PHENOTYPING OF H
iPSC-DERIVED NEURONS

PHENOTYPING OF HUMAN iPSC-DERIVED NEURONS

PATIENT-DRIVEN RESEARCH

Edited by

ELIZABETH D. BUTTERMORE
Human Neuron Core, Rosamund Stone Zander Translational Neuroscience Center, Boston Children's Hospital, Boston, MA, United States
Neurology Department, Boston Children's Hospital, Boston, MA, United States

ELSEVIER

ACADEMIC PRESS
An imprint of Elsevier

Academic Press is an imprint of Elsevier
125 London Wall, London EC2Y 5AS, United Kingdom
525 B Street, Suite 1650, San Diego, CA 92101, United States
50 Hampshire Street, 5th Floor, Cambridge, MA 02139, United States
The Boulevard, Langford Lane, Kidlington, Oxford OX5 1GB, United Kingdom

ISBN: 978-0-12-822277-5

For information on all Academic Press publications visit our website at https://www.elsevier.com/books-and-journals

Publisher: Nikki Levy
Acquisitions Editor: Joslyn T. Chaiprasert-Paguio
Editorial Project Manager: Susan Ikeda
Production Project Manager: Punithavathy Govindaradjane
Cover Designer: Mark Rogers

Typeset by TNQ Technologies

Dedication

Thanks to all of the members of the Cross-IDDRC Human Stem Cell Working Group, whose ideas and conversations helped shape this textbook, as well as the members of this group who wrote chapters included in the book. Thanks also to my colleagues (current and former) in the Rosamund Stone Zander Translational Neuroscience Center at Boston Children's Hospital who have fostered an environment of collaboration and innovation with the goal of contributing tangible improvements to preclinical research to identify novel therapeutics for patients with neurodevelopmental disorders.

Thanks to my family for supporting my scientific curiosity.

I hope this book serves as a valuable resource to anyone looking to enter the field of human iPSC-derived neuron modeling and stimulates conversations among colleagues for those already in the field.

Contents

5. Differentiation of Purkinje cells from pluripotent stem cells for disease phenotyping in vitro

Maria Sundberg

6. Brain organoids: models of cell type diversity, connectivity, and disease phenotypes

Madison R. Glass, Felix A. Kyere, Deborah L. French, Jason L. Stein, and Elisa A. Waxman

II

The use of iPSC-derived neurons to study neurological disorders

7. Human models as new tools for drug development and precision medicine

Alfred K. Dei-Ampeh, Manuj Shah, Gerarda Cappuccio, Damian W. Young, and Mirjana Maletic-Savatic

8. Use of cerebral organoids to model environmental and gene x environment interactions in the developing fetus and neurodegenerative disorders

Herbert M. Lachman

9. iPSC-derived models of autism: Tools for patient phenotyping and assay-based drug discovery

Gilles Maussion, Cecilia Rocha, Luisa Pimentel, and Thomas M. Durcan

10. Probing the electrophysiological properties of patient-derived neurons across neurodevelopmental disorders

Brooke Latour, Teun Klein Gunnewiek, and Nael Nadif Kasri

Contributors

Irene Antony Department of Developmental Biology, Washington University School of Medicine, Saint Louis, MO, United States

Elizabeth D. Buttermore Human Neuron Core, Rosamund Stone Zander Translational Neuroscience Center, Boston Children's Hospital, Boston, MA, United States; Neurology Department, Boston Children's Hospital, Boston, MA, United States

Gerarda Cappuccio Department of Pediatrics-Neurology, Baylor College of Medicine, Houston, TX, United States; Jan and Dan Duncan Neurological Research Institute, Texas Children's Hospital, Houston, TX, United States

Juan Cruz Casabona Instituto de Investigaciones Bioquímicas de Buenos Aires (IIBBA), Fundación Instituto Leloir, Buenos Aires, Argentina; Consejo Nacional de Investigaciones Científicas y Técnicas (CONICET), Buenos Aires, Argentina

Gareth Chapman Department of Developmental Biology, Washington University School of Medicine, Saint Louis, MO, United States

Ivy Pin-Fang Chen Tessera Therapeutics, Cambridge, MA, United States; Human Neuron Core, Boston Children's Hospital, Boston, MA, United States

Harue Chou Neuroregeneration and Stem Cell Programs, Institute for Cell Engineering, Johns Hopkins University School of Medicine, Baltimore, MD, United States; Department of Neurology, Johns Hopkins University School of Medicine, Baltimore, MD, United States

Laurence Daheron Harvard Stem Cell Institute, Harvard University, Cambridge, MA, United States

Alfred K. Dei-Ampeh Department of Pediatrics-Neurology, Baylor College of Medicine, Houston, TX, United States; Jan and Dan Duncan Neurological Research Institute, Texas Children's Hospital, Houston, TX, United States

Thomas M. Durcan Early Drug Discovery Unit (EDDU), Montreal Neurological Institute and Hospital, McGill University, Montreal, QC, Canada

Sandra J. Engle Biogen, Translational Biology, Cambridge, MA, United States

Patrick Faloon Vesalius Therapeutics, Cambridge, MA, United States; Biogen, Biotherapeutics and Medicinal Sciences, Cambridge, MA, United States

Kevin T. FitzGerald Department of Medical Humanities, Creighton University School of Medicine, Omaha, NE, United States; Pellegrino Center for Clinical Bioethics, Georgetown University Medical Center, Washington, DC, United States

Deborah L. French Department of Pathology and Laboratory Medicine, The Children's Hospital of Philadelphia, Philadelphia, PA, United States; Center for Cellular and Molecular Therapeutics, The Children's Hospital of Philadelphia, Philadelphia, PA, United States

James Giordano Pellegrino Center for Clinical Bioethics, Georgetown University Medical Center, Washington, DC, United States; Departments of Neurology and Biochemistry, Georgetown University Medical Center, Washington, DC, United States

Madison R. Glass Department of Genetics & UNC Neuroscience Center, University of North Carolina at Chapel Hill, Chapel Hill, NC, United States

Komal Kaushik Department of Developmental Biology, Washington University School of Medicine, Saint Louis, MO, United States

Robin J. Kleiman Biogen, Translational Biology, Cambridge, MA, United States; Alkermes Inc., Research, Waltham MA, United States

Teun Klein Gunnewiek Department of Human Genetics, Radboudumc, Donders Institute for Brain, Cognition and Behaviour, Nijmegen, Netherlands

Kristen L. Kroll Department of Developmental Biology, Washington University School of Medicine, Saint Louis, MO, United States

Felix A. Kyere Department of Genetics & UNC Neuroscience Center, University of North Carolina at Chapel Hill, Chapel Hill, NC, United States

Herbert M. Lachman Department of Psychiatry and Behavioral Sciences, Albert Einstein College of Medicine, Bronx, NY, United States; Department of Medicine, Albert Einstein College of Medicine, Bronx, NY, United States; Department of Genetics, Albert Einstein College of Medicine, Bronx, NY, United States; Dominick P. Purpura Department of Neuroscience, Albert Einstein College of Medicine, Bronx, NY, United States

Brooke Latour Department of Human Genetics, Radboudumc, Donders Institute for Brain, Cognition and Behaviour, Nijmegen, Netherlands

Mirjana Maletic-Savatic Department of Pediatrics-Neurology, Baylor College of Medicine, Houston, TX, United States; Jan and Dan Duncan Neurological Research Institute, Texas Children's Hospital, Houston, TX, United States; Center for Drug Discovery, Department of Pathology and Immunology, Baylor College of Medicine, Houston, TX, United States

Xiaobo Mao Neuroregeneration and Stem Cell Programs, Institute for Cell Engineering, Johns Hopkins University School of Medicine, Baltimore, MD, United States; Department of Neurology, Johns Hopkins University School of Medicine, Baltimore, MD, United States

Abigail Mariga Biogen, Biotherapeutics and Medicinal Sciences, Cambridge, MA, United States; Bristol Myers Squibb, Neuroscience Research and Early Development, Cambridge, MA, United States

Gilles Maussion Early Drug Discovery Unit (EDDU), Montreal Neurological Institute and Hospital, McGill University, Montreal, QC, Canada; Montreal Neurological Institute, Neurology and Neurosurgery, Montreal, QC, Canada

Kesavan Meganathan Department of Developmental Biology, Washington University School of Medicine, Saint Louis, MO, United States

Swati Mishra Laboratory Medicine and Pathology, University of Washington, Seattle, WA, United States; Institute for Stem Cell and Regenerative Medicine, University of Washington, Seattle, WA, United States

Nael Nadif Kasri Department of Human Genetics, Radboudumc, Donders Institute for Brain, Cognition and Behaviour, Nijmegen, Netherlands; Department of Cognitive Neuroscience, Radboudumc, Donders Institute for Brain, Cognition and Behaviour, Nijmegen, Netherlands

Andrew Petersen Madison, WI, United States

Luisa Pimentel Early Drug Discovery Unit (EDDU), Montreal Neurological Institute and Hospital, McGill University, Montreal, QC, Canada

Ramachandran Prakasam Department of Developmental Biology, Washington University School of Medicine, Saint Louis, MO, United States

Aiko Robert Center for Neurogenomics and Cognitive Research, Amsterdam Neuroscience, Alzheimer Center Amsterdam, Department of Neurology, Vrije Universiteit Amsterdam, Amsterdam UMC, Amsterdam, the Netherlands

Cecilia Rocha Early Drug Discovery Unit (EDDU), Montreal Neurological Institute and Hospital, McGill University, Montreal, QC, Canada

Manuj Shah Jan and Dan Duncan Neurological Research Institute, Texas Children's Hospital, Houston, TX, United States; Department of Neuroscience, Rice University, Houston, TX, United States

Jason L. Stein Department of Genetics & UNC Neuroscience Center, University of North Carolina at Chapel Hill, Chapel Hill, NC, United States

Maria Sundberg Department of Neurology, F.M. Kirby Center for Neurobiology, Boston Children's Hospital, Harvard Medical School, Boston, MA, United States

Rik Van Der Kant Center for Neurogenomics and Cognitive Research, Amsterdam Neuroscience, Alzheimer Center Amsterdam, Department of Neurology, Vrije Universiteit Amsterdam, Amsterdam UMC, Amsterdam, the Netherlands

Coen Van Deursen Center for Neurogenomics and Cognitive Research, Amsterdam Neuroscience, Alzheimer Center Amsterdam, Department of Neurology, Vrije Universiteit Amsterdam, Amsterdam UMC, Amsterdam, the Netherlands

Elisa A. Waxman Center for Cellular and Molecular Therapeutics, The Children's Hospital of Philadelphia, Philadelphia, PA, United States

Samantha Wu Pellegrino Center for Clinical Bioethics, Georgetown University Medical Center, Washington, DC, United States

Mingyao Ying Department of Neurology, Johns Hopkins University School of Medicine, Baltimore, MD, United States; Hugo W. Moser Research Institute at Kennedy Krieger, Baltimore, MD, United States

Damian W. Young Department of Pediatrics-Neurology, Baylor College of Medicine, Houston, TX, United States; Jan and Dan Duncan Neurological Research Institute, Texas Children's Hospital, Houston, TX, United States; Department of Pharmacology and Chemical Biology, Baylor College of Medicine, Houston, TX, United States; Center for Drug Discovery, Department of Pathology and Immunology, Baylor College of Medicine, Houston, TX, United States

Jessica E. Young Laboratory Medicine and Pathology, University of Washington, Seattle, WA, United States; Institute for Stem Cell and Regenerative Medicine, University of Washington, Seattle, WA, United States

SECTION I

Best practices and considerations when designing a new project

CHAPTER

1

iPSC culture: best practices from sample procurement to reprogramming and differentiation

Laurence Daheron[1] *and Ivy Pin-Fang Chen*[2,3]

[1]Harvard Stem Cell Institute, Harvard University, Cambridge, MA, United States; [2]Tessera Therapeutics, Cambridge, MA, United States; [3]Human Neuron Core, Boston Children's Hospital, Boston, MA, United States

In this chapter, we describe best practices and recommendations for patient sample collection, derivation and differentiation of human induced pluripotent stem cells (iPSCs) in a research laboratory setting. The purpose of this chapter is to set the stage for best practices that should be considered when one is implementing iPSC culture in their lab. Utility of iPSCs in specific disorders or applications, technical difficulties, and challenges are addressed in subsequent chapters, with additional requirements and recommendations for each topic.

Facility setup

Tissue culture room design

General guidelines and best practices for setting up a tissue culture room for iPSC culture are similar to those for any other mammalian cell culture system at Biological Safety Level 2 (BSL2).[1,2] However, iPSCs are typically cultured in the absence of antibiotics and additional measures should be taken to maintain sterility. Here we briefly touch on best practices for facility setup as well as equipment that comes in handy for culturing iPSCs. To reduce dust collection, it is ideal for a tissue culture room to have a balanced neutral or slightly positive airflow with respect to neighboring areas, unless biohazardous agents are used. Floors and walls must be durable and liquid tight and, although not required, epoxy coating can help maintain cleanliness. Similarly, smooth and cleanable drop ceiling tiles are ideal. Porous materials, such as fabric and wood, should be avoided wherever possible. A hand washing

Phenotyping of Human iPSC-derived Neurons
https://doi.org/10.1016/B978-0-12-822277-5.00008-0

sink (and eyewash station in some cases) is now a requirement in a tissue culture room due to safety concerns. However, still water and damp environment encourage microorganism growth and extensive measures—including leak proofing, antifungal foam spraying, waterproofing, and regular cleaning—should be taken to prevent the sink from turning into a source of contamination.

Tissue culture equipment

Basic equipment required includes storage racks, a refrigerator, a freezer, a vortex mixer, centrifuges, microscopes, a water/bead bath, an incubator, a biosafety cabinet, and a liquid nitrogen storage unit. For centrifuges, a swing bucket centrifuge (e.g., Eppendorf 5702, Cole–Parmer MS-3400) with biocontainment lids is essential and a mini centrifuge is nice to have for quick spins of microcentrifuge tubes. For microscopes, an entry-level inverted light microscope with 4×, 10×, and 20× phase contrast objectives (e.g., Leica DMi1, Zeiss PrimoVert, Nikon ECLIPSE Ts2) is recommended and a compact fluorescence microscope (e.g., ThermoFisher EVOS M5000, Leica DMi8) is helpful if using fluorescence reporters. A standard noncirculating 37°C water bath is commonly used to warm media and reagents, but we recommend replacing water with autoclavable metal beads due to contamination concerns mentioned above. If a water bath were to be used, weekly cleaning and water treatments (e.g., Spectrum Labs Clear Bath, Promo Cell Aquaguard, SP Scienceware Aqua-Clear) would be necessary to reduce growth of microorganisms. Similar practices should be done for the water pan in an incubator to maintain sterility. For incubators, self-sterilizing incubators are recommended (e.g., Panasonic MCO-170AICUVH, ThermoFisher Heracell 150i). Incubator temperature and gas level should be monitored and calibrated monthly, as well as certified annually. Complete sterilization, including replacing HEPA filter (if used), should be done annually at a minimum. Incubators should be placed away from foot traffic, doorways, or air vents to reduce exposure to contaminants. As vibration can cause cells to detach or distribute unevenly, it is not advised for incubators to be near or share one surface with a source of vibration, such as a large centrifuge. It is important to leave enough space around incubators for venting, tubing/cords, as well as maintenance and to ensure that the desired gas inlets (i.e., CO_2, N_2) are readily available nearby. For BSL2, class II type A2 biosafety cabinets are most commonly used, which should also be placed away from doorways or air vents to allow optimal laminar airflow. Similarly, it is important to leave enough space around biosafety cabinets for venting, tubing/cords, as well as vacuum. Biosafety cabinets should be certified annually for airflow and HEPA filter function.[3] A liquid nitrogen storage unit is required for long-term storage of iPSCs, which should be stored in vapor phase below −135°C.[4] Although manually filled liquid nitrogen storage dewars can be used, an automated cryopreservation system is highly recommended to have better control over liquid level and temperature, both of which are critical for iPSC viability post-thaw.

In addition to the equipment listed above, several pieces of specialized equipment are recommended for deriving, culturing, and differentiating iPSCs, as well as characterizing iPSCs and their derivatives. A dissection microscope (e.g., Nikon SMZ800, Zeiss Stemi305, Olympus SZX7) is necessary for processing skin biopsies to isolate and culture fibroblasts and for manually picking iPSC colonies during reprogramming or genome editing. To help maintain sterility, the dissection microscope can be put inside a biosafety cabinet or a laminar flow microscope enclosure with HEPA filtration (e.g., AirClean Systems AC632TLFUVMIC). A microscope objective marker (e.g., Nikon MBW10020), a stamp for marking colonies, is useful for clone isolation. For process standardization in tissue culture and treatments, a cell counter (e.g., ThermoFisher Countess, Bio-Rad TC20), electronic multichannel pipettes, and robotic liquid handlers are helpful, especially when working with a large number of small well formats, such as 96- and 384-well plates. In addition, automated imaging stations (e.g., Sartorius IncuCyte for live cell imaging, Molecular Devices IXM-c for high content imaging) are helpful for documenting reprogramming and/or differentiation progress. These platforms can also increase throughput for imaging and screening through the use of automation. Similarly, for increased throughput on functional readouts in iPSC-derived neurons, multi-electrode array recording devices (e.g., Axion BioSystems Maestro) and fluorescence/luminescence plate readers (e.g., Hamamatsu FDSS7000EX) can be used to measure neuronal network activity. Phenotyping equipment can be outfitted with robotic arms for streamlining and humidity control incubators to reduce edge effects in multi-well plates.

Primary sample collection

Somatic cells

The first human iPSC lines were generated from dermal fibroblasts,[5,6] but a large variety of cell types have been reprogrammed since, including different blood cells,[7] endothelial cells,[8] keratinocytes,[9] and adipose-derived stem cells.[10] The main consideration for the choice of somatic cells is the ease and safety of collection for the donors. Peripheral blood and skin biopsies are the most commonly used sources of cells for reprogramming. Skin biopsies are minimally invasive, but still more intrusive than a simple blood draw. The derivation of fibroblast culture from skin biopsy can be challenging and can take as long as 2–3 weeks. Moreover, the fibroblast proliferation rate is crucial for reprogramming success, as partially senescent fibroblasts have limited capacity for reprogramming. Mutational signatures associated with UV-exposure have been reported for iPSC lines derived from dermal fibroblasts,[11] raising concerns about the use of these cells in generating cGMP iPSC lines. However, fibroblasts remain a common source of cells for reprogramming as they are amenable to reprogramming using the three different non-integrative methods (Sendai virus, Episomal vector, and mRNA transfection).

From peripheral blood samples, different cell types can be isolated for culture. The first step (which should be done within 24 h post-draw to increase reprogramming efficiency) involves the isolation of peripheral blood mononuclear cells (PBMCs) by Ficoll gradient or equivalent. These cells can be frozen or cultured immediately for reprogramming. Hematopoietic progenitors (CD34 positive cells) purified from PBMCs show high reprogramming efficiency compared to more differentiated blood cells, but they are rare in the peripheral blood and therefore not often used as a starting material. Erythroblasts have become the most widely used cells for reprogramming.[12] By culturing in the presence of a combination of recombinant human cytokines (EPO, SCF and IL3) for 9 days, erythroblasts can be enriched to 90%–100% from the mixed population of cells in PBMC. T lymphocytes are another blood cell type that have been used for reprogramming.[13] These cells are abundant in PBMCs and are easy to activate in the presence of anti-CD3 and anti-CD8 antibodies, in addition to IL2. Within 3 days post blood draw, a high number of T lymphocytes can be ready for reprogramming. However, these cells are difficult to reprogram (low efficiency), and they have the TCR receptor rearrangement unique to each cell, which could be a problem for some applications. Erythroblasts and T lymphocytes can be reprogrammed using either the Sendai virus or the episomal vector methods. However, the mRNA transfection reprogramming method has been unsuccessful for any blood cell type so far. One alternative is to isolate endothelial progenitor cells (EPC) from fresh blood samples. EPCs are circulating cells that express a number of markers specific for vascular endothelial cells. They are rare and therefore difficult to expand from a small volume (less than 10 mL) of blood. EPCs are amenable to reprogramming with the mRNA method, but it takes approximately 3 weeks before they are ready for reprogramming. Urine samples are another convenient source of cells for reprogramming.[14] Cells are harvested by centrifugation of urine and two cell types can be expanded: epithelial-like cells with limited proliferation capacity and mesenchymal type cells with higher expansion capacity. Although urine is easy to collect, bacterial contamination of the culture is relatively high and the rate of failure for urine cell expansion is around 30%. Since urine sample collection is noninvasive, it is particularly suitable for children. The cGMP iPSC lines generated to date were derived from blood or fibroblast cells. For autologous transplantation, the use of T lymphocytes was proposed to take advantage of the TCR receptor rearrangement as a natural tag for the cells. This unique tag allows for identification of the transplanted cells post transplantation. For allogenic transplantation, obtaining somatic cells with limited mutational load is crucial. EPC or CD34+ cells isolated from cord blood could provide a safe source of cells for reprogramming.

Quality control of somatic cells

Mycoplasma testing should be performed on all somatic samples before starting the reprogramming process. For cells that have a greater risk of bacterial contamination (such as skin or urine cells), we recommend thoroughly washing the initial sample (skin biopsies and cell pellets from urine) with PBS containing antibiotics and initially expanding the cells in medium with antibiotics. Once established, it is good practice to culture somatic cells without antibiotics before reprogramming to reduce the chance of creating drug-resistant microorganisms. Viral testing (e.g., HIV, HBV, HCV, HTLV, etc.) is not commonly implemented on samples used for research grade iPSC lines. However, some stem cell banks or FACS sorting facilities might require viral testing before accepting new lines. For cGMP iPSC lines, a panel of relevant communicable disease agents or diseases (RCDADs) are tested in the somatic cells of origin. Karyotyping is rarely performed on somatic samples.

Reprogramming

In 2007, Shinya Yamanaka generated the first human iPSC lines by overexpressing Oct4, Sox2, Klf4, and c-Myc in fibroblasts using a retrovirus system.[5] This system offers three advantages: it efficiently transduces the fibroblast cells, allows high expression of the four transcription factors for a prolonged period of time, and results in silenced transgene expression in induced pluripotent stem cells, as retroviruses are known to be silenced upon reprogramming. However, since the retrovirus is randomly integrated into the genome, the resulting iPSC lines were not transgene free and therefore had limited applications. An optimal delivery system should provide the following advantages: (1) Efficient delivery: including strong expression of the reprogramming factors, sustained expression over time, and a high percentage of transfected/transduced cells; (2) Reliability: ability to deliver the transgenes in any somatic cell type; (3) Simplicity: one step delivery; (4) Limited toxicity: low cell death; (5) Safety: no integration into the genome and no residual expression of transgenes in iPSC lines. Several methods have been reported since the early study from Shinya Yamanaka. Currently, three non-integrating approaches are commonly used: the episomal vector, the Sendai virus and the mRNA transfection methods.

Pros and cons of each method

Episomal vector transfection

Episomal vectors are plasmids containing the *cis*-acting origin of replication (oriP) and *trans*-acting EBNA1 viral elements from the Epstein–Barr Virus. These two elements confer their ability to self-replicate in the cells without integration into the genome. The first iPSC lines generated with these vectors showed very low reprogramming efficiency and required the overexpression of Nanog, Lin28 and SV40 large T genes, in addition to the four original Yamanaka transcription factors.[6] Improvement in the efficiency of this method was described a few years later.[15] In this report, three episomal vectors were used: one with Oct4 and a short hairpin RNA (shRNA) for p53, a second that includes Klf4 and Sox2, and a third that includes L-Myc and Lin28. The p53 pathway acts as a roadblock to reprogramming somatic cells into pluripotency.[16] The downregulation of p53, which is described in this study, enhances the reprogramming efficiency. This method is simple since it requires only one transfection and can be used to reprogram a variety of cell types. One limiting factor is the transfection efficiency of the large episomal vectors (>10 Kb). Nucleofection is recommended over the lipofection system to ensure efficient delivery in the cells. The episomal method has been used to generate many research grade iPSC lines and most of the cGMP lines currently available. However, there are a few caveats to this method. The reprogramming efficiency (calculated based on the number of iPSC colonies generated per input cells) is the lowest of the three methods (around 0.01% for fibroblasts).[17] The addition of the shRNA against p53, a tumor suppressor, could increase the incidence of genetic abnormalities in the resulting iPSC lines. We reported that the percentage of iPSC lines with abnormal karyotype was the highest (11%) when using the episomal vectors. Last, integration of the episomal vectors into the genome of the iPSC lines is frequent.[17]

Sendai virus transduction

The sendai virus (SeV) is a rodent virus of the paramyxoviridae family causing respiratory infection. It is a single-stranded negative sense RNA virus. In 2009, it was used for the first time for the generation of iPSC lines from human fibroblasts.[18] Since this virus does not go through a DNA phase, there is no risk of integration into the host genome. This method offers several benefits. First, it provides an efficient system to deliver the Yamanaka factors. Even at low multiplicity of infection (MOI 1 to 3), Fusaki et al. showed close to 100% transfection efficiency in fibroblasts and high expression of the reprogramming factors. Next, it can be used to transduce a wide range of cells, as it binds to sialic acid found on glycoproteins at the surface of the cell membrane. Moreover, only one transduction is sufficient since the RNA polymerase from the Sendai virus maintains the RNA synthesis for an extended time. The first commercially available reprogramming Sendai virus kit (CytoTune-iPS 1.0) was launched in 2011. This method has been widely used by research labs and core facilities since its introduction. The Cytotune 1.0 consisted of four viral preparations, one for each of the reprogramming factors. An optimized version (CytoTune-iPS 2.0) with less cytotoxicity, higher efficiency and faster elimination of the SeV RNA in the iPSC lines was developed later. Besides the cost, there are two main concerns regarding this method: (1) it relies on a rodent virus which could hamper its use in countries or institutions with restrictive virus use policies, and (2) The SeV RNA is not eliminated rapidly in iPSC lines as the majority of the iPSC lines still express the transgenes at passage 5. A cGMP version of the reprogramming SeV kit is now available (CytoTune-iPS 2.1). It includes three viral preparations: one with Oct4, Sox2, and Klf4, one with L-Myc (instead of c-Myc) and a third with Klf4. Our group has successfully used this kit to generate cGMP iPSC lines from T cells from two patients with diabetes.

Both reprogramming methods described above may result in iPSC lines with transgene expression, either permanently for the episomal vectors (since it can be integrated into the genome) or transiently for the Sendai virus (until the SeV RNA virus is eliminated).

mRNA reprogramming method

A third approach eliminating this risk was reported in 2010 by the Rossi group.[19] It is based on the delivery of mRNA encoding reprogramming factors to the cells. mRNAs are swiftly degraded, so newly derived iPSC lines are instantly transgene free. However, due to the short-lived nature of mRNA, successive daily transfections are required to efficiently reprogram somatic cells. Moreover, the transfection of mRNA triggers an innate immune response in cells, often leading to cell death. To attenuate the immunogenicity of the mRNA, modified ribonucleotide bases need to be incorporated and B18R, an interferon response inhibitor, is required post transfection. This method enables a robust but transient expression of the reprogramming factors. In the first report, 15−17 successive transfections were necessary to obtain iPSC colonies. By adding microRNAs (including clusters miR-290-295 or miR-302-367) in the reprogramming cocktail, pluripotent reprogramming is promoted and the number of transfections can be reduced to 12. More recently, an optimized version of the mRNA/microRNA reprogramming method was described, allowing the reprogramming of somatic cells with only four transfections for fibroblasts and eight transfections for urine epithelial cells or endothelial progenitor cells.[20] Overall, the mRNA transfection method is the most efficient as it can reach 1% efficiency (1 iPSC colony per 100 input cells). However, the consecutive transfections can be challenging since the toxicity varies between samples. It is recommended to plate the cells at different densities to ensure successful reprogramming. Moreover, as stated before, this method is efficient for fibroblast reprogramming, but has not been demonstrated to reprogram blood cells. A comparison of iPSC lines generated by these three methods showed no difference in their differentiation potential.[17]

iPSC line characterization

The characterization and quality control of iPSC lines can be divided into five categories: sterility, pluripotency, transgene elimination, identity, and genetic stability.

Sterility

It is crucial to maintain sterility of cell culture during the different steps associated with deriving new iPSC lines, including reprogramming, picking iPSC clones, expanding, freezing and thawing newly derived iPSC lines. Good cell culture technique, appropriate set up (dissection microscope placed in a biosafety cabinet), and regular monitoring of the culture are key elements for human pluripotent stem cell (hPSC) culture. Bacterial, yeast, and fungal contaminations can easily be detected by observing the cell culture under the microscope on a regular basis. In contrast, mycoplasma contamination is not visible, but can negatively affect the cells and change their characteristics. It is important to test for mycoplasma contamination regularly or at strategic time-points, such as before freezing a new stock or before starting a new experiment. PCR-based or biochemical-based mycoplasma detection assays are commercially available. For the cGMP iPSC lines our group derived in collaboration with Dr. Ritz at Dana Farber, sterility was assessed by the BacT/alert system (BioMerieux, Marcy l'Etoile, France). Samples were inoculated directly into aerobic and anaerobic culture bottles and incubated for 14 days to assess microbial growth.

Pluripotency

The pluripotent state of newly derived iPSC lines is typically assessed by testing the expression of certain stem cell markers and, additionally, by evaluating the lines' trilineage differentiation potential. Cells that are fully reprogrammed reactivate stem cell associated markers, such as OCT4, NANOG, SOX2, and LIN28, to a level comparable to human embryonic stem cells.[21] This reactivation is evaluated by qRT-PCR using primers specific for the endogenous genes or by looking at the DNA methylation of their promoters through bisulfite sequencing. The somatic cells of origin are used as negative controls and hESCs are used as positive control. The protein expression of these markers can also be confirmed by immunocytochemistry or FACS analysis. Moreover, standardized assays are available to facilitate the characterization of iPSC lines. Pluritest is a bioinformatics assay allowing the comparison of newly derived iPSC lines to a large set of reference samples (more than 450 samples) based on gene expression profiles.[22] This assay provides both a pluripotency and a novelty score. The pluripotency score indicates how strong the pluripotency signature is in the new samples. The novelty score detects technical or biological variations in the data. The pluritest analysis tool is compatible with Illumina HT12V3/HT12V4 arrays, RNA-Seq data, and Affymetrix HG-U219 gene expression arrays. It is also supported by Thermo Fisher PrimeView Global Gene Expression profile assay. The Epi-Pluri score is another assay based on epigenetic biomarkers reported by Lenz et al.[23] It distinguishes pluripotent stem cells based on the DNA methylation level at three specific CpG sites. This test is offered by the company Cygenia. All of these aforementioned assays show the undifferentiated state of the cells, but the validation of their pluripotency state can only be demonstrated by differentiating the cells into the three germ layers. *In vitro* spontaneous differentiation is achieved using embryoid body (EB) formation or monolayer culture. Alternatively, iPSCs can be specifically differentiated into mesoderm, endoderm and ectoderm by using recombinant proteins that promote the emergence of specific lineages. This directed differentiation method provides a better yield, is faster, and is more consistent than EB formation. Following differentiation, cells from the three germ layers are detected by immunocytochemistry or by quantitative RT-PCR. The development of standardized quantitative assays, such as the scorecard assay, offers the possibility to compare differentiation potential of multiple iPSC lines. The scorecard was first established by monitoring the expression of 500 selected genes using the nanostring technology.[24] It was further simplified by assessing the expression of less than 100 genes with a TaqMan based method (TaqMan hPSC Scorecard Assay, Thermo Fisher).[25] Teratoma formation assay is the classic method used to evaluate differentiation capacity of iPSCs in vivo. When pluripotent stem cells are injected into immunodeficient mice, they give rise to a teratoma, a benign tumor comprising of cells from all three germ layers. Typically, teratomas are analyzed by performing histology sections and the pluripotency state is confirmed if the teratomas contain tissues from each of the three germ layers. Immunohistochemistry can also be applied to the teratomas to identify different cell types. In 2015, the Benvenisty group developed an algorithm to quantitatively assess the teratomas: TeratoScore.[26] The gene expression profile of the teratomas is compared with a set of references. This assay can be used to estimate the differentiation potential of each iPSC lines. The teratoma formation assay was considered the gold standard for pluripotency testing for many years. However, it is costly, time-consuming, and can be technically challenging depending on the injection site. In 2018, the ISCI group (International Stem Cell Initiative) reported their data on the comparison of the most commonly used methods to assess human pluripotent stem cells.[27] They showed that the Pluritest can predict, but not determine, the pluripotency state of an iPSC line. In addition, random or directed in vitro differentiation is sufficient to confirm the pluripotent differentiation potential of the cells. However, only the teratoma analysis (by histology or TeratoScore) can provide an evaluation of both pluripotency and malignant potential. Therefore, they concluded that in vitro differentiation is adequate for the characterization of research grade iPSC lines, but teratoma formation is recommended for clinical grade iPSC lines.

Transgene elimination

Depending on the reprogramming method, the iPSC lines may need to be tested for the residual expression of the Yamanaka factors. The first human iPSC lines were generated using either a retroviral or a lentiviral system and the transgenes were integrated into the genome. Despite silencing of the provirus during the reprogramming process, it was not uncommon to detect residual expression of the transgenes in these iPSC lines. Moreover, the transgene expression could be reactivated during the differentiation of the iPSC lines, leading to tumor formation after transplantation into mice.[28] However, with the non-integrative reprogramming methods, it is possible to obtain lines that are completely free of transgene expression. If the episomal vector method is used for reprogramming, it is important to test any potential integration of parts of the plasmids into the genome. Although the episomal vector replicates in the cytoplasm, we showed that 30% of the iPSC lines tested retained part of the plasmids over time.[17] A quantitative PCR for the EBNA sequence (found in all the episomal vectors) offers a simple screening method for determining if iPSC lines are free of the vectors. If the iPSC lines are generated using the Sendai virus method, the RNA virus will remain in the newly derived iPSC lines for several passages. The virus will be eliminated over time as the viral RNA polymerase protein is progressively diluted. The RNA virus can be monitored by quantitative RT-PCR using a probe specific for the Sendai virus sequence. We showed that the majority of the iPSC lines are free of transgene expression by passage 10. To accelerate the process, picking individual colonies during the first few passages and screening for SeV negative clones could be performed, but it would be a labor intensive process. Overall, the slow clearance of the SeV RNA is the main shortcoming of the Sendai virus method. The mRNA transfection reprogramming method generates iPSC lines that are instantly transgene free, eliminating the need for additional screening of the iPSC lines. However, as mentioned before, this method is labor intensive and it is not amenable to the reprogramming of blood cells.

Identity

To ensure that there was no swapping or mixing of iPSC lines during the reprogramming process or expansion of cell lines, a simple DNA fingerprinting test on genomic DNA extracted from both the iPSC lines and the somatic cells of origin should be performed. Both the UK stem cell bank and Wicell reported that some of the lines that were deposited in their banks were not the expected lines or were mixed with other cells. A short tandem repeat (STR) analysis, looking at the length of tandem repeats at 15 loci, is sufficient to differentiate lines from each donor. It is easy, inexpensive, and highly recommended for laboratories and facilities that are keeping multiple cell lines in culture at one time to perform STR analysis routinely or at strategic time-points, such as immediately after iPSC derivation, before freezing a new stock, or before starting a new experiment.

Genetic stability

The evidence that human pluripotent stem cells acquire genetic alterations during the reprogramming process, or the subsequent expansion of the lines, is well documented.[29,30] The most common chromosomal abnormalities are gains of whole chromosomes X and 8, gains of the long q arm of chromosomes 1, 17, and 20, and gains of the short p arm of chromosome 12.[31] These recurrent abnormalities provide a selective advantage by increasing the proliferation or survival of the cells. Interestingly, the relative frequencies of these alterations have changed over the years as culture conditions of the cells were modified (ISSCR poster presentation from Erik McIntire). Gains of the short arm of chromosome 12 were the most prevalent around 2010, while gains of chromosome 1 and 20 have increased recently, becoming the most predominant chromosomal alterations detected in human iPSCs now. Although chromosomal gains are more common, partial loss of chromosomes 7, 10, 18, and 22 are also recurrent in human iPSC cultures. Additionally, translocations involving chromosome 1 or chromosome 20 have been reported in multiple ESC and iPSC lines. G-Banding karyotyping is the most commonly used method to detect genetic instability in iPSC lines. However, the resolution of this method is only around 5 Mb. Higher resolution can be achieved using other methods, such as high-resolution array comparative genomic hybridization (aCGH) or single nucleotide polymorphism (SNP) microarrays. These methods can detect gain or loss of 0.05–0.4 Mb and the SNP microarray technology also allows for the detection of loss of heterozygosity (LOH). It was reported that small duplications of chromosome 20, undetectable by G-Banding, lead to additional copies of the gene BCL2L1 and increased expression of BCL-xL (the anti-apoptotic isoform of BCL2L1).[31] This genetic abnormality promotes cell survival upon loss of cell-to-cell contact. It is found at high frequency in genetically edited clones which have been through strong selective pressure during transfection and single cell cloning. Our group established a digital droplet PCR (ddPCR) assay to evaluate the copy number variation (CNV) of the gene *BCL2L1* (manuscript in preparation) to facilitate the screening of genetically edited clones. Unlike G-banding karyotype or aCGH/SNP microarrays, ddPCR is fast and inexpensive. Similarly, other qPCR- or ddPCR-based methods have been developed to identify the most common gain or loss of chromosome in human iPSCs. The hPSC Genetic Analysis kit (StemCell Technologies) can detect eight of the most common genomic alterations in hPSCs by qPCR, including gains of chromosomes 12, 17, 20, and X. A thorough analysis of all the reported genetic alterations in hPSCs led to the development of the iCS-digital PSC test (StemGenomics).[32] It is ddPCR-based and can detect more than 90% of recurrent genomic abnormalities using 24 probes or 76% using 12 probes. Besides the recurrent gain and loss of genetic material, single point mutations in the p53 gene were demonstrated to confer a selective advantage in hPSCs.[33] This was discovered by performing exome sequencing on hundreds of iPSC lines. This finding highlights the importance of carefully analyzing the genetic stability of hPSCs or their derivatives before their use in clinical trials. For cGMP iPSC lines, it is evident that G-banding karyotype, with its limited resolution, is not sufficient. In contrast, any mutation found through full genome sequencing or exome sequencing might be difficult to interpret, as a single mutation itself may not be tumorigenic. One approach to create safe cGMP PSC lines is to focus on oncogenes and tumor suppressor genes. Our group used the Oncopanel assay to detect CNV or SNV in 447 oncogenes. Based on this assay, we can eliminate iPSC lines with mutations that have a high risk of tumorigenic potential. Finally, since genetic alterations can occur at any point during in vitro culture, it is critical to verify the genome integrity of newly derived iPSC lines and to continuously monitor during expansion every 5 to 10 passages.[34]

Best practices prior to differentiation

The quality of starting iPSC culture is key for producing high quality neural culture. Differences in genetic, epigenetic, and pluripotent state can affect cell fate decision during differentiation and lead to variability between line-to-line and/or batch-to-batch. Therefore, when choosing an iPSC line for differentiation purposes, it is important to select a cell line that has as many documented characterizations (listed in the section above) as possible for quality control (QC). If the selected iPSC line comes from an external source, whether from a biorepository or a collaborator, it is important to keep cells in quarantine upon thawing and carry out mycoplasma test initially. Once proven negative for mycoplasma and other bacterial/fungal contaminants, the very first step is to expand iPSCs at low passage number to create a good number of cryovial stocks for cell banking, followed by iPSC line characterization discussed above. An example workflow is depicted in Fig. 1.1.

Cell banking

For good cell banking practices, a tiered banking system is recommended, including the initial token stocks, a master cell bank, and a working cell bank.[35] In general, the number of vials for each tier depends on the anticipated level of use. Token stocks are cryovials created upon in-house iPSC derivation or external purchasing and it is recommended to create 3–5 cryovials per iPSC line at this level. For iPSCs derived in-house, all iPSC QC, except reprogramming factor clearance, should be done at this stage. For externally sourced iPSCs that may or may not come with certificate of analysis, it is essential to assess parameters not characterized previously, but it is also a good idea to confirm reported data to ensure cell quality. It is worth noting that iPSC quality can be variable even when purchased from prestigious biobanks. Once the cells pass QC, one of the token stocks can be thawed and expanded to create a master cell bank, usually around 10–20 vials. Similarly, one vial from the master cell bank is thawed and expanded to create a working cell bank, typically around 20–50 vials for academic research purposes and 100–400 vials for therapeutic development. At a minimum, sterility, cell identity, and genomic stability should be tested after each expansion. The clearance of exogenous reprogramming agents should be confirmed around passage 10, ideally before the expansion for working cell bank. It is also important to record post-thawing cell morphology, survival rate, and spontaneous differentiation level. For protection from accidental loss, token stocks and/or several vials of the master cell bank should be stored apart from the working cell bank, ideally in a separate liquid nitrogen storage unit off-site.[36,37]

Culturing conditions

Traditionally, iPSCs are cultured on a mitotically inactivated mouse embryonic fibroblast (MEF) feeder layer, which provides growth factors and extracellular matrix (ECM) required to support proliferation and maintain pluripotency. However, it is labor intensive to prepare MEFs and the quality can be inconsistent between batch-to-batch, as evidenced by varying levels of Activin A in media, even when prepared properly.[38] In addition, technically challenging mechanical dissection or damage-prone enzymatic dissociation is required to separate iPSCs from the feeder layer during routine passaging.[39] Improved understanding in human embryonic stem cell (hESC) biology allowed the development of feeder-free culturing system with increasingly defined ECM, enzyme-free passaging reagents, and growth media.[40–43] Commonly used ECM for iPSCs includes murine-derived products, such as matrigel, geltrex, and cultrex, as well as recombinant human proteins, such as vitronectin and laminin-521. Recombinant proteins are often chosen for therapeutic development pipelines because they are xeno-free and defined subtrates. Commonly used enzyme-free

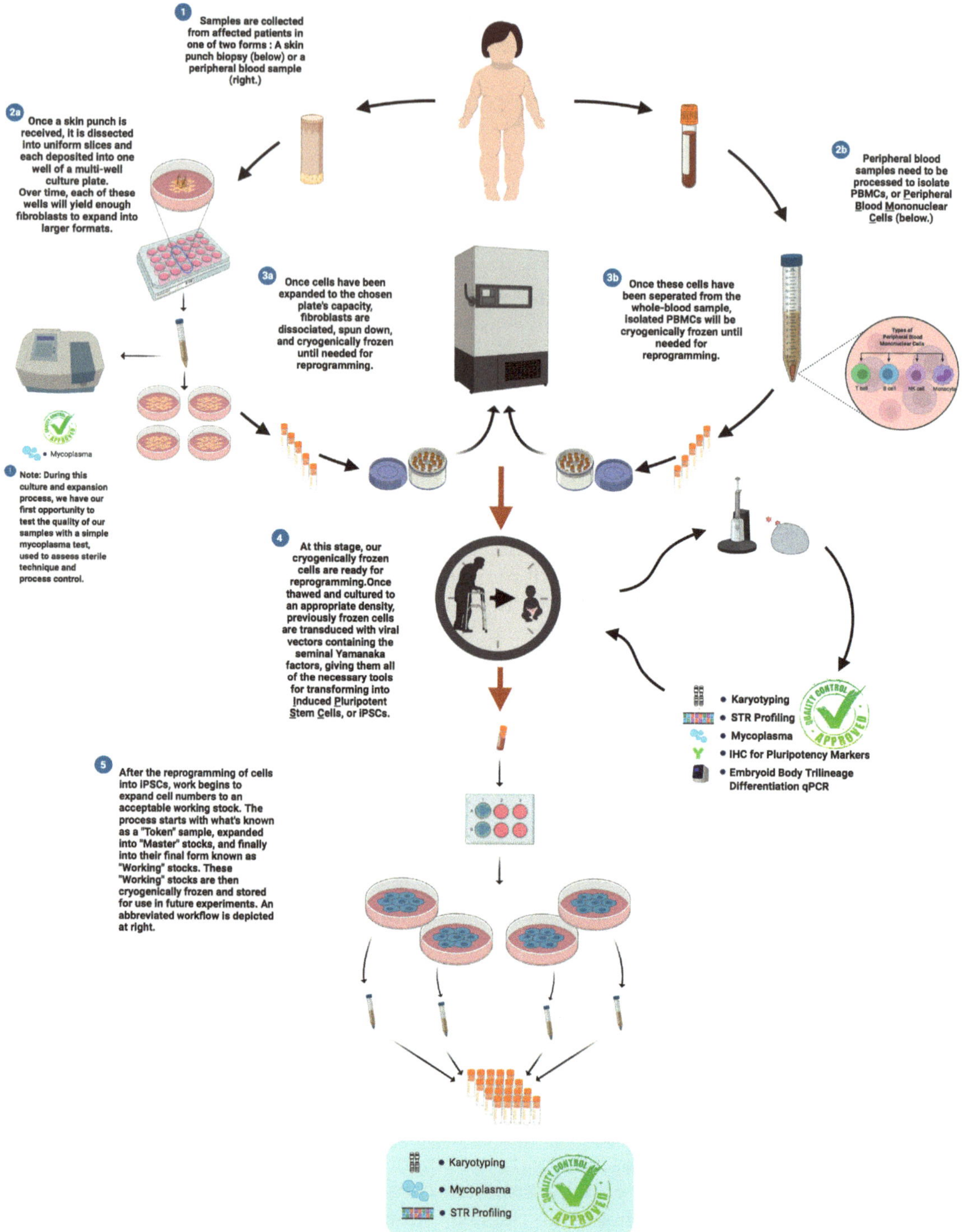

FIGURE 1.1 **Workflow for sample collection, reprogramming, cell banking, and quality control.** This is a brief overview of iPSC generation, with action point for each stage. Figure generated by Benjamin O'Kelly using Bio-Render software.

passaging reagents include EDTA, ReLeSR, and Gentle Cell Dissociation Reagent. These reagents allow for clump passaging of iPSC colonies, which is the preferred method of iPSC passaging to avoid chromosomal abnormalities. Accutase is a mixture of proteolytic and collagenolytic natural enzymes commonly used for single cell passaging in combination with Rho-associated kinase (ROCK) inhibitor,[44,45] which should not be used as a routine method for regular cell maintenance. Even with ROCK inhibitor present and dissociation time controlled, proteolytic enzyme activity damages cell membrane and imposes selective pressure on cell survival, which results in chromosomal abnormalities and rearrangements if done repeatedly.[46,47] Since it has a relatively forgiving incubation time for dissociation and does not require mechanical scraping or tapping to dislodge cells, Accutase is highly attractive for use with automation or large-scale iPSC culture.[48] To this end, a recent preprint article from National Institutes of Health (NIH) suggests that, in addition to ROCK inhibitor, several other small molecules are required during single cell dissociation to suppress stress responses in hPSCs, which enhances cell survival and genetic stability.[49] Based on this observation, if one must use Accutase for routine cell propagation, genomic stability in iPSCs, especially pathways that confer selective advantages, should be monitored rigorously and regularly. For growth media, a standard hPSC media used in feeder-dependent culture system can also be used to culture hESCs and iPSCs in a feeder-free culture system. However, this media lacks vital growth factors secreted by the feeder layer to support iPSCs in a feeder-free environment and needs to be "conditioned" on top of a feeder layer in advance.[38] Conditioning is commonly done by exposing the hPSC media to an MEF feeder layer overnight, collecting the media containing MEF-secreted factors the next day and storing the collected media at below −20°C if not used immediately. This method is cost effective, but can be inconsistent without stringent quality control due to variable raw material quality, feeder cell quality, incubation time, or storage condition. In 2007, the Thomson lab at the University of Wisconsin–Madison developed the first defined feeder-independent medium, mTeSR, for hESC culture.[43] Since then, a plethora of defined serum- and feeder-free medium has been commercialized, including commonly used Essential 8 medium, StemFlex medium, and mTeSR-Plus. Detailed recipes are available for in-house production of mTeSR and Essential 8 medium for cost reduction, but a stringent QC on raw materials is required to prevent inconsistency. It is important to note that StemFlex medium and mTeSR-Plus contain heat-stable basic fibroblast growth factor (bFGF) to reduce signaling fluctuation as a result of feeding cycle.[50] This technological advancement enhances cell proliferation and reduces spontaneous differentiation. This change also allowed for changes in iPSC feeding schedules, which were historically strict to every day feeds.

Differentiation

Experimental design

The success of a project largely relies on the experimental design. Since iPSC culture and neural differentiation require specialized equipment and are relatively time-consuming and labor-intensive, a well thought-out experimental design can be tremendously helpful in conserving financial, human, and physical resources. In order to design a project strategically, the first step is to identify and define a clear project goal. For example, an experimental design aimed at identifying a morphological phenotype for academic endeavor would be different from one aimed at screening novel compounds that modulates gene expression for therapeutic intervention. A project goal helps one define the appropriate scale of work based on the number of technical and biological replicates required to reach statistical significance. While the project goal casts an overall skeleton, the content of an experimental design is composed of multifaceted and interconnected aspects when using patient-derived iPSCs as an experimental system to study neurological disorders. Four broad aspects should be considered carefully and evaluated simultaneously in a holistic approach: (1) iPSC lines, (2) differentiation protocol, (3) readout/phenotype, and (4) stimulus/treatment. The last two aspects vary depending on diseases and project goals, which will be covered in other chapters. Here, we focus on the general considerations for selecting iPSC lines and differentiation protocols.

Cell line selection

When selecting patient-derived iPSC lines, it is important to consider the donor's clinical phenotype, genetic mutation(s), and to document demographics of the tissue donors, including biological sex, donor age, and race. While some of these parameters may have a bigger impact on one study over another, clinical phenotype, mutation information, and sex are the three most common and important factors. For genetically defined neurological disorders, one should validate reported variants in the gene of interest by Sanger sequencing the patient sample prior to reprogramming, or in patient-derived iPSCs if cells are readily available. For each clinically distinct disorder, different types of genetic aberration, such as copy number variations (CNVs), missense, nonsense, or frameshift mutations, often exist. On top of this, genetic mosaicism could exist if de novo mutation happens during embryo development[51] and random X-chromosome inactivation occurs in females with X-linked disorders.[52,53] For non-monogenic diseases, these genetic aberrations could be harbored in different genes in the same or, sometimes, different signaling pathways. These differences can affect patient clinical presentation independent of genetic background. In addition, gain-of-function and loss-of-function mutations can lead to closely associated sister disorders that have either complimentary or overlapping clinical phenotypes.[54] For non-genetically defined disorders, such as idiopathic autism or sporadic neurodegenerative disorders, the documented clinical phenotype of a patient donor is the only distinction from controls. If available, detailed clinical phenotyping can be useful for patient stratification. However, the utilization of iPSCs to study disorders without a clear genetic cause is limited until one establishes a baseline for natural variation in cellular phenotype from control individuals. In this case, iPSCs derived from patients with known genetic cause may be a good starting point to identify common pathological pathways.[55,56]

Patient clinical phenotypes often vary in severity. This heterogeneity within one disease could be a direct reflection of patient-specific genetic mutations or network disruptions, but patient genetic and epigenetic backgrounds, as well as environmental factors, most likely play a role as well.[57–60] Therefore, it is recommended to use iPSC lines derived from multiple controls and multiple patients that share similar clinical phenotypes wherever possible to distinguish disease-associated phenotypes from those stemming from individual background.[61,62] For genetically defined disorders, further patient stratification based on mutated genes or types of mutation may be necessary to achieve specific project goals. In the event that additional iPSC lines derived from multiple patients are not accessible, such as in the case of extremely rare disorders, it is recommended to use multiple clonal iPSC lines derived from the same patient, but careful control selection and statistical analysis are required to achieve confidence in the results.[61] Findings made with multiple clonal iPSC lines derived from a single individual are proof-of-concept and should be considered as personalized until further attested. When choosing an appropriate control iPSC line, it is recommended to match as many demographics categories as possible to the corresponding patient. This is especially important when studying nongenetically defined disorders or genetically defined diseases with large CNVs. One common strategy is to recruit an asymptomatic sex-matched familial control at the time of patient recruitment. Due to differences in sex chromosomes, changes in gene expression and cellular phenotype in iPSC derivatives have been reported between males and females.[63] It is still unclear whether donor age at the time of tissue collection or ancestral history has an effect on cellular phenotype after reprogramming. For patients with single nucleotide mutations or small insertions/deletions, an alternative strategy is to create an isogenic control from the patient iPSC line through genome editing. Patients with genetic mosaicism can be a powerful resource to create isogenic pairs without the need of genome manipulation.[64] In addition to using patient-derived experimental system described above, a complementary strategy is to create isogenic pairs by introducing a mutation to a hESC line or an iPSC line derived from an asymptomatic individual. Although there is no clinical significance by nature, this experimental system serves to prove that a mutation is sufficient to cause cellular phenotypes observed in the corresponding genetically defined disease regardless of genetic background.[56]

Differentiation protocol selection

Since relevant postmortem tissues (e.g., brain, spinal cord) are generally hard to access and vary in quality, iPSC derivatives are the closest surrogate for researchers to ethically study molecular mechanism of disease and develop therapies for neurological disorders in humans. Although the technology to differentiate neurons from iPSCs has progressed significantly in the past decade, the neural derivatives developed from human iPSCs do not reach the maturity and sophistication of a postnatal rodent brain and their developmental stage resembles that of the second trimester in humans.[65,66] Therefore, when selecting a differentiation protocol, it is important to consider limitations of the differentiation protocol, in addition to relevancy of the derived neuronal subtype to the clinical phenotype and target gene expression.

Current differentiation protocols often generate cells with generalized brain region identity (e.g., cortical, ventral) and/or identity defined by functional properties (e.g., GABAergic, dopaminergic). It is possible to narrow down to a specific brain region or certain cell identity that is likely to play a role in disease pathology by examining clinical phenotypes and/or postmortem human studies for a particular neurological disease. Transcriptomic data is often available for different differentiation protocols, which allows for selection of a protocol based on expression of particular genes of interest. In addition to the expression of appropriate cell identity markers (e.g., *TH* for dopaminergic neurons, *HB9* for motor neurons), it is important that the genes of interest are expressed at the developmental stage that one intends to conduct experiments. It is a good practice to utilize a time course to validate target gene expression at various differentiation time points before committing to a differentiation protocol. For protein-coding genes, this expression should be validated through Western blot and, ideally, immunostaining or fluorescence activated cell sorting (FACS) in combination with cell markers to verify that the cell population of interest indeed expresses the target protein.[62]

It is often hard to grasp the limitations of a specific differentiation protocol without first having hands-on experience. One can start by evaluating the scientific relevance, how compatible it is to the downstream experimental procedures, and equipment availabilities (e.g., a hypoxic incubator). In general, differentiation protocols can be divided into subcategories based on (1) culturing system and (2) method for inducing neurogenesis. The pros and cons for each subcategories are discussed in the following paragraphs.

Based on the culturing system, differentiation protocols can be categorized into monolayer versus 3-dimensional (3-D) protocols. Both monolayer and 3-D culturing systems are commonly used for differentiation of neurons, and some protocols even alternate between the two within a single protocol.[67] 3-D culture is accomplished by allowing singlized cells or dissociated cell clumps to spontaneously form globular aggregates in low attachment vessels. In some protocols, 3-D culture is utilized to reach high cell density required during differentiation, mimicking that of natural embryogenesis, which is hard to achieve with a monolayer culture.[68,69] In others, the act of lifting neural rosettes from a monolayer culture to form 3-D neurospheres allows the enrichment and expansion of neural progenitors in suspension.[67] There are significant differences in cell-cell interactions and microenvironment between monolayer and 3-D culturing systems.[69] One caveat of transiting cells back and forth between these two systems is the risk of substantial cell loss during each transition. More recently, advances in 3-D culture have been made to create neural spheroids on a large-scale, as well as advances in differentiation of brain organoids that develop structural complexity. Compared to brain organoids, neural spheroids are more homogeneous and smaller in size with no necrosis in the core regions.[70] On the other hand, brain organoids offer mixtures of cell population and complex organization in areas, which are in close resemblance to the human cortex and can be cultured indefinitely with labor-intensive care.[71–73] Traditionally, precursors and differentiated cells, whether cultured in a monolayer, 3-D environment, or both, go through a final dissociation to be plated as a monolayer for further maturation in a format suitable for terminal experimental assays. Although it remains challenging to interrogate morphological and functional cellular phenotypes in 3-D culture due to technical limitations, advances in microscopy, optical clearing, and multi-electrode array technology are well underway to reduce barriers and increase throughput.

Based on the method for inducing neurogenesis, differentiation protocols can be categorized into patterning-based directed differentiation protocols versus transcription-factor-based induced differentiation protocols. Directed differentiation protocols utilize growth factors or small molecules to direct neurogenesis through extrinsic signaling pathways and mimic in vivo embryonic brain development. As a result, directed differentiation protocols produce a mixed cell population, with various cell types that develop and mature alongside one another. Similar to normal brain development, patterning-based differentiation directs iPSCs into a neural ectoderm cell fate to produce neural progenitor cells (NPCs), which can then differentiate and mature into various neuronal and glial subtypes.[74,75] When patterned toward ectoderm, iPSCs tend to produce dorsal forebrain excitatory neurons;

however, when morphogenic cues are provided, NPCs can acquire specific regional or cellular identity and differentiate into other neural subtypes, including GABAergic, dopaminergic, spinal motor neurons, or astrocytes, just to name a few. It is worth noting that changes in NPCs have direct and often amplified effects on downstream differentiation, which can make it difficult to dissociate disease cellular phenotype from changes caused by technical and/or biological variability. Further, although researchers often refer to a directed differentiation protocol by one specific cell type it produces, it is important to keep in mind that the final cell population is heterogeneous by nature and cell sorting is often needed to enrich for the desired cell population. While the ability to reproduce diverse cell population similar to that of natural brain development is extremely valuable, especially when studying neurodevelopmental disorders, this heterogeneity can be a problem for sensitive platforms or assays that require uniform populations. Another notable caveat when using a directed differentiation protocol is that the final cell composition can vary between batch-to-batch. To reduce heterogeneity and variability, researchers developed gene-based induced differentiation protocols by ectopically expressing transcription factors and/or microRNAs to program iPSCs into final neural subtypes.[76–78] By compressing/bypassing the precursor stage, induced differentiation protocols also significantly shorten the time needed to reach later developmental stages compared to directed differentiation protocols. It is important to remember that, with a shortened progenitor stage, these protocols may not be suitable for studying changes that happen specifically during the NPC stage. The ectopic genes are commonly introduced in conjunction with antibiotic resistant genes through stable viral transduction in a system inducible by small molecules (e.g., doxycycline). Upon induction, cells are subjected to antibiotic selection to remove un-transduced cells, which aids in achieving homogeneity. Since each induced differentiation protocol aims at producing one specific cell type, unlike typical development, the cell derivatives do not develop and mature alongside with other cell types. As a result, co-culturing with other cell types may be necessary to improve cellular function and maturity, as well as to provide more biologically relevant cell populations for the disease being studied. In addition, many of the genes utilized for induced differentiations are transcription factors and master regulators of neurogenesis and the epigenome and, when exogenously expressed, may override and mask cellular phenotypes, especially when functioning in the same molecular pathway as the gene of interest. Lastly, if antibiotic selection is employed, it is important to consider its effect on cellular function and targeted phenotype. One could circumvent this issue by stably integrating inducible gene(s) into genomic safe harbors. Interestingly, induced differentiation techniques were first reported for the direct conversion of other terminally differentiated somatic cells, such as fibroblasts, into neural cells, a process termed transdifferentiation.[79] Transdifferentiation is especially useful for preserving patient epigenetic information, which could be important when studying age-related diseases, as the process of reprogramming into iPSCs has been reported to change epigenetic signatures.[80] This technology also holds promise as a therapeutic treatment for neurodegenerative disorders.[81] However, the pitfalls of transdifferentiation are that somatic cells like fibroblasts have limited cell expansion capacity before reaching senescence and that the current transdifferentiation protocols are still inefficient at converting human somatic cells into neural cells.

Best practices during differentiation

For studying neurological diseases using iPSCs, a primer by Engle et al. outlined best practices and workflows in detail and is a great resource for references.[62] It is recommended to culture paired diseased and control iPSC lines in parallel to ensure that all cells were subjected to the same culturing conditions. Although the appropriate number for individuals/cell lines used in a study heavily depends on experimental design, according to Germain et al., four patient/control pairs provide a reasonable statistical precision in general.[61] No matter which differentiation method is used, high quality iPSCs with minimal spontaneous differentiation that have passed stringent quality control stated above are essential. Since batch-to-batch variation is common for differentiation of neurons, each endpoint experiment should be conducted on cells produced from multiple differentiation batches and paired analysis should be done for statistics. For differentiation quality control, it is recommended to check cellular identity using immunocytochemistry or FACS with cell markers at various developmental stages (e.g., iPSCs, NPCs, neurons). Bulk quantitative polymerase chain reaction (qPCR) and bulk RNA sequencing are useful for getting an overview of cell population at each developmental stage, as well as for differential gene expression comparisons between control and experimental groups. Recent advancement in single cell qPCR and RNA sequencing technology allows researchers to look at gene expression profiles in individual cells. This granularity is helpful in understanding how iPSC-derived neural derivatives compare to brain cells in vivo and whether a specific cell population is more susceptible to a certain condition (e.g., mutation, compound treatment). Lastly, one should validate that the terminally differentiated cells demonstrate functional properties similar to those observed in vivo. For example, terminally differentiated neurons should exhibit spontaneous firing that is blocked by ion channel antagonists corresponding to the specific neuronal subtype.

If a lab is venturing into the world of iPSCs and neuronal differentiation for the first time, the best practices presented in this chapter will be critical for obtaining reliable and reproducible results.

References

1. Abraham D, et al. *Biosafety in Microbiological and Biomedical Laboratories*. 5th ed. U.S. Department of Health and Human Services, Public Health Service, Centers for Disease Control and Prevention, National Institutes of Health; 2009 (CDC)21-1112.
2. O'Connor R, O'Driscoll L. Setting up a cell culture laboratory. In: *Cell and Tissue Culture: Assorted Techniques*. (USA): Elsevier Science; 2005.
3. Gilpin RW. Biosafety Cabinetry: design, construction, performance, and Field certification. *NSF*. 2018;49(Annex E). NSF/ANSI 49-2018.
4. Hunt CJ. Cryopreservation of human stem cells for clinical application: a review. *Transfus Med Hemother*. 2011;38(2):107–123.
5. Takahashi K, Tanabe K, Ohnuki M, et al. Induction of pluripotent stem cells from adult human fibroblasts by defined factors. *Cell*. 2007;131(5):861–872.
6. Yu J, Vodyanik MA, Smuga-Otto K, et al. Induced pluripotent stem cell lines derived from human somatic cells. *Science*. 2007;318(5858):1917–1920.
7. Loh Y-H, Agarwal S, Park I-H, et al. Generation of induced pluripotent stem cells from human blood. *Blood*. 2009;113(22):5476–5479. https://doi.org/10.1182/blood-2009-02-204800.
8. Haase A, Olmer R, Schwanke K, et al. Generation of induced pluripotent stem cells from human cord blood. *Cell Stem Cell*. 2009;5(4):434–441. https://doi.org/10.1016/j.stem.2009.08.021.

9. Aasen T, Raya A, Barrero MJ, et al. Efficient and rapid generation of induced pluripotent stem cells from human keratinocytes. *Nat Biotechnol.* 2008;26(11):1276–1284. https://doi.org/10.1038/nbt.1503.
10. Sun N, Panetta NJ, Gupta DM, et al. Feeder-free derivation of induced pluripotent stem cells from adult human adipose stem cells. *Proc Natl Acad Sci USA.* 2009;106(37):15720–15725. https://doi.org/10.1073/pnas.0908450106.
11. D'Antonio M, Benaglio P, Jakubosky D, et al. Insights into the mutational burden of human induced pluripotent stem cells from an integrative multi-omics approach. *Cell Rep.* 2018;24(4):883–894. https://doi.org/10.1016/j.celrep.2018.06.091.
12. Sommer AG, Rozelle SS, Sullivan S, et al. Generation of human induced pluripotent stem cells from peripheral blood using the STEMCCA lentiviral vector. *J Vis Exp.* 2012;68. https://doi.org/10.3791/4327.
13. Loh Y-H, Hartung O, Li H, et al. Reprogramming of T cells from human peripheral blood. *Cell Stem Cell.* 2010;7(1):15–19. https://doi.org/10.1016/j.stem.2010.06.004.
14. Zhou T, Benda C, Dunzinger S, et al. Generation of human induced pluripotent stem cells from urine samples. *Nat Protoc.* 2012;7(12):2080–2089. https://doi.org/10.1038/nprot.2012.115.
15. Okita K, Matsumura Y, Sato Y, et al. A more efficient method to generate integration-free human iPS cells. *Nat Methods.* 2011;8(5):409–412. https://doi.org/10.1038/nmeth.1591.
16. Kawamura T, Suzuki J, Wang YV, et al. Linking the p53 tumour suppressor pathway to somatic cell reprogramming. *Nature.* 2009;460(7259):1140–1144. https://doi.org/10.1038/nature08311.
17. Schlaeger TM, Daheron L, Brickler TR, et al. A comparison of non-integrating reprogramming methods. *Nat Biotechnol.* 2015;33(1):58–63. https://doi.org/10.1038/nbt.3070.
18. Fusaki N, Ban H, Nishiyama A, Saeki K, Hasegawa M. Efficient induction of transgene-free human pluripotent stem cells using a vector based on Sendai virus, an RNA virus that does not integrate into the host genome. *Proc Jpn Acad Ser B Phys Biol Sci.* 2009;85(8):348–362.
19. Warren L, Manos PD, Ahfeldt T, et al. Highly efficient reprogramming to pluripotency and directed differentiation of human cells with synthetic modified mRNA. *Cell Stem Cell.* 2010;7(5):618–630. https://doi.org/10.1016/j.stem.2010.08.012.
20. Poleganov MA, Eminli S, Beissert T, et al. Efficient reprogramming of human fibroblasts and blood-derived endothelial progenitor cells using nonmodified RNA for reprogramming and immune evasion. *Hum Gene Ther.* 2015;26(11):751–766. https://doi.org/10.1089/hum.2015.045.
21. Adewumi O, Aflatoonian B, Ahrlund-Richter L, et al. Characterization of human embryonic stem cell lines by the International Stem Cell Initiative. *Nat Biotechnol.* 2007;25(7):803–816.
22. Müller F-J, Schuldt BM, Williams R, et al. A bioinformatic assay for pluripotency in human cells. *Nat Methods.* 2011;8(4):315–317. https://doi.org/10.1038/nmeth.1580.
23. Lenz M, Goetzke R, Schenk A, et al. Epigenetic biomarker to support classification into pluripotent and non-pluripotent cells. *Sci Rep.* 2015;5:8973. https://doi.org/10.1038/srep08973.
24. Bock C, Kiskinis E, Verstappen G, et al. Reference maps of human ES and iPS cell variation enable high-throughput characterization of pluripotent cell lines. *Cell.* 2011;144(3):439–452. https://doi.org/10.1016/j.cell.2010.12.032.
25. Tsankov AM, Akopian V, Pop R, et al. A qPCR ScoreCard quantifies the differentiation potential of human pluripotent stem cells. *Nat Biotechnol.* 2015;33(11):1182–1192. https://doi.org/10.1038/nbt.3387.
26. Avior Y, Biancotti JC, Benvenisty N. TeratoScore: assessing the differentiation potential of human pluripotent stem cells by quantitative expression analysis of teratomas. *Stem Cell Rep.* 2015;4(6):967–974. https://doi.org/10.1016/j.stemcr.2015.05.006.
27. Assessment of established techniques to determine developmental and malignant potential of human pluripotent stem cells. *Nat Commun.* 2018;9(1):1925. https://doi.org/10.1038/s41467-018-04011-3.
28. Okita K, Ichisaka T, Yamanaka S. Generation of germline-competent induced pluripotent stem cells. *Nature.* 2007;448(7151):313–317.
29. Yoshihara M, Hayashizaki Y, Murakawa Y. Genomic instability of iPSCs: challenges towards their clinical applications. *Stem Cell Rev Rep.* 2017;13(1):7–16. https://doi.org/10.1007/s12015-016-9680-6.
30. Martins-Taylor K, Xu R-H. Concise review: genomic stability of human induced pluripotent stem cells. *Stem Cells.* 2012;30(1):22–27. https://doi.org/10.1002/stem.705.
31. Baker D, Hirst AJ, Gokhale PJ, et al. Detecting genetic mosaicism in cultures of human pluripotent stem cells. *Stem Cell Rep.* 2016;7(5):998–1012. https://doi.org/10.1016/j.stemcr.2016.10.003.

32. Assou S, Girault N, Plinet M, et al. Recurrent genetic abnormalities in human pluripotent stem cells: definition and routine detection in culture supernatant by targeted droplet digital PCR. *Stem Cell Rep*. 2020;14(1):1–8. https://doi.org/10.1016/j.stemcr.2019.12.004.
33. Merkle FT, Ghosh S, Kamitaki N, et al. Human pluripotent stem cells recurrently acquire and expand dominant negative P53 mutations. *Nature*. 2017;545(7653):229–233. https://doi.org/10.1038/nature22312.
34. McIntire E, Taapken S, Leonhard K, Larson AL. Genomic stability testing of pluripotent stem cells. *Curr Protoc Stem Cell Biol*. 2020;52(1):e107. https://doi.org/10.1002/cpsc.107.
35. European Collection of Authenticated Cell Cultures, Sigma-Aldrich. Good cell banking practices. In: *Fundamental Techniques in Cell Culture*. 4th ed. Darmstadt, Germany: Merck KGaA; 2018.
36. Baghbaderani BA, Tian X, Neo BH, et al. cGMP-manufactured human induced pluripotent stem cells are available for pre-clinical and clinical applications. *Stem Cell Rep*. 2015;5(4):647–659. https://doi.org/10.1016/j.stemcr.2015.08.015.
37. Li Y, Ma T. Bioprocessing of cryopreservation for large-scale banking of human pluripotent stem cells. *BioResearch Open Access*. 2012;1(5):205–214. https://doi.org/10.1089/biores.2012.0224.
38. Jozefczuk J, Drews K, Adjaye J. Preparation of mouse embryonic fibroblast cells suitable for culturing human embryonic and induced pluripotent stem cells. *J Vis Exp*. 2012;64. https://doi.org/10.3791/3854.
39. Rebuzzini P, Zuccotti M, Redi CA, Garagna S. Chromosomal abnormalities in embryonic and somatic stem cells. *Cytogenet Genome Res*. 2015;147(1):1–9. https://doi.org/10.1159/000441645.
40. Navarro-Alvarez N, Soto-Gutierrez A, Yuasa T, et al. Long-term culture of Japanese human embryonic stem cells in feeder-free conditions. *Cell Transplant*. 2008;17(1–2):27–33.
41. Beers J, Gulbranson DR, George N, et al. Passaging and colony expansion of human pluripotent stem cells by enzyme-free dissociation in chemically defined culture conditions. *Nat Protoc*. 2012;7(11):2029–2040. https://doi.org/10.1038/nprot.2012.130.
42. Ohnuma K, Fujiki A, Yanagihara K, et al. Enzyme-free passage of human pluripotent stem cells by controlling divalent cations. *Sci Rep*. 2014;4:4646. https://doi.org/10.1038/srep04646.
43. Ludwig T, A Thomson J. Defined, feeder-independent medium for human embryonic stem cell culture. *Curr Protoc Stem Cell Biol*. 2007;2. https://doi.org/10.1002/9780470151808.sc01c02s2 (Chapter 1):Unit 1C.2.
44. Kim H. Splitting hESC/hiPSC lines on MEF using Accutase. *Stembook*. 2012. https://doi.org/10.3824/stembook.1.87.1.
45. Watanabe K, Ueno M, Kamiya D, et al. A ROCK inhibitor permits survival of dissociated human embryonic stem cells. *Nat Biotechnol*. 2007;25(6):681–686.
46. Garitaonandia I, Amir H, Boscolo FS, et al. Increased risk of genetic and epigenetic instability in human embryonic stem cells associated with specific culture conditions. *PloS One*. 2015;10(2):e0118307. https://doi.org/10.1371/journal.pone.0118307.
47. Bai Q, Ramirez J-M, Becker F, et al. Temporal analysis of genome alterations induced by single-cell passaging in human embryonic stem cells. *Stem Cells and Dev*. 2015;24(5):653–662. https://doi.org/10.1089/scd.2014.0292.
48. Katkov II, Kan NG, Cimadamore F, Nelson B, Snyder EY, Terskikh AV. DMSO-free programmed cryopreservation of fully dissociated and adherent human induced pluripotent stem cells. *Stem Cells Int*. 2011;2011:981606. https://doi.org/10.4061/2011/981606.
49. Chen Y, et al. A versatile Polypharmacology platform promotes cytoprotection and viability of human pluripotent and differentiated cells. *bioRxiv*. 2019;815761. https://doi.org/10.1101/815761.
50. Onuma Y, Higuchi K, Aiki Y, et al. A stable chimeric fibroblast growth factor (FGF) can successfully replace basic FGF in human pluripotent stem cell culture. *PloS One*. 2015;10(4):e0118931. https://doi.org/10.1371/journal.pone.0118931.
51. Spinner NB, Conlin LK. Mosaicism and clinical genetics. *Am J Med Genet C Semin Med Genet*. 2014;166C(4):397–405. https://doi.org/10.1002/ajmg.c.31421.
52. Hinz L, Hoekstra SD, Watanabe K, Posthuma D, Heine VM. Generation of isogenic controls for in vitro disease modelling of X-chromosomal disorders. *Stem Cell Rev Rep*. 2019;15(2):276–285. https://doi.org/10.1007/s12015-018-9851-8.
53. Bermejo-Alvarez P, Ramos-Ibeas P, Gutierrez-Adan A. Solving the "X" in embryos and stem cells. *Stem Cells Dev*. 2012;21(8):1215–1224. https://doi.org/10.1089/scd.2011.0685.
54. Niday Z, Tzingounis AV. Potassium Channel Gain of function in Epilepsy: an unresolved paradox. *Neuroscientist*. 2018;24(4):368–380. https://doi.org/10.1177/1073858418763752.

55. Aigner S, Heckel T, Zhang JD, Andreae LC, Jagasia R. Human pluripotent stem cell models of autism spectrum disorder: emerging frontiers, opportunities, and challenges towards neuronal networks in a dish. *Psychopharmacology*. 2014;231(6):1089–1104.
56. Hoffman GE, Schrode N, Flaherty E, Brennand KJ. New considerations for hiPSC-based models of neuropsychiatric disorders. *Mol Psychiatry*. 2019;24(1):49–66. https://doi.org/10.1038/s41380-018-0029-1.
57. Liu W, Deng Y, Liu Y, Gong W, Deng W. Stem cell models for drug discovery and toxicology studies. *J Biochem Mol Toxicol*. 2013;27(1):17–27. https://doi.org/10.1002/jbt.21470.
58. Rugg-Gunn PJ, Ferguson-Smith AC, Pedersen RA. Status of genomic imprinting in human embryonic stem cells as revealed by a large cohort of independently derived and maintained lines. *Hum Mol Genet*. 2007;16(Spec No. 2):R243–R251.
59. Churko JM, Lee J, Ameen M, et al. Transcriptomic and epigenomic differences in human induced pluripotent stem cells generated from six reprogramming methods. *Nat Biomed Eng*. 2017;1(10):826–837. https://doi.org/10.1038/s41551-017-0141-6.
60. Stephenson E, Ogilvie CM, Patel H, et al. Safety paradigm: genetic evaluation of therapeutic grade human embryonic stem cells. *J R Soc Interface*. 2010;7(Suppl 6):S677–S688. https://doi.org/10.1098/rsif.2010.0343.focus.
61. Germain P-L, Testa G. Taming human genetic variability: transcriptomic meta-analysis Guides the experimental design and interpretation of iPSC-based disease modeling. *Stem Cell Rep*. 2017;8(6):1784–1796. https://doi.org/10.1016/j.stemcr.2017.05.012.
62. Engle SJ, Blaha L, Kleiman RJ. Best practices for translational disease modeling using human iPSC-derived neurons. *Neuron*. 2018;100(4):783–797. https://doi.org/10.1016/j.neuron.2018.10.033.
63. Huo J, Wei F, Cai C, Lyn-Cook B, Pang L. Sex-related differences in drug-induced QT prolongation and Torsades de Pointes: a new model system with human iPSC-CMs. *Toxicol Sci*. 2019;167(2):360–374. https://doi.org/10.1093/toxsci/kfy239.
64. Cheung AYL, Horvath LM, Grafodatskaya D, et al. Isolation of MECP2-null Rett Syndrome patient hiPS cells and isogenic controls through X-chromosome inactivation. *Hum Mol Genet*. 2011;20(11):2103–2115. https://doi.org/10.1093/hmg/ddr093.
65. Velasco S, Kedaigle AJ, Simmons SK, et al. Individual brain organoids reproducibly form cell diversity of the human cerebral cortex. *Nature*. 2019;570(7762):523–527. https://doi.org/10.1038/s41586-019-1289-x.
66. Amiri A, Coppola G, Scuderi S, et al. Transcriptome and epigenome landscape of human cortical development modeled in organoids. *Science*. 2018;362(6420). https://doi.org/10.1126/science.aat6720.
67. Hu B-Y, Weick JP, Yu J, et al. Neural differentiation of human induced pluripotent stem cells follows developmental principles but with variable potency. *Proc Natl Acad Sci USA*. 2010;107(9):4335–4340. https://doi.org/10.1073/pnas.0910012107.
68. Efthymiou AG, Chen G, Rao M, Chen G, Boehm M. Self-renewal and cell lineage differentiation strategies in human embryonic stem cells and induced pluripotent stem cells. *Expert Opin Biol Ther*. 2014;14(9):1333–1344. https://doi.org/10.1517/14712598.2014.922533.
69. Sances S, Bruijn LI, Chandran S, et al. Modeling ALS with motor neurons derived from human induced pluripotent stem cells. *Nat Neurosci*. 2016;19(4):542–553. https://doi.org/10.1038/nn.4273.
70. Rigamonti A, Repetti GG, Sun C, et al. Large-scale production of mature neurons from human pluripotent stem cells in a three-dimensional suspension culture system. *Stem Cell Rep*. 2016;6(6):993–1008. https://doi.org/10.1016/j.stemcr.2016.05.010.
71. Marton RM, Pasca SP. Organoid and assembloid Technologies for investigating cellular Crosstalk in human brain development and disease. *Trends Cell Biol*. 2020;30(2):133–143. https://doi.org/10.1016/j.tcb.2019.11.004.
72. Lancaster MA, Renner M, Martin C-A, et al. Cerebral organoids model human brain development and microcephaly. *Nature*. 2013;501(7467):373–379. https://doi.org/10.1038/nature12517.
73. Qian X, Song H, Ming G-L. Brain organoids: advances, applications and challenges. *Development*. 2019;146(8). https://doi.org/10.1242/dev.166074.
74. Tcw J, Wang M, Pimenova AA, et al. An efficient platform for astrocyte differentiation from human induced pluripotent stem cells. *Stem Cell Rep*. 2017;9(2):600–614. https://doi.org/10.1016/j.stemcr.2017.06.018.
75. Chambers SM, Fasano CA, Papapetrou EP, Tomishima M, Sadelain M, Studer L. Highly efficient neural conversion of human ES and iPS cells by dual inhibition of SMAD signaling. *Nat Biotechnol*. 2009;27(3):275–280. https://doi.org/10.1038/nbt.1529.

76. Li X, Tao Y, Bradley R, et al. Fast generation of functional subtype Astrocytes from human pluripotent stem cells. *Stem Cell Rep*. 2018;11(4):998–1008. https://doi.org/10.1016/j.stemcr.2018.08.019.
77. Zhang Y, Pak C, Han Y, et al. Rapid single-step induction of functional neurons from human pluripotent stem cells. *Neuron*. 2013;78(5):785–798. https://doi.org/10.1016/j.neuron.2013.05.029.
78. Yang N, Chanda S, Marro S, et al. Generation of pure GABAergic neurons by transcription factor programming. *Nat Methods*. 2017;14(6):621–628. https://doi.org/10.1038/nmeth.4291.
79. Vierbuchen T, Ostermeier A, Pang ZP, Kokubu Y, Südhof TC, Wernig M. Direct conversion of fibroblasts to functional neurons by defined factors. *Nature*. 2010;463(7284):1035–1041. https://doi.org/10.1038/nature08797.
80. Lister R, Pelizzola M, Kida YS, et al. Hotspots of aberrant epigenomic reprogramming in human induced pluripotent stem cells. *Nature*. 2011;471(7336):68–73.
81. Smith DK, He M, Zhang C-L, Zheng JC. The therapeutic potential of cell identity reprogramming for the treatment of aging-related neurodegenerative disorders. *Prog Neurobiol*. 2017;157:212–229. https://doi.org/10.1016/j.pneurobio.2016.01.006.

CHAPTER

2

Phenotypic assay development with iPSC-derived neurons: technical considerations from plating to analysis

Elizabeth D. Buttermore[1,2]

[1]Human Neuron Core, Rosamund Stone Zander Translational Neuroscience Center, Boston Children's Hospital, Boston, MA, United States; [2]Neurology Department, Boston Children's Hospital, Boston, MA, United States

Introduction

The seminal discovery of four transcription factors (Oct3/4, Sox2, c-Myc, and Klf4) that can reprogram somatic cells (such as blood and skin) into human induced pluripotent stem cells (hiPSCs or iPSCs) drastically changed the landscape of preclinical disease modeling across many fields.[1] This is especially true for the field of neuroscience, where viable primary tissue is more challenging to obtain. In preclinical neuroscience studies, rodent and autologous cell models have failed to recapitulate the signaling mechanisms found in human neurons. Thus, many of the findings from these studies have failed to translate to the clinic. The use of human neurons for preclinical model systems is critical to increase confidence in targets for neurological disorders. In addition, the ability to obtain iPSCs from patients with genetically defined neurological disorders provides additional context for identifying mechanisms and pathways of importance in patient-derived neurons. While there have been advances in the protocols available to differentiate neurons, it's important to establish the best possible assay conditions for each biological question so that the assay readout is relevant to the patient or neurological disorder being investigated. This chapter presents some important technical tips and considerations when establishing in vitro assays with iPSC-derived neurons to study neurological disorders so that these studies have the best chance for success.

Phenotyping of Human iPSC-derived Neurons
https://doi.org/10.1016/B978-0-12-822277-5.00015-8

Establishing optimal conditions for phenotyping iPSC-derived neurons

When working with hiPSC-derived neurons, it is important to realize that they cannot be handled the same way as traditional cell lines, or even rodent neuronal cultures. Cell culture conditions for hiPSC-derived neurons will vary based on the subtype composition and maturation level of the neurons, which depends on the differentiation protocol used and the timing of differentiation when neurons are plated. As discussed in an earlier chapter, iPSCs should be clump passaged to reduce stress and karyotype changes that are more likely to occur when iPSCs are passaged as single cells. Similar to this concept, there are recommended conditions for the treatment of neurons derived from iPSCs to promote optimal neuronal health and assay development. This section explores considerations for culture conditions based on the neuronal differentiation protocol used and the assay end point of interest.

Differentiation protocol considerations

Optimal cell culture conditions for a chosen assay utilizing hiPSC-derived neurons depends heavily on the differentiation protocol selected and the timing of analysis. For example, neuronal progenitors will likely continue to proliferate over time and should potentially be plated at a lower density to allow for expansion and differentiation over time unless they will be processed immediately. However, plating mature post-mitotic neurons at too low of a density can be detrimental to the health of the neurons, as they prefer cell-to-cell contact to maintain optimal health. Other chapters will go into more detail about differentiation protocol options, including transcription factor-based protocols and small molecule, patterning-based protocols. Here, we highlight how the choices made will affect conditions for downstream assay development.

A common protocol that is used to generate cortical neurons is the transcription factor-based protocol that utilizes Neurogenin2 (NGN2) overexpression to rapidly produce a relatively pure population of cortical neurons.[2] Several groups have reported that NGN2 neurons benefit from being cocultured with astrocytes to maintain long-term cultures and to promote the formation of a functionally mature network.[3–5] Without the addition of astrocytes in coculture, NGN2-induced neurons tend to peel off of the plate by about 2–3 weeks into differentiation. This affects the timing that pure neuronal cultures can be collected for RNA or protein analysis if astrocyte signatures are unwanted for those analyses. One benefit to transcription factor-driven protocols is that the timeline from iPSCs to neurons is often much faster than patterning-based approaches. The NGN2 protocol, for example, derives neurons within 1 week of induction of NGN2 expression. Another advantage is that the transcription factor-based protocols tend to be more homogenous, which helps with reproducibility from well to well when using the neurons for drug screening. Interestingly, a recent paper showed that NGN2-induced neurons may not be as homogeneous as originally reported.[3] Although there seem to be more neuronal subtypes derived by NGN2 induction than previously thought, nearly all of the cells generated from NGN2 induction are neuronal. The homogeneity in transcription factor-based protocols may be due, in part, to the selection step made possible by introduction of the transcription factor vector. This neuronal purity improves reproducibility during assay development and screening.

Compared to transcription factor-based protocols, patterning-based protocols that use small molecules to follow developmental cues tend to take significantly more time, upwards of several months, to differentiate mature neurons.[6–8] Recent efforts have been made to identify small molecules that accelerate patterning-based differentiation timelines for neurons and astrocytes.[9,10] As will be described in other chapters in more detail, patterning-based approaches typically yield more heterogeneous mixtures of cells. This heterogeneity provides benefits of more physiologically relevant populations of neurons and glia, but also poses difficulties when scaling up for screening efforts or when a pure cell population is required. Patterning-based protocols generally do not require coculturing with astrocytes or other supportive materials to maintain long-term because supportive cells are frequently differentiated along with the neurons. However, astrocytes can still be added in coculture to help promote synapse formation and prevent neuronal clumping in 2D cultures. Patterning-based protocols may also contain progenitor-like cells that are not terminally differentiated and will continue to proliferate over time. Neurons can be enriched in these cultures with cell sorting techniques that utilize cell surface markers, such as Neural Cell Adhesion Molecule (NCAM).[11,12] As highlighted in Fig. 2.1, when deciding which protocol to use, it's essential to ask whether homogeneity is most important, reducing variability in culture conditions

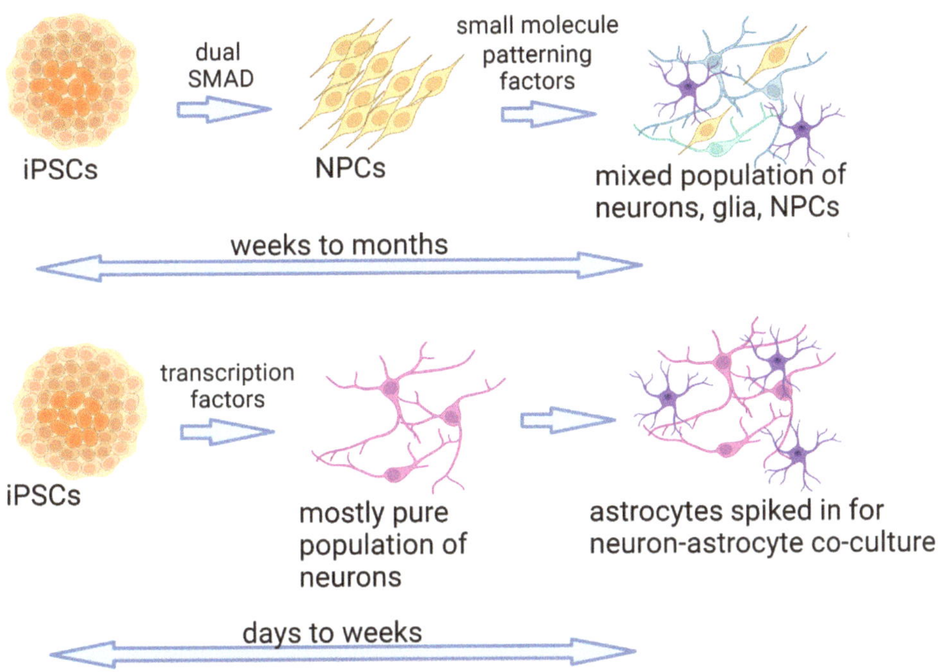

FIGURE 2.1 Small molecule-driven differentiation versus transcription factor-driven differentiation. Differentiating iPSCs into neuron can be achieved using small molecules to mimic neurodevelopmental cues over the course of weeks to months or with overexpression of single or combinations of transcription factors that are important for neural induction, which can induce formation of neurons in days to weeks. Small molecular-driven protocols tend to produce heterogenous cell populations while transcription factor-driven protocols produce more homogenous populations of neurons that typically require astrocyte coculture. *Created with BioRender.com.*

across wells, or whether it's more important to represent the physiologically relevant diversity of cellular subtypes seen in vivo. From there, the considerations below will allow for the optimization of assay conditions for various phenotypic endpoints.

Coating substrates

The coating materials used for rodent neuron cell culture also work well for iPSC-derived neurons. The substrate selected for a specific project is often chosen as the result of testing several combinations to determine what works best. In general, it is accepted that neurons, including iPSC-derived neurons, adhere better to plastic plates than to glass. However, confocal high-resolution imaging and electrophysiological end points require glass coverslips for optimal results. To plate neurons on glass, it's critical to first coat with a positively charged coating material, like poly-D-lysine (PDL) or poly-L-ornithine (Poly-O), which will enhance the adhesion of negatively charged cell membranes.[13,14] PDL is commonly used for its superior adhesion properties; however, it is toxic to cells, and therefore must be washed and dried before cells, or other coating materials, are added. As an alternative to PDL, Poly-O is sometimes chosen because it is also positively charged and is not toxic to the cells and therefore doesn't require thorough washing before adding other coatings or cells. However, the choice of coating material may also affect downstream differentiation efficiency and phenotyping, as Poly-O has been shown to support neural stem cell proliferation and differentiation more than poly-L-lysine or fibronectin under certain culture conditions.[15] Thus, when establishing a new assay, different coating materials should be compared before settling on the optimal coating combination.

Another alternative to PDL and Poly-O is polyetherimide (PEI). PEI has been shown to improve the distribution of cells across slides or wells, and in comparison to PDL and Poly-O was superior for maintaining an even distribution of neurons with reduced clumping over time.[16] Cell clumping can make downstream analysis more difficult for some assay endpoints and researchers are often looking for ways to minimize this clumping. Recent studies have also identified diaminopropane (DAP)[17] and dendritic polyglycerol amines (dPGAs)[13] as polymer coating materials that improve neuronal adherence to glass and plastic surfaces for long-term cultures and that may be useful to test for clumping issues.

After coating with PDL, Poly-O, or PEI, an additional coating may be added to enhance cell adhesion to slides or tissue culture plates. Extracellular matrix proteins and signaling molecules are commonly chosen materials for additional coating because they have been found to be important for supporting the health and survival of cells, especially neurons, in long-term cultures.[18–20] One of the most commonly used extracellular matrix glycoproteins for neurons is laminin. Laminin binds to extracellular matrix proteins and integrin receptors found on the cell membrane, promoting adhesion of cells to the surface of plates and dishes.[21,22] Another common extracellular matrix glycoprotein coating material is fibronectin. Importantly, substrates such as laminin and fibronectin have been shown to both support and promote neurite outgrowth in certain culture conditions.[23,24] Therefore, when examining neurite outgrowth in patient versus control neurons, or in treated versus untreated neurons, adding fibronectin or laminin may activate neurite growth pathways that interfere with the pathways being examined in the assay. Further, coating materials chosen for functional multi-electrode array (MEA) assays have also been shown to alter the results, with geltrex coating leading to the insulation of signal from electrodes and laminin promoting optimal signal relay.[25] These examples highlight the point that coating materials may change the behavior, morphology, or survival of cells that are plated on the chosen materials. Therefore, it is important to ensure that the selected coating materials do not interfere with the biological question being asked.

High content imaging (HCI)

Recent advances in imaging and analysis tools led to the development of automated imaging systems and software algorithms that can image cells and quantify phenotypes based on the presence, absence, or localization of fluorescent markers (Fig. 2.2). The use of customizable software algorithms to define unique shapes or cellular sub-compartments allows for complex biological questions to be quantified in a meaningful way. One example of this came from a group at Boston Children's Hospital, where they imaged and quantified cilia number and length in rodent control and *CDKL5* knockdown neurons and found longer cilia in neurons with loss of *CDKL5*.[26] This type of analysis was made possible across thousands of cells with the use of automated analysis algorithms. And while this publication highlights the elegance of this technology for scientific discovery research, it may be difficult to appreciate the amount of time and effort that went into optimizing precise conditions to allow for imaging and quantification of cilia across multiple batches of neuronal cultures. In addition to cilia, any other subcellular compartment or organelle can be quantified that is able to be uniquely labeled. For example, our group recently developed an algorithm to quantify synapses with high content imaging (Fig. 2.2), which will help with interpretation of cultures that exhibit functional differences.

When optimizing high content imaging and automated software analysis for a new project, it is critical to verify the specificity of the antibodies being used and imaging should be done carefully so as to not overexpose signals. In addition, to accurately measure different cell markers, cell culture conditions need to be optimized for the endpoint of interest. For measuring neurite outgrowth, there are numerous ways to establish the assay. In the early days of measuring neurite outgrowth, Sholl analysis was commonly used to identify changes in growth over time or across conditions.[27,28] In Sholl analysis, concentric circles are drawn around a neuronal cell body and the number of neurite crossings in each circle would provide information about the complexity and length of neurites for each neuron (Fig. 2.2). This analysis has improved with automation over the years, but has been replaced more recently with analysis algorithms that are able to trace every neurite and quantify the exact length and branch points of all neurites in every neuron imaged (Fig. 2.2). In a recent study, we have used this method to show that iPSC-derived neurons from patients with tuberous sclerosis complex (TSC) have increased neurite outgrowth compared to healthy control neurons due to increased mTOR activation.[29] It is clear that plating neurons at too high a density will make it more challenging to determine neurite outgrowth for individual neurons, but plating neurons at too low of a density can also affect neuronal health and result in decreased neurite outgrowth, unrelated to the biological question being addressed. Although neurite outgrowth is just one example, it highlights the importance for the neuronal cell density to be tested and optimized for each end point of interest.

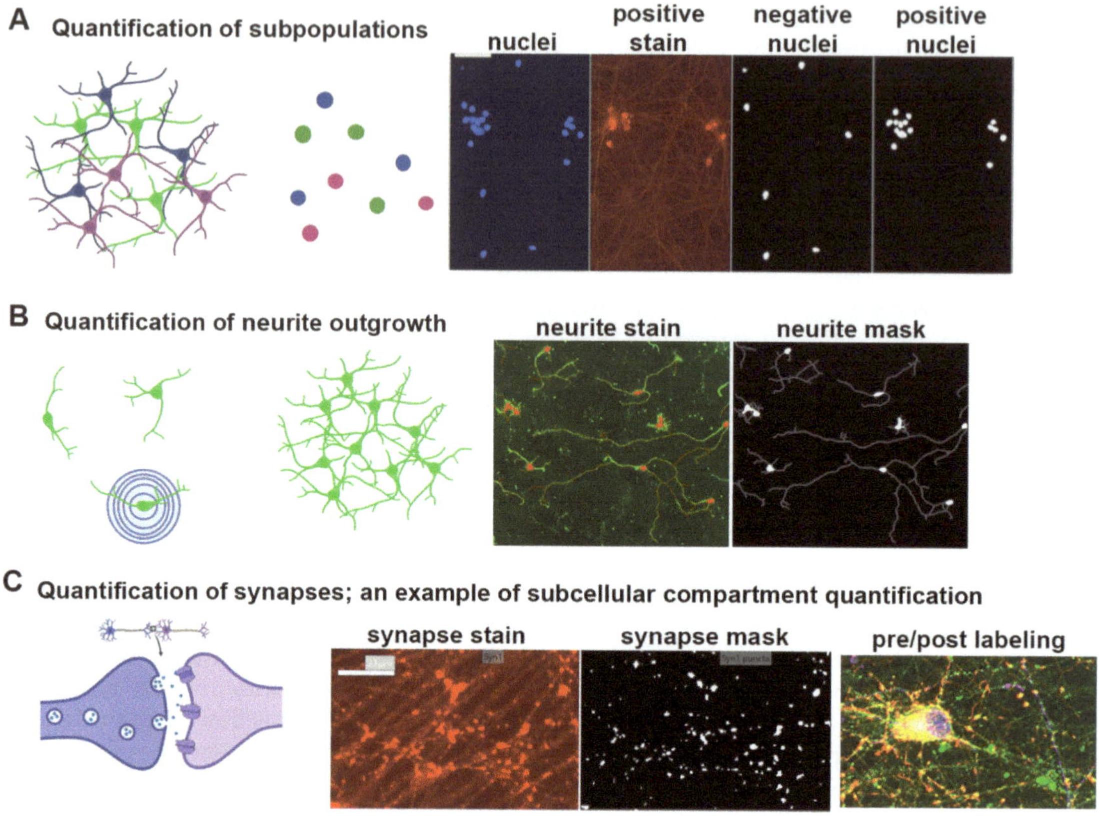

FIGURE 2.2 High content imaging for quantification of neuronal phenotypes. (A) High content imaging can be used to quantify percentages of neurons expressing markers of cellular subtypes. In this example, a Hoechst stain was used to label all nuclei (blue) and a second marker was use to stain positive cells (red). The algorithm created allows nuclei to be segmented as positive or negative for the stain in red. (B) Neurite outgrowth can be completed in low density cultures using Sholl analysis, exemplified by the concentric rings around a single neuron soma. Newer technology uses algorithms to trace all neurites present in the culture, as shown in the neurite tracing example on the right. (C) High content imaging can also be used to quantify subcellular structures and organelles, such as the synapse, which is labeled here with Synapsin1 (red) and PSD95 (green). The mask shows precise quantification of puncta, which can be colocalized with pre- and post-synaptic markers to quantify synapse number. *Created with BioRender.com.*

In addition to cell density, different coating conditions, astrocyte coculture, media components and growth factors should be optimized for each neuronal differentiation protocol and assay endpoint. For example, we have found that plating 20,000 neurons and 3000 astrocytes per well provides optimal conditions for long-term cultures of NGN2-induced neurons that are re-plated at day 6 of differentiation.[29] However, when utilizing a patterning-based protocol, astrocytes may not be necessary in coculture for long-term image analysis after replating.[7,8] This is likely because the patterning-based approaches also differentiate supportive cells within the same differentiation process.

There are many features to consider when deciding which high content imaging instrument will suit the needs for individual project goals. Recent advances in high content imaging instruments have allowed new features to keep pace with advances in manipulation of cell biology and in vitro assays.

The Carpenter lab at The Broad Institute of MIT and Harvard developed the Cell Painting assay, which utilizes high content imaging from six fluorescent dyes using five wavelengths to quantify hundreds of parameters in cell cultures.[30] Quantification with this method allows for the comparison of drug-treated neurons or neurons from different genetic indications with an output of hundreds of possible phenotypic differences. To operationalize an assay such as this, a high content imaging device needs to have at least five channels for imaging. Newer devices offer upgrades up to seven lasers, allowing for more precise control of imaging across multiple wavelengths in the same sample. The most common minimum recommended lasers to make sure a new high content imaging machine has are 405, 488, 561, and 640 nm, which will allow imaging of DAPI and other nuclear dyes in the shorter wavelengths and fluorescent secondary antibodies up to far-red wavelengths.

Beyond containing the option for imaging across multiple wavelengths, the instrument should be equipped with objectives ranging from 5× and 10× to image larger objects and cellular morphologies, to 20×, 40×, and 63×, which allow for more precise imaging of small subcellular compartments, such as synapses and cilia. With high magnification lenses, water-immersion objectives allow for sharper imaging of these details. The inclusion of high magnification water-based objectives with confocal capabilities enable investigators to image complex cellular models such as three-dimensional iPSC-derived spheroids and organoids. The imaging of these structures is often combined with clearing techniques, such as clarity, to allow penetration of antibodies into the inner regions of the three-dimensional shapes.[31]

Many questions can be answered by fixing and staining iPSC-derived neurons to understand the biological state of the neuronal cultures. However, it is becoming clear that live imaging provides a wealth of information that fixed cultures cannot reproduce. Imaging of live cultures allows for investigation of mitochondrial motility, pH changes across organelle membranes, and cell health changes.[32–35] In a recent study, live high content imaging technology was used to develop a screen for assessing mitochondrial motility in iPSC-derived cortical neurons from patients with SOD1 mutations, a recurrent variant found in patients with amyotrophic lateral sclerosis (ALS).[36] This screen was made possible by combining high content imaging and live cell chamber capabilities. Several companies provide dyes and kits for imaging these various live endpoints, but the instrument used to complete the imaging must be equipped with a live cell chamber that controls temperature, CO_2 levels, and humidity to assess these biological endpoints optimally.

In addition to multiple lasers, objectives, and a live cell chamber, it may be beneficial to consider whether multiple cameras that can acquire images simultaneously would be beneficial for your research studies. The ability to acquire two fluorophore channels simultaneously opens up a wealth of new experimental designs and reduces the overall scan time for more traditional image-based assays. During a typical imaging session with one camera, the camera will image each site in four, or more, wavelengths, one wavelength at a time before moving on to the next site. The ability to image two wavelengths simultaneously would cut imaging time in half for four wavelengths, allowing for imaging of more plates per day. In addition, functional assays that depend on rapid acquisition across a plate after treatment will be more feasible. The next section includes a discussion about imaging two wavelengths simultaneously, which enables functional analysis of multiple populations of neurons within the same site.

Perhaps the most important feature of any high content imaging device is the software that controls the device, including the analysis software that comes with it. There are dozens of available software packages that can be used to quantify anything with a fluorescent signal, including endless options for custom algorithms. However, the ease with which thresholds can be changed, changes can be visualized across images from different conditions, and data can be exported for downstream analysis, are critical for the effectiveness of these imaging platforms. In addition, some software systems require that certain cellular compartments, such as nuclei, be defined prior to other cellular compartments, but this strict requirement for an order of defining events can sometimes be restrictive when developing a novel analysis tool. Therefore, more-flexible systems that allow for customization will provide options for a greater number of assays. Further, new tools exist to combine machine learning for the classification of image data features to improve understanding of the complexity of big data generated, reducing hundreds of high-content multivariate features into a Z-value ranked data feature so that hits and interesting biology can be rapidly identified for follow up analysis in secondary screens or interrogated with orthogonal assays.[37,38]

Functional analysis

Analyzing the functional activity of iPSC-derived neurons and neuronal networks are excellent ways to understand how genetic or pharmacological perturbations affect neuronal behavior. Described below are a few of the ways the functional activity of neurons can be quantified, including multi-electrode array (MEA) recording, calcium imaging, and patch clamping (Fig. 2.3). Each of these methods should be carried out on a machine that allows for the regulation of temperature and CO_2 levels. Occasionally, devices will also regulate humidity, but the brevity of these recordings does not typically require this for success. While each method has its strengths and weaknesses, it is also important to remember that the conditions needed to generate optimal data readouts from each method may drive neuronal phenotypes. Thus, it is occasionally necessary to use multiple analysis modalities to have increased confidence in patient-specific or drug-induced phenotypes. In addition, functional readouts of neuronal activity can be easily manipulated with drug treatments or other perturbations, allowing for an even better understanding of how patient-derived neurons respond to stimuli, compared to healthy control neurons. Below are three potential methods for measuring neuronal activity using iPSC-derived neurons. More information about the ways functional readouts can be used for studying neurological disorders with iPSC-derived neurons can be found in a later chapter.

Multi-electrode array (MEA) recording

MEA machines measure the extracellular field potential from cells in culture, which changes as ions flux in or out of neurons during action potential propagation. Each electrode detects and reports changes in voltage over time, providing a readout of spontaneous or induced neuronal activity, reported as spikes over time (Fig. 2.3). With this spike information,

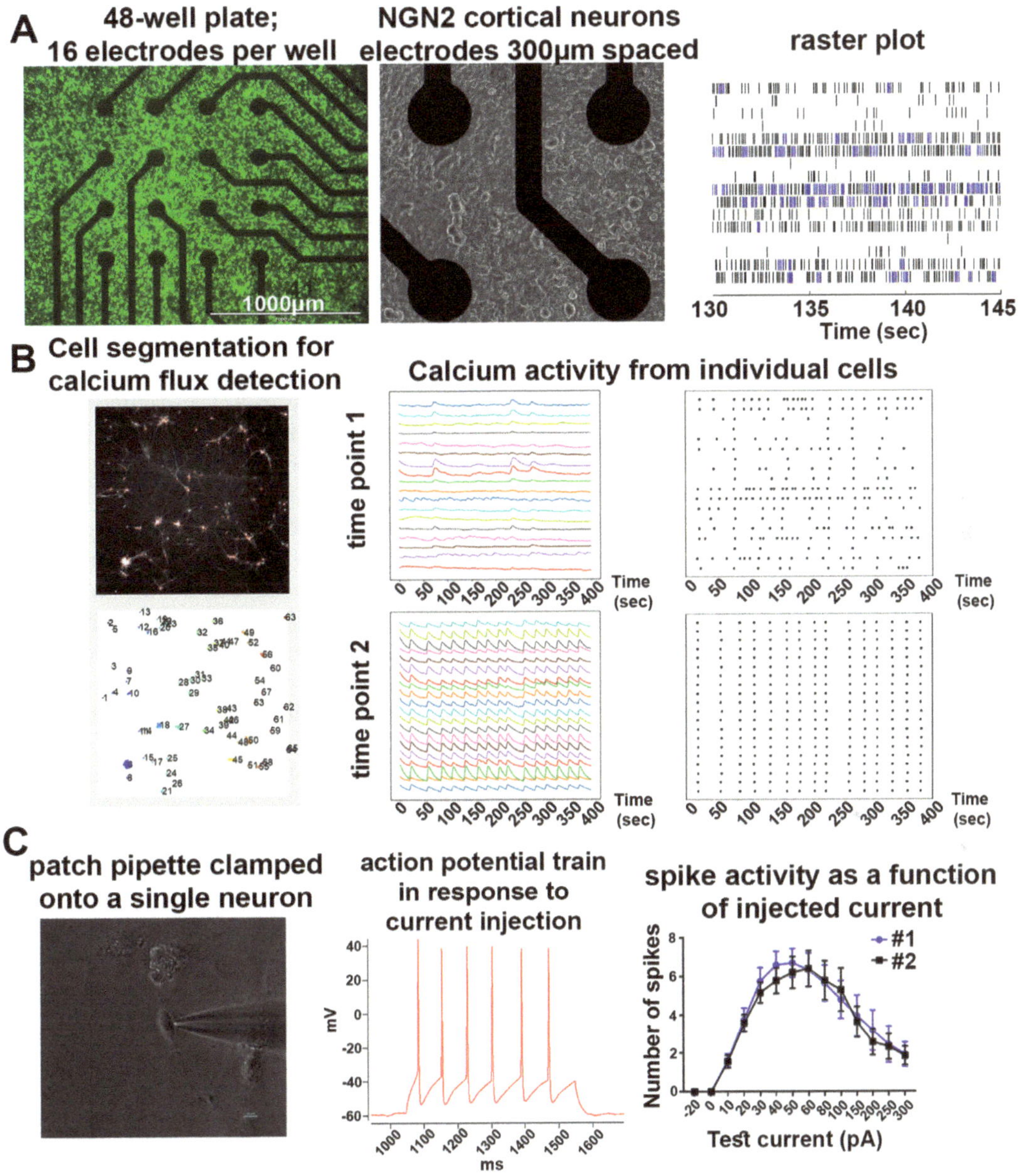

FIGURE 2.3 Methods to quantify functional activity in iPSC-derived neurons. Images in "A" from Hellen Jumo and Nina Makhortova. Images in "B" from Wardiya Afshar-Saber and Amanda Swanson. Images in "C" from Seungkyu Lee, Min-Joon Han, Clifford Woolf, and Mustafa Sahin. (A) Examples of neurons (GFP-labeled on the left and bright field in the middle) plated on low-density MEA plates (electrodes spaced 350 μm apart) shows the scale of individual neurons to electrode. The raster plot shows an example of a 15 s time window output from one well on a low-density MEA plate with each row representing one electrode. (B) Calcium imaging analysis is achieved by identification or segmenting cell bodies and measuring changes in fluorescence over time in these regions. Each colored trace shows the output from a single cell over time. The dots in the graphs on the right visualize when spikes are detected. The data from the second time point show how cultures become synchronous over time and how this is captured by calcium imaging. (C) Patch clamping allows for interrogation of one cell at a time, but can provide a lot of information about the activity of each neuron, including how many spikes occur with current injection steps, as plotted in this example.

users can quantify metrics such as mean firing rate, bursting rate, network bursting rate, and any iterative thereof, such as duration of bursts or inter-burst interval. Each of these metrics tells the story of the culture's neuronal activity profile and can be used to assess a disease or treatment state of the culture depending on how these parameters develop over time or how they compare at the time of culture maturity. The MEA is a common platform used in labs that study neurological disorders with iPSC-derived neurons due to the ease of use and lack of manipulation needed to obtain functional data from neuronal cultures. A recent report showed that this technology is robust and reproducible when certain aspects of the cell culture conditions are maintained across batches, such as the source of astrocytes.[4] Others have shown that the MEA can be used to detect changes in neuronal maturation over time[39] and several groups have used the technology to visualize network oscillations in neuronal organoids,[40,41] suggesting these cultures can be used as a model system for neurodevelopment. Our group used the MEA platform to visualize hyperexcitability in a TSC patient line with a second hit, which has been hypothesized to increase the risk of developing epilepsy.[29] These examples provide a brief window into the type of questions that can be asked with the MEA.

The parameters used to define a "spike" or a "burst" have the ability to change the data, so it's important to look closely at these parameters when establishing a new MEA pipeline. For example, for our iPSC-derived cortical neurons, we typically define a "spike" as any voltage change that is greater than six standard deviations above the background noise level and a "burst" is considered five spikes in 100 ms. Some MEA machines can stimulate neurons by releasing pulses of current from the electrodes in defined patterns[42,43] to see how different neuronal conditions or patient-derived lines respond to stimulation. Stimulation methods should be chosen carefully so as to not induce excitotoxicity, which can occur when neurons are hyperactive and may be more likely to occur in some patient-derived neurons.[44–46]

MEA machines have become more popular as companies developing the technology make the systems more user-friendly. In a broad sense, there are two main types of MEA machines: low density and high density devices. The density referred to is the density of electrodes per dish or well. Low-density machines utilize dishes with electrodes approximately 300 μm apart from each other, which allows multiple neurons to be in close proximity to each electrode, but also means that many neurons that are not recorded from, as shown in Fig. 2.3. On the other hand, high-density machines use plates with electrodes approximately 20 μm from each other (this varies depending on the chip technology used). To measure extracellular field potentials in 3D structures, such as spheroids and organoids, MEA systems also exist with pin electrodes that penetrate the three-dimensional structure, as was done to examine the functional activity of retinal organoids with light activation.[47]

In low-density MEA recordings, electrodes detect the signal from one or multiple neurons and those are compiled and reported as a single signal from each electrode. Therefore, it is typically not possible to determine if the signal is coming from one or multiple neurons using low density MEA machines. It can be challenging to maintain high reproducibility using low density MEA machines because neurons tend to migrate over time, resulting in cells moving away from electrodes and loss of signal. Strategies to improve reproducibility include testing different coating materials and coculturing with astrocytes. As mentioned previously, PEI has been shown to reduce the clumping of neuronal cultures over time and we have found it to be beneficial for this problem in MEA cultures.[16]

High density MEA recordings are unique because they not only provide information about firing rate and synchrony on the single-cell level, they also quantify action potential propagation speeds down neurites.[48] The structure of neurons can be visualized based on neuronal function on high density MEAs, allowing for structure and function to be elucidated from the same cultures.[49] In addition, high density MEA recordings provide reproducible single-cell spike detection accuracy for the classification of inhibitory versus excitatory waveforms.[50] As this technology continues to develop, researchers will be able to use it to understand how connectivity changes in treatment or disease states based on how different subtypes of cells communicate. The limitation of this technology is the throughput, which is currently orders of magnitude lower than low density MEAs.

There is currently no consensus for the density of neurons to astrocytes that should be cocultured to support phenotypic assay development on the MEA. In one study, astrocytes were plated 1:1 with iPSCs undergoing NGN2 differentiation at the start of differentiation.[51] In these cultures, the astrocytes support neuronal health and allow for control neuron firing rates to reach into the range of 3−5Hz.[4,51] In another study, astrocytes were plated at 15% of the total neuron count at day six of NGN2 differentiation.[29] In this study, control neurons were also supported to form a mature neuronal network, but control neuron firing rates only reached approximately 1 Hz.[29] It may be that different levels of astrocytes allow for varying levels of support of the neurons in the culture. Future experiments should be designed with these considerations to determine which coating conditions and neuron to astrocyte ratios allow for healthy control cultures and phenotypic differences to be visualized that are physiologically relevant. For low density plates, the density of neurons and ratio of astrocytes for each user should be defined at the start of a research study to form a monolayer of neurons and astrocytes covering the surface of the dish to try to reduce neuronal mobility and maintain neuron attachment by electrodes. These conditions will also need to be optimized based on the differentiation protocol chosen.

Calcium imaging

While MEA machines measure the extracellular field potential from neuronal cultures, calcium imaging measures changes in calcium ion concentrations and is also a powerful tool to infer neuronal activity.[52] In calcium imaging studies, a dye or a genetically encoded calcium indicator is used to measure calcium flux in active cells (Fig. 2.3). Intracellular calcium flux is an important readout of neuronal activity because many calcium-permeable channels and transporters are activated during neuron firing and calcium signaling plays an important role in both presynaptic and postsynaptic neuronal activity and signaling mechanisms.[53−57] The assay conditions necessary to optimize calcium imaging readout begin with selecting the calcium indicator. Dyes, such as Fura-2, are easily added to any cell culture medium, are uptaken by cells, and are rapidly available for measuring calcium signals.[58] Fura-2 is a commonly chosen dye because it is a ratiometric indicator, meaning it shifts its excitation when it binds to calcium, reducing artifacts in the imaging system. However, dyes such as Fura-2 are taken up by both neurons and astrocytes and can make it difficult to optimize analysis pipelines with high signal coming from astrocytes. To improve specificity, expression of genetically encoded calcium indicators, such as GCaMP6, can be driven by neuronal-specific promoters to visualize calcium flux in neurons only. Genetically encoded calcium indicators have been reviewed in the literature[59] and one downside is that they need to be introduced to the cells, which can be done with viral overexpression or genome editing techniques to introduce it into a safe harbor locus. In addition, recording of activity must wait until the calcium indicator is expressed, which can take several days if viral overexpression is used. On the flip side, once they are expressed, repeated recordings are possible for weeks after expression begins, allowing for long-term calcium interrogation without repeatedly adding dyes. Further, neurons utilized for calcium imaging can also be transduced with channelrhodopsins to utilize light to regulate neuronal activation and network connectivity.[60] Channelrhodopsins are a family of light-activated ion channels that allow neurons to become depolarized when light is shown in the correct wavelength.[61]

Culture conditions for calcium imaging depend on the differentiation protocol chosen and should be optimized to promote prolonged spacing of neurons so that individual neurons can be visualized and segmented (Fig. 2.3). Changes in fluorescent intensity over time can then be measured for each cell. As Fig. 2.3 shows, this technique is great for visualizing activity in individual cells as well as the population, as shown by increased synchrony across the population in later time points of recording. Image settings for calcium imaging should consider the toxicity of the light used to activate the calcium indicator, such as blue light used to activate GCaMP.[62] The specific recording duration and imaging frequency should be optimized for the culture conditions and biological question that are unique to each project. For example, calcium imaging has been used with continuous recording to visualize functional responses of neurons to the addition of different receptor agonists, which defines the subtype identity of the neurons.[63] In other studies, acute recording sessions of spontaneous or induced activity at different time points over the lifetime of a culture are used to visualize neuronal connectivity and changes between patient and control neuronal activity.[60,64] When transitioning to screening assays, calcium imaging is a lower cost method for visualizing neuronal activity in multi-well plates compared to MEA methods, which can be expensive due to costs of plates and the number of neurons needed for each well.

Patch clamping

Another method for interrogating neuronal activity at the single-cell level is patch clamping, which has been carefully reviewed in the literature[65] and allows for precise quantification of action potential activity and ion channel conductance. Patch clamping is typically achieved for one cell at a time, with a small glass pipette that is used to create a seal on the neuronal cell membrane, as shown in Fig. 2.3. Therefore, this technique is traditionally very time-consuming and low throughput. There are many variations in the way patch clamping is performed that allow the researcher to record from specific membrane sections, or the whole cell. These variations allow the researcher to apply different conditions through the patching pipette to better understand the function of the neuron, the ion channels present, and other information about the membrane composition.

For example, in whole cell patch, the ions and other compounds added to the patch pipette mix freely with the inside of the entire cell, allowing the manipulator to understand what channels are expressed in each cell.[65] With cell-attached methods of patch clamping, it's possible to use voltage-clamp or current-clamp methods. In voltage-clamp, the voltage, or membrane potential, of the cell is maintained at a constant level through injections of current.[65] This method allows the researcher to record the currents going through a cell and how the cell responds to those currents based on the amount of current that needs to be injected to maintain the membrane potential.[65] In current-clamp, the researcher controls the amount of current that is injected into the cell and the amplifier records the changes in membrane potential and how the cell reacts to the current injection.[65] A typical example of a current-clamp is the injection of current steps to see at which current injection level the cell will fire a single action potential or an action potential train (Fig. 2.3). The number of spikes elicited by each current step can also be compared across conditions, as shown in Fig. 2.3. This method allows the researcher to understand the excitability and ion channel content of the cell.

An advantage of patch clamping, in addition to precisely understanding the activity of the neuron that is interrogated, is that individual neurons can be collected into the pipette after recording. The collected cell can then be analyzed for gene expression, allowing for functional data to be tied to single-cell transcriptomic data in a technique now coined Patch-Seq.[66] Data from these experiments may help researchers understand how molecular changes in individual cells drive functional subtype differences.

Newer technologies are emerging that utilize automated patch clamp technology to increase throughput and make the technology accessible to more users.[67–69] However, operating and analyzing data from one of these machines still remains a technical process that requires expertise in the field of electrophysiology. Automated patch platforms differ in their capabilities and have been reviewed in the literature.[67] These instruments differ in their optimal cellular seal strength and throughput and particular projects or neuronal differentiation protocols may be better suited for one device over another based on these features. Continued advances in these technologies will allow for improved preclinical assays that better represent the biology of individual neurons across a population.

Live imaging

As described previously in the high content imaging section, many assays utilize live imaging of iPSC-derived neurons to understand the dynamic activity within neurons. These studies may help identify time points for other screening efforts using more static endpoints, or may be developed into assays for screening themselves. A few examples of the assays that are possible with live imaging include visualizing changes in the localization of internal signaling molecules, pH changes across internal cellular membranes, motility of mitochondria, migration of cells, cell survival, and proliferation. For each of these assays, unique kits are available with fluorescent dyes that change color or localization depending on the intended chemistry that is being detected. The optimal culture conditions for each endpoint are often described along with the kit when purchased. Considerations for live imaging machines should include the magnification required, throughput/speed of imaging, and environmental controls regulating temperature, CO_2, and humidity.

Fluorescent microplate assays

The last assay type discussed in this chapter is the microplate assay, which typically requires either collection of media that has been exposed to the cells for a defined period or lysing of the cells to readout a cell state. An example of such an assay that utilizes media collection is the lactate dehydrogenase (LDH) assay, which measures LDH released by cells in the well.[70] Since LDH is a cytoplasmic enzyme, the release of LDH to the media signals the death of a cell via plasma membrane damage. This assay depends on the ability of LDH to reduce NAD to NADH in the media solution, which is then able to interact with a probe designed by the company selling the kit to produce a color readout that can be measured at a specific absorbance. While the LDH assay provides a quantitative measure of the number of cells that have died, it does not provide a picture of the number of live cells that remain at any given time. To fill this gap, other assays can measure metabolic activity, such as the MTT (3-(4,5-dimethylthiazol-2-yl)-2,5-diphenyltetrazolium bromide) assay.[71] This assay depends on the ability of a live cell to metabolize MTT to formazan, which has a color that can be measured. Dead cells cannot metabolize MTT into formazan and do not produce any signal. The amount of color quantifies the number of metabolically active cells. This readout can help determine the effects of drug treatments on cell health, but the cells cannot be kept in culture after the assay is run.

There are no specific cell culture requirements in terms of coating for a successful microplate assay. However, cell density is critical to regulate and proper controls should be in place to understand the levels read out by the assay. Any plate reader that can shake, incubate at set temperatures, and readout a defined absorbance will be sufficient for these assays.

Assay development for screening

Once a phenotype has been identified that characterizes an important difference between patient and control neurons, that phenotype can be optimized into an assay for a phenotypic screen. There are many considerations for determining whether an assay is amenable to screening. While the order of these considerations may depend on the resources available, the first consideration may be the cost of running the assay of choice. For example, if a functional difference is identified, such as hyperexcitability of patient neurons compared to healthy control neurons, it is likely more cost effective to optimize an assay using calcium imaging compared to MEA because far fewer neurons are needed per well for calcium imaging and the plates used are significantly less expensive. The MEA, on the other hand, serves as a great secondary assay. Another consideration that contributes to the overall cost is the protocol used to generate neurons. New protocols have been developed that utilize 3D spheroid differentiation methods in large bioreactors to allow for the formation of hundreds of millions of neurons in a single differentiation batch, reducing costs and variability associated with differentiating multiple batches of neurons to obtain enough for screening.[72] However, if a homogeneous neuron population is required for the assay endpoint, then transcription factor-based approaches, such as the NGN2 differentiation protocol,[2] may be the better choice. The NGN2 differentiation protocol does not utilize a progenitor phase of differentiation, so if signaling mechanisms during this phase of neuronal development are important for your phenotype of interest, that should be considered when selecting the differentiation protocol.

As described above, all of the culture conditions that were optimized for phenotype identification can also be re-addressed during assay optimization, including coating materials, neuronal plating density, the decision to include astrocyte coculture (and at what level), and timing of the assay, among other considerations specific to each phenotypic endpoint. During this phase, automation can be used to reduce the variability introduced by human error. Typical steps that can be automated in a phenotypic screen includes plate coating and washing, cell plating and feeding, addition of compounds, fixatives, and antibodies during staining. Different liquid handlers are available that specialize in each modality and vary in their ability to regulate pipetting speeds, media volumes that can be exchanged, including dead volumes (which are especially important when plating valuable cells), whether different plate formats are compatible, and whether the device can fit inside of a biosafety cabinet. Automation will not only improve the reproducibility of a screen, it will also enhance throughput by not relying on individual pipetting to complete each task.

Most importantly, an assay will be deemed ready for screening when the difference in phenotype between positive and negative controls is significant enough and the variation across replicates within positive and negative controls is small enough.[73] The best way to confirm this is to run an assay plate with half of the plate treated as positive controls, and half treated as negative controls. The strength of the assay can then be quantified using the Z-prime statistic equation, which is $= 1 - (3$ (standard deviation of positive controls) $+ 3$(standard deviation of negative controls))/[mean of positive control-mean of negative control]. This equation quantifies how different the phenotype is, based on the mean of positive and negative controls, but in the context of the distribution of positive and negative controls. Typical autologous cell-based assays can reach the optimal range of 0.5 to 1 for the Z-prime factor calculation, but this is more difficult for phenotypic assays using iPSC-derived neurons, usually because the variability across wells is higher. A Z-prime factor between 0 and 0.5 is seen as acceptable for such phenotypic screens.[74] A Z-prime factor value less than 0 means there is either too much variability or the positive and negative control effects are not different enough for a high throughput screen.

Conclusion

This chapter offers advice on ways to optimize phenotypic assays for iPSC-derived neurons. This technology holds the promise to improve the efficacy of therapeutics identified through preclinical research in the neurosciences by creating patient-relevant in vitro systems with human neurons. While animals studies are important for understanding how whole organisms and interconnected biological systems are affected by different genetic indications or drug treatments, human iPSC-derived neurons provide a model system for understanding human neuronal-specific responses to these genetic changes and drug treatments. There is great promise for the future of iPSC-derived neurons to move new therapeutic targets for neuroscience to the clinic, but it starts with high-quality assay design and development.

References

1. Takahashi K, Yamanaka S. Induction of pluripotent stem cells from mouse embryonic and adult fibroblast cultures by defined factors. *Cell*. 2006;126(4):663–676.
2. Zhang Y, Pak C, Han Y, et al. Rapid single-step induction of functional neurons from human pluripotent stem cells. *Neuron*. 2013;78(5):785–798. https://doi.org/10.1016/j.neuron.2013.05.029.
3. Lin H-C, He Z, Ebert S, et al. NGN2 induces diverse neuron types from human pluripotency. *Stem Cell Rep*. 2021;16(9):2118–2127. https://doi.org/10.1016/j.stemcr.2021.07.006.
4. Mossink B, Verboven AHA, van Hugte EJH, et al. Human neuronal networks on micro-electrode arrays are a highly robust tool to study disease-specific genotype-phenotype correlations in vitro. *Stem Cell Rep*. 2021;16(9):2182–2196. https://doi.org/10.1016/j.stemcr.2021.07.001.
5. Shih P-Y, Kreir M, Kumar D, et al. Development of a fully human assay combining NGN2-inducible neurons co-cultured with iPSC-derived astrocytes amenable for electrophysiological studies. *Stem Cell Res*. 2021;54:102386. https://doi.org/10.1016/j.scr.2021.102386.
6. Muratore CR, Srikanth P, Callahan DG, Young-Pearse TL. Comparison and optimization of hiPSC Forebrain cortical differentiation protocols. *PLOS ONE*. 2014;9(8):e105807. https://doi.org/10.1371/journal.pone.0105807.
7. Maroof AM, Keros S, Tyson JA, et al. Directed differentiation and functional maturation of cortical interneurons from human embryonic stem cells. *Cell Stem Cell*. 2013;12(5):559–572. https://doi.org/10.1016/j.stem.2013.04.008.
8. Maury Y, Côme J, Piskorowski RA, et al. Combinatorial analysis of developmental cues efficiently converts human pluripotent stem cells into multiple neuronal subtypes. *Nat Biotechnol*. 2015;33(1):89–96. https://doi.org/10.1038/nbt.3049.
9. Shen L, Yuan F, Hong Y, Xu M, Hu Y, Liu Y. Identification of small molecules for accelerating the differentiation of GABA interneurons from human pluripotent stem cells. *J Mol Cell Biol*. 2020;12(3):245–248. https://doi.org/10.1093/jmcb/mjaa002.
10. Zhang L, Yin JC, Yeh H, et al. Small molecules efficiently reprogram human astroglial cells into functional neurons. *Cell Stem Cell*. 2015;17(6):735–747. https://doi.org/10.1016/j.stem.2015.09.012.
11. Bowles KR, Julia TCW, Qian L, Jadow BM, Goate AM. Reduced variability of neural progenitor cells and improved purity of neuronal cultures using magnetic activated cell sorting. *PLoS ONE*. 2019;14(3). https://doi.org/10.1371/journal.pone.0213374.
12. Kim D-S, Lee DR, Kim H-S, et al. Highly pure and expandable PSA-NCAM-positive neural precursors from human ESC and iPSC-derived neural rosettes. *PloS One*. 2012;7(7):e39715. https://doi.org/10.1371/journal.pone.0039715.
13. Clément JP, Al-Alwan L, Glasgow SD, et al. Dendritic polyglycerol amine: an enhanced substrate to support long-term neural cell culture. *ASN Neuro*. 2022;14. https://doi.org/10.1177/17590914211073276.
14. McKeehan WL, Ham RG. Stimulation of clonal growth of normal fibroblasts with substrata coated with basic polymers. *J Cell Biol*. 1976;71(3):727–734. https://doi.org/10.1083/jcb.71.3.727.
15. Ge H, Tan L, Wu P, et al. Poly-L-ornithine promotes preferred differentiation of neural stem/progenitor cells via ERK signalling pathway. *Sci Rep*. 2015;5. https://doi.org/10.1038/srep15535.
16. Overland AC, Rauch JN, Oupicka L, Rock DM, Appledorn DM. *Application Note: Quantitative Live-Cell Analysis for Optimization of Culture Conditions and Evaluation of Cell Health in Human Induced Pluripotent Stem Cell-Derived Neurons*. 2017.
17. Milky B, Zabolocki M, Al-Bataineh SA, et al. Long-term adherence of human brain cells in vitro is enhanced by charged amine-based plasma polymer coatings. *Stem Cell Rep*. 2022;17(3):489–506. https://doi.org/10.1016/j.stemcr.2022.01.013.
18. Chambers SM, Fasano CA, Papapetrou EP, Tomishima M, Sadelain M, Studer L. Highly efficient neural conversion of human ES and iPS cells by dual inhibition of SMAD signaling. *Nat Biotechnol*. 2009;27(3):275–280. https://doi.org/10.1038/nbt.1529.
19. Perrier AL, Tabar V, Barberi T, et al. Derivation of midbrain dopamine neurons from human embryonic stem cells. *Proc Natl Acad Sci USA*. 2004;101(34):12543–12548. https://doi.org/10.1073/pnas.0404700101.
20. Shi Y, Kirwan P, Livesey FJ. Directed differentiation of human pluripotent stem cells to cerebral cortex neurons and neural networks. *Nat Protoc*. 2012;7(10):1836–1846. https://doi.org/10.1038/nprot.2012.116.

21. Lam D, Enright HA, Cadena J, et al. Tissue-specific extracellular matrix accelerates the formation of neural networks and communities in a neuron-glia co-culture on a multi-electrode array. *Sci Rep*. 2019;9(1). https://doi.org/10.1038/s41598-019-40128-1.
22. Paul C. *I'm Sticking With You: Four Coatings To Help Cells Stick To Microscopy Slides*; 2013. https://bitesizebio.com/13452/im-sticking-with-you-four-coatings-to-help-cells-stick-to-microscopy-slides/.
23. Siddiqui AM, Brunner R, Harris GM, et al. Promoting neuronal outgrowth using ridged scaffolds coated with extracellular matrix proteins. *bioRxiv*. October 2019 https://doi.org/10.1101/788539.
24. Tonge DA, de Burgh HT, Docherty R, Humphries MJ, Craig SE, Pizzey J. Fibronectin supports neurite outgrowth and axonal regeneration of adult brain neurons in vitro. *Brain Res*. 2012;1453:8–16. https://doi.org/10.1016/j.brainres.2012.03.024.
25. LaFosse P, Zamponi M, Mollica P, et al. Effect of substrate coating material on spontaneous activity of human-induced pluripotent stem cell-derived neuronal stem cells. *Front Cell Neurosci*. 2018;12. https://doi.org/10.3389/conf.fncel.2018.38.00115.
26. Di Nardo A, Rühmkorf A, Award P, Brennecke A, Fagiolini M, Sahin M. Phenotypic characterization of Cdkl5-knockdown neurons establishes elongated cilia as a functional assay for CDKL5 Deficiency Disorder. *Neurosci Res*. 2022;176:73–78. https://doi.org/10.1016/j.neures.2021.10.001.
27. Gensel JC, Schonberg DL, Alexander JK, McTigue DM, Popovich PG. Semi-automated Sholl analysis for quantifying changes in growth and differentiation of neurons and glia. *J Neurosci Methods*. 2010;190(1):71–79. https://doi.org/10.1016/j.jneumeth.2010.04.026.
28. Ye E-A, Chawla SS, Khan MZ, Sakaguchi DS. Bone marrow-derived mesenchymal stem cells (MSCs) stimulate neurite outgrowth from differentiating adult hippocampal progenitor cells. *Stem Cell Biol Res*. 2016;3(1):3. https://doi.org/10.7243/2054-717X-3-3.
29. Winden KD, Sundberg M, Yang C, et al. Biallelic mutations in TSC2 lead to abnormalities associated with cortical tubers in human iPSC-derived neurons. *J Neurosci*. 2019;39(47):9294–9305. https://doi.org/10.1523/JNEUROSCI.0642-19.2019.
30. Bray M-A, Singh S, Han H, et al. Cell Painting, a high-content image-based assay for morphological profiling using multiplexed fluorescent dyes. *Nat Protoc*. 2016;11(9):1757–1774. https://doi.org/10.1038/nprot.2016.105.
31. Du H, Hou P, Zhang W, Li Q. Advances in CLARITY-based tissue clearing and imaging. *Exp Ther Med*. 2018;16(3):1567–1576. https://doi.org/10.3892/etm.2018.6374.
32. Boecker CA, Olenick MA, Gallagher ER, Ward ME, Holzbaur ELF. ToolBox: live Imaging of intracellular organelle transport in induced pluripotent stem cell-derived neurons. *Traffic*. 2020;21(1):138–155. https://doi.org/10.1111/tra.12701.
33. Dittgen T, Nimmerjahn A, Komai S, et al. Lentivirus-based genetic manipulations of cortical neurons and their optical and electrophysiological monitoring in vivo. *Proc Natl Acad Sci USA*. 2004;101(52):18206–18211.
34. Little D, Luft C, Mosaku O, Ketteler R, Devine MJ, Gissen P. High-content analysis of mitochondrial function in iPSC-derived neurons. In: *Methods in Molecular Biology*. United Kingdom: Humana Press Inc.; 2019:175–184; Vol 1994.
35. Ma L, Ouyang Q, Werthmann GC, Thompson HM, Morrow EM. Live-cell microscopy and fluorescence-based measurement of luminal pH in intracellular organelles. *Front Cell Dev Biol*. 2017;5. https://doi.org/10.3389/fcell.2017.00071.
36. Shlevkov E, Basu H, Bray M-A, et al. A high-content screen Identifies TPP1 and Aurora B as Regulators of axonal mitochondrial transport. *Cell Rep*. 2019;28(12):3224–3237.e5. https://doi.org/10.1016/j.celrep.2019.08.035.
37. Scheeder C, Heigwer F, Boutros M. Machine learning and image-based profiling in drug discovery. *Curr Opin Syst Biol*. 2018;10:43–52. https://doi.org/10.1016/j.coisb.2018.05.004.
38. Yang SJ, Lipnick SL, Makhortova NR, et al. Applying deep neural network analysis to high-content image-based assays. *SLAS Discov*. 2019;24(8):829–841. https://doi.org/10.1177/2472555219857715.
39. Odawara A, Katoh H, Matsuda N, Suzuki I. Physiological maturation and drug responses of human induced pluripotent stem cell-derived cortical neuronal networks in long-term culture. *Sci Rep*. 2016;6:26181. https://doi.org/10.1038/srep26181.
40. Samarasinghe RA, Miranda OA, Buth JE, et al. Identification of neural oscillations and epileptiform changes in human brain organoids. *Nature Neuroscience*. 2021;24(10):1488–1500. https://doi.org/10.1038/s41593-021-00906-5.
41. Trujillo CA, Gao R, Negraes PD, et al. Complex oscillatory Waves emerging from cortical organoids model early human brain network development. *Cell Stem Cell*. 2019;25(4):558–569.e7. https://doi.org/10.1016/j.stem.2019.08.002.

42. Jimbo Y, Kasai N, Torimitsu K, Tateno T, Robinson HPC. A system for MEA-based multisite stimulation. *IEEE Trans Bio-Med Eng*. 2003;50(2):241–248.
43. Ronchi S, Fiscella M, Marchetti C, et al. Single-cell electrical stimulation using CMOS-based high-density microelectrode arrays. *Front Neurosci*. 2019;13. https://doi.org/10.3389/fnins.2019.00208.
44. Abiega O, Beccari S, Diaz-Aparicio I, et al. Neuronal hyperactivity disturbs ATP microgradients, impairs microglial motility, and reduces phagocytic receptor expression triggering apoptosis/microglial phagocytosis uncoupling. *PLoS Biol*. 2016;14(5):e1002466. https://doi.org/10.1371/journal.pbio.1002466.
45. Olloquequi J, Cornejo-Córdova E, Verdaguer E, et al. Excitotoxicity in the pathogenesis of neurological and psychiatric disorders: therapeutic implications. *J Psychopharmacol*. 2018;32(3):265–275. https://doi.org/10.1177/0269881118754680.
46. Velasco M, Quintero JR, Castillo MC, et al. Excitotoxicity: an organized Crime at the cellular level. *J Neurol Neurosci*. 2017;08(03). https://doi.org/10.21767/2171-6625.1000193.
47. Quadrato G, Nguyen T, Macosko EZ, et al. Cell diversity and network dynamics in photosensitive human brain organoids. *Nature*. 2017;545(7652):48–53. https://doi.org/10.1038/nature22047.
48. Bakkum DJ, Frey U, Radivojevic M, et al. Tracking axonal action potential propagation on a high-density microelectrode array across hundreds of sites. *Nat Commun*. 2013;4. https://doi.org/10.1038/ncomms3181.
49. Torsten B, Radivojevic MHSD. Hierlemann andreas; Frey Urs T. Large-scale mapping of axonal Arbors using high-density microelectrode arrays. *Front Cell Neurosci*. 2019;13:1662–5102.
50. Mita T, Bakkum D, Frey U, Hierlemann A, Kanzaki R, Takahashi H. Classification of inhibitory and excitatory neurons of dissociated cultures based on action potential waveforms on high-density CMOS microelectrode arrays. *IEEJ Trans Electron Inf Syst*. 2019;139(5):615–624. https://doi.org/10.1541/ieejeiss.139.615.
51. Frega M, Linda K, Keller JM, et al. Neuronal network dysfunction in a model for Kleefstra syndrome mediated by enhanced NMDAR signaling. *Nat Commun*. 2019;10(1):4928. https://doi.org/10.1038/s41467-019-12947-3.
52. Carafoli E, Krebs J. Why calcium? How calcium Became the best communicator. *J Biol Chem*. 2016;291(40):20849–20857.
53. Berridge MJ, Bootman MD, Roderick HL. Calcium signalling: dynamics, homeostasis and remodelling. *Nat Rev Mol Cell Biol*. 2003;4(7):517–529. https://doi.org/10.1038/nrm1155.
54. Berridge MJ, Lipp P, Bootman MD. The versatility and universality of calcium signalling. *Nat Rev Mol Cell Biol*. 2000;1(1):11–21. https://doi.org/10.1038/35036035.
55. Grienberger C, Konnerth A. Imaging calcium in neurons. *Neuron*. 2012;73(5):862–885. https://doi.org/10.1016/j.neuron.2012.02.011.
56. Neher E, Sakaba T. Multiple roles of calcium ions in the regulation of neurotransmitter release. *Neuron*. 2008;59(6):861–872. https://doi.org/10.1016/j.neuron.2008.08.019.
57. Zucker RS. Calcium- and activity-dependent synaptic plasticity. *Curr Opin Neurobiol*. 1999;9(3):305–313. https://doi.org/10.1016/S0959-4388(99)80045-2.
58. Barreto-Chang OL, Dolmetsch RE. Calcium imaging of cortical neurons using Fura-2 AM. *J Vis Exp*. 2009:23. https://doi.org/10.3791/1067.
59. Oh J, Lee C, Kaang B-K. Imaging and analysis of genetically encoded calcium indicators linking neural circuits and behaviors. *Korean J Physiol Pharmacol*. 2019;23(4):237–249. https://doi.org/10.4196/kjpp.2019.23.4.237.
60. Afshar Saber W, Gasparoli FM, Dirks MG, Gunn-Moore FJ, Antkowiak M. All-optical assay to study biological neural networks. *Front Neurosci*. 2018;12:451. https://doi.org/10.3389/fnins.2018.00451.
61. Boyden ES, Zhang F, Bamberg E, Nagel G, Deisseroth K. Millisecond-timescale, genetically targeted optical control of neural activity. *Nat Neurosci*. 2005;8(9):1263–1268.
62. Nakai J, Ohkura M, Imoto K. A high signal-to-noise Ca(2+) probe composed of a single green fluorescent protein. *Nat Biotechnol*. 2001;19(2):137–141.
63. Wainger BJ, Buttermore ED, Oliveira JT, et al. Modeling pain in vitro using nociceptor neurons reprogrammed from fibroblasts. *Nat Neurosci*. 2015;18(1):17–24. https://doi.org/10.1038/nn.3886.
64. Culotta L, Penzes P. Exploring the mechanisms underlying excitation/inhibition imbalance in human iPSC-derived models of ASD. *Mol Autism*. 2020;11(1):32. https://doi.org/10.1186/s13229-020-00339-0.
65. Hill CL, Stephens GJ. An introduction to patch clamp recording. *Methods Mol Biol*. 2021;2188:1–19. https://doi.org/10.1007/978-1-0716-0818-0_1.

66. van den Hurk M, Erwin JA, Yeo GW, Gage FH, Bardy C. Patch-seq protocol to analyze the electrophysiology, morphology and transcriptome of whole single neurons derived from human pluripotent stem cells. *Front Mol Neurosci*. 2018;11. https://doi.org/10.3389/fnmol.2018.00261.
67. Bell DC, Dallas ML. Using automated patch clamp electrophysiology platforms in pain-related ion channel research: insights from industry and academia. *Br J Pharmacol*. 2018;175(12):2312–2321. https://doi.org/10.1111/bph.13916.
68. Billet A, Froux L, Hanrahan JW, Becq F. Development of automated patch clamp technique to investigate CFTR Chloride Channel function. *Front Pharmacol*. 2017;8. https://doi.org/10.3389/fphar.2017.00195.
69. Obergrussberger A, Rinke-Weiß I, Goetze TA, et al. The suitability of high throughput automated patch clamp for physiological applications. *J Physiol*. 2022;600(2):277–297. https://doi.org/10.1113/jp282107.
70. Kaja S, Payne AJ, Naumchuk Y, Koulen P. Quantification of lactate dehydrogenase for cell viability testing using cell lines and primary cultured astrocytes. *Curr Protoc Toxicol*. 2017;72:2.26.1–2.26.10. https://doi.org/10.1002/cptx.21.
71. Ghasemi M, Turnbull T, Sebastian S, Kempson I. The MTT assay: utility, limitations, pitfalls, and interpretation in bulk and single-cell analysis. *Int J Mol Sci*. 2021;22(23). https://doi.org/10.3390/ijms222312827.
72. Rigamonti A, Repetti GG, Sun C, et al. Large-scale production of mature neurons from human pluripotent stem cells in a three-dimensional suspension culture system. *Stem Cell Rep*. 2016;6(6):993–1008. https://doi.org/10.1016/j.stemcr.2016.05.010.
73. Bray M, Carpenter A, Markossian S, et al. *Advanced Assay Development Guidelines for Image-Based High Content Screening and Analysis*. Assay Guidance Manual Eli Lilly & Company and the National Center for Advancing Translational Sciences; 2004.
74. Bar H, Zweifach A. Z′ does not need to be > 0.5. *SLAS Discov*. 2020;25(9):1000–1008. https://doi.org/10.1177/2472555220942764.

CHAPTER

3

Derivation of cortical interneurons from human pluripotent stem cells to model neurodevelopmental disorders

Kesavan Meganathan, Ramachandran Prakasam*, Komal Kaushik*, Irene Antony*, Gareth Chapman and Kristen L. Kroll*

Department of Developmental Biology, Washington University School of Medicine, Saint Louis, MO, United States

Introduction

The cerebral cortex consists of two main categories of neurons: excitatory neurons (cExNs) that convey information to distant neurons within and outside the cortex and GABAergic cortical interneurons (cINs) that provide local inhibitory inputs to modulate excitatory neuronal activity.[1,2] Disruption of cIN development alters the balance between neuronal excitation and inhibition, which can contribute to neurodevelopmental disorders (NDDs), including autism spectrum disorder, intellectual disability, and epilepsy.[1–5] The approaches described in this chapter enable in vitro modeling of human cIN development, by directed differentiation of human pluripotent stem cells (hPSCs; including human embryonic stem cells (hESCs) and human subject-derived induced pluripotent stem cells (iPSCs)). These approaches can be effectively used, both to identify the underlying regulatory networks that control human cIN development as well as to determine how these are disrupted to contribute to NDDs. Below, we first outline key aspects of human cortical development, some of which these protocols mimic during directed hPSC differentiation. We then describe

* These authors contributed equally to this work.

Phenotyping of Human iPSC-derived Neurons
https://doi.org/10.1016/B978-0-12-822277-5.00007-9

the development of protocols for cIN derivation from hPSCs, along with key features and major protocol variants that have been tested or are in common use. We also provide a detailed step-by-step protocol for deriving cINs from human pluripotent stem cells and describe common methods for cellular phenotyping that can be used in combination with this protocol to assess NDD-associated alterations of cIN development. Most experimental evaluation of NDDs is currently conducted only in iPSC-derived cExNs, as these can be more readily and rapidly generated from hPSCs. Therefore, use of these cIN models, including the use of cExN and cIN neurons or cocultures, both expands the utility of iPSC models for identifying NDD-associated phenotypes and provides another disease-relevant neuronal platform for molecular and pharmacologic screening to reverse NDD-associated phenotypes, enabling the development of therapeutic interventions for individuals with NDDs.

Development of the human cortex

Initial specification of neural tissue that will form the central nervous system occurs during early embryogenesis, when pluripotent embryonic cells are allocated to the ectodermal, mesodermal, or endodermal germ layers in response to differential exposure to growth factor signaling. During this process, antagonism of bone morphogenetic protein (BMP) and Activin/Nodal-type TGF-β growth factor signaling specifies dorsal ectoderm as neural tissue, generating the neural plate by the third post-conception week of human embryogenesis;[6–8] the lateral edges of the neural plate subsequently elevate and fuse dorsally to create the hollow, dorsal neural tube. Along the anterior-posterior axis of the neural tube, specification of the brain and spinal cord occurs through differential, low versus high, exposure to posteriorizing growth factor signaling. The dorsal-ventral axis of the neural tube is likewise patterned by differential growth factor signaling: dorsal Wnt and BMP signaling and ventral Sonic hedgehog (SHH) signaling activate restricted expression of transcription factors to establish distinct regional neuronal identities along the neural tube's dorsal-ventral axis.[6–8] Within the telencephalon of the developing forebrain, this dorsal BMP and Wnt signaling specifies cortical cExN progenitors that express the transcription factor PAX6 and line the ventricles of the developing brain. As these progenitors differentiate, they migrate radially inward from the ventricular surface to form six layers of cExNs, which are further elaborated to form the cortex.[6–8] Below the developing cortex, subcortical neural tissue in the ventral telencephalon is specified as three progenitor domains (the medial, lateral, and caudal ganglionic eminences; MGE/LGE/CGE) that give rise to distinct neural types. One of these, the MGE, gives rise to a majority of cINs that express the neurotransmitter GABA. These cINs undergo a long-range tangential migration from the MGE, the subcortical progenitor domain where they are born, to the cortex, where they establish synaptic connections with cExNs and modulate their activity. Together, these developmental processes establish balanced excitatory and inhibitory activities of synapses and circuits required for normal brain function.[1,2]

Modeling human cortical interneuron development in vitro

Understanding the etiology of human NDDs is particularly challenging because the human embryonic and fetal brain is not accessible or able to be experimentally manipulated during neurodevelopment. To bypass this obstacle, in vitro models that recapitulate various aspects of human cortical development have been developed, with directed differentiation of cortical neuronal cell types from human subject-derived iPSC models serving as a particularly useful tool for identifying how perturbation of neurodevelopment leads to NDDs (Fig. 3.1).

These approaches are useful because genetically engineered murine models often do not recapitulate NDD-associated human phenotypes, due to differences in both human genetics and brain development. For example, there are significant differences between human and rodent cIN development. During primate evolution, GABAergic interneurons increased in complexity and proportion relative to cortical excitatory neurons,[9,10] comprising 25%–34% of cortical neurons in humans, but only 15%–20% or less in mice.[11,12] Human cIN development is highly susceptible to perturbation that leads to NDDs,[4] in part because it occurs throughout a long embryonic and fetal timeframe encompassing human post-conception weeks 10–~35.[9,10,13] iPSC-derived neuronal models recapitulate many neurodevelopmental processes that occur during fetal brain development and thus can be used to define NDD-associated alterations of cIN development and function that may not be detected in rodent NDD models.[1,14,15] Human iPSC models can also recapitulate the complex and often human-specific genetic contributors that underlie most NDD risk, including genetic interactions between pathogenic mutations and the genetic background of individual human subjects.[16–18] As such, creating human subject-specific human NDD models enables cellular,

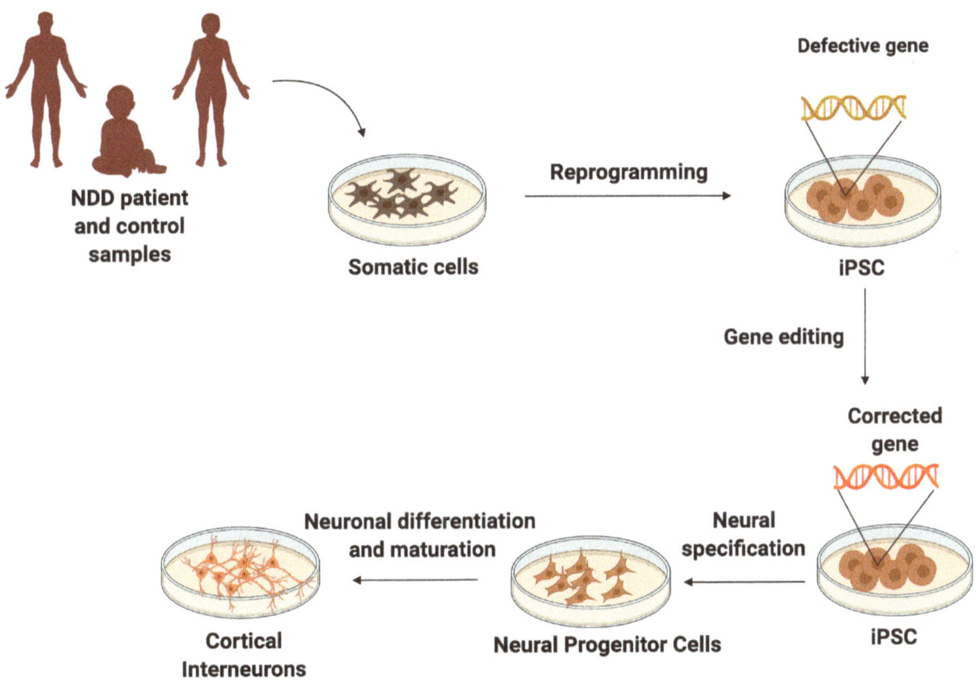

FIGURE 3.1 Overview: modeling neurodevelopmental disorders (NDDs) with human subject-derived induced pluripotent stem cells (iPSCs). Somatic cells from control subjects and probands with NDDs are isolated and reprogrammed to generate iPSC models. To assess pathogenicity, a suspected gene mutation is corrected by genome editing in the proband-derived iPSCs, and neural specification, differentiation, and function are subsequently assessed in these models (schematic generated in BioRender).

molecular, and functional anomalies to be identified that are associated with and may contribute to NDDs. These models then serve as a platform for targeted or high throughput screening to identify chemical or molecular treatments that can reverse phenotypes and that may have utility for therapeutic intervention.

The development of protocols for cortical interneurons from human pluripotent stem cells (hPSCs) to model neurodevelopmental disorders

Patient-derived iPSC lines are now readily obtained from individual human subjects by reprogramming various somatic cell types from patients, including skin fibroblasts,[19] renal epithelial cells,[20] blood,[21] and dental pulp;[22] this is usually accomplished by transient expression of four transcription factors (OCT4, SOX2, KLF4, and c-MYC). Derivation of cExNs from these iPSC models is the most commonly utilized approach to model NDDs.[23–26] This can be readily accomplished by using so called "dual SMAD inhibition" to drive iPSC differentiation: small molecule antagonists of both BMP signaling through SMAD1/5/8 and of Activin/Nodal-type TGF-β signaling through SMAD2/3 are used to generate neural progenitor cells (NPCs) of dorsal telencephalic character, which are then differentiated into cExNs.[27] However, many NDDs, including autism spectrum disorder, intellectual disability, epilepsy, schizophrenia, and bipolar disorder, are likely to be associated with altered development or imbalanced activity of either or both cExNs and cINs in the cerebral cortex.[28–30] Therefore, the development of protocols for effectively deriving MGE-like neural progenitors and GABAergic cINs from hPSCs represents an important advance. Most prior studies using iPSC models have characterized deficits in only cExNs,[23–25,31,32] while the contribution of disrupted cIN development to NDDs has been largely unexplored.

While protocols for deriving cIN-like cells from mouse embryonic stem cells were previously described, a first effort to develop these protocols for human ESCs was made in 2007.[33] The authors used protein inhibitors of WNT, Nodal, and BMP signaling (DKK1, LeftyA, and BMPR1A-Fc, respectively) to generate PAX6-expressing cells with a dorsal telencephalic, cortical regional character; they then ventralized these neuroectodermal cells with Sonic Hedgehog (SHH) addition, generating NKX2-1-expressing ventral forebrain progenitors as floating cellular aggregates.[33] To assess the efficiency of this differentiation protocol, Goulburn and colleagues generated a transgenic hESC line with knock-in of GFP at the NKX2-1 locus; this reagent has been used in subsequent studies for protocol optimization and to purify populations of MGE-like progenitors.[34] Multiple subsequent studies have likewise shown that specifying formation of PAX6-expressing neural progenitors while augmenting SHH signaling, with or without blocking WNT signaling, yields >80% NKX2-1-expressing MGE-like progenitors.[35,36] Both of the previously described protocols[35,36] used the formation of 3-D neural aggregates for neural specification and then plated the progenitors as a monolayer for differentiation, which was driven by addition of either addition of BDNF, GDNF, and IGF1, or by treatment with valproic acid.

In 2013, two studies used directed hESC differentiation to obtain NKX2-1 positive MGE-like progenitors and then further differentiated these into mature cortical interneurons expressing markers of the two major mature MGE-derived cIN subtypes, somatostatin (SST), and parvalbumin (PV).[37,38] Maroof et al. initiated differentiation in DMEM/F-12 plus N2 basal media in monolayer culture, with addition of small molecule antagonists of BMP, WNT, and TGF-beta signaling (LDN 193189, XAV 939, and SB 431542). Nicholas et al. instead initiated differentiation using Neurobasal-A plus B-27 basal media, with specification occurring in three dimensional aggregates, and utilized both small molecules and recombinant proteins to block the same pathways (a BMPR1A blocking antibody, the WNT antagonist DKK1, and the TGF-beta antagonist SB 431542). Both groups generated >90% NKX2-1-expressing progenitors with these distinct methods. Another major difference between these protocols was that Maroof et al. added the SHH agonist purmorphamine only during days 10–18 of the protocol, while Nicholas et al. added purmorphamine from initiation of specification on day 0. For progenitor differentiation and maturation, Maroof et al. used BDNF, cAMP, and ascorbic acid, while Nicholas et al. used DAPT, a gamma-secretase inhibitor, in the presence of Neurobasal-A media. Both methods drove efficient differentiation and maturation of NKX2-1-expressing MGE-like progenitors into SST and PV expressing mature cIN subtypes that exhibited electrophysiological activity.

A recent review by Anderson et al. provides a detailed cross-comparison of prior and current protocols for deriving cINs.[39] Currently, most protocols similarly modulate BMP and Activin/Nodal TGF-beta signaling and Wnt and SHH signaling,[31,32,40–45] although some exclude use of Wnt inhibition.[24,46–48] However, current protocols continue to vary in the time window during which SHH agonists are added, with some protocols modulating SHH signaling throughout cIN specification, while others utilize specific timeframes for SHH agonist exposure.[40,42,43,45] Modulation of growth factor signaling is achieved either through recombinant proteins or small molecule agonists or antagonists. For example, some protocols use recombinant Noggin or BMPR1A Fc chimeric proteins to antagonize BMP signaling or DKK1 to antagonize Wnt signaling,[40,46,49] while other protocols instead suppress BMP signaling by using the small molecule inhibitors LDN 193189 or Dorsomorphin, or antagonize Wnt signaling with a small molecule tankyrase inhibitor, XAV 939.[32,41–43,50–54] SHH signaling agonists utilized include recombinant SHH protein[23,24,46,48,49,55] or the small molecule agonists Purmorphamine[23,40–43,48] or SAG.[31,32,44,45]

Basal media and supplement composition used throughout experiments also varies among protocols, with some protocols using DMEM-F12 as the basal media, with addition of either KnockOut serum replacement (KSR),[56] N-2,[54,57–59] or N-2 plus B27 supplements.[25,27,50,52,60] While cIN specification and differentiation protocols that supplement media with either B27 or with N-2 plus B27 appear to have the highest success rates, protocols that use Neurobasal-A[23,40,42,45,55] and DMEM-F12[24,31,32,41,43,47] as basal media are also commonly used. Another major variable in these protocols involves whether specification and differentiation utilizes monolayer culture (ML) or formation of three-dimensional cellular aggregates referred to here as embryoid bodies (EBs). Both ML and EB methods appear effective in specifying high percentages of cells as NKX2-1 expressing MGE-like progenitors,[37,38] but there is not yet a clear consensus as to whether utilizing monolayer versus 3D aggregate-based methods during specification and/or differentiation achieves higher frequencies of GABAergic, functionally mature cINs.[37,38,43,57,61,62] Notably, many authors continue to use the NKX2-1-eGFP transgenic reporter line to optimize differentiation protocols and/or to purify the NKX2-1 expressing cell population.[34,37,38,57,63]

Below, we provide a detailed step-by-step protocol that reproducibly results in the specification of 80%–90% of the starting iPSCs as NKX2-1-expressing MGE-like progenitors, and subsequent differentiation of most of these progenitors into mature cINs, with additional approaches used to purify differentiated and mature, functional cINs.[52] We also describe protocol variants and troubleshooting tips and provide protocols for several assays that are commonly used for cellular phenotyping to identify NDD-associated alterations of cIN development. Additional assays for characterization of NDD-associated phenotypes are described in Refs. 52,64,65.

A protocol for cortical interneuron derivation from human pluripotent stem cells (hPSCs)

Equipment and supplies

1. Biosafety cabinet
2. 5% CO_2 37°C tissue culture incubator
3. Standard bacterial petri plates (60 mm) for suspension culture
4. Standard bacterial petri plates (100 mm) for suspension culture
5. Centrifuge outfitted with adaptors for spinning both 15 and 50 mL tubes and tissue culture plates, for example, the Thermo Sorvall Legend XTR Refrigerated Centrifuge including 15 and 50 mL adaptors and Swinging bucket rotor for 96 well plate (#75-003-670) or equivalent.
6. Orbital shaker (placed inside the 5% CO_2 tissue culture incubator)
7. Tissue culture dishes (60 mm)
8. Water bath or dry bead bath, set at 37°C
9. 5 mL serological pipettes
10. 8 well chamber slides (Nunc Lab-Tek Chamber Slide System)
11. 15 mL polystyrene conical tubes
12. Untreated 96 well V-bottom plates and lids
13. 200 and 1000 μL wide bore Pipetteman tips
14. 200 μL Pipetteman tips
15. Aggrewell-800 24-well plates (Stemcell Technologies 34811)
16. Sterile cell strainer 70 μm (Corning 431751)
17. Sterile cell strainer 40 μm (Corning 431750)
18. Multi-well cell culture plates 6 well (USA Scientific CC7682-7506)
19. Multi-well cell culture plates 12 well (USA Scientific CC7682-7512)
20. Multi-well cell culture plates 24 well (USA Scientific CC7682-7524)
21. Multi-well cell culture plates 48 well (USA Scientific CC7682-7548)

Reagents

1. Accutase (Gibco A11105-01)2
2. BDNF (PeproTech 450-02)
3. B-27 supplement without vitamin A (Gibco 12587-010)
4. cAMP (Millipore Sigma D0627)
5. DAPT (Tocris Bioscience 2634)

6. DMEM/F12 (Gibco 11330-032)
7. DPBS, no calcium, no magnesium (Gibco, 14190144)
8. Glutamax (Gibco 35050-061)
9. hESC qualified matrigel matrix (Corning 354277)
10. Laminin (Millipore Sigma L2020)
11. L-ascorbic acid (Millipore Sigma A4403)
12. LDN 193189 (Tocris Bioscience 6053)
13. MEM nonessential amino acids (NEAA) (Gibco 11140050)
14. mTESR-1 (Stemcell Technologies 05852/05851)
15. mFreSR (Stemcell Technologies 05854)
16. Neurobasal-A (Life technologies 10888022)
17. Neural Progenitor Freezing Medium (Stemcell Technologies 05838)
18. Neural Rosette Selection Reagent (Stemcell Technologies 05832)
19. Penicillin-Streptomycin (Gibco 15070063)
20. SAG (Tocris Bioscience 4366)
21. SB 431542 (Tocris Bioscience 1614)
22. XAV 939 (Calbiochem 575545)
23. Y-27632 (Tocris Bioscience 1254)
24. 2-mercaptoethanol (Gibco 21985023)
25. CultureOne supplement (Gibco A3320201)
26. PD0332991 (Tocris Bioscience 4786)
27. GABA (Tocris Bioscience 0344)
28. Anti-Adherence Rinsing Solution (Stemcell Technologies 07010)
29. AggreWell EB Formation Medium (Stemcell Technologies 05893)

Preparation of reagents

Accutase cell detachment solution

Thaw Accutase overnight at 4°C or at room temperature for several hours. Aliquot 12 mL into each 15 mL conical tube and freeze at −20°C. After thawing, Accutase can be stored at 4°C for up to 2 weeks. Note: Thaw Accutase at room temperature.

B-27 supplement (50×) minus vitamin A

Thaw B-27 supplement overnight at 4°C or at room temperature, add an appropriate volume to the tissue culture media, and aliquot the remaining solution into sterile 1.5 mL microfuge tubes and store at −20°C.

Preparing matrigel

The day before aliquoting, place sterile microfuge tubes, sterile 100–200 μL tips, and tube racks needed at −20°C or −80°C overnight. Thaw the 5 mL Matrigel solution overnight at 4°C in an ice bucket filled with ice. The next day, in a tissue culture hood, arrange the prechilled tubes into the prechilled tube rack. Open the Matrigel bottle carefully, aliquot 100 μL per tube using the cold tips, and immediately place the tubes containing Matrigel

TABLE 3.2 Neurobasal complete medium (NBM) composition.

Reagent	Stock concentration	Final concentration	Volume per 100 mL
Neurobasal A	1×	1×	95.0 mL
Pen strep	100×	1×	1.0 mL
GlutaMAX	100×	1×	1.0 mL
B27 without vitamin A	50×	1×	2.0 mL
NEAA	100×	1×	1.0 mL
2-Mercaptoethanol	1000×	1×	100 μL

Reagents, their stock and final concentrations, and the quantity needed to derive 100 mL of media are shown.

TABLE 3.3 Maturation media (MM) composition.

Reagents name	Stock concentration	Final concentration	For 100 mL
BrainPhys Media	1×	1×	
B27 without vitamin A	50×	1×	2.0 mL
Pen strep	100×	1×	1.0 mL
BDNF	20 μg/mL	20 ng/mL	100 μL
cAMP	200 mM	200 μM	100 μL
Ascorbic acid	200 mM	200 μM	100 μL

Reagents, their stock and final concentrations, and the quantity needed to derive 100 mL of media are shown.

Specification of cortical interneuron progenitors from hPSCs

1. On day 0, remove any spontaneously differentiating colonies from the plate containing the hPSCs, wash this 60 mm petri dish with 3 mL of 1× PBS (without Ca^{2+}/Mg^{2+}), add 1.5 mL of Accutase to dissociate the colonies into single cells, and incubate the plate in a 37°C, 5% CO_2 incubator for 5 min. Pipette the cell slurry up and down and check that the colonies have been dissociated into single cells and then neutralize the Accutase by adding 3 volumes of mTeSR media per dish. Collect the cell slurry in a 15 mL conical tube and centrifuge at 200g for 5 min, then aspirate the supernatant and resuspend the cell pellet in Neural induction medium (NIM) plus 10 μM Rock inhibitor. Count cells and seed at a density of 30,000–40,000 cells per well into each well of a V-bottom 96 well plate, spin this plate at 200g for 5 min at room temperature, and incubate the plate in a 37°C, 5% CO_2 incubator for 48 h without disruption. This enables the formation of 3D aggregates, referred to here as embryoid bodies (EBs).
2. On day 2, remove 50 μL medium from each well without touching the EBs by using a multichannel pipette. Add back 50 μL of fresh NIM with 10 μM Rock inhibitor without disrupting the EBs.

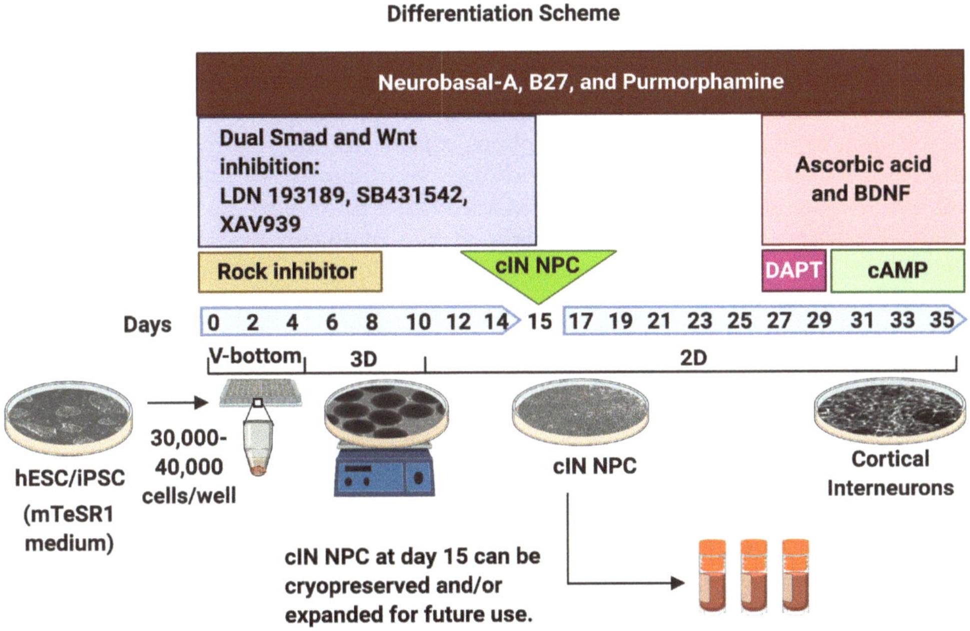

FIGURE 3.2 Differentiation of human pluripotent stem cells into cortical interneurons. Human embryonic stem cells (hESCs) or induced pluripotent stem cells (iPSCs) are maintained in mTeSR medium and then seeded into V-bottom 96-well plates at a cell density of 30,000–40,000 cells/well in neural induction medium with the indicated small molecules to form embryoid bodies (EBs). These are then cultured for 3 days (D) on an orbital shaker in the media indicated. On day 10, EBs are transferred to Matrigel-laminin coated dishes and cultured as a monolayer (2-D). The cortical interneuron progenitors (cIN NPC) generated by day 15 of this protocol are then either cryopreserved, expanded as NPCs, or undergo continued differentiation and maturation through day 35, in the presence of DAPT, Ascorbic acid, cAMP, and BDNF to generate cortical interneurons (cINs) (schematic generated in BioRender).

3. On day 4, set a multichannel pipette to draw 80 μL medium from each well of the V-bottom plate and to eject it back directly onto the EBs, to lift them from the V-bottom and enable their transfer. Pick up all EBs with wide bore 200 μL pipette tips and collect them in a 100 mm nonadherent petri dish (e.g., a petri plate for bacterial culture) and then transfer EBs into a 15 mL conical tube with a 5 mL serological pipet. Spin the EBs at 200g for 2 min at room temperature, aspirate the media carefully, and wash the EBs once with 3 mL of DMEM/F-12 by pipetting onto the wall of the tube, to avoid disrupting the EBs. Resuspend the EB pellet with 4 mL of NIM plus 10 μM Rock inhibitor, and transfer approximately 90–100 EBs into a 60 mm bacteriological plate (ideally transferring the contents of one 96 well plate to one 60 mm petri dish).
4. Incubate plates on an orbital shaker at 80 rpm at 37°C and 5% CO_2 from days 4–10.
5. On day 6, remove the petri dish from the incubator and perform a gentle orbital rotation by hand (e.g., swirl gently) to bring all EBs to the center of the plate. Remove ~3.5 mL of NIM, leaving enough media so that the EBs remain submerged and do not dry out, and immediately add back 4 mL of fresh NIM plus 10 μM Rock inhibitor.

6. On day 8, replenish the NIM without Rock inhibitor, performing a media change as described at step 4.
7. On day 10, prepare Matrigel-laminin coated plates (see Appendix for details). Aspirate the excess Matrigel-laminin coating medium from the plate and add 4 mL of NIM without Rock inhibitor. Removing the plates containing the EBs from the incubator, perform a gentle orbital rotation by hand (e.g., swirl gently) to bring the EBs to the center of the plate, and use a P1000 PIPETMAN with a wide bore tip to move the EBs to a Matrigel-Laminin coated 60 mm dish. One 60 mm plate can accommodate approximately 50 EBs. If using larger plates for culture, scale media quantities and numbers of EBs plated accordingly.
8. On day 12 and day 14, gently remove ~3 mL of NIM from the petri dish and replenish with 4 mL of NIM.
9. By day 15, cIN neural progenitors (NPCs) have been specified and are ready for subsequent applications. If mature interneurons are required, proceed to the interneuron differentiation protocol as described in Section "Interneuron differentiation and maturation from cIN neural progenitor cells."

Maintenance and expansion of cIN NPCs

Interneuron progenitors can be also expanded and maintained in NIM with passaging every 5–6 days or when the plate becomes 100% confluent. For this purpose, do not expose the cells to differentiation medium (DM), as described in the Section "Interneuron differentiation and maturation from cIN neural progenitor cells." Instead, re-suspend the pellet in NIM plus Rock inhibitor for cell counting. Then plate the cells to be maintained as NPCs at a density of 1×10^6 per 60 mm plate. During every passaging, seed at least $0.5–1 \times 10^6$ cells into a 60 mm petri dish using NIM. This will yield a plate of progenitors at >50% confluence after 24 h. The characteristics of NPCs (e.g., regional character and marker expression) can drift during passaging, so it is recommended that NPC maintenance be limited to a maximum of five passages (with fewer being optimal). These interneuron progenitors can be used for various applications, including molecular manipulations via either transfection or lentiviral transduction of plasmid constructs. These progenitors can also be cryopreserved for future use.

Cryopreservation of cIN neural progenitor cells

Dissociate the EBs on day 15 by aspirating the media and washing once with PBS (free of Ca^{2+}/Mg^{2+}) and then adding 1.5 mL of Accutase. Incubate plates for 5 min at 37°C in a 5% CO_2 incubator and then dissociate the cells by pipetting up and down with a P1000 PIPETMAN several times, until the EBs are fully dissociated into single cells. Neutralize the Accutase by adding 3 volumes of NIM and centrifuging at 200 g for 5 min. Aspirate the supernatant and resuspend the cell pellet with Neural progenitor freezing medium (Stem Cell Technologies) at a concentration of $1.5–2.0 \times 10^6$ cells per mL, filling each cryovial with 1 mL of cell slurry. Move the vials to a Mr. Frosty Freezing Container, or similar cryopreservation equipment enabling a controlled −1°C per minute cooling rate. Store the freezing container with the cryovials for a minimum of 24 h at −80°C and then move the vials to a liquid nitrogen tank for long term storage. To obtain mature interneurons from these interneuron progenitor cells (cIN-NPCs), revive the NPCs as described in Section "Revival and maintenance of cryopreserved cIN neural progenitor cells" and follow the interneuron differentiation and maturation protocol described in Section "Interneuron differentiation and maturation from cIN neural progenitor cells."

Revival and maintenance of cryopreserved cIN neural progenitor cells

To recover and expand cryopreserved cIN-NPCs, thaw a cryovial that has been removed from liquid nitrogen storage in a 37°C water bath (or a dry thermal bead bath at the same temperature). Once the vial is largely thawed, with only a small ice crystal visible, wipe the surface of the vial with 70% ethanol and gently remove the cells with a serological pipette. Add the cell solution dropwise to a 15 mL conical tube containing 7 mL of NIM medium plus 10 μM ROCK inhibitor. Centrifuge the cells at 200 g for 5 min. Resuspend the cell pellet with 4 mL of cold NIM with 10 μM ROCK inhibitor and plate on a Matrigel-Laminin coated dish. If the cIN NPCs were cryopreserved as described above, the vial should be readily recovered into one 60 mm Petri dish. After 24 h, the plate should be at least 50% confluent. While cIN NPCs derived from some hPSC lines may initially exhibit a lower survival rate, within a few days they should also be growing well. As needed, enable the cells additional time to reach 80%–90% confluency. As the plate of cells reaches 100% confluency, passage them as described above (Section "Maintenance and expansion of cIN NPCs"), using a seeding density of at least 1×10^6 cells per 60 mm dish.

Interneuron differentiation and maturation from cIN neural progenitor cells

1. A schematic for the workflow of revival and differentiation of cIN NPCs is shown in Fig. 3.3.
2. Begin differentiation once the plate of NPCs reaches confluency. To initiate cIN differentiation, on differentiation day 0 wash the 60 mm petri dish containing the NPCs with 3 mL of PBS (without Ca^{2+}/Mg^{2+}), add 1.5 mL Accutase, and incubate the plates for 5 min at 37°C in a 5% CO_2 incubator. Dissociate the cells by pipetting up and down with a P1000 PIPETMAN and then neutralize the Accutase by adding 3 volumes of NBM per dish. Collect the cells in a 15 mL conical tube, centrifuge at 200 g for 5 min, aspirate the supernatant, and resuspend the cell pellet in 5 mL of NBM plus 10 μM ROCK inhibitor. Count the cells and, using NBM plus 10 μM ROCK inhibitor, bring the volume up to a density of 30,000–40,000 cells per well in a 96-well V-bottom plate, centrifuge the plate at 200g for 5 min and place the plate into a 37°C, 5% CO_2 incubator. Do not disturb the plate for the next 48 h.
3. On differentiation day 2, remove 50 μL medium from each well and add 50 μL of NB plus 10 μM ROCK inhibitor, using a multichannel pipette and without disturbing the neurospheres.
4. On differentiation day 4, use a multichannel pipette to collect 80 μL of media from each well of the V-bottom plate and eject it onto the neurospheres, causing them to detach from the well bottom. Pick the neurospheres by using a multichannel pipette outfitted with wide bore tips (200 μL) and collect them into a 100 mm bacteriological (nonadherent) plate. Use a serological pipette to collect the neurospheres and the media they are in and transfer this to a 15 mL conical tube. Centrifuge this tube for 2 min at 200 g, remove the supernatant, and wash the neurospheres by ejecting 2 mL of DMEM/F-12

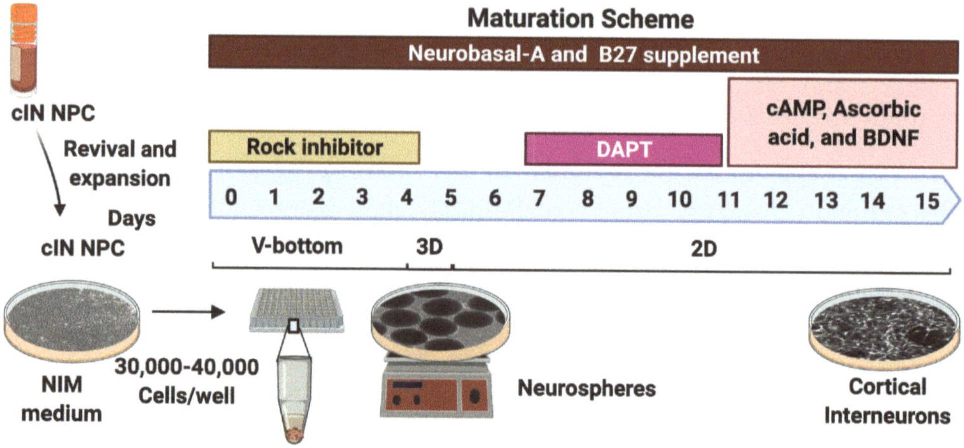

FIGURE 3.3 Differentiation and maturation of cryopreserved cIN NPCs into cortical interneurons. The cIN NPCs that were cryopreserved at day 15 can be revived and expanded in NIM, followed by neurosphere formation at day 4 of differentiation. Neurospheres can then be differentiated in monolayer culture in the presence of DAPT, Ascorbic acid, cAMP, and BDNF to generate differentiated cINs (schematic generated in BioRender).

onto them. Allow the tube to sit undisturbed for ~30–60 s to allow the neurospheres to settle to the tube bottom. Remove most of the supernatant and transfer ~90 neurospheres to each 60 mm bacteriological plate containing 4.5 mL NBM without ROCK inhibitor. Place these plates on an orbital shaker set to 80 rpm and incubate at 37°C in a 5% CO_2 incubator for 1 day. ROCK inhibitor is not necessary from this step onward.

5. On differentiation day 5, remove the plates from the incubator and bring the neurospheres to the plate center by gentle orbital rotation (swirling) of the plate by hand. Move ~50 neurospheres to each 60 mm Matrigel-Laminin coated plate containing 4 mL of DM.
6. On differentiation day 6, neurite extension from the attached neurospheres should begin to occur and can be monitored over time by morphometric analysis and quantitation by acquiring brightfield images with a light microscope.
7. On differentiation days 7 and 9 aspirate and replenish the media with 4 mL of NBM plus DAPT (10 μM). All media for these steps can be prepared at the same time.
8. On differentiation days 11 and 13 aspirate 3.5 mL of medium and add 4 mL of NBM plus cAMP (200 μM), Ascorbic acid (200 μM), and BDNF (20 ng/mL). Neurites will continue to extend and acquire greater length and complexity during this period, so take care to avoid disrupting these during media changes, ensuring that the solution is never aspirated completely and handling the plate gently to avoid detaching the neurospheres. Add 5 μg/mL of Laminin if the neurospheres appear to be nonadherent, to encourage reattachment to the dish bottom. Day 11 and 13 media can be prepared at the same time and stored at 4°C. On day 15 the differentiated cINs will be ready for downstream applications, such as immunocytochemistry for markers or molecular analysis (e.g., by ChIP-seq or RNA-seq).

Enrichment and purification of cIN neural progenitor cells and neurons

Enrichment for post-mitotic cINs with neural rosette selection reagent

As shown in Fig. 3.4A–C, an hESC line with eGFP knock-in at the NKX2-1 locus[34] can be used to assess the efficiency with which cIN NPCs are specified and subsequently differentiate into cINs. At day 35, neural rosette selection can be used to enrich the culture for differentiated and maturing post-mitotic cINs and to remove remaining cIN progenitors (Fig. 3.4D–G). To perform neural rosette selection, wash cells with 3 mL of 1× PBS (without Ca^{2+}/$Mg2^{+}$), add 4 mL of Neural Rosette Selection Reagent (NRS; Stem Cell Technologies), and incubate for 60 min at 37°C. After 60 min, gently remove the NRS reagent by pipetting and wash once with 2 mL of DMEM/F-12. To enrich the cell population for differentiated cINs, use a P1000 PIPETMAN to eject DMEM/F-12 directly onto the center of the neurospheres to lift them off of the plate and move them away from less differentiated cells at the neurosphere periphery. After collecting these interneuron containing neurospheres from the plate with a P1000 pipet tip, move them to a new plate, pipetting the mixture up and down several times to dissociate the cells. Prior to Neural Rosette Selection, interneurons will be a mixed population of differentiated post-mitotic interneurons (cells with small, round cell bodies) and proliferating progenitors (cells with larger, flatter cell bodies) (Fig. 3.4F). After Neural Rosette selection, the culture will be enriched for the differentiated interneurons, with ~80% expressing GABA, GAD65, and other markers of differentiated GABAergic neurons (Fig. 3.4G). These can be used for further analyses, as described below.

Purification of post-mitotic cINs with NCAM bead selection

Purification with Neural Rosette Selection reagent, as described above, enriches the culture for cINs sufficiently for many further applications. However, some residual undifferentiated progenitors and younger, less mature neurons will remain. Proliferating interneuron progenitors expressing Ki-67, a marker of actively dividing cells, represent ~5% of cells in the culture at this stage. For some downstream applications, such as engraftment into the murine brain or functional analysis by electrophysiology, removal of these proliferating progenitors is essential. In this case, differentiated, post-mitotic interneurons can be purified away from these residual proliferating progenitors based upon their expression of the differentiated neuronal marker NCAM, expressed only by post-mitotic neurons, by using the Anti-PSA-NCAM MicroBead Kit (Miltenyi Biotech) according to the manufacturer's instructions.

Critical steps and troubleshooting

hPSC lines can exhibit differences in their rate of proliferation and sensitivity to some experimental manipulations. These differences may relate either to differences in their genetic background and/or the conditions under which they have been derived or maintained, including their passage number. This variability can cause differences in the cellular response to handling when these specification and differentiation protocols are conducted. For example, we have found cIN NPC derivation to be more challenging with the H1 versus

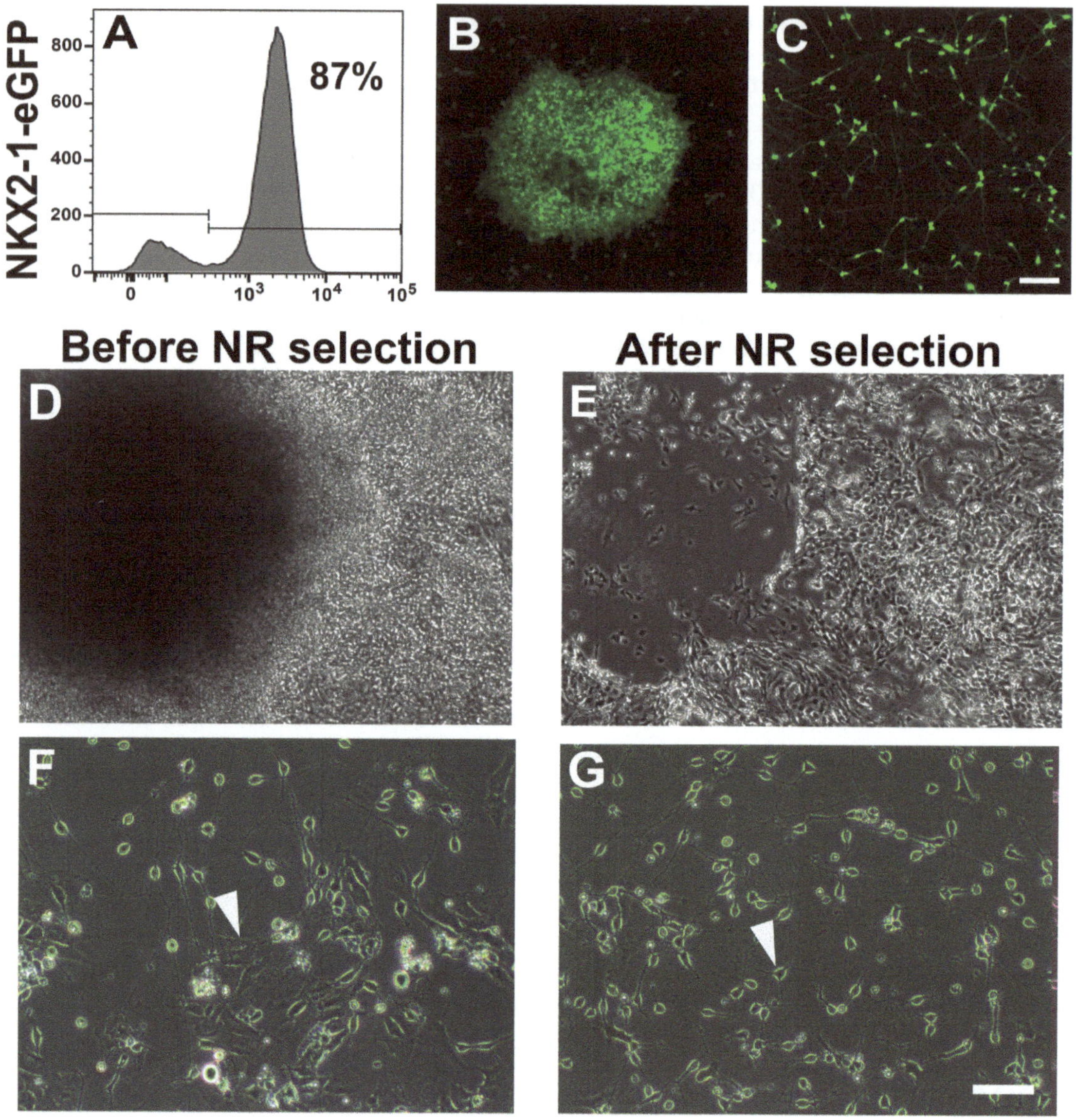

FIGURE 3.4 Isolation and characterization of cIN NPCs. (A–C) An hESC line with eGFP knock-in at the NKX2-1 locus (34) can be used to assess the efficiency with which cIN NPCs are specified and subsequently differentiate into cINs. (A) Flow cytometry can be used to define the fraction of cIN NPCs and neurons expressing NKX2-1-eGFP. (B–C) NKX2-1-eGFP expression can also be detected by immunofluorescence in (B) embryoid bodies with expressing cIN NPCs or (C) in plated cINs. Scale bar for B–C = 50 μm. (D–G) Neural rosette (NR) selection reagent can be used to isolate NRs, enriching the isolated cIN population for differentiated, post-mitotic cells. (D) Prior to and (E) after isolation of each neural rosette (NR) from the culture. (F) Prior to NR selection, plated cINs are heterogeneous and contain many dividing progenitors with larger cell bodies (one highlighted with arrowhead). (G) The replated cINs isolated after NR rosette isolation using the NR selection reagent. The culture is enriched for post-mitotic neurons with small round cell bodies and neurite extension (one highlighted with arrowhead). Scale bar in F–G = 200 μm.

the H9 hESC line, due to this line's greater propensity to undergo cell death in response to handling or media changes, including Accutase treatment or addition of the SHH agonist SAG to cultures. In working with such lines, several tips and a modified protocol for EB formation are provided below. This includes growing hESC lines in mTeSR medium, with care taken to remove colonies undergoing spontaneous differentiation whenever they arise. EBs are derived when the plates are ~75%–80% confluent, by washing the cells once with 1× PBS (without Ca^{2+} and Mg^{2+}). Colonies are dissociated by adding 1.5 mL of Accutase and incubating the plate for 5 min at 37°C in a 5% CO_2 incubator.

Tip: Do not exceed an incubation time of >10 min in Accutase, as this causes cell death and will reduce the efficiency of later cIN specification and differentiation. Add three volumes of 10% FBS in DMEM/F-12 to inactivate Accutase when the colonies are dissociated into single cells.

Tip: Handle EBs gently using a 5 mL serological pipet or a wide bore P1000 PIPETTEMAN tip rather than a 10 mL serological pipet, as the opening of the latter is too narrow for pipetting EBs.

Tip: Long term maintenance of post-mitotic cINs is challenging as cells have a tendency to lose adherence to culture plates over time. To avoid this loss of cells, feed using an automated pipet using the slowest possible speed and avoid unnecessary movement of the plates. Furthermore, using multi-well plates with smaller surface area per well (i.e., switching to 24 or 48 well plates) can reduce the chances of cells lifting off of the plate. Regular supplementation with Laminin is also recommended.

Tip: Passaging post-mitotic neurons (day 30) can result in an enrichment of astrocytes, as they are more resistant to manual manipulations. To avoid this, pre-warm Accutase to 37°C before use and do not triturate cells to encourage dissociation. Longer incubation with Accutase can be performed, but Rock inhibitor should be added to neurons 1 h before dissociation to prevent cell death. Finally, straining cells through a 40 μM cell strainer can aid in removal of astrocytes. While these modifications should achieve a more homogeneous neuronal population and limit numbers of astrocytes in the cultures, there will likely be a significant increase in cell death. Therefore, the starting material should be increased to compensate.

Cellular phenotyping of hPSC-derived cINs

Using immunocytochemistry to benchmark hPSC-derived cINs and to assess NDD-related alterations of neurodevelopment

Immunocytochemistry using the antibodies shown in can be used to assess the efficiency with which NPCs of a forebrain (e.g., FOXG1), ganglionic eminence (e.g., DLX2/5), or medial ganglionic eminence (e.g., NKX2-1) character are generated, and to assess the efficiency with which these differentiate into GABAergic cINs (e.g., GABA, GAD65/67) expressing markers of medial ganglionic eminence-derived cIN subtypes (e.g., PV/SST). Common antibodies, sources, and dilutions are listed in Table 3.4.

Morphometric analysis of neurite extension and length

To measure neurite length in cINs from day 15 neurospheres, dissociate neurospheres and seed 1–1.5 × 10^6 cells/well into Matrigel-Laminin coated chambers of an 8 well chamber slide. Incubate the cells for 24–48 h at 37°C in a 5% CO_2 incubator, and stain with MAP2 antibody as described below. Acquire images using a spinning-disk (Quorum) confocal microscope and an Axiovision inverted microscope. Process the images with Imaris software (Bitplane), measuring neurospheres length using the filament tracer application, and

TABLE 3.4 Antibodies for immunocytochemistry of cIN NPCs and neurons.

#	Antibody name	Company name	Source	Dilution used
1	DLX2	Millipore	Rb	1 in 200
2	DLX5	Abcam	Rb	1 in 400
3	FOXG1	Abcam	Goat	1 in 100
4	TUBB3	Covance	Ms	1 in 1000
5	NKX2-1	Abcam	Rb	1 in 100
6	ASCl1	Millipore	Ms	1 in 50
7	OLIG2	Millipore	Rb	1 in 100
8	DCX	Santa cruz	Goat	1 in 200
9	CALB	Swant	Rb	1 in 200
10	GAD65/67	Millipore	Rb	1 in 100
11	GABA	Sigma	Rb	1 in 200
12	CXCR4	BD	Ms	1 in 100
13	PV	Swant	Rb	1 in 200
14	SST	Peninsula	Rb	1 in 200
15	VGAT	Synaptic System	Ms	1 in 200
16	VGLUT1	Synaptic System	Rb	1 in 100
17	Ki67	Abcam	Rb	1 in 1000
18	Map2	Millipore	Ms	1 in 500
19	Map2	Abcam	Rb	1 in 250
21	Nestin	Abcam	Ms	1 in 250

This collection of antibodies can be used for immunocytochemistry to assess the fraction of hPSCs that have been specified as cIN NPCs or differentiated into cINs, as well as assessing other aspects of the system such as the fraction that differentiate into the PV- versus SST-expressing cIN subtypes. For each antibody, the supplier, source species, and suggested dilution for use are shown. *Rb*, rabbit; *Ms*, mouse.

normalizing to the number of DAPI stained nuclei in the neurospheres, as measured with the particle application in Imaris.

On day 6 of cIN NPC differentiation, transfer the neurospheres to a Matrigel-Laminin coated 60 mm dish or an 8 well chamber (3–4 neurospheres/well). Incubate the cells for 24 h at 37°C in a 5% CO_2 incubator. The next day, acquire images using an inverted light microscope (examples in A and Ref. 65). Neurite extension length can be measured from neurospheres using the ImageJ program, as the distance between two circles drawn at the border of the plated neurospheres and at the tips of neurites extending from that neuroid. See Fig. 3.5.

Neuronal migration assay

To study the migration of cExN and cIN neurons, we developed an approach that utilizes fused coculture of two neurospheres consisting of cExNs and cINs, respectively.[65] cExN and cIN NPCs are generated as described (64) and are transduced with a lentiviral Synapsin-eGFP or Synapsin-RFP expression construct, respectively. 72 h post-transduction, cExN and cIN NPCs are separately dissociated with Accutase, spun down at 200 g, and each pellet is resuspended in NBM containing 10 μM ROCK inhibitor. Count the cells and bring the volume up to a density of 3×10^5 cells per ml. Aliquot 100 μL/well of each cell type separately into a 96-Well V-bottom plate, centrifuge the plate at 200 g for 5 min, and place the dish into a 37°C, 5% CO_2 incubator. Do not disturb the plate for the next 48 h. On day 2, replace 50% medium with new NBM plus 10 μM ROCK inhibitor without disturbing the spheres, to generate cExN and cIN containing neurospheres. On day 4, select cExN and cIN neurospheres with wide bore P1000 tips and place one cExN neurosphere and one cIN neurosphere side by side in each U bottom well. Placement of the neurospheres in close juxtaposition will cause them to undergo fusion without further manipulation, enabling assessment of neuronal migration. On Day 6, move the fused neurospheres with wide bore P1000 tips to a coverslip placed in a 30 mm Petri dish coated with Matrigel-Laminin. On day 10, acquire images using a spinning-disk (Quorum) confocal microscope and an Axiovision inverted microscope, and perform image processing using ImageJ. An example of this assay is shown in Fig. 3.5B.

Measurement of synaptic puncta

To quantify VGAT- and VGLUT-expressing synaptic punctain cocultured cINs and cExNs, dissociate and plate neuroids generated as described above and perform VGAT and VGLUT immunocytochemistry, using the antibodies shown in Table 3.4. Acquire images using a spinning-disk (Quorum) confocal microscope and Axiovision inverted microscope and measure puncta using the synaptic counter plugin in ImageJ. An example of this assay is shown in Fig. 3.6.

Alternate protocol for derivation of cIN NPCs from hPSCs

This protocol variant differs from the version described above in that the formation of smaller EBs from fewer hPSCs is conducted by using AggreWell plates (rather than V bottom

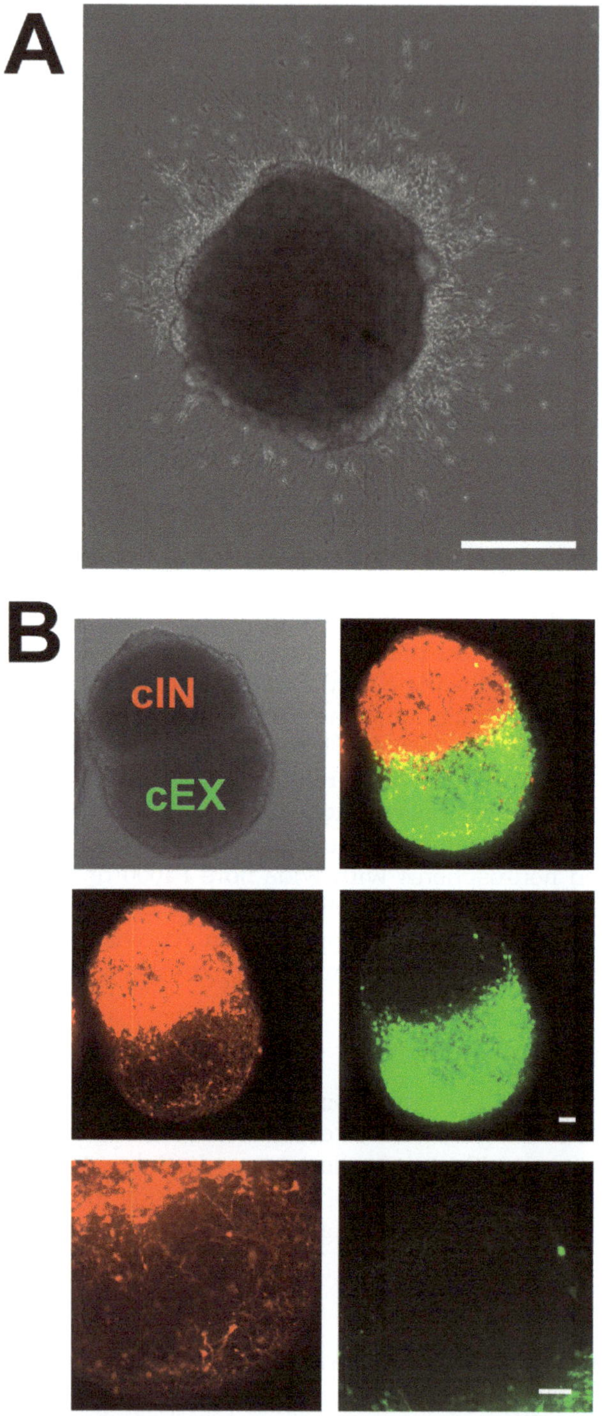

FIGURE 3.5 Neurite extension and neuronal migration assays. (A) Neurite extension: after 10 days of cIN differentiation, cINs in the plated EB exhibit robust neurite extension. Scale bar = 200 μm. (B) Neuronal migration can be assessed by separately deriving cINs and cortical excitatory neurons (cExNs) as described (64), transducing these with lentiviral expression constructs that encode Synapsin-red fluorescent protein (RFP) and Synapsin-GFP, respectively, and placing these EBs in apposition. After EB fusion, red cINs that have migrated into the cExN territory and green cExNs that have migrated into the cIN territory are visible. Scale bars in B = 50 μm; lower panels provide higher magnification views of the center panel coculture interfaces.

96 well plates), with EB formation in mTeSR rather than neural induction media. This alternative may be preferable for working with hPSC lines that are more challenging to work with, because the cells are kept under less stressful conditions during EB formation, and fewer hPSCs are required to initiate EB formation.

1. On day 0, remove any spontaneously differentiating colonies from the plate containing the hPSCs, wash this 60 mm petri dish with 3 mL of 1× PBS (without Ca^{2+}/Mg^{2+}), add 1.5 mL of Accutase to dissociate the colonies into single cells, and incubate the plate in a 37°C, 5% CO_2 incubator for 5 min.
2. While cells are dissociated, begin to prepare the desired number of wells on an AggreWell plate (ideally using the 24-well variant). To do this, add 500 μL Anti-Adherence Rinsing Solution to each well and centrifuge the plate at 1300 g for 5 min.

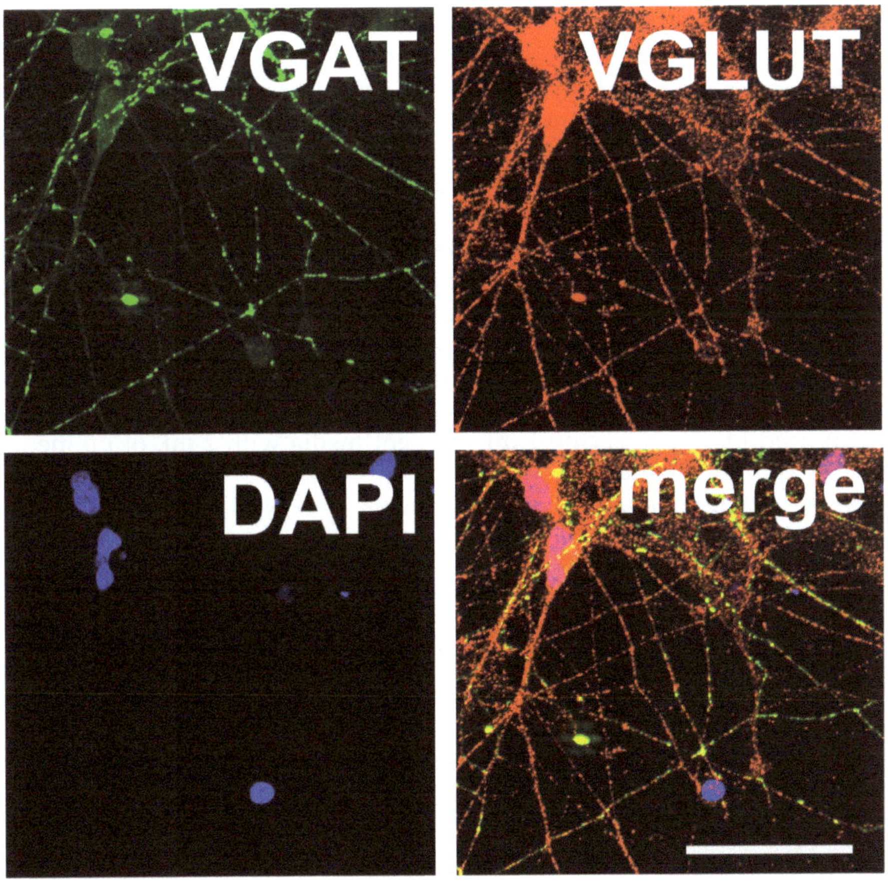

FIGURE 3.6 Visualization of synaptic puncta is performed by immunocytochemistry for VGAT and VGLUT, vesicular transporters for the neurotransmitters GABA and glutamate, respectively, in cIN and cExN cocultures. The presence of vesicular puncta expressing these markers indicates capacity of neurons for neurotransmitter release at synapses. DAPI staining indicates the locations of neuronal nuclei within the field shown. Scale bar = 50 μm.

3. Pipette the cell slurry up and down and check that the colonies have been dissociated into single cells. Then neutralize the Accutase by adding 3 volumes of mTeSR media per dish. Collect the cell slurry in a 15 mL conical tube and centrifuge at 200 g for 5 min.
4. While cells are centrifuged, finish preparing the AggreWell plate by aspirating the Anti-Adherence Rinsing Solution and washing each well with 1 mL DMEM/F12 media, before adding 1 mL of EB formation media supplemented with 20 μM Rock inhibitor to each well.
5. After centrifugation, aspirate the supernatant and re-suspend the cells in 1 mL EB formation media, count cells and adjust the density of cells to a concentration of 1.5×10^6 per mL (concentration can be adjusted ±10% if necessary); then add 1 mL to each prepared well on the AggreWell plate.
6. Centrifuge the AggreWell plate at 200 g for 5 min and then incubate for 24 h in a 37°C, 5% CO_2 incubator.
7. Observe the formation of EBs in the AggreWell plate for 1–3 days and replace 50% of the media with fresh EB formation media supplemented with 10 μM Rock inhibitor daily (make sure media is removed and added slowly so as not to disturb the EB formation). Once EBs are formed and have a clearly defined edge, proceed to step 8. If EBs have not formed after 4 days, then the iPSCs used were likely not of sufficient quality to produce EBs and the process should be restarted. Normally EBs are ready to proceed after 48 h.
8. Remove the EBs from the AggreWell plate using a wide bore 200 μL pipette tips and transfer the media with EBs into a 60 mm nonadherent petri dish (e.g., an untreated petri plate for bacterial culture). Wash the well with a further 1 mL of EB media (containing 10 μM Rock inhibitor), collect the wash in the same nonadherent petri dish. Repeat the wash with a further 1 mL EB media (containing 10 μM Rock inhibitor) and incubate plate on an orbital shaker at 80 rpm at 37°C and 5% CO_2.
9. After 24 h, feed EBs by replacing 1 mL of spent media with 2 mL of EB media (containing 10 μM Rock inhibitor).
10. After a further 24 h replace 2 mL of spent media with NIM plus 10 μM Rock inhibitor. This constitutes day 0.
11. On day 2, remove the petri dish from the incubator and perform a gentle orbital rotation by hand (e.g., swirl gently) to bring all EBs to the center of the plate. Remove ~3.5 mL of NIM, leaving enough media so that the EBs remain submerged and do not dry out, and immediately add back 4 mL of fresh NIM plus 10 μM Rock inhibitor.
12. On days 4 and 6, replenish the NIM with Rock inhibitor, performing a media change as described at step 11.
13. On day 8, replenish the NIM without Rock inhibitor, performing a media change as described at step 4.
14. On day 10, prepare Matrigel-laminin coated plates (see Appendix for details). Aspirate the excess Matrigel-laminin coating medium from the plate and add 4 mL of NIM without Rock inhibitor. Removing the plates containing the EBs from the incubator, perform a gentle orbital rotation by hand (e.g., swirl gently) to bring the EBs to the center of the plate, and use a P1000 PIPETMAN with a wide bore tip to move the EBs to a Matrigel-Laminin coated 60 mm dish. One 60 mm plate can accommodate all EBs

from one 60 mm dish (~150 EBs). If using larger plates for culture, scale media quantities and numbers of EBs plated accordingly.

15. On day 12 and day 14, gently remove ~3 mL of NIM from the petri dish and replenish with 4 mL of NIM.
16. By day 15, cIN neural progenitors (NPCs) have been specified and are ready for subsequent applications. If mature interneurons are required, proceed to the interneuron differentiation protocol.

Alternate protocol for differentiation of cIN NPCs into interneurons

This alternate protocol for differentiation of cIN NPCs into interneurons involves the use of additional reagents (Cytarabine, GABA, and BrainPhys) to facilitate further maturation of the cINs that are produced. It may be better suited to preparing neurons for functional analysis by electrophysiology or multi-electrode array.

1. Begin differentiation once the plate of NPCs reaches confluency. To initiate cIN differentiation, on differentiation day 0 wash the 60 mm petri dish containing the NPCs with 3 mL of PBS (without Ca^{2+}/Mg^{2+}), add 1.5 mL Accutase, and incubate the plates for 5 min at 37°C in a 5% CO_2 incubator.
2. While cells are dissociated, begin to prepare the desired number of wells on an AggreWell plate (ideally using the 24-well variant). Add 500 μL Anti-Adherence Rinsing Solution to each well and centrifuge the plate at 1300 g for 5 min.
3. Pipette the cell slurry up and down and check that the cells have been dissociated into single cells and then neutralize the Accutase by adding 3 volumes of NBM per dish. Collect the cell slurry in a 15 mL conical tube and centrifuge at 200 g for 5 min.
4. While cells are centrifuged, finish preparing the AggreWell plate by aspirating the Anti-Adherence Rinsing Solution and washing each well with 1 mL DMEM/F12 media, before adding 1 mL of NBM supplemented with 20 μM Rock inhibitor to each well.
5. After centrifugation, aspirate the supernatant and re-suspend the cells in 1 mL NBM, count cell numbers, and adjust the density of cells to a concentration of 1.7×10^6 per ml (concentration can be adjusted ±10% if necessary); then add 1 mL to each prepared well on the AggreWell plate.
6. Centrifuge the AggreWell plate at 200 g for 5 min and then incubate for 24 h in a 37°C, 5% CO_2 incubator.
7. On day 16, feed neurospheres by replacing 1 mL of media with fresh NBM supplemented with 10 μM Y-27632. Make sure media is removed and added slowly so as not to disturb the neurospheres.
8. On day 17, remove the neurospheres from the AggreWell plate using a wide bore 200 μL pipette tips and transfer the media with neurospheres into a 60 mm nonadherent petri dish (e.g., a petri plate for bacterial culture) and wash the well with a further 1 mL of NBM (containing 10 μM Rock inhibitor), collecting the wash in the same nonadherent petri dish. Repeat the wash with a further 1 mL NBM (containing

10 μM Rock inhibitor) and incubate plate on an orbital shaker at 80 rpm at 37°C and 5% CO_2.

9. On day 18, prepare Matrigel-Laminin coated plates by adding 12 mL of cold DMEM/F-12, 100 μL Matrigel (thawed), and 60 μL (5 μg/mL final concentration) laminin to a 15 mL tube. Dispense 1 mL per well on a 12-well plate (or equivalent amount scaled for other multi-well plates or chamber slides). Incubate overnight at 4°C.
10. On Day 19, plate the neurospheres onto Matrigel-Laminin coated culture plates or chamber slides. A single well from an AggreWell plate can be plated into $^1/_2$ a multi-well culture plate. Neurospheres should be plated in NBM (1 mL per well on a 12 well plate or equivalent).
11. On day 22, feed the plated neurospheres by removing 250 μL (or equivalent) of spent media and replacing it with 750 μL of fresh NBM.
12. On days 24 and 26, feed the plated neurons by removing 750 μL (or equivalent) of spent media and replacing it with 750 μL of NBM supplemented with BDNF (20 ng/mL) and Culture One Supplement (1×).
13. On day 28 feed the plated neurons by removing 750 μL (or equivalent) of spent media and replacing it with 750 μL of NBM supplemented with BDNF (20 ng/mL), ascorbic acid (200 μM), DAPT (10 μM) and PD0332991 (2 μM).
14. On day 30 of differentiation, cells can be maintained or can be passaged to achieve desired density for immunocytochemistry/immunofluorescence or for functional assessments.
    a. For maintenance, feed the plated neurons by removing 750 μL (or equivalent) of spent media and replacing it with 750 μL of NBM supplemented with BDNF (20 ng/mL), ascorbic acid (200 μM), DAPT (10 μM), PD0332991 (2 μM) and Laminin (2–2.5 μg/mL).
    b. For passaging, do not wash cells with DPBS as this can cause cells to lift. Instead aspirate media and immediately add Accutase solution. Dissociation should be carried out at 37°C for 8–12 min and no longer than 15 min. After 8 min of dissociation, cell suspension can be gently triturated with a 1000 μL pipet tip to improve cell dissociation. Cell suspension should be collected and Accutase solution inactivated by 1:5 dilution in DMEM/F12. Cells can be collected by centrifugation at 200 g for 5 min. If necessary, neurons can be suspended in 1 mL NBM and passed through a sterile (70 μm) cell strainer to achieve a single cell suspension. Desired density is dependent on endpoint assay, but ~50,000 cells/cm^2 is preferable for functional and morphological assessments that require distinct cells or small cell clumps, ~200–400 cells/cm^2 is preferable for assays that require dense cell cultures and abundant cell-cell connections. Cells should be plated in 1.5 mL (or equivalent) of NBM supplemented with BDNF (20 ng/mL) and ascorbic acid (200 μM) in 12-well Matrigel-Laminin coated plates.
15. On days 32 and 34, feed the plated neurons by removing 750 μL (or equivalent) of spent media and replacing it with 750 μL of NBM supplemented with BDNF (20 ng/mL), ascorbic acid (200 μM) and cAMP (200 μM).
16. On day 36, feed the plated neurons by removing 750 μL (or equivalent) of spent media and replacing it with 750 μL of Maturation media (MM).

17. On day 38, feed the plated neurons by removing 750 μL (or equivalent) of spent media and replacing it with 750 μL of MM supplemented with Cytarabine (1 μM).
18. On day 40, feed the plated neurons by removing 750 μL (or equivalent) of spent media and replacing it with 750 μL of MM supplemented with Cytarabine (1 μM) and Laminin (2–2.5 μg/mL).
19. On day 42, feed the plated neurons by removing 750 μL (or equivalent) of spent media and replacing it with 750 μL of MM supplemented with GABA (300 μM).
20. Neurons can be maintained by feeding with $^1/_2$ media replacements every 4 days using MM supplemented with GABA (300 μM).
    a. For improved cell adhesion, Laminin can be added every 10 days (i.e., day 50, 60 etc.)
    b. For improved maturation, BrainPhys media can be preconditioned by 48h incubation with human, rat or mouse astrocytes (note: media must be sterile filtered after removal from astrocytes and should only be supplemented after filtering).

Acknowledgments

We thank the family for providing biomaterials used in the examples shown in Fig. 3.5. We thank the Genome Engineering and iPSC Center (GEiC) at Washington University School of Medicine (WUSM) for deriving the iPSC lines used here. We thank the Alvin J. Siteman Cancer Center at WUSM, which is supported in part by NCI Cancer Center Support Grant #P30 CA091842, for use of their Flow Cytometry Core (data in Fig. 3.4). We thank WUSM Cytogenetics & Molecular Pathology for karyotyping services. We also thank A.G. Elefanty, E.G. Stanley, and S.A. Anderson for providing our laboratory with the NKX2-1-eGFP knock-in hESC line used in Fig. 3.4.[34]

Approvals

Subjects for example data shown in Fig. 3.5 were consented for iPSC line generation by the Washington University Institutional Review Board of the Human Research Protection Office under human studies protocol #201409091 (Dr. John Constantino). Consent to publish data was provided by all subjects. Human embryonic stem cell work was approved by the ESCRO board at Washington University under protocol ESCRO#12-002 (Dr. Kristen Kroll).

Funding

This work was supported by R01NS114551 and R01MH124808, and U01HG007530 (NINDS/NIH Common Fund), and U54HD087011 (administrative supplement), and the M-CM Network, the Jakob Gene Fund, and the Washington University (WU) Center for Regenerative Medicine, WU Children's Discovery Institute (CDI-LI-2019-819), the Engelhart Family Foundation, and the WU Institute for Clinical and Translational Sciences to KLK and P50 HD129966 to John Constantino and Christina Gurnett (KLK, PI, core director).

References

1. Arber C, Li M. Cortical interneurons from human pluripotent stem cells: Prospects for neurological and psychiatric disease. *Front Cell Neurosci*. 2013. https://doi.org/10.3389/fncel.2013.00010.
2. Kessaris N, Magno L, Rubin AN, Oliveira MG. Genetic programs controlling cortical interneuron fate. *Curr Opin Neurobiol*. 2014;26:79–87. https://doi.org/10.1016/j.conb.2013.12.012.
3. Kepecs A, Fishell G. Interneuron cell types are fit to function. *Nature*. 2014;505(7483):318–326. https://doi.org/10.1038/nature12983.

4. Powell EM. Interneuron development and epilepsy: early genetic defects cause long-term consequences in seizures and susceptibility. *Epilepsy Current*. 2013;13(4):172–176. https://doi.org/10.5698/1535-7597-13.4.172.
5. Tyson JA, Anderson SA. GABAergic interneuron transplants to study development and treat disease. *Trends Neurosci*. 2014;37(3):169–177. https://doi.org/10.1016/j.tins.2014.01.003.
6. Breedlove SM. *Foundations of Neural Development*. Sunderland, MA: Sinauer Associates, Inc.; 2017.
7. Patestas MA, Gartner LP. *A Textbook of Neuroanatomy*. 2016.
8. Stiles J, Jernigan TL. The basics of brain development. *Neuropsychol Rev*. 2010;20(4):327–348. https://doi.org/10.1007/s11065-010-9148-4.
9. Clowry GJ. An enhanced role and expanded developmental origins for gamma-aminobutyric acidergic interneurons in the human cerebral cortex. *J Anat*. 2015;227(4):384–393. https://doi.org/10.1111/joa.12198.
10. Letinic K, Zoncu R, Rakic P. Origin of GABAergic neurons in the human neocortex. *Nature*. 2002;417(6889):645–649. https://doi.org/10.1038/nature00779.
11. Defelipe J. The evolution of the brain, the human nature of cortical circuits, and intellectual creativity. *Front Neuroanat*. 2011;5.
12. Jones EG. The origins of cortical interneurons: mouse versus monkey and human. *Cerebr Cortex*. 2009;19(9):1953–1956. https://doi.org/10.1093/cercor/bhp088.
13. Arshad A, Vose LR, Vinukonda G, et al. Extended production of cortical interneurons into the third trimester of human gestation. *Cerebr Cortex*. 2016;26(5):2242–2256. https://doi.org/10.1093/cercor/bhv074.
14. Xiang Y, Tanaka Y, Patterson B, et al. Fusion of regionally specified hPSC-derived Organoids models human brain development and interneuron migration. *Cell Stem Cell*. 2017;21(3):383–398. https://doi.org/10.1016/j.stem.2017.07.007. e7.
15. Xiang Y, Yoshiaki T, Patterson B, et al. Generation and fusion of human cortical and medial ganglionic eminence brain Organoids. *Curr Protoc Stem Cell Biol*. 2018;47(1). https://doi.org/10.1002/cpsc.61.
16. Ardhanareeswaran K, Mariani J, Coppola G, Abyzov A, Vaccarino FM. Human induced pluripotent stem cells for modelling neurodevelopmental disorders. *Nat Rev Neurol*. 2017;13(5):265–278. https://doi.org/10.1038/nrneurol.2017.45.
17. Kim YS, Leventhal BL. Genetic epidemiology and insights into interactive genetic and environmental effects in autism spectrum disorders. *Biol Psychiatr*. 2015;77(1):66–74. https://doi.org/10.1016/j.biopsych.2014.11.001.
18. Tick B, Bolton P, Happé F, Rutter M, Rijsdijk F. Heritability of autism spectrum disorders: a meta-analysis of twin studies. *J Child Psychol Psychiatry Allied Discip*. 2016;57(5):585–595. https://doi.org/10.1111/jcpp.12499.
19. Takahashi K, Tanabe K, Ohnuki M, et al. Induction of pluripotent stem cells from Adult human fibroblasts by defined factors. *Cell*. 2007;131(5):861–872. https://doi.org/10.1016/j.cell.2007.11.019.
20. Zhou T, Benda C, Dunzinger S, et al. Generation of human induced pluripotent stem cells from urine samples. *Nat Protoc*. 2012;7(12):2080–2089. https://doi.org/10.1038/nprot.2012.115.
21. Loh YH, Agarwal S, Park IH, et al. Generation of induced pluripotent stem cells from human blood. *Blood*. 2009;113(22):5476–5479. https://doi.org/10.1182/blood-2009-02-204800.
22. Tamaoki N, Takahashi K, Tanaka T, et al. Dental pulp cells for induced pluripotent stem cell banking. *J Dent Res*. 2010;89(8):773–778. https://doi.org/10.1177/0022034510366846.
23. Chailangkarn T, Trujillo CA, Freitas BC, et al. A human neurodevelopmental model for Williams syndrome. *Nature*. 2016;536(7616):338–343. https://doi.org/10.1038/nature19067.
24. Griesi-Oliveira K, Acab A, Gupta AR, et al. Modeling non-syndromic autism and the impact of TRPC6 disruption in human neurons. *Mol Psychiatr*. 2015;20(11):1350–1365. https://doi.org/10.1038/mp.2014.141.
25. Marchetto MCN, Carromeu C, Acab A, et al. A model for neural development and treatment of rett syndrome using human induced pluripotent stem cells. *Cell*. 2010;143(4):527–539. https://doi.org/10.1016/j.cell.2010.10.016.
26. Muratore CR, Rice HC, Srikanth P, et al. The familial alzheimer's disease APPV717I mutation alters APP processing and Tau expression in iPSC-derived neurons. *Hum Mol Genet*. 2014;23(13):3523–3536. https://doi.org/10.1093/hmg/ddu064.
27. Chambers SM, Fasano CA, Papapetrou EP, Tomishima M, Sadelain M, Studer L. Highly efficient neural conversion of human ES and iPS cells by dual inhibition of SMAD signaling. *Nat Biotechnol*. 2009;27(3):275–280. https://doi.org/10.1038/nbt.1529.
28. Donovan APA, Basson MA. The neuroanatomy of autism – a developmental perspective. *J Anat*. 2017;230(1):4–15. https://doi.org/10.1111/joa.12542.
29. Oblak AL, Gibbs TT, Blatt GJ. Decreased GABAB receptors in the cingulate cortex and fusiform gyrus in Autism. *J Neurochem*. 2010;114(5):1414–1423. https://doi.org/10.1111/j.1471-4159.2010.06858.x.

30. Zikopoulos B, Barbas H. Altered neural connectivity in excitatory and inhibitory cortical circuits in autism. *Front Hum Neurosci*. 2013. https://doi.org/10.3389/fnhum.2013.00609.
31. Marchetto MC, Belinson H, Tian Y, et al. Altered proliferation and networks in neural cells derived from idiopathic autistic individuals. *Mol Psychiatr*. 2017;22(6):820–835. https://doi.org/10.1038/mp.2016.95.
32. Ross PJ, Zhang WB, Mok RSF, et al. Synaptic dysfunction in human neurons with autism-associated deletions in PTCHD1-AS. *Biol Psychiatr*. 2020;87(2):139–149. https://doi.org/10.1016/j.biopsych.2019.07.014.
33. Watanabe K, Ueno M, Kamiya D, et al. A ROCK inhibitor permits survival of dissociated human embryonic stem cells. *Nat Biotechnol*. 2007;25(6):681–686. https://doi.org/10.1038/nbt1310.
34. Goulburn AL, Alden D, Davis RP, et al. A targeted NKX2.1 human embryonic stem cell reporter line enables identification of human basal forebrain derivatives. *Stem Cell*. 2011;29(3):462–473. https://doi.org/10.1002/stem.587.
35. Li XJ, Zhang X, Johnson MA, Wang ZB, LaVaute T, Zhang SC. Coordination of sonic hedgehog and Wnt signaling determines ventral and dorsal telencephalic neuron types from human embryonic stem cells. *Development*. 2009;136(23):4055–4063. https://doi.org/10.1242/dev.036624.
36. Ma L, Hu B, Liu Y, et al. Human embryonic stem cell-derived GABA neurons correct locomotion deficits in quinolinic acid-lesioned mice. *Cell Stem Cell*. 2012;10(4):455–464. https://doi.org/10.1016/j.stem.2012.01.021.
37. Maroof AM, Keros S, Tyson JA, et al. Directed differentiation and functional maturation of cortical interneurons from human embryonic stem cells. *Cell Stem Cell*. 2013;12(5):559–572. https://doi.org/10.1016/j.stem.2013.04.008.
38. Nicholas CR, Chen J, Tang Y, et al. Functional maturation of hPSC-derived forebrain interneurons requires an extended timeline and mimics human neural development. *Cell Stem Cell*. 2013;12(5):573–586. https://doi.org/10.1016/j.stem.2013.04.005.
39. Anderson NC, Chen P-F, Meganathan K, et al. Balancing serendipity and reproducibility: pluripotent stem cells as experimental systems for intellectual and developmental disorders. *Stem Cell Rep*. 2021;16(6):1446–1457. https://doi.org/10.1016/j.stemcr.2021.03.025.
40. Ahn S, Kim TG, Kim KS, Chung S. Differentiation of human pluripotent stem cells into medial ganglionic eminence vs. Caudal ganglionic eminence cells. *Methods*. 2016;101:103–112. https://doi.org/10.1016/j.ymeth.2015.09.009.
41. Amit M, Carpenter MK, Inokuma MS, et al. Clonally derived human embryonic stem cell lines maintain pluripotency and proliferative potential for prolonged periods of culture. *Dev Biol*. 2000;227(2):271–278. https://doi.org/10.1006/dbio.2000.9912.
42. Kirmizi-Alsan E, Bayraktaroglu Z, Gurvit H, Keskin YH, Emre M, Demiralp T. Comparative analysis of event-related potentials during Go/NoGo and CPT: decomposition of electrophysiological markers of response inhibition and sustained attention. *Brain Res*. 2006;1104(1):114–128. https://doi.org/10.1016/j.brainres.2006.03.010.
43. Liu Y, Liu H, Sauvey C, Yao L, Zarnowska ED, Zhang SC. Directed differentiation of forebrain GABA interneurons from human pluripotent stem cells. *Nat Protoc*. 2013;8(9):1670–1679. https://doi.org/10.1038/nprot.2013.106.
44. Thomas CA, Tejwani L, Trujillo CA, et al. Modeling of TREX1-dependent Autoimmune disease using human stem cells Highlights L1 accumulation as a source of neuroinflammation. *Cell Stem Cell*. 2017;21(3):319–331. https://doi.org/10.1016/j.stem.2017.07.009. e8.
45. Thomson JA. Embryonic stem cell lines derived from human blastocysts. *Science*. 1998;282(5391):1145–1147. https://doi.org/10.1126/science.282.5391.1145.
46. Inzunza J, Gertow K, Strömberg MA, et al. Derivation of human embryonic stem cell lines in serum replacement medium using postnatal human fibroblasts as feeder cells. *Stem Cell*. 2005;23(4):544–549. https://doi.org/10.1634/stemcells.2004-0201.
47. Stojkovic P, Lako M, Przyborski S, et al. Human-serum matrix supports undifferentiated growth of human embryonic stem cells. *Stem Cell*. 2005;23(7):895–902. https://doi.org/10.1634/stemcells.2004-0326.
48. Xu RH, Peck RM, Li DS, Feng X, Ludwig T, Thomson JA. Basic FGF and suppression of BMP signaling sustain undifferentiated proliferation of human ES cells. *Nat Methods*. 2005;2(3):185–190. https://doi.org/10.1038/nmeth744.
49. Ludwig T, A Thomson J. Defined, feeder-independent medium for human embryonic stem cell culture. *Curr Protoc Stem Cell Biol*. 2007;1(1C.2).

50. Cunningham M, Cho JH, Leung A, et al. hPSC-derived maturing GABAergic interneurons ameliorate seizures and abnormal behavior in epileptic mice. *Cell Stem Cell*. 2014;15(5):559–573. https://doi.org/10.1016/j.stem.2014.10.006.
51. Hannoun Z, Fletcher J, Greenhough S, et al. The comparison between conditioned media and serum-free media in human embryonic stem cell culture and differentiation. *Cell Reprogr*. 2010;12(2):133–140. https://doi.org/10.1089/cell.2009.0099.
52. Meganathan K, Lewis EMA, Gontarz P, et al. Regulatory networks specifying cortical interneurons from human embryonic stem cells reveal roles for CHD2 in interneuron development. *Proc Natl Acad Sci USA*. 2017;114(52):E11180–E11189. https://doi.org/10.1073/pnas.1712365115.
53. Vogt J, Traynor R, Sapkota GP. The specificities of small molecule inhibitors of the TGFß and BMP pathways. *Cell Signal*. 2011;23(11):1831–1842. https://doi.org/10.1016/j.cellsig.2011.06.019.
54. Zhang SC, Wernig M, Duncan ID, Brüstle O, Thomson JA. In vitro differentiation of transplantable neural precursors from human embryonic stem cells. *Nat Biotechnol*. 2001;19(12):1129–1133. https://doi.org/10.1038/nbt1201-1129.
55. Muratore CR, Srikanth P, Callahan DG, Young-Pearse TL. Comparison and optimization of hiPSC forebrain cortical differentiation protocols. *PLoS One*. 2014;9(8). https://doi.org/10.1371/journal.pone.0105807.
56. Kim TG, Yao R, Monnell T, et al. Efficient specification of interneurons from human pluripotent stem cells by dorsoventral and rostrocaudal modulation. *Stem Cell*. 2014;32(7):1789–1804. https://doi.org/10.1002/stem.1704.
57. Chen CY, Plocik A, Anderson NC, et al. Transcriptome and in vitro differentiation profile of human embryonic stem cell derived NKX2.1-positive neural progenitors. *Stem Cell Rev Rep*. 2016;12(6):744–756. https://doi.org/10.1007/s12015-016-9676-2.
58. Shi Y, Kirwan P, Livesey FJ. Directed differentiation of human pluripotent stem cells to cerebral cortex neurons and neural networks. *Nat Protoc*. 2012;7(10):1836–1846. https://doi.org/10.1038/nprot.2012.116.
59. Vadodaria KC, Ji Y, Skime M, et al. Altered serotonergic circuitry in SSRI-resistant major depressive disorder patient-derived neurons. *Mol Psychiatr*. 2019;24(6):808–818. https://doi.org/10.1038/s41380-019-0377-5.
60. Hu BY, Weick JP, Yu J, et al. Neural differentiation of human induced pluripotent stem cells follows developmental principles but with variable potency. *Proc Natl Acad Sci USA*. 2010;107(9):4335–4340. https://doi.org/10.1073/pnas.0910012107.
61. Birey F, Andersen J, Makinson CD, et al. Assembly of functionally integrated human forebrain spheroids. *Nature*. 2017;545(7652):54–59. https://doi.org/10.1038/nature22330.
62. Chi L, Fan B, Zhang K, et al. Targeted differentiation of regional ventral Neuroprogenitors and related neuronal subtypes from human pluripotent stem cells. *Stem Cell Rep*. 2016;7(5):941–954. https://doi.org/10.1016/j.stemcr.2016.09.003.
63. Germain ND, Banda EC, Becker S, Naegele JR, Grabel LB. Derivation and isolation of NKX2.1-positive basal forebrain progenitors from human embryonic stem cells. *Stem Cell Dev*. 2013;22(10):1477–1489. https://doi.org/10.1089/scd.2012.0264.
64. Lewis EMA, Meganathan K, Baldridge D, et al. Cellular and molecular characterization of multiplex autism in human induced pluripotent stem cell-derived neurons. *Mol Autism*. 2019;10(1). https://doi.org/10.1186/s13229-019-0306-0.
65. Meganathan K. *Alterations in Neuronal Physiology, Development, and Function Associated with a Common Duplication of Chromosome 15 Involving CHRNA7*. 2020. https://doi.org/10.1101/2020.01.28.922187v2.

CHAPTER

4

Development of transcription factor-based strategies for neuronal differentiation from pluripotent stem cells

Harue Chou[1,2], *Xiaobo Mao*[1,2] *and Mingyao Ying*[2,3]

[1]Neuroregeneration and Stem Cell Programs, Institute for Cell Engineering, Johns Hopkins University School of Medicine, Baltimore, MD, United States; [2]Department of Neurology, Johns Hopkins University School of Medicine, Baltimore, MD, United States; [3]Hugo W. Moser Research Institute at Kennedy Krieger, Baltimore, MD, United States

Introduction

Transcription factors (TFs) are proteins serving as essential epigenetic regulators and key determinants of cell identity. TFs bind to specific DNA sequences to modulate gene transcription and the transcriptional landscape, which determine cellular functions and identity. The human genome encodes approximately 1500–2000 TF proteins[1] that form the single largest family of human proteins. It is common for multiple TFs to work in a coordinated fashion to direct various cellular events, such as cell proliferation, death, and differentiation. During the process of stem cell differentiation to lineage-specific progenitor/precursor cells and finally to functionally mature cell progenies, a number of master TFs are known to initiate and maintain specific transcriptional regulatory networks driving this process. The over 1500 TFs from the human genome provide a gold mine to develop efficient strategies for shifting cell identity, such as differentiation of human pluripotent stem cells (PSCs), including induced pluripotent stem cells (iPSCs) and embryonic stem cells (ESCs), to different types of neurons.

Stem cell biologists have demonstrated that TFs are powerful tools for shifting cell identity. Pioneering work led by Shinya Yamanaka in 2006 revealed that induction of a combination of four TFs (myc proto-oncogene (Myc), octamer-binding transcription factor 3/4 (Oct3/4), sex-

Phenotyping of Human iPSC-derived Neurons
https://doi.org/10.1016/B978-0-12-822277-5.00003-1

determining region Y-box 2 (Sox2), and Krüppel-like factor 4 (KLF4)), also known as the Yamanaka factors, was sufficient to revert a somatic cell into an embryonic pluripotent stem cell state,[2] now widely known as iPSCs. The discovery of iPSCs first proved that TFs can be used not only for changing cell identity, but also for controlling the epigenetic age of a cell. Scientists have attempted to change cell identity using TFs since long before the discovery of iPSCs. In 1986, Harold M. Weintraub and colleagues first discovered that overexpression of a single transcription factor, myoblast determination protein 1 (MyoD1), can directly convert mouse fibroblasts to myocytes,[3,4] suggesting that overexpression of lineage-specific TFs can induce transdifferentiation of somatic cells into desired cell types even if they are very distantly related. Inspired by these results, biologists have attempted to directly convert fibroblasts into neuronal cells. In 2010, Vierbuchen and colleagues discovered that induction of three TFs (POU class 3 homeobox 2 (POU3F2, also known as BRN2), achaete-scute family bHLH transcription factor 1 (Ascl1), and myelin transcription factor 1 like (Myt1l)) can efficiently convert mouse fibroblasts and PSCs into functional induced neuronal (iN) cells, indicating that direct lineage conversions are possible even between distantly related cell types if correct combination of TFs are used.[5] Furthermore, overexpression of the same three combinations of neuronal lineage-determining TFs (BRN2, ASCL1, and MYT1L) has been shown to convert human somatic cells and PSCs into functional neurons in 8 days.[6] The rapid conversion of PSCs into neuronal fate implies that TF-induced cells can bypass both the stem cell stage and sequential neural developmental stages.[7]

During the process of brain development, how can TFs instruct a vast number of undifferentiated cells to acquire neuronal fate? The early stages of neural development is regulated by morphogen gradients that provide instructive cues for activation of neurogenic TFs. Interestingly, multiple TFs function sequentially and/or cooperatively to drive the differentiation of multipotent stem cells to lineage-specific precursors, and finally to functionally mature cells in the brain, including multiple subtypes of neurons and glial cells, such as astrocytes, oligodendrocytes, and microglia.

When the three germ layers are established during vertebrate embryogenesis, ectodermal cells differentiate into either the epidermis layer of the skin or the neural ectodermal layer for the nervous system, depending on the activity of specific morphogen gradients. Ectoderm becomes epidermis in the presence of bone morphogen protein (BMP) and Wnt morphogens. If these two neurogenic signals are inhibited, ectoderm differentiates into neuroepithelium.[8,9] Neuroepithelial cells are the stem cells of the central nervous system, which are also referred to as neural stem cells (NSCs). NSCs in the neuroectoderm can self-renew or differentiate into neural progenitor cells (NPCs) or glial progenitor cells. NPCs give rise to specific subtypes of neurons, which is a process defined as neurogenesis. On the other hand, gliogenesis is the process by which glial progenitor cells differentiate into glial cells, such as astrocytes and oligodendrocytes. The resulting cell types in both neurogenesis and gliogenesis are determined by TFs.

Neuronal and glial cell specification in the neuroepithelium is associated with the spatial pattern of a group of TFs, most of which belong to the basic helix-loop-helix (bHLH) TF family. The bHLH family are evolutionally conserved TFs that play important roles in tissue development and tissue maintenance.[10] One of the subfamilies of this bHLH TFs, known as proneural TFs, are required and sufficient to initiate neural lineage determination and neural differentiation.[11] These bHLH TFs cooperate to form a transcriptional regulatory network to regulate key processes in neural induction, including neural patterning, neuronal maturation, and migration. The timing of their expression is precisely controlled by specific gradients of morphogens.[12] For example, the morphogens, fibroblast growth factor 8 (FGF8), Sonic hedgehog (SHH), and Wnt Family Member 1 (Wnt1), induce the expression of bHLH TFs, forkhead box A2 (FoxA2), LIM homeobox transcription factor 1 alpha (Lmx1a), and LIM homeobox transcription factor 2 alpha (Lmx2a), which control the differentiation of midbrain dopaminergic (DA) neurons in the floor plate region of the mesencephalon.[13] Similarly, the differentiation of forebrain cortical neurons requires cooperation of two proneural bHLH TFs (neurogenin 2 (Ngn2) and neuronal differentiation 4 (NeuroD4/Math3)), to accelerate cortical gene transcription.[14] These bHLH TFs are capable of directly instructing embryonic ectoderm cells to acquire neural identity. Neurodevelopmental studies completed in the 1970s using drosophila and vertebrate models have shown evidence to support that bHLH TFs are capable of inducing neural lineage commitment.[11,15] For example, ectopic expression of the bHLH TF NeuroD in Xenopus directly converts both ectoderm and epidermal cells into neurons.[16] Based on our knowledge of bHLH TFs in brain development, it is feasible to apply these TFs to guide PSC differentiation to neurons and other neural cells.

Traditional methods of PSC differentiation are designed based on the principle of using chemical compounds, growth factors, and/or stromal feeder cell coculture to induce a TF signature defining a given cell type. Alternatively, ectopic TF expression is capable of directly driving PSC differentiation, which is potentially more straightforward and efficient. In this chapter, we focus on TF-driven strategies for the differentiation of PSCs to various neuron subtypes, as well as other important brain cells. Current applications and future development of these highly efficient TF-based differentiation strategies will facilitate neuron production from PSCs, especially patient-derived iPSCs, for disease modeling, drug discovery and cell replacement therapy. We will also discuss strategies to identify key TF drivers of a cell type of interest and efficient methods for TF delivery, toward the goal of maximizing the applications of TF-based technologies in neuroscience research and regenerative medicine.

Neuron differentiation driven by transcription factors

TF-driven neuronal conversion is a feasible and efficient strategy for neuron production from various cell resources, including PSCs and somatic cells. In this section, we will discuss currently available TF-based strategies for differentiating PSCs and transdifferentiating somatic cells into major neuron subtypes and glia. These strategies are also summarized in Table 4.1.

Dopaminergic (DA) neurons

In recent years, researchers have extensively investigated methods for the generation of functional midbrain DA neurons from patient-derived iPSCs, in order to develop an effective cell replacement therapy for Parkinson's disease (PD). Midbrain DA neurons constitute approximately 75% of total DA neurons in the brain, with largest populations located in the substantia nigra pars compacta (SNc) and the ventral tegmental area (VTA).[17] These midbrain DA neurons progressively degenerate in PD, resulting in cognitive-motor

TABLE 4.1 Transcription factor-driven strategies for the conversion of pluripotent stem cells and somatic cells to neurons and other neural cells.

Cell source	Neuron subtype	Transcription factor	Delivery	Induction time (Days)	Neuron purity	Functional testing	References
Fibroblast (H)	DA	Ascl1, Brn2, Mytl1, Lmx1a, FoxA2	LV	12	95% (MAP2+) 90% (synaptophysin+)	Electrophysiology	Pfisterer et al.[20]
Fibroblast (H and M)	DA	Ascl, Lmx1a, Nurr1	LV	6	85%(TH+/TUJ1+)	Electrophysiology, dopamine release, engraftment in the brain	Caiazzo et al.[21]
ESC, iPSC (H)	DA	Ascl, Lmx1a, Nurr1	LV	6	51% (TUJ1+) 65% (TH+/TUJ1+)	Electrophysiology, dopamine release, engraftment in the brain	Theka et al.[22]
ESC, iPSC (H)	DA	Atoh1	LV	5	80% (TUJ1+/MAP2+)	Electrophysiology, dopamine release	Sagal et al.[23]
NPC (R)	DA	Nurr1, FoxA2	mRNA	10	55% (NURR1+) 28% (FOXA2+)	Electrophysiology, dopamine release	Kim et al.[25]
ESC, iPSC (H)	DA	Atoh1, Ngn2	mRNA	5	>90% (TUJ1+) >97% (TH+/TUJ1+)	Electrophysiology, dopamine release	Xue et al.[29]
iPSC (H)	DA	Lmx1a, Nurr1, Pitx3	AAV	10	30% (TH+/DAPI+) 50% (TH+/β-tubulin)	Dopamine release, α-synuclein neurotoxicity	Mahajani et al.[30]
ESC (M)	Glu	Ngn2	Plasmid	10	40% (TUJ1+)	Electrophysiology, network formation with primary neurons	Thoma et al.[37]
ESC, iPSC (H)	Glu	Ngn2 or NeuroD1	LV	7	~100% (vGlut2+), 20% (vGlut1+)	Electrophysiology, calcium imaging, engraftment in the brain	Zhang et al.[39]
iPSC, NPC (H)	Glu	Ngn2	LV	4	>90% (MAP2+, AB+)	Electrophysiology	Ho et al.[41]
ESC (H)	Glu/ GABA	Ngn2, Satb2, Fezf2	LV	7	Ngn2: 83% (MAP2+) Ngn2 + Fezf2: 64% (MAP2+) Ngn2 + Satb2: 49% (MAP2+) Ngn2: 55% (vGlut2+) Ngn2 + Fezf2: 67% (vGlut2+)	Electrophysiology, transplantation into organotypic slice cultures of human cortex	Miskinyte et al.[38]

TABLE 4.1 Transcription factor-driven strategies for the conversion of pluripotent stem cells and somatic cells to neurons and other neural cells.—cont'd

Cell source	Neuron subtype	Transcription factor	Delivery	Induction time (Days)	Neuron purity	Functional testing	References
iPSC, fibroblast (H)	GABA	Ascl1, Sox2, FoxG1, Dlx5, Lhx6	LV	14	>30% GABAergic neuron	Electrophysiology, engraftment in the brain	Colasante et al.[43]
ESC (H)	GABA	Ascl1, DLX2, LHX6, miR-9/9*, miR-124	LV	10	81% (MAP2+) 85%(GABA+/MAP2+)	Electrophysiology, engraftment in the brain	Sun et al.[44]
ESC (H)	GABA	Ascl1, Dlx2	LV	14	89.1% (GABA+/MAP2+) FOXG1 (93.5%) DLXs (88.5%) GAD 65/67 (94.4%)	Electrophysiology, engraftment in the brain	Yang et al.[47]
ESC iPSC (H)	Motor neuron	Ngn2, Isl1, Lhx3	AAV	11	49–72% (HB9+/CHAT+)	Electrophysiology, neuromuscular junction formation	Hester et al.[53]
ESC (M)	Motor neuron	Isl1–Lhx3 fusion protein	Genome editing	6	77% (HB9+/TUJ1+)	Neuromuscular junction formation	Lee et al.[54]
ESC (M)	Motor neuron	Isl1–Lhx3 fusion protein	Genome editing	6			Lee et al.[55]
ESC (M)	Cranial motor neuron	Ngn2, Isl1, Phox2A	Genome editing	6	99.77% (NeuN+) 99.03% (Phox2b+) 0.11% (HB9+)	Electrophysiology	Mazzoni et al.[56]
ESC (M)	Motor neuron	Ngn2, Isl1, Lhx3	Genome editing	2			Velasco et al.[57]
iPSC (H)	Motor neuron	Lhx3, Ngn2, Isl1,	SeV	14	>90% (TUJ1+) >90% (HB9+)	Electrophysiology, neuromuscular junction formation	Goto et al.[59]
iPSC (H)	Motor neuron	Ngn1 Ngn2 Ngn3 NeuroD1 NeuroD2	mRNA	7	98.2% (TUJ1+) 83.2%(MAP2+/TUJ1+) 94.3% (HB9+/TUJ1+) 97.7% (ChAT+/TUJ1+) 100% (ISL1+/TUJ1+)	Electrophysiology, calcium imaging	Goparaju et al.[60]
Fibroblast (M)	Astrocyte	NfiA, NfiB, Sox9	LV	12–16	15% (GFAP+) 20% (S100B+) 10% (GLT-1+)	Electrophysiology, calcium imaging	Caiazzo et al.[78]
NPC (M)	Astrocyte	NfiA, Atf3, Runx2	Plasmid	21		Calcium imaging	Tiwari et al.[79]

(Continued)

TABLE 4.1 Transcription factor-driven strategies for the conversion of pluripotent stem cells and somatic cells to neurons and other neural cells.—cont'd

Cell source	Neuron subtype	Transcription factor	Delivery	Induction time (Days)	Neuron purity	Functional testing	References
ESC, iPSC (H)	Astrocyte	NfiA or NfiA plus Sox9	CRISPR/ Cas9	28	day 52: 70%–80% (GFAP+) day 60: 80% –90% (GFAP+)	Neurite outgrowth, calcium imaging	Li et al.[80]
ESC (H)	Astrocyte	NfiB or NfiB plus Sox9	LV	7	90% (GFAP+) 95% (S100B+) 90% (VIM+)	Calcium imaging, glutamate uptake, cytokine stimulation, electrophysiology, engraftment in the brain	Canals et al.[81]
PSC-derived NSC (H)	Astrocyte	NfiA	LV	5	Day 56: >60% (GFAP+) Day 77: ~100% (GFAP+)	Cytokine stimulation, calcium imaging, glutamate uptake, electrophysiology, engraftment in the brain	Tchieu et al.[82]
Primary NPC (H)	Olig	Sox10	LV	14	>60% (NG2+/ A2B5+)	Engraftment in the brain	Wang et al.[86]
iPSC-derived NPC (H)	Olig	Sox10, Olig2, Nkx6.2	LV	4	70% (O4+)	Myelination in vitro, engraftment in the brain	Ehrlich et al.[87]
iPSC (M)	RGC	Math5	Plasmid	8	15% (MATH5+) 10% (BRN3B+) 5% (THY+)	Engraftment	Chen et al.[68]
iPSC (H)	RGC	Math5	Plasmid	16	23% (BRN3B+)	Calcium imaging	Deng et al.[69]

This table summarizes publications that report transcription factor-driven strategies for the differentiation of pluripotent stem cells or the transdifferentiation of somatic cells to various neuron subtypes as well as other neural cells. Abbreviations: *AAV*, adeno-associated virus; *DA*, dopaminergic; *GABA*, GABAergic; *Glu*, glutamatergic; *H*, human; *LV*, lentivirus; *M*, mouse; *Olig*, oligodendrocyte; *RGC*, retinal ganglion cell; *R*, rat; *SeV*, sendai virus.

dysfunctions, and currently there are no effective therapies. Patterning-based methods for generating midbrain DA neurons from PSCs involve dual-SMAD inhibition and Wnt signaling activation to induce coexpression of FoxA2 and Lmx1a, which leads to the development of floor-plate precursor cells that give rise to midbrain DA neurons.[18,19] In order to shorten the lengthy patterning-based differentiation processes that take months, it is attractive to develop TF-driven rapid differentiation methods for midbrain DA neurons that have wide applications in autologous cell replacement therapy for PD, as well as disease modeling and drug discovery.

In 2011, Pfisterer and colleagues provided the first report showing that the combination of two TFs (LMX1A and FOXA2) involved in DA neuron development in addition to three other TFs (BRN2, ASCL1, and MYT1L) responsible for driving neuronal conversion could induce differentiation of human fibroblast cells into functional DA neurons.[20] In the same year, Caiazzo and colleagues identified a set of three TFs (ASCL1, nuclear receptor 4A2 (NURR1), and LMX1A) that rapidly induce functional DA neurons directly from mouse and human fibroblast cells.[21] These findings led to the establishment of rapid methods of generating mature and functional human iPSC-derived DA neurons by overexpressing three TFs (ASCL1, NURR1, and LMX1A) as reported by Theka and colleagues.[22] With this method, functionally mature DA neurons can be efficiently generated from iPSCs in 21 days. Although the purity and maturity of iPSC-derived DA neurons are slightly lower than those produced from somatic cells, these iPSC-derived DA neurons express all of the essential DA-specific markers (e.g., tyrosine hydroxylase (TH), vesicular monoamine transporter (VMAT), dopamine transporter (DAT), dihydroxyphenylalanine decarboxylase (DDC), G-protein-regulated inward-rectifier potassium channel 2 (GIRK2)), and show electrophysiological activity similar to bona fide DA neurons (e.g., Na+/K+ currents, multiple evoked action potential, and spontaneous pace-making activity). This method showed for the first time that mature DA neurons could be made efficiently in less than 1 month, compared with traditional methods that take 2–3 months and involve embryoid body formation. In addition to the discovery of the three potent TFs for DA neuron conversion (ASCL1, NURR1, and LMX1A), researchers later showed that overexpressing one single TF is sufficient to efficiently convert human iPSCs and ESCs to midbrain DA neurons. In 2013, Sagal and colleagues reported that the proneural TF, ATOH1, is upregulated during the conversion of human iPSCs into midbrain DA neurons, and ectopic ATOH1 expression in PSCs mediated by a Tet-On lentiviral vector drives efficient DA neuron conversion in 6 days.[23] ATOH1-induced DA neurons express midbrain DA neuron markers, including FOXA2, NURR1, LMX1A, orthodenticle homeobox 2 (OTX2), TH, and DAT,[23] and also exhibit spontaneous spiking, a key electrophysiological feature of primary midbrain DA neurons. ATOH1 induction in PSCs was found to activate multiple downstream neurogenic TFs involved in DA neuron conversion, such as NGN2 and NEUROD1.[24] The expression of these crucial TFs was sustained even after the exogeneous ATOH1 expression was silenced, indicating that the neuronal differentiation program initiated by ectopic ATOH1 expression can later become self-supporting and ATOH1-independent.[23] It is important to reduce the number of TFs used in differentiation process to make it more feasible. Compared to neuronal induction using multiple TFs that have functional redundancy, delivering a single TF is simpler and has been widely utilized these days.

Although DA neurons can be rapidly generated by expressing DA conversion TFs using viral vectors, chromosomal abnormalities caused by genome-integrating viruses have been a major concern for translating such technology into clinical applications. In contrast, synthetic mRNA-based gene delivery approaches avoid this problem and therefore have gained attention in the development of integration-free differentiation strategies for neuron production. In 2017, Kim and colleagues first reported that rat forebrain neural precursor cells could be differentiated into DA neurons by transfecting mRNA encoding two key TFs, Nurr1 and FoxA2. In this protocol, the efficiency of DA neuron generation was further enhanced by the addition of the small molecule dibutyryl cyclic AMP (db-cAMP).[25] db-cAMP is a derivative of cAMP that has been shown to enhance protein kinase A (PKA) signaling, contributing to gene expression activation.[26–28] Therefore, with the addition of db-cAMP in mRNA-transfected rat neural precursor cells, the expression of DA neuron markers (Nurr1 and FoxA2) can be efficiently increased. Following the development of mRNA-based methods to generate DA neurons from rat NPCs, Xue and colleagues in 2019 showed that human PSCs sequentially transfected with mRNA coding for two TFs (ATOH1 and NGN2) were rapidly differentiated into midbrain DA neurons with more than 90% purity in 5 days.[29] Proteomic analysis revealed that ATOH1 interacts with the non-muscle myosin II motor protein complex in the nucleus, which promotes the transcriptional activity of ATOH1 and enhances the efficiency of ATOH1-driven DA neuron conversion.[29]

In addition to synthetic mRNAs, non-integrating adeno-associated virus (AAV) has also been used for TF delivery to drive PSC differentiation. AAV-mediated transient expression of three TFs (LMX1A, NURR1, and pituitary homeobox 3 (PITX3)) led to generation of DA neurons from multiple human iPSC lines.[30] Among these three TFs, LMX1A was identified as a master regulator of this differentiation process.[30] Moreover, an AAV vector harboring a reporter gene under the control of the neuron-specific human Synapsin1 promoter has been used as a surrogate marker to monitor DA neuron maturation, providing a strategy to more precisely evaluate the efficiency of neuronal conversion, as well as the status of neuron maturation.[30] If being optimized, TF induction via transient expression or integrative vectors can achieve similar conversion efficiency as summarized in Table 4.1.

Here, we described the development of differentiation strategies for generating DA neurons from PSCs, as well as somatic cells. Conversion efficiency from human fibroblast to DA neuron is ~16%.[20] Generally, neuronal conversion starting from PSCs is more efficient as it is possible to achieve >80% conversion efficiency.[23,29] Acquisition of a generic midbrain DA neuron fate requires the activation of several key TFs, such as ASCL1, NGN2, NURR1, LMX1A, and FOXA2. Compared to conventional patterning-based neuron differentiation methods, TF-directed differentiation strategies enable rapid production of highly pure DA neurons with less variability. It is also noteworthy to mention that patterning-based and TF-based methods can be used in combination to further enhance the overall efficiency of DA neuron differentiation by precisely steering neural induction and DA lineage specification. For example, the ATOH1-driven protocol also uses two morphogens, SHH and FGF8b, to promote DA lineage specification.[23] Here, we have also described the disadvantage of using lentivirus-mediated TF delivery due to integrative gene transfer which likely leads to long-term expression of the TF transgenes in differentiated progenies. Long-term TF expression may interfere with cellular functions of mature DA neurons.[31] In contrast, transient TF gene expression mediated by synthetic mRNA or AAV does not involve genome integration, thus lowering the risk of insertional mutagenesis, and providing a safer alternative to generate midbrain DA neurons for cell replacement therapy. Currently, scalable TF-driven differentiation procedures still need to be developed for generating sufficient number of clinical-grade PSC-derived DA neurons for clinical use.

Glutamatergic neurons

Cortical glutamatergic neurons originate from the dorsal telencephalon and comprise the major excitatory network in the central nervous system. Glutamatergic neurons play critical roles in controlling cognitive and motor function, and defects of cortical glutamatergic neurons have been implicated in many neurological diseases, such as Alzheimer's disease and Huntington's disease.[32] Traditional patterning-based methods to differentiate PSCs into cortical neurons are based on dual-SMAD inhibition and embryoid body formation method.[33] In comparison, TF-induced generation of glutamatergic neurons from PSCs provides a more rapid and less cumbersome method to facilitate various applications of this important neuron subtype in disease modeling, mechanistic studies and drug discovery.

Over the years, scientists have attempted to screen potent TFs driving the acquisition of glutaminergic neuron fate during neurogenesis. Neurodevelopmental studies revealed that Ngn2 expressed in the dorsal telencephalic progenitors is the prominent bHLH TF that controls the development of glutaminergic neurons in the nervous system.[34–36] Based on this knowledge, many studies have attempted to identify a minimal set of TFs to convert PSCs into cortical glutamatergic neurons. In 2012, Thoma and colleagues demonstrated that ectopic expression of Ngn2 alone is sufficient to convert mouse ESCs into functionally mature glutamatergic neurons without additional extrinsic factors.[37,38] Remarkably, Ngn2-induced glutamatergic neurons exhibited action potentials and formed connections with mouse hippocampal primary neurons in coculture. Following successful glutamatergic neuron generation using mouse ESCs, in 2013, rapid conversion of human PSCs into homogeneous cortical glutamatergic neurons was correspondingly achieved by inducible lentiviral expression of NGN2 alone.[39,40] Importantly, integration into existing synaptic networks was observed when these NGN2-induced neurons were transplanted into the mouse brain. Moreover, in 2018, it was shown that three different combinations of TFs, NGN2 only, NGN2 plus forebrain embryonic zinc finger-like protein 2 (FEZF2), and NGN2 plus special AT-rich sequence-binding protein 2 (SATB2) can convert human ESCs into both upper and deeper layer cortical excitatory pyramidal neurons with functional electrophysiological properties.[38] SATB2 promotes upper layer cortical neuron identity by inhibiting subcerebral features,[40] and FEZF2 drives deep-layer cortical neuron development.[36] This finding indicates that a TF cocktail may more precisely convert PSCs to layer-specific cortical neurons with distinct area identity, which strategy will facilitate studies of the neuronal network in the cortex and various brain disorders associated with the cortical network, such as Alzheimer's disease and autism spectrum disorder.

Several challenges still remain to achieve efficient production of cortical glutamatergic neurons. Due to time and labor involved in culturing and expanding human iPSCs as the starting source for differentiation, efficient protocols need to be developed to allow scalable and reproducible glutamatergic neuron production from PSCs, which is essential for high-throughput drug screening. In addition, the dependence of current protocols on repeated iPSC expansion may result in variation depending on the passage number of iPSCs at the time of differentiation. Instead of starting from iPSCs, an adapted Ngn2-induced neuronal induction protocol has been developed that starts with human iPSC-derived NPCs.[41] As this protocol is still new to the field, it will be interesting to compare Ngn2-induced glutamatergic neurons from iPSCs and NPCs to determine which strategy is optimal for individual applications.

GABAergic neurons

GABAergic (γ-aminobutyric acid-secreting) interneurons (INs) are inhibitory neurons located in the cerebral cortex and hippocampus. GABAergic INs arise from precursors located in the medial and caudal ganglionic eminence (MGE and CGE) from where they migrate out to integrate into the circuitry of the developing striatum, cortex, and hippocampus.[42] The imbalance between the excitatory and inhibitory tone in the cerebral neuronal network has been shown to be associated with various neurological disorders, such as epilepsy. Interestingly, transplantation of GABAergic INs has been shown to improve long-term function in animal models of epilepsy.[43] Due to the importance of GABAergic INs in disease modeling and drug discovery, as well as their potential application in cell replacement therapy, TF-driven strategies to improve the differentiation efficiency of INs from PSCs have been extensively explored.

In 2015, Colasante and colleagues reported that overexpression of five TFs, forkhead box G1 (FOXG1), SOX2, ASCL1, distal-less homeobox 5 (DLX5), and LIM homeobox 6 (LHX6), can convert human fibroblasts and iPSCs into functional GABAergic INs possessing molecular signatures of telencephalic INs with approximately 30% purity.[43] It was found that SOX2 and FOXG1 cooperate with ASCL1 to activate a transcriptional network required for GABAergic fate specification both from fibroblasts and PSCs. It is noteworthy to mention that ectopically expressed ASCL1 in mouse fibroblast cells without SOX2 and FOXG1 induces glutamatergic but not GABAergic neurons.[5] These GABAergic INs successfully integrated into mouse hippocampal circuitry and triggered inhibition of granule neurons.[43] In 2016, a study by Sun and colleagues showed that highly pure and functional GABAergic INs could be generated from human PSCs by expressing microRNA-9/9* and microRNA-124 (miRNA-9/9*-124) together with three TFs, ASCL1, distal-less homeobox 2 (DLX2), and LHX6.[44] During GABAergic IN development, these TFs are abundantly expressed in the MGE to control GABAergic neurogenesis.[45] miRNA-9/9* and miRNA-124 are two highly conserved miRNAs expressed in the mammalian nervous system and involved in regulating neuronal development.[46] By synergizing miRNA-9/9*-124 with these three TFs, cortical GABAergic INs were produced with more than 80% purity. Following transplantation in the mouse brain, these neurons effectively formed functionally mature synapses with the host neural circuits.[44] Later in 2017, it was reported that transient expression of the combination of two TFs, ASCL1 and DLX2, is sufficient to convert human PSCs into GABAergic INs with nearly 90% purity that belong to the telencephalon and diencephalon neuronal subtype.[47] This protocol is now widely adapted by groups working to understand the role of GABAergic neurons in neurological diseases. In addition, these relatively pure INs can be cocultured with NGN2-induced glutamatergic neurons to understand the roles of each neuronal subtype in disease-relevant phenotypes. These studies also demonstrated that conversion efficiency and purity of GABAergic INs starting from iPSCs are relatively higher than that from fibroblasts.

Overall, these studies suggest that the combination of ASCL1 and DLX2 may be a minimal set of TFs required for GABAergic neuron differentiation from human PSCs. Patterning-based differentiation of GABAergic INs is extremely challenging compared to other neuron subtypes. The establishment of TF-driven protocols for GABAergic IN generation provides a simple and robust approach for generating and applying this important neuron subtype in modeling and treating human neurological disorders. For future development and optimization, it will be critical to develop integration-free and ideally virus-free methods for delivering these TF drivers to drive efficient generation of GABAergic INs.

Cholinergic motor neurons

Motor neurons (MNs) are a subtype of neurons that control a variety of downstream effector organs, particularly muscles and glands. The cell bodies of MNs reside in the cerebral cortex, brainstem or spinal cord, with their axons sending output electrical signals by synapsing with target muscles. There are two subtypes of motor neurons, upper and lower MNs. The upper MNs originate in the brainstem, cerebral cortex, and spinal cord, and carry electrical information down to connect with the lower MNs that reside in the brainstem and spinal cord. The upper MNs release the neurotransmitter glutamate, while the lower MNs use acetylcholine to control muscle movements. Dysfunction or injury of these MNs affects the motor circuits in the nervous system and results in various MN disorders, such as spinal muscular atrophy, amyotrophic lateral sclerosis, and spinal cord injury. To understand molecular mechanisms associated with these disorders and develop effective therapies for MN-associated diseases, researchers have now established TF-based differentiation methods to efficiently generate MNs from human PSCs. Traditional patterning-based MN generation methods use dual-SMAD inhibitors and WNT activators in combination with growth factors to induce MNs in 14–28 days.[48] In comparison, TF-based protocols allow rapid MN generation within a week.

In the 2000s, scientists identified key transcriptional networks that control spinal MN specification. During embryonic development, MN progenitor (pMN) cells in the spinal cord firstly differentiate into spinal MNs and then switch to produce oligodendrocytes in a sequential manner.[49,50] Thus, whether the population of pMN cells retain their progenitor cell identity or differentiate into motor neurons or oligodendrocytes depends on the unique spatial and temporal gene expression patterns in the developing spinal cord. Recent studies have demonstrated that the bHLH TF, oligodendrocyte transcription factor 1/2 (Olig1/2), is a crucial TF driver determining pMN cell fate. Olig1/2-deficient mice were found to lack spinal MNs and oligodendrocytes.[51] At the initial stage of embryonic spinal cord development, retinoic acid (RA) and SHH signaling induces early expression of Olig2 in pMN cells. At this early stage, Olig2 expression in pMN cells functions to maintain their precursors state by repressing MN genes (e.g., motor neuron and pancreas homeobox 1 (Hb9)) selectively expressed in post-mitotic MNs.[52] At the later stage in development, Olig2 activates expressions of downstream bHLH TFs (Ngn2, NeuroM (no official name found/NeuroD)) for neurogenesis, and homeodomain factors (homeobox gene (Mnr2), LIM homeobox 3 (Lhx3), Islet-1 (Isl1)) for the specification of the MN lineage.[52] When Ngn2 is activated, it represses the function of Olig2 and induces a post-mitotic MN lineage commitment, suggesting that high Ngn2 expression in replacement of Olig2 in pMN cells favors the conversion of pMN cells into post-mitotic MNs.[52]

With this background, Hester and colleagues in 2011 developed an AAV-based gene delivery system to express NGN2, ISL1, and LHX3, and successfully converted human PSCs to functional spinal MNs with 60%–70% efficiency.[53] Compared to conventional patterning-based MN differentiation protocols, which took around 60 days at that time, this approach enabled rapid and efficient production of spinal MNs from human PSCs in 11 days. In addition to using AAV vectors, in 2012 Lee and colleagues reported that the MN hexamer complex, a fusion protein consisting of Isl1 and Lhx3, can efficiently induce the conversion of mouse ESCs into spinal MNs in 6 days while suppressing V2-interneuron production.[54] This LIM-complex collaborates with two other TFs (signal transducer and activator of transcription 3 (Stat3) and Ngn2) to directly activate genes specifying a terminal MN fate, while repressing the expression of V2-interneuron genes by inducing the motor neuron TF, Hb9, suggesting a TF-mimicking complex may be useful to precisely

number of key TFs controlling astrocyte specification. The nuclear factor I TF family and the SRY (sex determining region Y)-box 9 (Sox9) TF are known to be induced during the initiation of astrogliogenesis in the developing spinal cord and in the cerebral cortex.[74–77] During gliogenesis, Sox9 induces nuclear factor I/A (Nfia) to form a Sox9-Nfia complex that regulates the expression of adenomatosis polyposis coli down-regulated 1 (Apcdd1), monocyte to macrophage differentiation associated 2 (Mmd2), and zinc finger CCHC-type containing 24 (Zcchc24), which are essential for glial maturation, migration and metabolism.[77] In 2015, it was demonstrated that coexpression of Nfia, nuclear factor I/B (Nfib), and Sox9 enabled direct conversion of mouse fibroblasts into astrocytes, suggesting their instructive role in specifying the astrocytic phenotype.[78] Based on this study, Tiwari and colleagues in 2018 successfully converted mouse NPCs into astrocytes by overexpressing Nfia, activating transcription factor 3 (Atf3), and runt related transcription factor 2 (Runx2), revealing that these TFs are capable of driving NPCs to the astroglial lineage by interfering with the neurogenesis program.[79] Similarly, Li and colleagues demonstrated that CRISPR-Cas9 mediated transgenic expression of NfiA alone or NfiA plus SOX9 efficiently induced functional astrocytes from human PSCs in 4–7 weeks, and these TF-induced astrocytes exhibited the hallmarks of mature astrocytes, including glutamate reuptake, a key functional hallmark of bona fide astrocytes.[80] This astrocyte differentiation process can be further accelerated by SOX9 expression, consistent with the fact that both NfiA and SOX9 pattern early neuroepithelial cells to generate astrocytes.[80] Instead of using NfiA, overexpressing the TF NfiB, can also induce astrocyte conversion from human PSCs. Canals and colleagues showed that NfiB alone or NfiB plus SOX9 can rapidly generate homogeneous populations of astrocytes from human PSCs in 7 days.[81] These NfiB/SOX9-induced astrocytes from PSCs exhibited major astrocytic functions, including glutamate uptake, supporting synaptogenesis, and forming functional gap junctions.[81] Moreover, transient expression of NfiA together with extrinsic glial-promoting factors has also been shown to be sufficient to drive astrocyte conversion from human PSC-derived NSCs.[82] It was found that NfiA acts as a molecular switch to induce human glial competency by increasing the chromatin accessibility for particular genes in response to extrinsic gliogenic factors, such as leukemia inhibitory factor (LIF) and BMP. Taken together, these results support that NfiA, NfiB and SOX9 are key TF drivers that work collaboratively to drive astrogliogenesis. For future development, optimizing their delivery in PSCs is necessary for establishing more robust methods to generate PSC-derived astrocytes for disease modeling and cell replacement therapy.

A major function of oligodendrocytes is to form myelin sheath around neuronal axons in the central nervous system, which provides electrical insulation allowing for saltatory conduction of action potentials. Each single oligodendrocyte can extend to approximately 50 axons with its processes to myelinate axons and create nodes of Ranvier, which are the gaps in myelin along an axon that allow action potentials to jump from node to node to increase the speed of action potential propagation. Oligodendrocyte dysfunction or damage to myelin sheath result in demyelinating diseases, such as multiple sclerosis. Although remyelination in the nervous system can occur to restore myelin sheath damages, it often fails to fully recover. Transplantation of human oligodendrocyte progenitor cells (OPCs) has been considered as a potential approach to promote remyelination.[83,84] To fulfill the need of PSC-derived OPCs for cell transplantation, researchers have been searching for optimal TF-induced differentiation protocols for OPC production to treat demyelinating diseases. During embryonic development, radial glial cells in the embryonic ventricular zone give rise to OPCs, which migrate to different regions of the developing brain and spinal cord. When these OPCs are fully populated, they differentiate into mature myelin-forming oligodendrocytes, regulated by the expression of master TFs. In the spinal cord VZ, the bHLH TF Olig2 is expressed in the pMN domain to control sequential differentiation of MN and oligodendrocyte.[51,52] Olig2 expressed in the pMN domain is a crucial driver determining pMN and oligodendrocyte cell fate. Coexpression of Olig2 and Nkx2.2 in the spinal cord induces ectopic oligodendrocyte differentiation.[85] In addition, it was shown that Sox10 expression is restricted in the myelinogenic oligodendrocytes in the brain to directly regulate myelin gene expression.[74] Knowledge of these key TFs for oligodendrocyte fate specification has led to successful OPC differentiation from human iPSC-derived NPCs. In 2014, Wang and colleagues showed that Sox10 overexpression alone is sufficient to induce OPCs from human NPCs.[86] Later in 2017, Ehrlich and colleagues demonstrated that overexpression of three TFs, SOX10, OLIG2, and NKX6.2, in iPSC-derived neural progenitor cells, generated O4-positive oligodendrocytes with up to 70% purity within 28 days.[87] Together, these discoveries support that a set of three TFs, SOX10, OLIG2 and NKX6.2, are sufficient to drive efficient oligodendrocyte conversion from human iPSC-derived NPCs. These TF-induced oligodendrocytes, with validated myelin-forming function, provide a valuable source for the development of cell replacement therapies for demyelinating diseases, and also for establishing patient-derived models of these diseases. However, it is noteworthy to mention that, for generating clinical-grade OPCs suitable for cell replacement therapy, these TFs need to be delivered using integration-free methods.

Microglia are the resident macrophages in the central nervous system (CNS) known for their roles in immune regulation and protecting the brain from injury or infection.[88] They are also critical for proper brain development through the regulation of synapse pruning and formation, establishing the blood-brain barrier and maintaining brain structures.[89,90] When neuronal injury occurs, microglia are activated to release proinflammatory cytokines, such as IL-1β, IL-6 and TNF-α. It is also known that excessive secretion of these proinflammatory factors could be toxic to neurons and glia, which is an important mechanism underlying neurodegeneration. Microglia-mediated chronic neuroinflammation has been linked to various neurodegenerative diseases, including Alzheimer's disease and Parkinson's disease.[91] Microglia are known to originate from hematopoietic primitive erythro-myeloid precursor cells in the extra-embryonic yolk sac and migrate to the CNS during early embryonic development.[92] It is also known that migration of microglia to the CNS takes place before the generation of other glial cells.[93]

A number of patterning-based microglia differentiation methods have been developed from human PSCs. However, TF-based microglia differentiation protocols have yet to be established. This might be due to the fact that the development of microglia involves multiple developmental stages, thus it is challenging to use only TFs to accurately drive this developmental processes. In patterning-based microglia differentiation methods, chemical compounds and growth factors are used to induce iPSCs to undergo a differentiation process that resembles yolk sac-hematopoiesis in the developing embryo. First, iPSCs are induced into hemangioblasts, then into erythro-myeloid progenitors that give rise to primitive macrophages, then to CNS tissue resident macrophages, and lastly to functionally mature microglia.[89] Several studies have provided evidence that Pu.1 is a critical TF for specifying the microglia lineage and also for regulating their functional maturation.[94–96] More studies are required to identify TF drivers of microglia conversion from PSCs, which may contain multiple TF drivers corresponding to the multi-step processes of microglia development. These studies will guide the development and optimization of TF-based strategies for robust microglia generation from PSCs.

Transcription factor-driven differentiation: considerations when designing a new protocol

TF-driven shift of cell identity has been proven as a robust platform for PSC differentiation to neurons and various other cell progenies. However, to build an optimal TF-driven differentiation strategy, proper design, optimization and quality control are critical. Here, we will discuss two essential elements in this process, including selection of TF drivers and methods for efficient TF delivery.

Design a cocktail of transcription factors

The principle of developing a successful TF-driven differentiation method is to accurately design a TF cocktail containing both "driver" and "helper" TFs that can collaboratively direct PSC differentiation into a neuronal type of interest. In general, the "driver" TF triggers neurogenesis by activating a cell-intrinsic program for neuronal conversion. The "helper" TF may enhance differentiation efficiency through promoting neuronal conversion, neuronal lineage specification and/or progenitor/precursor cell survival. Multiple lines of evidence support that proneural genes encoding the bHLH family of TFs (e.g., ASCL1, NGN2, ATOH1, and NEUROD1) are core "driver" TFs. These proneural TFs are activated during the neuronal conversion of human PSCs, which initiate and sustain a neurogenic transcriptional network,[18,23] and have been successfully used to differentiate PSCs into neurons and also transdifferentiate somatic cells into neurons.[20,21,37,39,97,98]

Nonetheless, it is still possible that a single "driver" TF might be insufficient to induce highly efficient production of a neuron subtype of interest. Overexpression of a single TF could also lead to a mixed population of different neuron subtypes, as the same proneural TF could be involved in the development of different neuron subtypes. Therefore, additional "helper" TFs are needed to more precisely direct differentiation into a specific neuronal subtype. For example, differentiation of human PSCs into DA and GABAergic neurons both uses the proneural TF ASCL1 as a master driver for neuronal conversion. In this setting, adding Nurr1 and Lmx1a promotes the lineage specification to DA neurons. In contrast, DLX2 and LHX6 have been successfully used with ASCL1 for GABAergic neuron differentiation (Table 4.1). Similarly, glutamatergic neurons and MNs both share the NGN2 TF for neuronal induction. Adding ISL1 and LHX3 has been shown to direct the neuronal fate to spinal cord MNs (Table 4.1). Several studies indicated that each of these proneural TFs can activate a neurogenic TF network that over time becomes self-supporting. For example, our laboratory previously showed that two neurogenic TFs (NEUROD1 and NGN2) are induced by ectopic expression of the ATOH1 TF, and their expression was sustained after silencing the transgenic expression of ATOH1.[23] This demonstrates that an optimal TF-driven differentiation strategy should initiate a neurogenic program that can become self-supporting and independent of exogenous TF expression. Moreover, it is also necessary to design a TF cocktail with a minimal number of TFs that do not have functional redundancy, as overexpressing multiple TFs is more challenging than delivering fewer TFs. If mRNA and AAV vectors are used for TF delivery, it may be feasible to deliver members of the defined TF cocktail sequentially rather than simultaneously. Our laboratory has shown that DA neuron differentiation is more efficient when we use a strategy including daily delivery of mRNAs coding ATOH1 for 3 days followed by mRNAs coding NGN2 for 1 day, as compared to using ATOH1 or NGN2 mRNAs alone or together. This design of sequential TF delivery may more closely recapitulate the sequential activation of proneural TFs during DA development. Finally, it is noteworthy to mention that, after we gain more knowledge of protein sequences and domain structures associated with the proneural function of bHLH TFs, it will be feasible to design artificial TFs, e.g., a hybrid TF with optimized domains from different wild-type TFs, in order to achieve more efficient neuron production from PSCs.

Transcription factor delivery

When an optimal TF cocktail is established, efficient TF delivery is the next critical step toward successful PSC differentiation. A number of methods have been established for ectopic TF expression in PSCs, including genome-integrating and non-integrating viral vectors, plasmids, synthetic mRNAs, as well as genome editing. Here, we will discuss the application of these methods in PSC differentiation.

Genome integrating vectors

TF delivery by genome integrating viruses, including retrovirus and lentivirus, results in integration of the viral vector to the genome of PSCs and the expression of the TF encoded by the viral vector. Retroviruses and lentiviruses can induce stable expression of the TF transgene, both of which are suitable for TF delivery in PSCs. The main differences between retroviruses and lentiviruses include transfection efficiency and viral tropism. Generally, lentivirus has higher transfection efficiency in PSCs compared to retrovirus, which may also lead to higher level of TF expression.[99] Regarding tropism, retrovirus can only infect dividing cells, while lentivirus can deliver TFs into both proliferating and nondividing cells, making them advantageous for multi-step protocols where lentivirus expression might be targeted to immature neurons as a way to promote maturation.

Promoters in the viral vector for driving TF expression, especially in the lentiviral vector, deserve careful selection for optimal TF induction in PSCs and differentiated progenies. Norrman and colleague have performed quantitative comparison of a panel of constitutive promoters in lentiviral vectors, including the human β-actin (ACTB), cytomegalovirus (CMV), elongation factor-1α, (EF1α), phosphoglycerate kinase (PGK), and ubiquitinC (UbC) promoters, based on their capability to drive high and stable expression of a GFP transgene in human ESCs as well as differentiated progenies.[100] The ACTB, EF1α, and PGK promoters maintain stable activities in human ESCs, and the EF1α promoter was found as the most stable promoter during ESC differentiation. It is noteworthy to mention that these promoters showed various activities in different cell lineages during ESC differentiation, suggesting that it may be necessary to determine if the promoter of choice drives TF expression in both PSCs and early neural progenitors, when we focus on neuronal differentiation. A lentiviral vector with the Doxycycline-inducible promoter may also be a good choice for regulatable TF expression in PSCs. Our laboratory has successfully used a single-cassette Tet-On lentiviral vector to express the ATOH1 TF and drive efficient DA neuron differentiation from human iPSCs and ESCs.[23] This vector also has a puromycin selection marker for generating stable cell lines that allow us to rapidly induce or silence TF transgene expression after doxycycline treatment or withdrawal, respectively.

Beside viral vectors, genome-editing methods have also been used to generate neurons and glia cells from PSCs. Li and colleagues have used the CRISPR/Cas9 technique to establish human PSCs with doxycycline inducible expression of the gliogenic transcription factors (NFIA alone or NFIA plus SOX9) for rapid and efficient astrocyte generation from PSCs.[80] This procedure can be great when studying one or a few lines, but can become labor intensive if the goal is to compare cells across multiple patient-derived iPSC lines.

Overall, these genome-integrating viral or nonviral strategies have been proven as efficient methods for TF delivery. However, their genome integration feature brings a risk of insertional mutagenesis and may significantly increase the risk of tumorigenicity when PSC-derived cell progenies are used for cell transplantation. This drawback has been considered as a major hurdle for PSC-based cell replacement therapies.

Non-genome integrating viral vectors

To overcome the limitation of genome-integrating viruses, non-integrating viral vectors as well as nonviral vectors (e.g., synthetic mRNA) have been developed for TF delivery to PSCs. Sendai virus (SeV) is an enveloped single-stranded RNA virus that belongs to the family of Paramyxoviridae. Unlike retrovirus, SeV vectors only replicate in the cytoplasm without integrating into the host genome.[101] SeV vectors have been successfully used to express reprogramming TFs and are capable of reprogramming human skin and blood cells with equal or higher efficiency when compared to other reprogramming methods (e.g., lentiviral and episomal vectors), as described in more detail in an earlier chapter. As an efficient vehicle for TF delivery, SeV vectors encoding three TFs, LHX3, NGN2, and ISL1, have been shown to rapidly induce MNs from iPSCs derived from normal individuals and ALS patients.[59] Now, SeV-driven differentiation kits for generating a panel of neuron subtypes from human iPSCs are also commercially available .

Beside SeV, another feasible non-integrating viral system is adeno-associated virus (AAV), which is a non-enveloped single-stranded DNA virus belonging to the *genus Dependovirus* in the family of *Parvoviridae*.[102] Once entering host cells, AAV vectors reside in the nucleus without integrating into the host genome. AAV vectors have also been widely used in gene therapy as they are nonpathogenic and have low to mild immunogenicity. However, since the AAV genome is relatively short, it has relatively small packaging capacity compared to other types of viral vectors. It may be necessary to deliver multiple AAV vectors, if a TF cocktail is required for PSC differentiation. It is also noteworthy to mention that both SeV and AAV can deliver transgenes into quiescent and dividing cells, thus allowing sequential delivery of TFs to drive neuronal conversion and possibly neuron maturation, when a TF of interest needs to be expressed in nondividing neuronal precursors.

Synthetic mRNA

Synthetic mRNA is a practical tool for expressing TF drivers to change cell identity, and an mRNA-based strategy has been widely used in cell reprogramming through delivery of synthetic mRNAs coding for the four Yamanaka factors. Compared to viral vectors, lipid-mediated mRNA delivery offers several advantages. Since mRNAs function in the cytoplasm to produce proteins, this strategy does not involve genome integration and avoids the risk of insertional mutagenesis.[103] Large-scale production of synthetic mRNAs through in vitro transcription is also easy, safe, and cost-effective. Furthermore, TFs can be sequentially or simultaneously delivered via synthetic mRNAs to induce transient gene expression in both dividing and nondividing cells. Goparaju and colleagues reported a novel mRNA-based strategy using the cotransfection of five TFs to generate iPSC-derived MNs.[60] In addition, Xue and colleagues have successfully used mRNAs coding two proneural TFs, ATOH1 and NGN2, to drive highly efficient DA neuron differentiation from human iPSCs and ESCs.[29] Notably, they showed that modifications to the mRNA sequence to block protein phosphorylation of ATOH1 and NGN2 prevented their degradation and enhanced protein expression and function to drive neuronal conversation of human PSCs.[29] This report provides a feasible and widely applicable strategy to enhance the efficiency of mRNA-driven PSC differentiation, as phosphorylation is a key regulator of protein stability and transcriptional activity of various TFs.[29] Overall, mRNA-based TF delivery is a safer and more flexible strategy for PSC differentiation, and is also suitable for generating clinical-grade cell progenies from PSCs for cell replacement therapies.

Summary and future directions

In conclusion, TF-driven differentiation provides a powerful and robust tool to enable rapid and efficient generation of neurons and other cells in the nervous system from human PSCs, as well as somatic cells. Current advances in stem cell and developmental biology have defined a panel of TF drivers for differentiating PSCs to individual subtypes of neurons and other brain cells. TF-driven differentiation will produce reliable cell resources for disease modeling, drug discovery and cell replacement therapy for various brain disorders. We believe that emerging technologies, including TF protein sequence optimization by artificial intelligence and robust TF delivery methods (e.g., self-replicating mRNAs), will significantly accelerate the development of TF-driven strategies for PSC differentiation and possibly trans-differentiation of somatic cells to finally achieve our goal to shift a cell's identity freely by TFs.

Acknowledgement

This work was supported by grants from the NIH (NS099460, NS106407, NS101400, NS107318, AG073291, AG071820, NS125592, AG056841 and NS125559), the Maryland Stem Cell Research Fund (2017-MSCRFD-3714, 2019-MSCRFD-4292), and the Parkinson's Foundation (PF-JFA-1933).

References

1. Wenger AM, Clarke SL, Guturu H, et al. PRISM offers a comprehensive genomic approach to transcription factor function prediction. *Genome Res*. 2013;23(5):889–904. https://doi.org/10.1101/gr.139071.112.

2. Takahashi K, Tanabe K, Ohnuki M, et al. Induction of pluripotent stem cells from adult human fibroblasts by defined factors. *Cell*. 2007;131(5):861–872.
3. Tapscott SJ, Davis RL, Thayer MJ, Cheng PF, Weintraub H, Lassar AB. MyoD1: a nuclear phosphoprotein requiring a Myc homology region to convert fibroblasts to myoblasts. *Science*. 1988;242(4877):405–411.
4. Lassar AB, Paterson BM, Weintraub H. Transfection of a DNA locus that mediates the conversion of 10T1/2 fibroblasts to myoblasts. *Cell*. 1986;47(5):649–656.
5. Vierbuchen T, Ostermeier A, Pang ZP, Kokubu Y, Südhof TC, Wernig M. Direct conversion of fibroblasts to functional neurons by defined factors. *Nature*. 2010;463(7284):1035–1041. https://doi.org/10.1038/nature08797.
6. Pang ZP, Yang N, Vierbuchen T, et al. Induction of human neuronal cells by defined transcription factors. *Nature*. 2011;476(7359):220–223. https://doi.org/10.1038/nature10202.
7. Oh Y, Jang J. Directed differentiation of pluripotent stem cells by transcription factors. *Mol Cell*. 2019;42(3):200–209. https://doi.org/10.14348/molcells.2019.2439.
8. Sasai Y, Lu B, Steinbeisser H, Geissert D, Gont LK, De Robertis EM. Xenopus chordin: a novel dorsalizing factor activated by organizer-specific homeobox genes. *Cell*. 1994;79(5):779–790.
9. Lee H-K, Lee H-S, Moody SA. Neural transcription factors: from embryos to neural stem cells. *Mol Cell*. 2014;37(10):705–712. https://doi.org/10.14348/molcells.2014.0227.
10. Tutukova S, Tarabykin V, Hernandez-Miranda LR. The role of Neurod genes in brain development, function, and disease. *Front Mol Neurosci*. 2021;14:662774. https://doi.org/10.3389/fnmol.2021.662774.
11. Bertrand N, Castro DS, Guillemot F. Proneural genes and the specification of neural cell types. *Nat Rev Neurosci*. 2002;3(7):517–530.
12. Guillemot F. Spatial and temporal specification of neural fates by transcription factor codes. *Development (Camb)*. 2007;134(21):3771–3780.
13. Hegarty SV, Sullivan AM, O'Keeffe GW. Midbrain dopaminergic neurons: a review of the molecular circuitry that regulates their development. *Dev Biol*. 2013;379(2):123–138. https://doi.org/10.1016/j.ydbio.2013.04.014.
14. Mattar P, Langevin LM, Markham K, et al. Basic helix-loop-helix transcription factors cooperate to specify a cortical projection neuron identity. *Mol Cell Biol*. 2008;28(5):1456–1469.
15. Shirasaki R, Pfaff SL. Transcriptional codes and the control of neuronal identity. *Annu Rev Neurosci*. 2002;25:251–281.
16. Lee JE, Hollenberg SM, Snider L, Turner DL, Lipnick N, Weintraub H. Conversion of Xenopus ectoderm into neurons by NeuroD, a basic helix-loop-helix protein. *Science*. 1995;268(5212):836–844.
17. Wallén A, Perlmann T. Transcriptional control of dopamine neuron development. *Ann N Y Acad Sci*. 2003;991:48–60.
18. Kriks S, Shim J-W, Piao J, et al. Dopamine neurons derived from human ES cells efficiently engraft in animal models of Parkinson's disease. *Nature*. 2011;480(7378):547–551. https://doi.org/10.1038/nature10648.
19. Chambers SM, Fasano CA, Papapetrou EP, Tomishima M, Sadelain M, Studer L. Highly efficient neural conversion of human ES and iPS cells by dual inhibition of SMAD signaling. *Nat Biotechnol*. 2009;27(3):275–280. https://doi.org/10.1038/nbt.1529.
20. Pfisterer U, Kirkeby A, Torper O, et al. Direct conversion of human fibroblasts to dopaminergic neurons. *Proc Natl Acad Sci USA*. 2011;108(25):10343–10348. https://doi.org/10.1073/pnas.1105135108.
21. Caiazzo M, Dell'Anno MT, Dvoretskova E, et al. Direct generation of functional dopaminergic neurons from mouse and human fibroblasts. *Nature*. 2011;476(7359):224–227. https://doi.org/10.1038/nature10284.
22. Theka I, Caiazzo M, Dvoretskova E, et al. Rapid generation of functional dopaminergic neurons from human induced pluripotent stem cells through a single-step procedure using cell lineage transcription factors. *Stem Cells Transl Med*. 2013;2(6):473–479. https://doi.org/10.5966/sctm.2012-0133.
23. Sagal J, Zhan X, Xu J, et al. Proneural transcription factor Atoh1 drives highly efficient differentiation of human pluripotent stem cells into dopaminergic neurons. *Stem Cells Transl Med*. 2014;3(8):888–898. https://doi.org/10.5966/sctm.2013-0213.
24. Klisch TJ, Xi Y, Flora A, Wang L, Li W, Zoghbi HY. In vivo Atoh1 targetome reveals how a proneural transcription factor regulates cerebellar development. *Proc Natl Acad Sci USA*. 2011;108(8):3288–3293. https://doi.org/10.1073/pnas.1100230108.
25. Kim S-M, Lim M-S, Lee E-H, et al. Efficient generation of dopamine neurons by synthetic transcription factor mRNAs. *Mol Ther*. 2017;25(9):2028–2037. https://doi.org/10.1016/j.ymthe.2017.06.015.

26. Morris BJ, Adams DJ, Beveridge DJ, van der Weyden L, Mangs H, Leedman PJ. cAMP controls human renin mRNA stability via specific RNA-binding proteins. *Acta Physiol Scand*. 2004;181(4):369–373.
27. Dong Y, Aronsson M, Gustafsson JA, Okret S. The mechanism of cAMP-induced glucocorticoid receptor expression. Correlation to cellular glucocorticoid response. *J Biol Chem*. 1989;264(23):13679–13683.
28. Cornelius P, Marlowe M, Call K, Pekala PH. Regulation of glucose transport as well as glucose transporter and immediate early gene expression in 3T3-L1 preadipocytes by 8-bromo-cAMP. *J Cell Physiol*. 1991;146(2):298–308.
29. Xue Y, Zhan X, Sun S, et al. Synthetic mRNAs drive highly efficient iPS cell differentiation to dopaminergic neurons. *Stem Cells Transl Med*. 2019;8(2):112–123. https://doi.org/10.1002/sctm.18-0036.
30. Mahajani S, Raina A, Fokken C, Kügler S, Bähr M. Homogenous generation of dopaminergic neurons from multiple hiPSC lines by transient expression of transcription factors. *Cell Death Dis*. 2019;10(12):898. https://doi.org/10.1038/s41419-019-2133-9.
31. Chung J, Scherer LJ, Gu A, et al. Optimized lentiviral vectors for HIV gene therapy: multiplexed expression of small RNAs and inclusion of MGMT(P140K) drug resistance gene. *Mol Ther*. 2014;22(5):952–963. https://doi.org/10.1038/mt.2014.32.
32. Colasante G, Rubio A, Massimino L, Broccoli V. Direct neuronal reprogramming reveals Unknown functions for known transcription factors. *Front Neurosci*. 2019;13:283. https://doi.org/10.3389/fnins.2019.00283.
33. Qi Y, Zhang X-J, Renier N, et al. Combined small-molecule inhibition accelerates the derivation of functional cortical neurons from human pluripotent stem cells. *Nat Biotechnol*. 2017;35(2):154–163. https://doi.org/10.1038/nbt.3777.
34. Parras CM, Schuurmans C, Scardigli R, Kim J, Anderson DJ, Guillemot F. Divergent functions of the proneural genes Mash1 and Ngn2 in the specification of neuronal subtype identity. *Genes Dev*. 2002;16(3):324–338.
35. Mattar P, Britz O, Johannes C, et al. A screen for downstream effectors of Neurogenin2 in the embryonic neocortex. *Dev Biol*. 2004;273(2):373–389.
36. Chen J-G, Rasin M-R, Kwan KY, Sestan N. Zfp312 is required for subcortical axonal projections and dendritic morphology of deep-layer pyramidal neurons of the cerebral cortex. *Proc Natl Acad Sci USA*. 2005;102(49):17792–17797.
37. Thoma EC, Wischmeyer E, Offen N, et al. Ectopic expression of neurogenin 2 alone is sufficient to induce differentiation of embryonic stem cells into mature neurons. *PLoS One*. 2012;7(6):e38651. https://doi.org/10.1371/journal.pone.0038651.
38. Miskinyte G, Grønning Hansen M, Monni E, et al. Transcription factor programming of human ES cells generates functional neurons expressing both upper and deep layer cortical markers. *PLoS One*. 2018;13(10):e0204688. https://doi.org/10.1371/journal.pone.0204688.
39. Zhang Y, Pak C, Han Y, et al. Rapid single-step induction of functional neurons from human pluripotent stem cells. *Neuron*. 2013;78(5):785–798. https://doi.org/10.1016/j.neuron.2013.05.029.
40. McKenna WL, Ortiz-Londono CF, Mathew TK, Hoang K, Katzman S, Chen B. Mutual regulation between Satb2 and Fezf2 promotes subcerebral projection neuron identity in the developing cerebral cortex. *Proc Natl Acad Sci USA*. 2015;112(37):11702–11707. https://doi.org/10.1073/pnas.1504144112.
41. Ho S-M, Hartley BJ, Julia TCW, et al. Rapid Ngn2-induction of excitatory neurons from hiPSC-derived neural progenitor cells. *Methods*. 2016;101:113–124. https://doi.org/10.1016/j.ymeth.2015.11.019.
42. Marín O, Rubenstein JL. A long, remarkable journey: tangential migration in the telencephalon. *Nat Rev Neurosci*. 2001;2(11):780–790.
43. Colasante G, Lignani G, Rubio A, et al. Rapid conversion of fibroblasts into functional forebrain GABAergic interneurons by direct genetic reprogramming. *Cell Stem Cell*. 2015;17(6):719–734. https://doi.org/10.1016/j.stem.2015.09.002.
44. Sun AX, Yuan Q, Tan S, et al. Direct induction and functional maturation of forebrain GABAergic neurons from human pluripotent stem cells. *Cell Rep*. 2016;16(7):1942–1953. https://doi.org/10.1016/j.celrep.2016.07.035.
45. Kessaris N, Magno L, Rubin AN, Oliveira MG. Genetic programs controlling cortical interneuron fate. *Curr Opin Neurobiol*. 2014;26:79–87. https://doi.org/10.1016/j.conb.2013.12.012.
46. Xue Q, Yu C, Wang Y, et al. miR-9 and miR-124 synergistically affect regulation of dendritic branching via the AKT/GSK3β pathway by targeting Rap2a. *Sci Rep*. 2016;6:26781. https://doi.org/10.1038/srep26781.
47. Yang N, Chanda S, Marro S, et al. Generation of pure GABAergic neurons by transcription factor programming. *Nat Methods*. 2017;14(6):621–628. https://doi.org/10.1038/nmeth.4291.

48. Du Z-W, Chen H, Liu H, et al. Generation and expansion of highly pure motor neuron progenitors from human pluripotent stem cells. *Nat Commun*. 2015;6:6626. https://doi.org/10.1038/ncomms7626.
49. Richardson WD, Smith HK, Sun T, Pringle NP, Hall A, Woodruff R. Oligodendrocyte lineage and the motor neuron connection. *Glia*. 2000;29(2):136–142.
50. Rowitch DH, Lu QR, Kessaris N, Richardson WD. An "oligarchy" rules neural development. *Trends Neurosci*. 2002;25(8):417–422.
51. Zhou Q, Anderson DJ. The bHLH transcription factors OLIG2 and OLIG1 couple neuronal and glial subtype specification. *Cell*. 2002;109(1):61–73.
52. Lee S-K, Lee B, Ruiz EC, Pfaff SL. Olig2 and Ngn2 function in opposition to modulate gene expression in motor neuron progenitor cells. *Genes Dev*. 2005;19(2):282–294.
53. Hester ME, Murtha MJ, Song S, et al. Rapid and efficient generation of functional motor neurons from human pluripotent stem cells using gene delivered transcription factor codes. *Mol Ther*. 2011;19(10):1905–1912. https://doi.org/10.1038/mt.2011.135.
54. Lee S, Cuvillier JM, Lee B, Shen R, Lee JW, Lee S-K. Fusion protein Isl1-Lhx3 specifies motor neuron fate by inducing motor neuron genes and concomitantly suppressing the interneuron programs. *Proc Natl Acad Sci USA*. 2012;109(9):3383–3388. https://doi.org/10.1073/pnas.1114515109.
55. Lee S, Cuvillier JM, Lee B, Shen R, Lee JW, Lee S-K. Fusion protein Isl1-Lhx3 specifies motor neuron fate by inducing motor neuron genes and concomitantly suppressing the interneuron programs. *Proc. Natl. Acad. Sci. USA*. 2012 Feb 28;109(9):3383–3388. https://doi.org/10.1073/pnas.1114515109. PMID: 22343290.
56. Mazzoni EO, Mahony S, Closser M, et al. Synergistic binding of transcription factors to cell-specific enhancers programs motor neuron identity. *Nat Neurosci*. 2013;16(9):1219–1227. https://doi.org/10.1038/nn.3467.
57. Velasco S, Ibrahim MM, Kakumanu A, et al. A multi-step transcriptional and chromatin state cascade Underlies motor neuron programming from embryonic stem cells. *Cell Stem Cell*. 2017;20(2):205–217. https://doi.org/10.1016/j.stem.2016.11.006. e8.
58. Toch M, Harris A, Schakman O, et al. Onecut-dependent Nkx6.2 transcription factor expression is required for proper formation and activity of spinal locomotor circuits. *Sci Rep*. 2020;10(1):996. https://doi.org/10.1038/s41598-020-57945-4.
59. Goto K, Imamura K, Komatsu K, et al. Simple derivation of spinal motor neurons from ESCs/iPSCs using sendai virus vectors. *Mol Ther Methods Clin Dev*. 2017;4:115–125. https://doi.org/10.1016/j.omtm.2016.12.007.
60. Goparaju SK, Kohda K, Ibata K, et al. Rapid differentiation of human pluripotent stem cells into functional neurons by mRNAs encoding transcription factors. *Sci Rep*. 2017;7:42367. https://doi.org/10.1038/srep42367.
61. Dennis DJ, Han S, Schuurmans C. bHLH transcription factors in neural development, disease, and reprogramming. *Brain Res*. 2019;1705:48–65. https://doi.org/10.1016/j.brainres.2018.03.013.
62. Cherry TJ, Wang S, Bormuth I, Schwab M, Olson J, Cepko CL. NeuroD factors regulate cell fate and neurite stratification in the developing retina. *J Neurosci*. 2011;31(20):7365–7379. https://doi.org/10.1523/JNEUROSCI.2555-10.2011.
63. Castro DS, Martynoga B, Parras C, et al. A novel function of the proneural factor Ascl1 in progenitor proliferation identified by genome-wide characterization of its targets. *Genes Dev*. 2011;25(9):930–945. https://doi.org/10.1101/gad.627811.
64. Brown NL, Patel S, Brzezinski J, Glaser T. Math5 is required for retinal ganglion cell and optic nerve formation. *Development (Camb)*. 2001;128(13):2497–2508.
65. Riesenberg AN, Le TT, Willardsen MI, Blackburn DC, Vetter ML, Brown NL. Pax6 regulation of Math5 during mouse retinal neurogenesis. *Genesis*. 2009;47(3):175–187. https://doi.org/10.1002/dvg.20479.
66. A bHLH transcriptional network regulating the specification of retinal ganglion cells. *Development*. 2005;132:3907–3921. https://doi.org/10.1242/dev.01960.
67. Riazifar H, Jia Y, Chen J, Lynch G, Huang T. Chemically induced specification of retinal ganglion cells from human embryonic and induced pluripotent stem cells. *Stem Cells Transl Med*. 2014;3(4):424–432. https://doi.org/10.5966/sctm.2013-0147.
68. Chen M, Chen Q, Sun X, et al. Generation of retinal ganglion-like cells from reprogrammed mouse fibroblasts. *Invest Ophthalmol Vis Sci*. 2010;51(11):5970–5978. https://doi.org/10.1167/iovs.09-4504.
69. Deng F, Chen M, Liu Y, et al. Stage-specific differentiation of iPSCs toward retinal ganglion cell lineage. *Mol Vis*. 2016;22:536–547.

70. Miller RH, Raff MC. Fibrous and protoplasmic astrocytes are biochemically and developmentally distinct. *J Neurosci*. 1984;4(2):585–592.
71. Tabata H. Diverse subtypes of astrocytes and their development during corticogenesis. *Front Neurosci*. 2015;9:114. https://doi.org/10.3389/fnins.2015.00114.
72. Bellot-Saez A, Kékesi O, Morley JW, Buskila Y. Spatial buffering. *Neurosci Biobehav Rev*. 2017;77:87–97. https://doi.org/10.1016/j.neubiorev.2017.03.002.
73. Chen KC, Nicholson C. Spatial buffering of potassium ions in brain extracellular space. *Biophys J*. 2000;78(6):2776–2797.
74. Stolt CC, Lommes P, Sock E, Chaboissier M-C, Schedl A, Wegner M. The Sox9 transcription factor determines glial fate choice in the developing spinal cord. *Genes Dev*. 2003;17(13):1677–1689.
75. Cebolla B, Vallejo M. Nuclear factor-I regulates glial fibrillary acidic protein gene expression in astrocytes differentiated from cortical precursor cells. *J Neurochem*. 2006;97(4):1057–1070.
76. Deneen B, Ho R, Lukaszewicz A, Hochstim CJ, Gronostajski RM, Anderson DJ. The transcription factor NFIA controls the onset of gliogenesis in the developing spinal cord. *Neuron*. 2006;52(6):953–968.
77. Kang P, Lee HK, Glasgow SM, et al. Sox9 and NFIA coordinate a transcriptional regulatory cascade during the initiation of gliogenesis. *Neuron*. 2012;74(1):79–94. https://doi.org/10.1016/j.neuron.2012.01.024.
78. Caiazzo M, Giannelli S, Valente P, et al. Direct conversion of fibroblasts into functional astrocytes by defined transcription factors. *Stem Cell Rep*. 2015;4(1):25–36. https://doi.org/10.1016/j.stemcr.2014.12.002.
79. Tiwari N, Pataskar A, Péron S, et al. Stage-specific transcription factors drive astrogliogenesis by remodeling gene regulatory landscapes. *Cell Stem Cell*. 2018;23(4):557–571. https://doi.org/10.1016/j.stem.2018.09.008. e8.
80. Li X, Tao Y, Bradley R, et al. Fast generation of functional subtype Astrocytes from human pluripotent stem cells. *Stem Cell Rep*. 2018;11(4):998–1008. https://doi.org/10.1016/j.stemcr.2018.08.019.
81. Canals I, Ginisty A, Quist E, et al. Rapid and efficient induction of functional astrocytes from human pluripotent stem cells. *Nat Methods*. 2018;15(9):693–696. https://doi.org/10.1038/s41592-018-0103-2.
82. Tchieu J, Calder EL, Guttikonda SR, et al. NFIA is a gliogenic switch enabling rapid derivation of functional human astrocytes from pluripotent stem cells. *Nat Biotechnol*. 2019;37(3):267–275. https://doi.org/10.1038/s41587-019-0035-0.
83. Wang S, Bates J, Li X, et al. Human iPSC-derived oligodendrocyte progenitor cells can myelinate and rescue a mouse model of congenital hypomyelination. *Cell Stem Cell*. 2013;12(2):252–264. https://doi.org/10.1016/j.stem.2012.12.002.
84. Windrem MS, Schanz SJ, Guo M, et al. Neonatal chimerization with human glial progenitor cells can both remyelinate and rescue the otherwise lethally hypomyelinated shiverer mouse. *Cell Stem Cell*. 2008;2(6):553–565. https://doi.org/10.1016/j.stem.2008.03.020.
85. Zhou Q, Choi G, Anderson DJ. The bHLH transcription factor Olig2 promotes oligodendrocyte differentiation in collaboration with Nkx2.2. *Neuron*. 2001;31(5):791–807.
86. Wang J, Pol SU, Haberman AK, Wang C, O'Bara MA, Sim FJ. Transcription factor induction of human oligodendrocyte progenitor fate and differentiation. *Proc Natl Acad Sci USA*. 2014;111(28):E2885–E2894. https://doi.org/10.1073/pnas.1408295111.
87. Ehrlich M, Mozafari S, Glatza M, et al. Rapid and efficient generation of oligodendrocytes from human induced pluripotent stem cells using transcription factors. *Proc Natl Acad Sci USA*. 2017;114(11):E2243–E2252. https://doi.org/10.1073/pnas.1614412114.
88. Tay TL, Savage JC, Hui CW, Bisht K, Tremblay M-È. Microglia across the lifespan: from origin to function in brain development, plasticity and cognition. *J Physiol*. 2017;595(6):1929–1945. https://doi.org/10.1113/JP272134.
89. Takata K, Kozaki T, Lee CZW, et al. Induced-pluripotent-stem-cell-derived primitive macrophages provide a platform for modeling tissue-resident macrophage differentiation and function. *Immunity*. 2017;47(1):183–198. https://doi.org/10.1016/j.immuni.2017.06.017. e6.
90. da Fonseca ACC, Matias D, Garcia C, et al. The impact of microglial activation on blood-brain barrier in brain diseases. *Front Cell Neurosci*. 2014;8:362. https://doi.org/10.3389/fncel.2014.00362.
91. Subhramanyam CS, Wang C, Hu Q, Dheen ST. Microglia-mediated neuroinflammation in neurodegenerative diseases. *Semin Cell Dev Biol*. 2019;94:112–120. https://doi.org/10.1016/j.semcdb.2019.05.004.
92. Ginhoux F, Prinz M. Origin of microglia: current concepts and past controversies. *Cold Spring Harbor Perspect Biol*. 2015;7(8):a020537. https://doi.org/10.1101/cshperspect.a020537.

93. Squarzoni P, Oller G, Hoeffel G, et al. Microglia modulate wiring of the embryonic forebrain. *Cell Rep.* 2014;8(5):1271–1279. https://doi.org/10.1016/j.celrep.2014.07.042.
94. Davies LC, Jenkins SJ, Allen JE, Taylor PR. Tissue-resident macrophages. *Nat Immunol.* 2013;14(10):986–995. https://doi.org/10.1038/ni.2705.
95. Holtman IR, Skola D, Glass CK. Transcriptional control of microglia phenotypes in health and disease. *J Clin Invest.* 2017;127(9):3220–3229. https://doi.org/10.1172/JCI90604.
96. Yeh H, Ikezu T. Transcriptional and epigenetic regulation of microglia in health and disease. *Trends Mol Med.* 2019;25(2):96–111. https://doi.org/10.1016/j.molmed.2018.11.004.
97. Liu M-L, Zang T, Zou Y, et al. Small molecules enable neurogenin 2 to efficiently convert human fibroblasts into cholinergic neurons. *Nat Commun.* 2013;4:2183. https://doi.org/10.1038/ncomms3183.
98. Qiang L, Fujita R, Abeliovich A. Remodeling neurodegeneration: somatic cell reprogramming-based models of adult neurological disorders. *Neuron.* 2013;78(6):957–969. https://doi.org/10.1016/j.neuron.2013.06.002.
99. Mali S. Delivery systems for gene therapy. *Indian J Hum Genet.* 2013;19(1):3–8. https://doi.org/10.4103/0971-6866.112870.
100. Norrman K, Fischer Y, Bonnamy B, Wolfhagen Sand F, Ravassard P, Semb H. Quantitative comparison of constitutive promoters in human ES cells. *PLoS One.* 2010;5(8):e12413. https://doi.org/10.1371/journal.pone.0012413.
101. Li HO, Zhu YF, Asakawa M, et al. A cytoplasmic RNA vector derived from nontransmissible Sendai virus with efficient gene transfer and expression. *J Virol.* 2000;74(14):6564–6569.
102. Daya S, Berns KI. Gene therapy using adeno-associated virus vectors. *Clin Microbiol Rev.* 2008;21(4):583–593. https://doi.org/10.1128/CMR.00008-08.
103. Yamamoto A, Kormann M, Rosenecker J, Rudolph C. Current prospects for mRNA gene delivery. *Eur J Pharm Biopharm.* 2009;71(3):484–489. https://doi.org/10.1016/j.ejpb.2008.09.016.

CHAPTER

5

Differentiation of Purkinje cells from pluripotent stem cells for disease phenotyping in vitro

Maria Sundberg

Department of Neurology, F.M. Kirby Center for Neurobiology, Boston Children's Hospital, Harvard Medical School, Boston, MA, United States

Development of the cerebellum

Although the development of the cerebellum begins early during brain formation, the cerebellum is one of the last brain regions to become completely mature. During early embryogenesis, the cerebellum forms from the anteriormost dorsal hindbrain after midbrain/hindbrain boundary development in the neural plate. The isthmic organizer regulates this development via secretion of several growth factors and signaling molecules, such as fibroblast growth factor 8 (FGF8) and Wnt signaling.[1,2] The neural precursor cells that originate from the cerebellar ventricular zone (VZ) are GABAergic Purkinje cells, cerebellar nuclei, and cerebellar interneuron precursors.[3,4] The neural precursor cells that originate from the cerebellar rhombic lip (RL) give rise to glutamatergic granule neurons, which proliferate extensively to form the external granule layer (EGL). Mature granule neurons migrate from the EGL zone through the Purkinje cell layer to form the internal granule layer (IGL), giving rise to the overall structure of the cerebellum[5–7] (Fig. 5.1).[8,9] The mouse cerebellum starts to develop around embryonic day 10 and is completed after birth, at approximately P21. In humans, the development of the cerebellum continues for 2 years after birth (Fig. 5.1).[8,9]

Gene expression cascades in the midbrain/hindbrain boundary guide cerebellar development and differentiation of Purkinje cells and granule neurons. Of these genes, the expression of *FGF8* regulates the expression of *orthodenticle homeobox 2* (*OTX2*) and *gastrulation brain homeobox 2* (*GBX2*). *OTX2* expression is specific for midbrain and forebrain development, whereas *GBX2* is specific for the midbrain/hindbrain boundary and posterior central nervous

https://doi.org/10.1016/B978-0-12-822277-5.00002-X

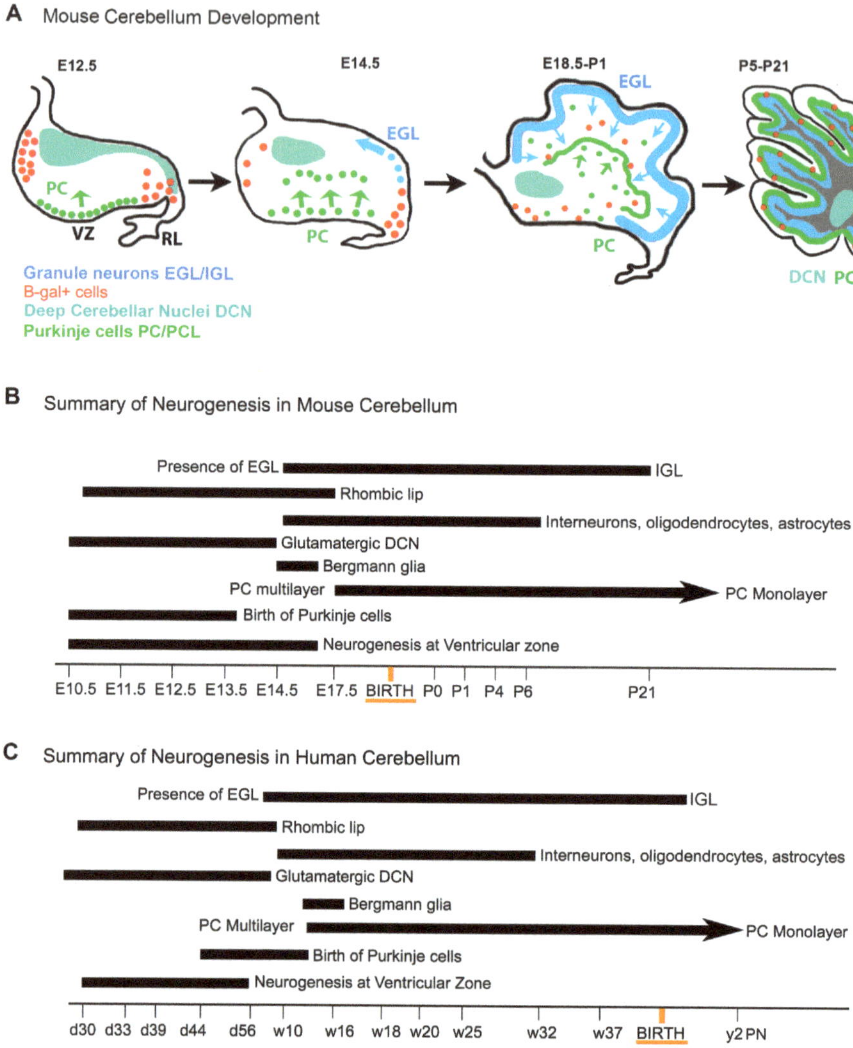

FIGURE 5.1 **Cerebellum development.** A. Schematic representation of mouse cerebellum development.[8] B. Schematic time scale of the differentiation of the neuronal subpopulations in the mouse cerebellum during development.[9] C. Schematic time scale of the differentiation of neuronal subpopulations in the human cerebellum during development.[9] *Modified from refs. Selvadurai H. J. and Mason J. O., Wnt/b-catenin Signalling Is Active in a Highly Dynamic Pattern during Development of the Mouse Cerebellum. Plos One, 2011, https://doi.org/10.1371/journal.pone.0023012, and Haldipur P., Dang D., Millen K.J., Embryology. Elsevier. 2018, DOI: 10.1016/B978-0-444-63956-1.00002-3.*

system development. The homeobox protein, HOXA2, is expressed in the caudal central nervous system at the boundary of rhombomere 1 and 2 expression, defining the outer limit of cerebellum.[6,10]

In the rhombic lip, the basic helix-loop-helix (bHLH) transcription factor, *atonal homologue 1* (*ATOH1*), induces gene cascades that regulate downstream targets, such as *PAX6*, which enhances neural precursor development in the rhombic lip and differentiation of granule neuron precursors. During development, granule neurons migrate out of the rhombic lip to form the EGL. Development of the EGL alongside granule neuron proliferation leads to cerebellar volume growth. During cerebellum development, FGF8 enhances WNT1 signaling at the midbrain/hindbrain boundary, and this feedback loop activates PAX2 and EN2 expression in the area. *EN1* and *EN2* work together to regulate granule neuron progenitor cell development, along with the bone morphogenic proteins, which dorsalize the cells within this region. The dorsal genes, *ZIC1/2* and *MATH1*, also enhance differentiation and migration of the granule neurons. From the EGL, the granule neurons migrate through the glial scaffolds, which is regulated by group of molecules called astrotactins, and specifically astrotactin2 (ACTN2). Migration of granule neurons through the Purkinje cell layer is regulated by Sonic Hedgehog (SHH) signaling from the Purkinje cells, leading to formation of the internal granule layer in the cerebellum,[6] as presented in Fig. 5.1.

The cerebellar ventricular zone gives rise to Purkinje cells, where the bHLH transcription factor, pancreas transcription factor1a (PTF1A), and OLIG2 regulate Purkinje cell precursor development. Expression of nephrin-like protein 3 (NEPH3, also known as KIRREL2 in humans) also enhances development of Purkinje cells. Purkinje cell precursors express CORL2 (also known as SKOR2 in humans) and E-cadherin when they become post-mitotic, which occurs in mice at embryonic day 12. After differentiation, the mature Purkinje cells express L7 or Purkinje cell specific protein-2 (PCP2) and calbindin (CALB).[11–14]

The ventricular zone also gives rise to other GABAergic inhibitory interneurons that populate the cerebellum. These interneurons originate from PAX2-positive inhibitory interneuron progenitors and include the stellate, lugaro, Golgi, candelabrum, globular, and basket interneuron subtypes. The ventricular zone of the cerebellum is also the origin of the glial cells of the cerebellum, including Bergmann glia, oligodendrocytes, and cerebellar astrocytes (Fig. 5.1). The development of cerebellar glial cells is regulated by Notch signaling and expression of Achaete-Scute homolog 1 (ASCL1). Oligodendrocytes arise from the OLIG2-positive precursor cells, and Bergman glia development is regulated by SHH signaling from the Purkinje cells.[6]

Although the cerebellum is made up of these numerous subpopulations of neurons, in this chapter, we will focus on the description of the development of Purkinje cells. Next, we will describe the establishment of differentiation protocols for the generation of Purkinje cells from pluripotent stem cells, and disease phenotyping of these neurons that are derived from different patient populations.

Differentiation of pluripotent stem cells into Purkinje cells

Cerebellar organoids derived from iPSCs and ESCs in 3D cultures

In 2010, Muguruma and colleagues published one of the first methods for generating mouse embryonic stem cell (ESC)-derived Purkinje cells in 3D cultures.[13] In this protocol, the ESCs were initially dissociated into single cells and reaggregated on 96-well plates to form clusters for neuronal differentiation in suspension cultures.[13] To guide the midbrain-hindbrain boundary development from the pluripotent stem cells, the cells were cultured in the presence of growth factor-free, chemically defined medium (gfCDM), which is a combination of Iscove's modified Dulbecco's Medium plus Ham's F12 medium with glutamax, bovine serum albumin, lipids, transferrin, and monothioglycerol from days 0–11.[15] After this, the cells were cultured in the DMEM/F12 with N2-supplement and FBS from days 11–19. In addition to this, a high concentration of insulin and basic fibroblast growth factor (bFGF) were used to induce Wnt-signaling and expression of Fgf8b in the cells. Cyclopamine was also added for the precursor cells at day 7 to inhibit Shh signaling and to induce ventralization of the cultures during the embryoid body (EB) differentiation process. These factors also efficiently enhanced Corl2 and Lhx5 expression in the developing Purkinje cell precursor cells compared to addition of Bmp4 or Shh.[13] When the EBs were dissociated and plated as adherent cultures to form neural rosettes, 25%–35% of these cells expressed Neph3+ at day 13 of differentiation.[13] The Neph3+ cells were sorted for enrichment of Purkinje cells in the differentiation process. These sorted Neph3+ cells differentiated into a high percentage of Corl2+ cells; 83% from total cell population. To enhance the survival of purified Purkinje cell-progenitor cultures, the sorted cells were cultured with mouse granule neurons that were isolated from the rhombic lip at E15.5.[13] After 79 days of co-culturing the ESC-derived cerebellar cells with mouse granule neurons, the differentiating cells matured to L7-positive Purkinje cells that expressed the Purkinje cell-specific marker, glutamate receptor subunit d2 (GRID2)[13] (Fig. 5.2). The construction of this protocol is a perfect example for how the signaling pathways known to take place during the development of Purkinje cells in vivo were applied to stem cells in vitro to mimic the developmental stages and lead to mature Purkinje cells for research purposes (Fig. 5.3).

Similar to mouse ESC-derived Purkinje cell derivation,[13] human ESC and iPSC-derived Purkinje cells have been successfully derived using this 3D-culture protocol, with minor modifications[12] (Fig. 5.2). In this protocol, sequential addition of growth inducing factors were introduced to human PSC cell cultures, which produced multilayered cerebellum tissue formation in vitro.[12] The stem cell colonies were first dissociated into single cells and then reaggregated to form small cellular clusters in gfCDM. A high concentration of insulin, ROCK-inhibitor (Y-27632) was used to inhibit cell death, and a transforming growth factor b-receptor blocker SB-431542 was used to induce neural induction of PSCs. On day 2, recombinant bFGF was added, which not only enhanced *FGF8* and *WNT1* signaling in the cerebellar progenitors but also reduced the expression of forebrain specific markers *OTX2* and *SIX3*. From day 21 forward, the cell culture medium comprised Neurobasal medium with N2-supplement containing transferrin, insulin, transferrin, putrescine, selenite, and progesterone to support neuronal differentiation of the cerebellar precursor cells.[12] At day 14 to day 21 of differentiation these PSC-derived Purkinje cell-precursor cells expressed markers *EN2* and *GBX2*. Addition of the bFGF in the cultures increased *WNT1* and *FGF8* signaling, indicating midbrain-hindbrain boundary development in the cell populations. The KIRREL2+ precursor cells were sorted with FACS at day 35 of differentiation, to enrich for Purkinje cell differentiation in the cultures. The KIRREL2 percentage was approximately 27% in cell culture at that point of the differentiation. The Purkinje cell precursor cells were characterized by expression of PTF1A, EN2, and these cells were also proliferative expressing Ki67. Post-mitotic Purkinje cell precursors were SKOR2 positive and located

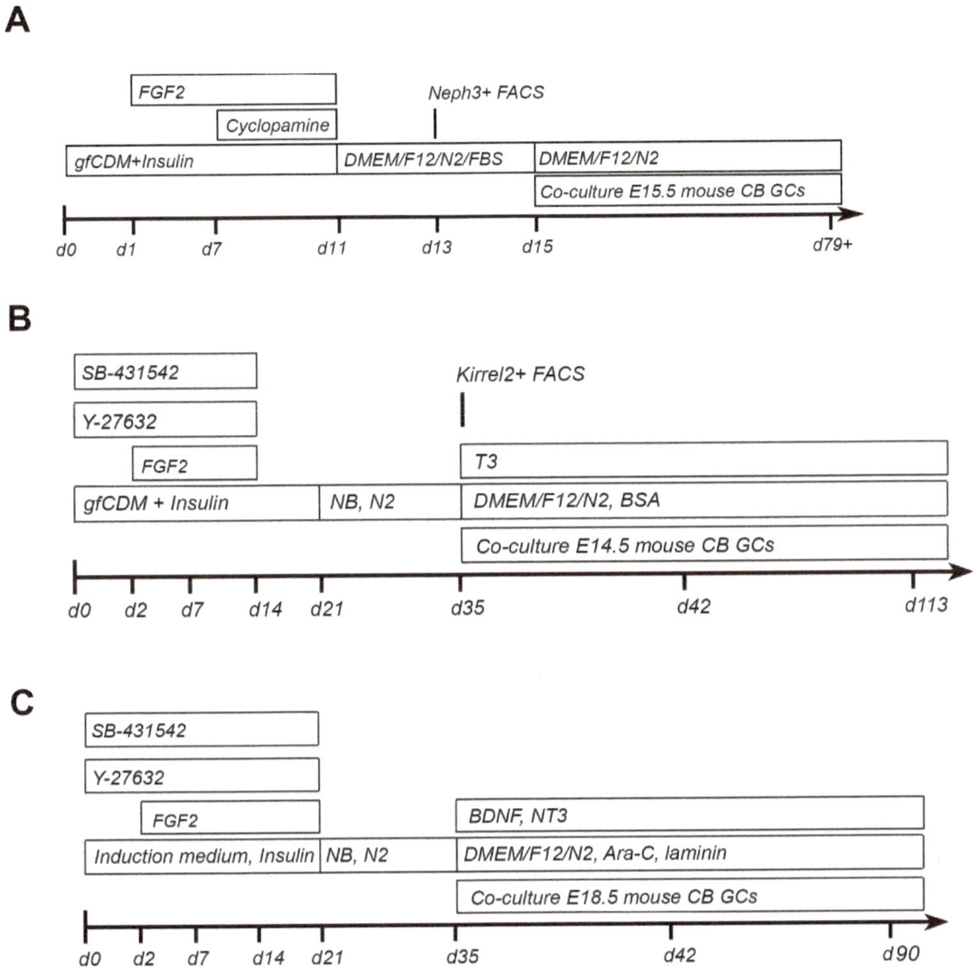

FIGURE 5.2 **Differentiation of pluripotent stem cells into Purkinje cells.** A. Schematic representation of the timeline, growth factors, and small molecules used to differentiate mouse ESCs into Purkinje cells [13]. B. Schematic representation of the timeline, growth factors, and small molecules used to differentiate human ESCs into Purkinje cells [12]. C. Schematic representation of the timeline, growth factors, and small molecules used to differentiate human iPSCs into Purkinje cells [17]. *Modified from refs. Muguruma K, Nishiyama A, Ono Y, et al. Ontogeny-recapitulating generation and tissue integration of ES cell-derived Purkinje cells. Nat Neurosci. 2010;13(10):1171–1180. doi:10.1038/nn.2638, Muguruma K, Nishiyama A, Kawakami H, Hashimoto K, Sasai Y. Self-organization of polarized cerebellar tissue in 3D culture of human pluripotent stem cells. Cell Rep. 2015;10(4):537–550. doi:10.1016/j.celrep.2014.12.051, Watson LM, Wong MMK, Vowles J, Cowley SA, Becker EBE. A simplified method for generating Purkinje cells from human-induced pluripotent stem cells. Cerebellum. 2018;17(4):419–427. doi:10.1007/s12311-017-0913-2.*

next to proliferative precursor cells in these cultures. After 113 days of culturing the stem cell derived Purkinje cell expressed PCP2, Calbindin, and LHX5. In addition, complex branching of the dendrites was detected in the mature Purkinje cells in culture with spines that expressed GRID2 (Fig. 5.3). Calbindin is a member of a calcium-binding superfamily that has been described to be a specific marker of Purkinje cells in the cerebellum.[16] Purkinje

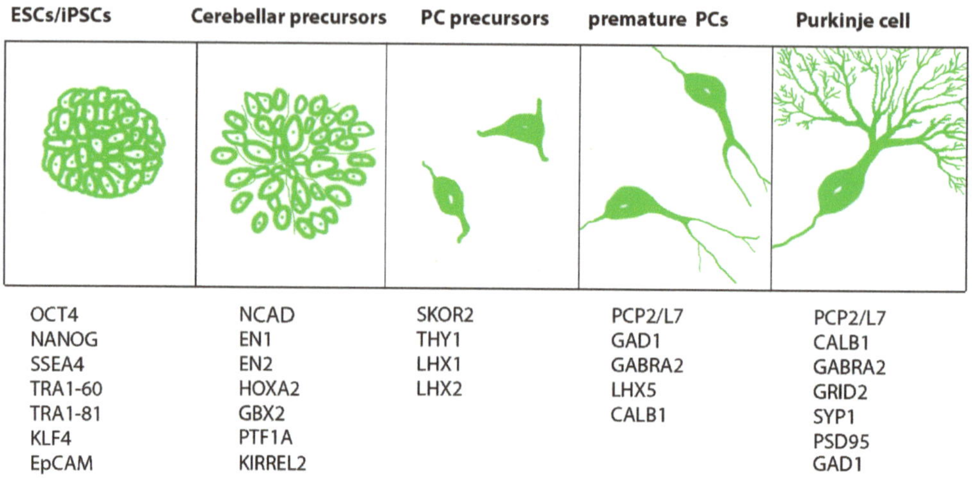

FIGURE 5.3 **Characterization of pluripotent stem cell derived Purkinje cells in vitro.** Schematic presentation of the morphological development of pluripotent stem cells toward cerebellar precursor cells, Purkinje cell precursors, premature Purkinje cells, and mature Purkinje cells. Lists of specific proteins detected during different cellular stages of Purkinje cell development.

cell protein-2 (PCP2) has also been identified as Purkinje cell specific marker in the human cerebellum (Fig. 5.3).

Interestingly, these methods have been difficult to reproduce across labs and with different stem cell lines . The reason for this may be in the differences of the origin of the studied pluripotent stem cell lines (e.g., ESCs versus iPSCs). In addition, the differences in the genetic background of the pluripotent stem cell lines, or the different reprogramming methods that were used to generate iPSC lines may affect in the differentiation efficiency of these cell lines. Also, different culture conditions that have been used to maintain the PSCs, for example, feeder-free cultures on Matrigel/Geltrex/Vitronectin-coated plates versus fibroblast-supported co-cultures. In addition, cell passaging methods with enzymatic or mechanical dissociation of the PSCs may affect their phenotype and differentiation capacity, as well as the different cell culture media compositions that have been used to maintain the pluripotency of the stem cell lines. All of these factors may contribute to the success of the differentiation into cerebellar lineage and subpopulations, including Purkinje cells. For this reason, we suggest that published neuronal differentiation protocols may require some level of optimization and adjustments before adapting them to individual cell lines, different laboratories, and cell culture conditions.

In light of this discussion, other groups have worked to optimize the Purkinje cell differentiation protocol,[12] including a protocol that is completed in approximately 90 days.[17] In this protocol, the iPSCs were first cultured on mTESR1 medium and then exposed to neural inducing factors SB-432154 and high concentration of insulin, and bFGF [17], similarly to Muguruma's protocol.[12] The main difference in this protocol was that the cells were not sorted with KIRREL2 prior to co-culturing of the human iPSC-derived cerebellar cells with E18.5 mouse cerebellar granule neurons [17] Fig. 5.2. The authors described development of human iPSC-derived cerebellar progenitors that expressed EN1, EN2, and GB2 at day 21 of differentiation. At day 35, the cells expressed LHX5, OLIG2, and granule cell marker, ATOH1. In addition, PTF1A and KIRREL2 were detected in these cell clusters. After 70 days of differentiation, 10% of the human cells in the co-cultures with mouse cerebellar cells expressed CALB and FOXP2. On day 90, the human cells expressed PCP4, CALB and FOXP2 in the co-cultures with mouse granule neurons. However, the morphology of these cells were immature at this stage of differentiation, and complex dendritic trees with multiple branches and spines were not detected in this study.[17]

In addition, a recently published method describes the derivation of human cerebellar neurons and Purkinje cells from iPSCs without cell sorting and without co-culturing of the human Purkinje cells on mouse granule neurons.[18] In this method, the iPSCs were first plated as single cells with ROCK-inhibitor on low-attachment plates to form aggregates in suspension cultures. The cells were then guided toward neural fate with the same factors used in the original protocol; bFGF and SB-431542 in growth factor-free, chemically-defined medium.[12] Cerebellar progenitors were derived by the addition of FGF19 from day 14 until day 21 [18]. Stromal cell derived factor1 (SDF1) was used to support multi-layered self-organization in the cerebellar spheroids in Neurobasal media. After 35 days of differentiation, the cells were transferred to BrainPhys-media supplemented with brain-derived neurotrophic factor (BDNF) and glial-cell line-derived neurotrophic factor (GDNF) [18]. Immunocytochemical characterization of the cells revealed that at the early stages of the differentiation process, the cerebellar spheroids contained a mixed populations of immature neural progenitor cells (NPCs) that were positive for Nestin, NCAD, SOX2, PAX6, and PAX2. Further, neural precursors were positive for OLIG2, CORL2 (SKOR2), and KIRREL2 at day 35 of differentiation.[18] At day 35, the cell aggregates were dissociated and the neural cells were plated onto laminin-coated dishes to allow the Purkinje cells to mature. At day 50, 10%–20% of the total cell population expressed SKOR2 (CORL2). Individual CALB-positive Purkinje Cells were also detected at day 80.[18] However, the morphology of these cells was also quite immature, consistent with previous methods.[17], [18]Without the cell sorting step, the Purkinje cells were not enriched in this protocol, which resulted in a low percentage of Purkinje cells produced from the total cell population.[18] Also, many other neuronal cell types were detected in these cell cultures, including glutamatergic neurons, Golgi cells, GABAergic interneurons, and astrocytes. In addition, the lack of co-culturing with mouse granule neurons led to the development of immature Purkinje cells with low dendritic branching or spine development.[18] These results underline the importance of cell sorting methods and the establishment of co-cultures with correct cerebellar sub-populations that can support Purkinje cell development in vitro.

Human iPSC- and ESC-derived Purkinje cell differentiation in 2D co-cultures with mouse cerebellar cells

Several research groups have developed and optimized human pluripotent stem cell differentiation into cerebellar precursor cells and pre-mature Purkinje cells with small molecule- and growth factor-based protocols.[11,12,14,17,18] Initially, the identification of the timing and content of small molecule and growth factor cocktails for the induction of Purkinje cell differentiation were optimized by analyzing the expression profiles of cerebellar lineage-inducing transcription factors, and by comparisons of the gene expression profiles to mouse Purkinje development in vivo.[11,14] In these methods, the initial patterning of the pluripotent stem cells toward neural lineage have been achieved with NOGGIN and SB-431542[14] or NOGGIN by itself[11] in the basal media with N2-supplement, B27-supplement and nicotinamide (Fig. 5.4). For induction of cerebellar neural precursor cells and for midbrain/hindbrain boundary development from human iPSC or ESC cultures, CHIR99021, and FGF8b have been used,[11,14] and FGF2 has been utilized for KIRREL2+ precursor cell induction.[12] After

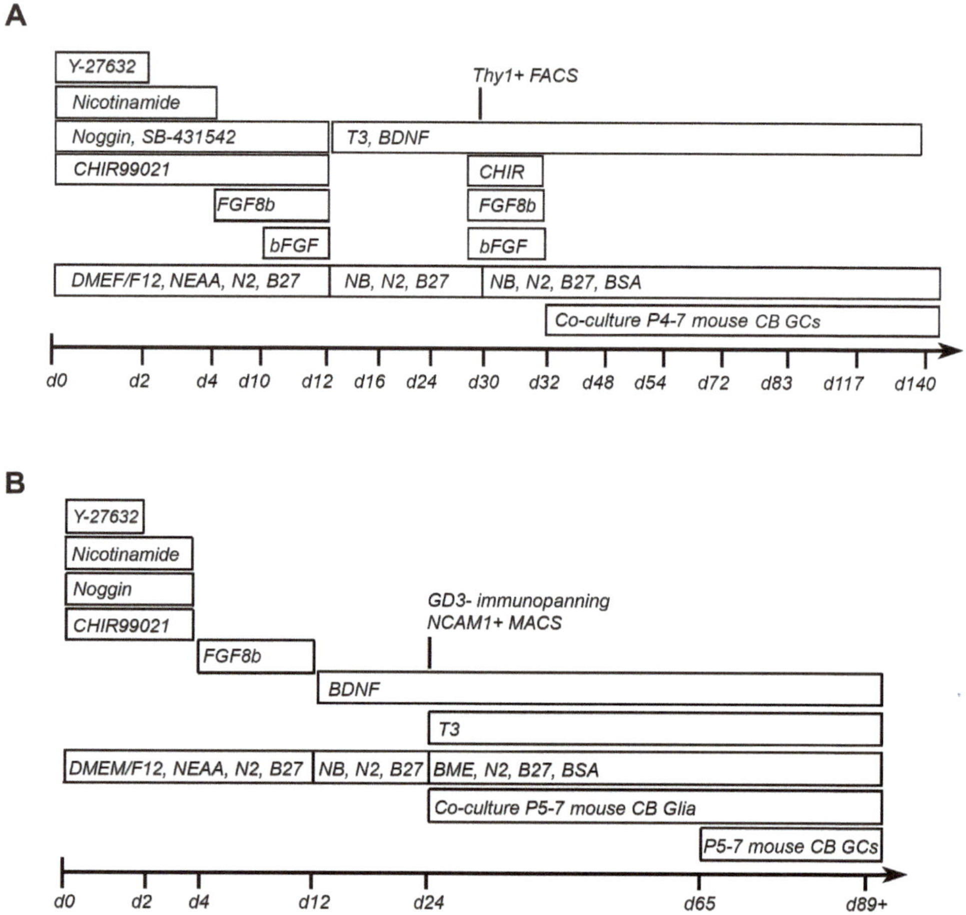

FIGURE 5.4 **Differentiation of human pluripotent stem cells into Purkinje cells in 2D co-cultures with mouse cerebellar cells.** (A). Schematic representation of human iPSCs differentiation into Purkinje cells in 2D co-cultures with mouse cerebellar granule cells [14]. (B). Schematic representation of human ESCs differentiation into Purkinje cells in 2D co-cultures with mouse astrocytes and cerebellar granule cells [11]. *Modified from refs. Sundberg M, Tochitsky I, Buchholz DE, et al. Purkinje cells derived from TSC patients display hypoexcitability and synaptic deficits associated with reduced FMRP levels and reversed by rapamycin. Mol Psychiatr. 2018;23(11):2167–2183. doi:10.1038/s41380-018-0018-4, and Buchholz DE, Carroll TS, Kocabas A, et al. Novel genetic features of human and mouse Purkinje cell differentiation defined by comparative transcriptomics. Proc Natl Acad Sci USA. 2020;117(26):15085–15095. doi:10.1073/pnas.2000102117.*

initial spheroid culturing from 6 to 8 days in culture, these cells were plated as monolayers on top of laminin to enhance neural rosette formation and expression of KIRREL2, LHX1A, PITF1a, and SKOR2. At this point of differentiation, the precursor cells express cerebellar lineage markers EN1, EN2, HOXA2, and GBX2[11,14] (Figs. 5.3 and 5.4).

The normal secretion of thyroid hormone L-3,3′,5-triiodothyronine (T3) is essential for the development of Purkinje cells in the cerebellum.[19–21] Several studies have shown that absence of T3 during cerebellum development negatively affects Purkinje cell branching, spine formation, and synaptic development.[21] T3 also affects granule cell migration and development of synaptic connections with Purkinje cells.[19,20] In mouse Purkinje cell cultures, the addition of T3 enhanced dendritic arborization via activation of the TRa1 receptor in Purkinje cells.[22] Neurotrophins, like BDNF, have also been shown to have significant effect on Purkinje cell differentiation and maturation.[23,24] In a mouse model of loss of BDNF ($BDNF^{-/-}$), significant deficits were detected in cerebellar development and Purkinje cell maturation, which indicates that BDNF is important for initial differentiation of stem cells toward neurons during embryonal cerebellum formation.[24] Previous studies have also shown that the addition of external BDNF or NT-4 in the cell culture media of purified mouse postnatal day 1 Purkinje cells, increased survival of Purkinje cells.[25] However, in co-cultures with mouse granule neurons, the addition of BDNF decreased Purkinje cell survival,[25] which indicates that the cell culture environment and cellular interactions with granule neurons alter Purkinje cell adaptation and survival after the Purkinje cells have been differentiated. Interestingly, late addition of BDNF in the Purkinje cell cultures has also been shown to enhance spine development.[25] BDNF also affects dendritic branching in neurons,[23] and axonal development in cortical neurons, optic neurons, and in retinal cells.[26–28] Taken together, BDNF signaling is important for early cerebellum development and Purkinje cell development from neural progenitors, and during later stages of differentiation it enhances dendritic formation and spine development. In the iPSC- and ESC-derived Purkinje cell cultures, BDNF is added to the cell cultures to enhance neural progenitor cell development toward neurons and Purkinje cells at day 12 of differentiation, and continues until the end of the differentiation at day 95,[11] or day 125.[14] Also, T3 was added to enhance Purkinje cell differentiation from the iPSC-derived cerebellar progenitors at day 12[14] or at day 24[11] of differentiation, and kept in culture until day 95[11] or day 125 of the differentiation[14] Fig. 5.4.

The morphological development of Purkinje cells continues postnatally in both rodents and humans. Rodent studies have shown that premature Purkinje cells begin to extend a few long processes around embryonic day E16-18. After this time, the Purkinje cells develop an oval shaped soma and extend short peri-somatic processes at E17 to postnatal day 3 (P3). From P3 to P10, the Purkinje cells start to develop thicker primary branches with short secondary branches. From P10 to adulthood, the Purkinje cells form a tree-like shape and branches that are complex and contain short spines on their dendrites.[29] Immunocytochemical characterization of developing Purkinje cells in the developing mouse and rat cerebellum show expression of Thy1.2, which is specific for Purkinje cells.[29] Also, the staining of Thy1.2-positive cerebellar cultures reveals high levels of CALB-positive cells.[29] To follow in vivo development of cerebellar Purkinje cells, the iPSC- or ESC-derived Purkinje cellsshow expression of PCP2 starting at days 24–30 in vitro. The iPSC- or ESC-derived Purkinje cell protocols result in approximately 22% PCP2-positive precursor cells from the total cell population in the culture at days 24–30 of differentiation.[11,14] At this stage, the cellular morphology is still undeveloped and the cells have round soma with thin processes.[11,14] Co-culturing of the iPSC- or ESC-derived Purkinje cells with mouse granule neurons up to differentiation days 95–125 allowed the Purkinje cell morphology to develop further, including oval shaped soma and more complex dendritic branching.[11,14] At this final stage of the differentiation, the Purkinje cells express synaptophysin, post-synaptic density 95 (PSD95) and Purkinje cell-specific marker, GRID2[11,14] (Figs. 5.3 and 5.4).

To enhance the number of developing Purkinje cells in the iPSC- or ESC-derived cerebellar cell cultures, several cell sorting methods have been utilized. For example, Purkinje cells can be enriched with positive sorting of THY1+ cells via immunopanning and fluorescence-activated cell sorting (FACS).[14] Negative selection can also be used to remove nonspecific cells, including immunopanning using a b-series ganglioside (GD3) marker, followed by magnetic cell sorting with neural cell adhesion molecular (NCAM)-positive selection.[11] During initial development of these methods, THY1 was selected for the enrichment of Purkinje cells because THY1 is a cell surface glycoprotein that is specifically expressed in Purkinje cells in the cerebellum.[22,30] Previous studies have also indicated that THY1 is present in undifferentiated pluripotent ESCs and it is strongly expressed in neuronal precursor cells derived from ESCs.[31] Thy1 immunopanning has also been used to isolate Purkinje cells from mouse cerebellum.[22] In the pluripotent stem cell-derived cerebellar cultures, the THY1+ selection is used to enrich the premature Purkinje cell precursors that will eventually develop into PCP2- and CALB-positive Purkinje cells with further culturing *in vitro*.[14] In line with this, NCAM is a commonly used pan-neuronal cell surface marker that is present in several subtypes of neurons and neural precursor cells,[31,32] including dopaminergic neuron precursor cells,[33,34] and fetal cortical neural precursor cells.[35] NCAM has also been shown to regulate Purkinje cell migration in the cerebellum by counteracting the effects of GFRa1 on migration.[36] GD3 is a ganglioside glycolipid, and it is expressed in several neuronal classes of the cerebellum. Antibodies against GD3 stain cells in the granule layer and Purkinje cell layer, including the cell bodies of the granule cells and Purkinje cells, with intra-cytoplasmic and vesicular localization.[29,37] In the pluripotent stem cell-derived cerebellar cultures, the negative selection against GD3 is used to remove cerebellar granule cells, and the NCAM + selection was used to isolate the premature Purkinje neurons.[11] These different cell sorting methods of the ESC/iPSC-derived cerebellar cultures at day 25 or at day 30 of the differentiation have been shown to enrich the number of PCP2-positive Purkinje cells up to 84% of the human cell population at day 95 of differentiation[11] or up to 60%–91% PCP2+/huNuc + at day 125 of differentiation.[14]

Cerebellar granule neurons play an important role in the development of Purkinje cells, by providing growth factors and synaptic inputs to enhance Purkinje cell maturation.[29] This is why several co-culturing protocols have been established to enhance iPSC-derived Purkinje cell differentiation in vitro. Mouse granule neurons and cerebellar cells have been isolated from P4–P7 mouse cerebellum and co-cultured with sorted human iPSC-derived Purkinje cell precursor cells for 95 days in vitro to support the branching and synaptic development of the Purkinje cells.[14] In addition, several studies have shown that astroglial cells support dendritic outgrowth of neurons. However, astroglial cells do not enhance the neurotransmission signaling or the complex branching patterning or spine formation in the co-cultures with Purkinje cells.[29] Despite this, it has been described that co-culturing of isolated mouse astrocytes with $GD3^-/NCAM^+$ human iPSC-derived Purkinje cells for 65 days supports Purkinje cell survival. Also, in this protocol, the addition of purified mouse granule cells was required for an additional 24 days to enhance human iPSC-derived Purkinje cell maturation and synaptic marker expression[11] (Fig. 5.4).

Functional characterization of human pluripotent stem cell-derived Purkinje cells in vitro and in vivo

Electrophysiological characterization of neurons and neuronal networks is crucial for understanding the complexity of brain development and brain disorders. To model neurophysiological disease phenotypes in vitro, several electrophysiological technologies have been established to measure neuronal network firing properties or single cell synaptic properties in a dish, e.g., patch-clamping, calcium imaging, and micro-electrode array (MEA) technologies. Next, we describe how these methods have been used to characterize human iPSC- or ESC-derived Purkinje cells in vitro.

Patch-clamp recordings are an ideal method for single cell characterization of ion channel properties, as well as neuronal membrane conductance, capacitance, and resistance information.[38] Our group and others have used patch clamp to characterize the functional properties of iPSC-derived Purkinje cells.[12,14,39] Selection of which cells to patch is conventionally done by morphological assessment of the cells by their size, soma shape and branching properties. Also, expression of virally transduced Purkinje neuron-specific markers with fluorescent tags, like PCP2-EGFP or GAD-GFP, allows for easy identification of Purkinje cells for patch clamping. Voltage-clamp mode can provide data on the cellular excitability, rheobase, capacitance, ionic flux, and spontaneous activity, whereas current-clamp mode can be used to measure synaptic activity, such as spontaneous excitatory post-synaptic currents (sEPSC) and inhibitory post-synaptic currents (sIPSC) and the resting membrane potential of cells.[40,41] These detailed characterizations provide a unique window to cellular functionality at the single cell level.

To characterize the functional properties of iPSC-derived Purkinje cells that were differentiated with the 3D method, the cells were dissociated and plated in 2D with mouse granule neurons and analyzed with whole cell patch clamping at differentiation days 108–113.[12] This characterization showed that the Purkinje cells displayed a depolarizing sag after injecting cells with hyperpolarizing current pulses, which is consistent with rodent Purkinje cells.[42] The human ESC-derived Purkinje cells also had lower input resistance compared to mouse Purkinje cells, likely due to the larger surface area of human Purkinje cells. [12] To verify that the human Purkinje cells had synaptic activity that was specific for glutamatergic and GABAergic synapses, the spontaneous inward and outward currents in the Purkinje cells were blocked with NBQX, RCPP, and bicuculline. The miniature excitatory postsynaptic currents (EPSCs) in human ESC-derived Purkinje cells were silenced by the addition of NBQX, which suggested that AMPA receptors were mainly involved in the excitatory synaptic activity in these cultures. These results concluded that this cell culture protocol was sufficient to generate human Purkinje cells with morphological and electrophysiological properties similar to rodent Purkinje cells in the cerebellum. In addition, this protocol generated two other neuronal subtypes, which were CALB1/parvalbumin-positive interneurons and Golgi cells. [12]

Previous studies have also shown that iPSC-derived Purkinje cells that were cultured on top of human fetal cerebellum slices exhibit resting membrane potentials from −36.6 mV to −58.7 mV during 2–4 weeks of culturing and recordings.[39] These iPSC-derived Purkinje cells also displayed action potentials after 2 weeks of culturing on the slices. Moreover, after 4 weeks of co-culturing, the majority of the iPSC-derived Purkinje cells (90%) displayed action potentials and 40% of these cells showed frequent spontaneous action potentials. In addition, these cells formed synaptic connections to the cerebellar cells, which were verified by detection of inhibitory-post synaptic currents (IPSCs).[39] These results indicate that this method was sufficient to generate human Purkinje cells with functional properties that were comparable to in vivo Purkinje cell development. [39] Although patch clamping can be useful for identification of neuronal subtypes and ion channel properties, it is a labor intensive method and doesn't provide network level functional data about the neuronal cultures.

Calcium imaging can provide data on the network level of human ESC-derived Purkinje cells in co-cultures with granule neurons and astroglial cells.[11] This characterization was done by transfecting the human ESC-derived Purkinje cells with the genetically encoded calcium indicator jRGECO1a and PCP2-mGFP reporter.[11] The calcium imaging revealed that the human ESC-derived Purkinje cells formed functional connections with mouse granule neurons and ~58% of the Purkinje cells displayed calcium transients $\Delta F/F_0 > 0.3$ that are typical of human neurons in cell culture.[11] The downside of this method is that the repeated imaging of the cell cultures under the fluorescent microscope may damage the cells and decrease viability. Calcium imaging also requires access to a specific microscope setup with optimized detection filters and channels, which requires time and expertise to set it up in different laboratories.

The final method described here is the MEA, which involves the recording of extracellular voltage changes in cell cultures that are plated on specialized plates with multiple electrodes in each well. This technology allows for the monitoring of neuronal networks with millisecond time scale resolution.[43] The most widely used MEAs are low density MEAs, which are not able to resolve activity from individual neurons, rather recording from the extracellular field potential of multiple neurons at each electrode. The advantage of the low density MEA platform is that they are typically in a multi-well format, allowing for simultaneous recording of neural network from several wells, with different cellular lines or treatment conditions. These MEA platforms have been widely used to characterize the network level functionality of developing iPSC-derived neuronal cultures.[44–48]

In addition to low density MEAs, high-density MEAs have been developed to record the neuronal activity in the networks at sub-cellular resolution. High density MEAs can be used to analyze network connectivity and axonal velocity, synchronization and bursting patterns at subcellular level, and visualizing the morphological development of the neurons with fluorescent imaging and stimulation capacities.[49–51] High density MEA platforms facilitate single cell characterization of the neuronal network and spike sorting based on individual cell type waveforms.[52] Previous studies have shown that human iPSC-derived neuronal networks can be recorded with the high density MEA chips for over 1 month.[46] In addition, 3D neuronal brain organoids have been differentiated and recorded with high density MEAs.[53] The high density MEA recordings facilitate more in depth, single cell level analyses of the network development and neuronal features compared with conventionally used low density MEA platforms. In the future, characterization of the iPSC-derived Purkinje cells in co-cultures with granule neurons with high density MEA will provide important information about the cerebellar network formation in a dish. This platform also facilitates disease phenotyping of different patient-derived Purkinje cell cultures.

Challenges in the differentiation of human Purkinje cells in 2D- and 3D-cell cultures

The development of 2D and 3D methods for the differentiation of human pluripotent stem cells to Purkinje cells has presented several challenges. To develop these differentiation protocols, researchers have taken advantage of data from mouse developmental studies. Evaluation of the factors driving mouse cerebellar development guided the selection of efficient growth factors and small molecules that induce the activation of the molecular pathways that are important for cerebellar development and specialization of Purkinje cells. However, this is not always a straightforward process, since timing of mouse cerebellum and Purkinje cell development is different from human cerebellar development. Thus, optimization of the timing to add factors from mouse development to human cell differentiation in cell culture is quite challenging. In addition, the current average time to differentiate mature human PSC-derived Purkinje cells in vitro varies between 90 and 140 days, which increases challenges to keep the cell cultures free from bacterial contaminations or detachment of the neurons from the culture plates. To prevent this problem, addition of penicillin-streptomycin, normocin, or primocin to cultures have been used to prevent bacterial, fungal or mycoplasma contaminations.[11,14]

Conversion of embryonic development from in vivo to in vitro is a challenging endeavor. Developing Purkinje cells require support from extracellular matrix proteins that support cell attachment and differentiation. Laminin can be added to the cell culture medium to support iPSC-derived Purkinje cell co-cultures with mouse granule cells.[14] Others have tried to embed $NEPH3^{+}$ iPSC-derived Purkinje cell precursors and mouse granule neurons into Matrigel to try to enhance the morphological development of the Purkinje cells.[39] However, the viability of the cells inside Matrigel was poor, and the cells couldn't be detected after 1 week of culturing.[39] In addition to extracellular matrix signaling, tight cell to cell connections are also important for Purkinje cell development. As described previously, Purkinje cells grow in dense layers in the cerebellum with close contact to each other. Mimicking this environment in vitro has proven to be challenging, and it has required the development of demanding co-culture paradigms with various different compositions of mouse cerebellar granule neurons, mouse astrocytes, mouse cerebellar sections, or even human fetal cerebellum sections.[11,12,14,39] Quality of co-cultures depends on the dissection of the mouse cerebellum and selection of proper developmental time points for isolation of the cerebellar cells from mice, e.g., embryonic versus postnatal cells, which have differences in viability and phenotype. Also, challenges arise from the optimization of suitable media compositions that supports both human and rodent cells in co-cultures.[11,12,14,39]

In addition, patterning-based differentiation protocols that use growth factors and small molecules to guide differentiation tend to lead to rather heterogenous neuronal cell cultures.[11,12,14,39] The use of cell sorting methods to isolate developing human Purkinje cells from PSC-derived cultures is a crucial step for optimizing the yield from these cultures and reducing heterogeneity. Cell sorting can be achieved with immunopanning, magnetic-activated cell sorting (MACS), and fluorescence-activated cell sorting (FACS), using

KIRREL2+ selection for Purkinje precursor cells, THY1+ for premature Purkinje cells, and a combination of GD3-/NCAM+ selection for premature Purkinje cell isolation.[11,12,14,39] The success of cell sorting depends on the quality of internal controls and gating strategies used for FACS and MACS. Also, the viability of the neuronal cells may decrease during prolonged cell sorting and processing time. Addition of a cell death inhibitor in the media, such as ROCK-inhibitor, will enhance the survival of the sorted neurons post-sorting.

In the future, viral transduction of transcription factors that enhance Purkinje cell differentiation from iPSCs could be beneficial to increase efficiency of the differentiation. As described in the previous chapter, similar methods have been developed for the generation of pure glutamatergic neurons with Neurogenin2 overexpression[54] or induced GABAergic neurons with ASCL1/DLX2 overexpression[55,56] from PSCs. The advantage of viral induced neuronal differentiations of PSCs is that the time required for generation of these cell populations is significantly shorter, compared with conventionally used small molecule and growth factor based methods. With transcription factor overexpression, electrophysiologically active excitatory glutamatergic neurons can be generated in 2 weeks and functional inhibitory GABAergic neurons in 5–6 weeks. However, viral transduction of pluripotent stem cells may cause damage to the genome with viral integration, which could increase chromosomal abnormalities in the stem cells. This will require extensive quality control for the transduced cell lines with single nucleotide polymorphism (SNP) arrays or whole genome sequencing technologies.

Each of these factors contributes to the challenges of developing an efficient protocol for the differentiation of human Purkinje cells from PSCs. These protocols require precise culture techniques from highly specialized researchers. The establishment of standard operating procedures with detailed information for each step and process is of uttermost importance for the generation of Purkinje cells from human pluripotent stem cells.

Disease phenotyping of Purkinje cells

Purkinje cells in cerebellar ataxia

Mouse Purkinje cell models of cerebellar ataxia

In cerebellar ataxia, there are three main signs of motor deficits, including ataxia, increased muscle passivity, and tremor. Degeneration of the cerebellum and deficits in Purkinje cell function result in various deficits for patients, including deficits in movement regulation, eyeball movement, ataxic posture and gate, difficulties in goal-directed movements, and abnormal speech development. In humans, cerebellar degeneration can be treated mainly with rehabilitation and motor coordination training. Currently, there are no curative treatments for cerebellar degeneration disorders.

Mouse models of ataxia are essential for studies of cerebellar degeneration, since these animals can be monitored with various behavioral and motor tasks in the laboratory, and the cellular morphologies of the cerebellum are relatively conserved across species and can be detected with immunocytochemical characterizations of the cerebellum sections. Several ataxia models have been created, including Scrambler mice, mutant *Agtpbp1 pcd/J* mice, Nervous mice, Weaver mice, Reeler mice, Staggerer mice, and *SCA1, SCA2, SCA3, SCA6, SCA7,*

SCA23, DRPLA mutant mice. The most common cellular and molecular deficit in these animals are the loss of Purkinje cells in the cerebellum, overall cell death, cerebellum atrophy and deficits in the axonal connections between the cerebellum and other areas of central nervous system.[57] Model systems such as these can be useful for testing treatment therapies aimed at cell survival or Purkinje neuron replacement.

Human iPSC-derived NPCs and Purkinje cells in cerebellar ataxia

Previously, researchers have reported the generation of iPSCs from patients with Spinocerebellar ataxia type 6 (SCA6).[58] SCA6 is characterized by loss of Purkinje cells in the cerebellum. Trinucleotide repeat expansion of CAG in *CACNA1A*, a gene which encodes a subunit of the calcium channel $Ca_V2.1$, leads to development of SCA6. SCA6 is a dominantly inherited disorder and there is currently no cure. In this study, the iPSC-derived Purkinje cells from patients with SCA6 showed vulnerability to thyroid hormone depletion.[58] Quantification of Purkinje cell number showed an overall decrease in the number of cells in the heterozygous and homozygous loss of SCA6 lines, compared to healthy control cells. In addition, the arborization of the dendrites were deficient in the SCA6 mutant Purkinje cells and decreased dendritic areas were detected, compared to control cells. Importantly, this study showed that these Purkinje cell deficits were rescued by treating the cells with thyroid releasing hormone and Riluzole.[58] Although the results of this study were promising, more detailed follow-up studies are needed to determine the long term safety of these compounds in cell culture and in animal models. Overall, this study showed that the advantage of patient iPSC-derived Purkinje cell cultures is the possibility to test different drugs, compounds and small molecules in cell culture, and potentially find new treatment options for severe neurodegenerative disorders that do not have curative treatments available.

To rescue degeneration of the cerebellum and the loss of Purkinje cells in the mouse models of ataxia, researchers have also studied the effects of human iPSC-derived NPC transplantations in the cerebellum. The focus of these NPC transplantation was to determine if the cells could restore the function of remaining cells in the cerebellum or protect the animals from Purkinje cell degeneration further.[59] This study described that human NPCs, which were injected bilaterally into the cerebellum, were viable, were able to migrate toward host Purkinje cell layer, and they secreted protective factors that slowed the death and degeneration of the host Purkinje cells in the cerebellum. After NPC transplantation, behavioral improvements were detected in these animals with improved motor function, increased lifetime, and body weight of the animals were maintained longer, compared to mice with ataxia that did not receive the cell transplantations.[59] These are promising results that indicate that human NPCs are a powerful source for the secretion of supporting factors after transplantation to the cerebellum. However, one restriction for wider use of NPC transplantation for regenerative purposes of central nervous system disorders is the remaining risk of NPCs' proliferation, which possess a great risk for graft overgrowth.[35] This would be difficult to control post-transplantation in the cerebellum. Thus, development of more specialized cells from ESCs or iPSCs would be a safer option for cell therapy development for cerebellar ataxia, such as Purkinje precursor cells that are FACS or MACS sorted before transplantation.[11,12,14,39]

Purkinje cells in Tuberous Sclerosis Complex (TSC)

Mouse Purkinje cell models of TSC

Tuberous sclerosis complex (TSC) is a multi-organ disorder that manifests with benign tumor formation in different organs. TSC patients also have neurological deficits, including epilepsy (~80–90% of patients) and autism (~50% of patients).[60] Previous studies have shown that heterozygous and homozygous loss of Tsc1 in mouse cerebellar Purkinje cells leads to hypoexcitable Purkinje cells with enlarged soma size and reduced spine densities, compared to control Purkinje cells.[61] Behavioral studies of these mice with Tsc1 mutant Purkinje cells revealed autistic-like behavioral deficits, including repetitive behaviors and abnormal social interactions, compared to control mice.[61] These deficits were shown to be mTOR-dependent, since rapamycin, a specific inhibitor of mTOR-pathway activation, rescued the cellular disease phenotypes and behavioral deficits in the Tsc1 mutant mice.[61] More recently, studies have suggested that there is a critical period for utilizing rapamycin to rescue the ASD-related behavioral deficits and Purkinje cell dysfunctions in Tsc1 mutant mice.[62] The authors found that when rapamycin treatment was initiated in Tsc1-deficient mice in early adulthood, starting at the 6 weeks of age, rapamycin was able to rescue motor and social deficits in the Tsc1-mutant mice. This social behavioral rescue effect was also correlated with Purkinje cell function and pathology.[62] However, administration of rapamycin at a later timepoint, starting at the 10 weeks of age, did not rescue the social or behavioral deficits or the cellular dysfunctions seen in the Tsc1 mutant mice.[62] These findings indicate that inhibition of the mTOR pathway during sensitive periods of brain development could reduce the neurological deficits in TSC.

TSC patient iPSC-derived Purkinje cells

Interestingly, human iPSC-derived Purkinje cells from patients with TSC display phenotypes that are similar to those seen in Tsc1-deficient mouse Purkinje cells.[14] The TSC2-deficient iPSC-derived Purkinje cells have enlarged soma and increased number of neurites. In addition, the TSC2-deficient Purkinje cells are hypoexcitable and have reduced synaptic marker expression, compared to Purkinje cells derived from healthy control individuals.[14] In this study, we examined three TSC-patient derived iPSC-lines with heterozygous mutation of TSC2, as well as an iPSC line with TALEN-induced biallelic mutation of TSC2, and a CRISPR-cas9 corrected TSC-patient derived iPSC line.[14] The line with the biallelic mutation of TSC2 and the CRISPR-corrected line were generated from the same patient line so that they are all isogenic. Healthy control iPSC lines that were derived from family members of the TSC patients were also included. RNA sequencing of Purkinje cells derived from these lines revealed downregulation of fragile X mental retardation protein (FMRP) and its targets in TSC2-deficient iPSC-derived Purkinje cells, compared with healthy controls. These experiments also revealed a decrease in expression of GRID2 and pre- and post-synaptic markers in the TSC2-deficient Purkinje cells, compared to control cells.[14]

In addition, patch clamping was used in this study to identify functional disease phenotypes of TSC patient iPSC-derived Purkinje cells compared to healthy control cells.[14] In this study, the patched cells were characterized by their capacitance, resting membrane potential (mV), action potential threshold (mV), Rheobase (pA), and spontaneous firing rate.[14] Increasing current injections into the Purkinje cells revealed that the TSC2-deficient

Purkinje cells were hypoexcitable compared to control Purkinje cells. In addition, the specific mTOR inhibitor, rapamycin, rescued these deficits *in vitro*,[14] similarly as seen with Tsc1 mutant mice Purkinje cells.[61] This study also confirmed that the Purkinje cells formed synaptic connections with mouse cerebellar granule neurons in co-cultures, shown by characterization of frequency (Hz) and amplitude (pA) of mini EPSCs.[14] These studies highlight the utility of patient-specific iPSC-derived Purkinje cells as a powerful tool for discovering cellular and molecular deficits and molecular mechanisms driving disease development. Disease phenotyping of iPSC-derived Purkinje cells can also facilitate screening compounds for development of new treatment options for patients with specific genetic mutations in the future.

Future perspectives for stem cell-derived Purkinje cells in translational medicine

Cell transplantation for treatment of cerebellar degeneration

A proof of concept study for Purkinje cell transplantation was recently undertaken in a mouse model of Purkinje cell degeneration. In this study, iPSC-derived NEPH3+ sorted Purkinje cell precursors were transplanted into the mouse cerebellum.[39] This study showed that transplanted iPSC-derived Purkinje cell somas aligned with the mouse Purkinje cell layer and formed complex dendritic branches in the cerebellum of severe combined immunodeficient (SCID) mice after 4 weeks. These cells differentiated into L7-positive Purkinje cells, which suggested that the progenitors injected had been patterned to the correct cell lineage.[39] In line with this study, research from Muguruma and colleagues showed that after transplantation, mouse ESC-derived Purkinje cells integrated in the embryonic mouse cerebellum, and after 4 weeks of transplantation the GAD-GFP-labeled Purkinje cells formed complex branches in the Purkinje cell layer and expressed Purkinje cell-specific markers, GRID2, CALB, L7, and VGLUT2.[13] These studies indicated that PSC-derived Purkinje cells have potential to be used for cellular transplantation experiments and studies of cerebellum degeneration in vivo. In the future, optimization of these transplantations methods into mouse models of cerebellar ataxia or Tuberous Sclerosis Complex would facilitate behavioral assessments of these animal models that have loss of Purkinje cells in the cerebellum. These animal studies could also provide answers to whether iPSC or ESC-derived Purkinje cells could be safely and effectively used for treatment of neurodegenerative and neurodevelopmental disorders in the future.

To generate Purkinje cells for transplantation therapies, protocols need to be established that do not rely on co-culture with rodent granule cells. With this in mind, researchers have developed a protocol that generates both Purkinje cells and cerebellar granule neurons from human iPSCs in 3D-suspension aggregate cultures.[18] This method produced up to 10%–20% of CORL2(SKOR2)-positive Purkinje cell precursor cells at day 50.[18] In addition, the cell populations included immature neural precursor cells, which were positive for PAX6 and negative for MAP2. Also, 46% of the cells were positive for the glutamatergic granule neuron marker, BARHL1, at day 50. Neurogranin and PAX2-positive Golgi cells were also detected, as well as GABAergic interneurons that were negative for CALB. At day 80, a small percentage of the cells were positive for mature Purkinje cell markers,

CALB, L7/PCP2 and LHX5. After 80 days of differentiation, GFAP-positive astrocytes were also detected in the cultures.[18] Although this method did not require co-culturing of the cells with mouse granule cells or cerebellar slides, the capacity of this protocol to generate human Purkinje cells for transplantation is currently inefficient. For cell replacement therapies where a specific subpopulation of neuronal cells are needed, the heterogeneity of this protocol would not be ideal. However, this method may be useful for the generation of patient-specific cerebellar populations that can be used for preclinical drug screening efforts. In addition, cell sorting could potentially aid in purification of Purkinje cell populations for transplantation efforts from protocols such as this. To overcome the need for mouse co-cultures for differentiation of Purkinje cells in vitro, researchers have also recently developed a rapid method to generate human PSC derived excitatory granule neuron precursor cells in a cell culture.[63] In this method the ATOH1 cells were generated in 16 days in vitro by sequential addition of six factors, and the final maturation to excitatory cerebellar granule neurons was verified by transplantation of these cells into mice cerebellum.[63] In the future, establishment of these protocols for co-culturing of the human PSC derived granule neurons with human PSC derived Purkinje cells will enhance the understanding of human cerebellum development in more detail, and facilitate the development of cell therapies for cerebellum degeneration disorders.

In addition to these methodological requirements and advances, there are many safety considerations that need to be implemented for generation of human pluripotent stem cell-derived Purkinje cells for transplantation therapies. These include control of the cell grafts possibility to over proliferate, form tumors, cause toxicity, or bacterial or viral infections to host tissues. First and foremost, validated and quality-controlled cell manufacturing processes must be established according to good manufacturing practices (GMP). This includes the optimization and writing of standard operation procedures (SOPs) for cell differentiation and quality control of all processes in cell culture. For the differentiation procedures, all reagents need to be traceable, validated and GMP level, and each differentiation batch needs to be precisely recorded, quality controlled, and traceable. Cell production in GMP facilities also includes regular internal and external audits of the cell manufacturing processes. With these controls in place, the management of errors is highly controlled, and the cell transplantation therapies could be safely developed to treat patients.[64]

Drug screening with pluripotent stem cell-derived Purkinje cells

One strategy that has more recently been implemented in neuroscience drug discovery is to identify molecular mechanisms responsible for cellular phenotypes using phenotypic screens. These screens can use a well-annotated libraries or libraries with compounds already approved by the Food and Drug Administration (FDA) or European Medicines Agency (EMA). These libraries have been widely used in phenotypic screens against various neurological diseases in PSC-derived neurons. These studies included the identification of treatments for zika virus,[65] Alzheimer's disease,[66,67] and Amyotrophic Lateral Sclerosis.[68] Phenotypic drug screening platforms are often developed using high-throughput automated cellular imaging analyses methods. As an example, neurons plated on 96-well plates can be readily imaged by automated machines, such as the ImageXpressMicro-Confocal (IXM-C,

Molecular Devices) or the Arrayscan XTI (Cellomics, Thermo Fisher Scientific).[69,70] Phenotypes developed for screening may include examination of cells with a specific expression profile, expression of a specific protein, phosphorylation of a specific protein, or loss of expression of a specific protein, which can be detected with antibody staining. In addition, localization of target proteins to different subcellular regions can be utilized for assay development. Morphological examination of neurons from patient-derived iPSCs is also a common phenotypic endpoint. Small molecule libraries can then be used to treat these neurons to identify compounds that rescue the patient-specific cellular phenotypes with the image-based outcome measures. Identification of hit compounds may elucidate the molecular pathways involved in disease pathogenesis and can be further studied in downstream assays including functional MEA[71] and calcium imaging assays[72] as previously described, or western blots, or other biochemical methods such as mitochondrial function assays, or metabolic analyses assays.[69,70,73] If network activity can be rescued with a specific compound, then more detailed functional assays can be used to verify the functionality of the synapses and excitability of the cells with patch clamping of the drug treated iPSC-derived Purkinje cells versus vehicle-treated iPSC-derived Purkinje cells.[14] In addition to this, automated patch clamping methods are available for high throughput analyses of individual neurons that are treated with different compounds.[72] Selected compounds should also be tested in animal models of cerebellar degenerative diseases to study their safety, toxicity, efficacy, optimal dosing, and behavioral effects in animals.

In summary, although there are several challenges facing the development of cell transplantation therapies for treatment of cerebellum degeneration, current research efforts are aimed at making this possible in the future. Moreover, development of human iPSC-derived Purkinje cells from patients with cerebellar degenerative diseases for drug screening efforts will allow researchers to screen novel compounds that could enhance the survival of the Purkinje cells or granule neurons. Although these research methods are challenging, the continued improvement of these protocols will establish new methods for studying cerebellar diseases and will lead to improved treatments of these devastating neurodegenerative disorders. Strategies such as these can also be applied to the development of patterning-based protocols for other specific neuronal subtypes.

Acknowledgments

The schematic pictures and illustrations to this chapter were made by Miss Samantha Hasan and Dr. Sundberg. Dr. Sundberg has received funding from the Rosamund Stone Zander Translational Neuroscience Center, Tommy Fuss Foundation for Neuropsychiatric Research and Boston Children's Hospital, and National Ataxia Foundation.

References

1. Harada H, Sato T, Nakamura H. Fgf8 signaling for development of the midbrain and hindbrain. *Dev Growth Differ*. 2016;58(5):437–445. https://doi.org/10.1111/dgd.12293.
2. Sato T, Joyner AL. The duration of Fgf8 isthmic organizer expression is key to patterning different tectal-isthmo-cerebellum structures. *Development*. 2009;136(21):3617–3626. https://doi.org/10.1242/dev.041210.
3. Green MJ, RJT W. Developmental origins of diversity in cerebellar output nuclei. *Neural Dev*. 2014;9(1). https://doi.org/10.1186/1749-8104-9-1.

4. Butts T, Green MJ, RJT W. Development of the cerebellum: simple steps to make a 'little brain. *Development*. 2014;141(21):4031–4041. https://doi.org/10.1242/dev.106559.
5. Leto K, Arancillo M, Becker EBE, et al. Consensus paper: cerebellar development. *Cerebellum*. 2016;15(6):789–828. https://doi.org/10.1007/s12311-015-0724-2.
6. Marzban H, Del Bigio MR, Alizadeh J, Ghavami S, Zachariah RM, Rastegar M. Cellular commitment in the developing cerebellum. *Front Cell Neurosci*. 2015;8:1–26. https://doi.org/10.3389/fncel.2014.00450.
7. Millen KJ, Gleeson JG. Cerebellar development and disease. *Curr Opin Neurobiol*. 2008;18(1):12–19. https://doi.org/10.1016/j.conb.2008.05.010.
8. Selvadurai H, Mason J. Wnt/β-catenin signalling is active in a highly dynamic pattern during development of the mouse cerebellum. *Plos One*. 2011;6(8), e23012. https://doi.org/10.1371/journal.pone.0023012.
9. Haldipur P, Dang D, Millen KJ. Embryology. *Handb Clin Neurol*. 2018;154:29–44. https://doi.org/10.1016/B978-0-444-63956-1.00002-3.
10. Gibbs H, Chang-Gonzalez A, Hwang W, Yeh A, Lekven A. Midbrain-hindbrain boundary morphogenesis: at the intersection of Wnt and Fgf signaling. *Front Neuroanat*. 2017;11:64. https://doi.org/10.3389/fnana.2017.00064.
11. Buchholz DE, Carroll TS, Kocabas A, et al. Novel genetic features of human and mouse Purkinje cell differentiation defined by comparative transcriptomics. *Proc Natl Acad Sci USA*. 2020;117(26):15085–15095. https://doi.org/10.1073/pnas.2000102117.
12. Muguruma K, Nishiyama A, Kawakami H, Hashimoto K, Sasai Y. Self-organization of polarized cerebellar tissue in 3D culture of human pluripotent stem cells. *Cell Rep*. 2015;10(4):537–550. https://doi.org/10.1016/j.celrep.2014.12.051.
13. Muguruma K, Nishiyama A, Ono Y, et al. Ontogeny-recapitulating generation and tissue integration of ES cell-derived Purkinje cells. *Nat Neurosci*. 2010;13(10):1171–1180. https://doi.org/10.1038/nn.2638.
14. Sundberg M, Tochitsky I, Buchholz DE, et al. Purkinje cells derived from TSC patients display hypoexcitability and synaptic deficits associated with reduced FMRP levels and reversed by rapamycin. *Mol Psychiatr*. 2018;23(11):2167–2183. https://doi.org/10.1038/s41380-018-0018-4.
15. Johansson BM, Wiles MV. Evidence for involvement of activin A and bone morphogenetic protein 4 in mammalian mesoderm and hematopoietic development. *Mol Cell Biol*. 1995;15(1):141–151. https://doi.org/10.1128/MCB.15.1.141.
16. Bastianelli E. Distribution of calcium-binding proteins in the cerebellum. *Cerebellum*. 2003;2(4):242–262. https://doi.org/10.1080/14734220310022289.
17. Watson LM, Wong MMK, Vowles J, Cowley SA, Becker EBE. A simplified method for generating Purkinje cells from human-induced pluripotent stem cells. *Cerebellum*. 2018;17(4):419–427. https://doi.org/10.1007/s12311-017-0913-2.
18. Silva TP, Bekman EP, Fernandes TG, et al. Maturation of human pluripotent stem cell-derived cerebellar neurons in the absence of Co-culture. *Front Bioeng Biotechnol*. 2020;8. https://doi.org/10.3389/fbioe.2020.00070.
19. Lauder JM. Effects of early hypo- and hyperthyroidism on development of rat cerebellar cortex. IV. The parallel fibers. *Brain Res*. 1978;142(1):25–39. https://doi.org/10.1016/0006-8993(78)90174-9.
20. Nicholson JL, Altman J. The effects of early hypo- and hyperthyroidism on the development of the rat cerebellar cortex. II. Synaptogenesis in the molecular layer. *Brain Res*. 1972;44(1):25–36. https://doi.org/10.1016/0006-8993(72)90363-0.
21. Vincent J. Effects of thyroid hormone on synaptogenesis in the molecular layer of the developing rat cerebellum. *J Physiol*. 1982;(8):729–738.
22. Heuer H, Mason CA. Thyroid hormone induces cerebellar Purkinje cell dendritic development via the thyroid hormone receptor α1. *J Neurosci*. 2003;23(33):10604–10612. https://doi.org/10.1523/jneurosci.23-33-10604.2003.
23. McAllister AK, Lo DC, Katz LC. Neurotrophins regulate dendritic growth in developing visual cortex. *Neuron*. 1995;15(4):791–803. https://doi.org/10.1016/0896-6273(95)90171-X.
24. Schwartz PM, Borghesani PR, Levy RL, Pomeroy SL, Segal RA. Abnormal cerebellar development and foliation in BDNF(-/-) mice reveals a role for neurotrophins in CNS patterning. *Neuron*. 1997;19(2):269–281. https://doi.org/10.1016/S0896-6273(00)80938-1.
25. Morrison ME, Mason CA. Granule neuron regulation of Purkinje cell development: striking a balance between neurotrophin and glutamate signaling. *J Neurosci*. 1998;18(10):3563–3573. https://doi.org/10.1523/jneurosci.18-10-03563.1998.

26. Cohen-Cory S, Fraser SE. Effects of brain-derived neurotrophic factor on optic axon branching and remodelling in vivo. *Nature*. 1995;378(6553):192–196. https://doi.org/10.1038/378192a0.
27. Ghosh A, Carnahan J, Greenberg ME. Requirement for BDNF in activity-dependent survival of cortical neurons. *Science*. 1994;263(5153):1618–1623. https://doi.org/10.1126/science.7907431.
28. Sawai H, Clarke DB, Kittlerova P, Bray GM, Aguayo AJ. Brain-derived neurotrophic factor and neurotrophin-4/5 stimulate growth of axonal branches from regenerating retinal ganglion cells. *J Neurosci*. 1996;16(12):3887–3894. https://doi.org/10.1523/jneurosci.16-12-03887.1996.
29. Baptista CA, Hatten ME, Blazeski R, Mason CA. Cell-cell interactions influence survival and differentiation of purified Purkinje cells in vitro. *Neuron*. 1994;12(2). https://doi.org/10.1016/0896-6273(94)90268-2.
30. Morris RJ, Beech JN, Barber PC, Raisman G. Early stages of Purkinje cell maturation demonstrated by Thy-1 immunohistochemistry on postnatal rat cerebellum. *J Neurocytol*. 1985;14(3):427–452. https://doi.org/10.1007/BF01217754.
31. Sundberg M, Jansson L, Ketolainen J, et al. CD marker expression profiles of human embryonic stem cells and their neural derivatives, determined using flow-cytometric analysis, reveal a novel CD marker for exclusion of pluripotent stem cells. *Stem Cell Res*. 2009;2(2):113–124. https://doi.org/10.1016/j.scr.2008.08.001.
32. Pruszak J, Sonntag KC, Moe HA, Sanchez-Pernaute R, Isacson O. Markers and methods for cell sorting of human embryonic stem cell-derived neural cell populations. *Stem Cell*. 2007;25(9):2257–2268. https://doi.org/10.1634/stemcells.2006-0744.
33. Hargus G, Cooper O, Deleidi M, et al. Differentiated Parkinson patient-derived induced pluripotent stem cells grow in the adult rodent brain and reduce motor asymmetry in Parkinsonian rats. *Proc Natl Acad Sci USA*. 2010;107(36):15921–15926. https://doi.org/10.1073/pnas.1010209107.
34. Sundberg M, Bogetofte H, Lawson T, et al. Improved cell therapy protocols for Parkinson's disease based on differentiation efficiency and safety of hESC-, hiPSC-, and non-human primate iPSC-derived dopaminergic neurons. *Stem Cell*. 2013;31(8):1548–1562. https://doi.org/10.1002/stem.1415.
35. Sundberg M, Andersson PH, kesson EA, et al. Markers of pluripotency and differentiation in human neural precursor cells derived from embryonic stem cells and CNS tissue. *Cell Transplant*. 2011;20(2):177–191. https://doi.org/10.3727/096368910X527266.
36. Sergaki MC, Ibáñez CF. GFRα1 regulates Purkinje cell migration by counteracting NCAM function. *Cell Rep*. 2017;18(2):367–379. https://doi.org/10.1016/j.celrep.2016.12.039.
37. Molander M, Berthold CH, Persson H, Fredman P. Immunostaining of ganglioside GD1b, GD3 and GM1 in rat cerebellum: cellular layer and cell type specific associations. *J Neurosci Res*. 2000;60(4):531–542. https://doi.org/10.1002/(SICI)1097-4547(20000515)60:4<531::AID-JNR12>3.0.CO;2-6.
38. Wood C, Williams C, Waldron GJ. Patch clamping by numbers. *Drug Discov Today*. 2004;9(10):434–441. https://doi.org/10.1016/S1359-6446(04)03064-8.
39. Wang S, Wang B, Pan N, et al. Differentiation of human induced pluripotent stem cells to mature functional Purkinje neurons. *Sci Rep*. 2015:9232. https://doi.org/10.1038/srep09232.
40. Perkins KL. Cell-attached voltage-clamp and current-clamp recording and stimulation techniques in brain slices. *J Neurosci Methods*. 2006;154(1–2):1–18. https://doi.org/10.1016/j.jneumeth.2006.02.010.
41. Yang Z, Santamaria F. Purkinje cell intrinsic excitability increases after synaptic long term depression. *J Neurophysiol*. 2016;116(3):1208–1217. https://doi.org/10.1152/jn.00369.2016.
42. Williams SR, Christensen SR, Stuart GJ, Häusser M. Membrane potential bistability is controlled by the hyperpolarization-activated current/H in rat cerebellar Purkinje neurons in vitro. *J Physiol*. 2002;539(2):469–483. https://doi.org/10.1113/jphysiol.2001.013136.
43. Obien MEJ, Deligkaris K, Bullmann T, Bakkum DJ, Frey U. Revealing neuronal function through microelectrode array recordings. *Front Neurosci*. 2015;9:423. https://doi.org/10.3389/fnins.2014.00423.
44. McConnell ER, McClain MA, Ross J, LeFew WR, Shafer TJ. Evaluation of multi-well microelectrode arrays for neurotoxicity screening using a chemical training set. *Neurotoxicology*. 2012;33(5):1048–1057. https://doi.org/10.1016/j.neuro.2012.05.001.
45. Negri J, Menon V, Young-Pearse TL. Assessment of spontaneous neuronal activity in vitro using multi-well multi-electrode arrays: implications for assay development. *eNeuro*. 2020;7(1). https://doi.org/10.1523/ENEURO.0080-19.2019.

46. Sundberg M, Pinson H, Smith RS, et al. 16p11.2 deletion is associated with hyperactivation of human iPSC-derived dopaminergic neuron networks and is rescued by RHOA inhibition in vitro. *Nat Commun*. 2021;12(1). https://doi.org/10.1038/s41467-021-23113-z.
47. Wainger BJ, Kiskinis E, Mellin C, et al. Intrinsic membrane hyperexcitability of amyotrophic lateral sclerosis patient-derived motor neurons. *Cell Rep*. 2014;7(1):1–11. https://doi.org/10.1016/j.celrep.2014.03.019.
48. Winden KD, Sundberg M, Yang C, et al. Biallelic mutations in TSC2 lead to abnormalities associated with cortical tubers in human ipsc-derived neurons. *J Neurosci*. 2019;39(47):9294–9305. https://doi.org/10.1523/JNEUROSCI.0642-19.2019.
49. Ballini M, Muller J, Livi P, et al. 1024-channel CMOS microelectrode array with 26,400 electrodes for recording and stimulation of electrogenic cells in vitro. *IEEE J Solid State Circ*. 2014;49(11):2705–2719. https://doi.org/10.1109/JSSC.2014.2359219.
50. Frey U, Sedivy J, Heer F, et al. Switch-matrix-based high-density microelectrode array in CMOS technology. *IEEE J Solid State Circ*. 2010;45(2):467–482. https://doi.org/10.1109/JSSC.2009.2035196.
51. Müller J, Ballini M, Livi P, et al. High-resolution CMOS MEA platform to study neurons at subcellular, cellular, and network levels. *Lab Chip*. 2015;15(13):2767–2780. https://doi.org/10.1039/c5lc00133a.
52. Shein-Idelson M, Pammer L, Hemberger M, Laurent G. Large-scale mapping of cortical synaptic projections with extracellular electrode arrays. *Nat Methods*. 2017;14(9):882–890. https://doi.org/10.1038/nmeth.4393.
53. Sharf T, van der Molen T, Guzman E, et al. Intrinsic network activity in human brain organoids. *bioRxiv preprint*. 2021. https://doi.org/10.1101/2021.01.28.428643.
54. Zhang Y, Pak CH, Han Y, et al. Rapid single-step induction of functional neurons from human pluripotent stem cells. *Neuron*. 2013;78(5):785–798. https://doi.org/10.1016/j.neuron.2013.05.029.
55. Sun AX, Yuan Q, Tan S, et al. Direct induction and functional maturation of forebrain GABAergic neurons from human pluripotent stem cells. *Cell Rep*. 2016;16(7):1942–1953. https://doi.org/10.1016/j.celrep.2016.07.035.
56. Yang N, Chanda S, Marro S, et al. Generation of pure GABAergic neurons by transcription factor programming. *Nat Methods*. 2017;14(6):621–628. https://doi.org/10.1038/nmeth.4291.
57. Cendelin J. From mice to men: lessons from mutant ataxic mice. *Cerebellum Ataxias*. 2014;1.
58. Ishida Y, Kawakami H, Kitajima H, et al. Vulnerability of Purkinje cells generated from spinocerebellar ataxia type 6 patient-derived iPSCs. *Cell Rep*. 2016;17(6):1482–1490. https://doi.org/10.1016/j.celrep.2016.10.026.
59. Nuryyev RL, Uhlendorf TL, Tierney W, et al. Transplantation of human neural progenitor cells reveals structural and functional improvements in the spastic Han-Wistar rat model of ataxia. *Cell Transplant*. 2017;26(11):1811–1821. https://doi.org/10.1177/0963689717723637.
60. Sundberg M, Sahin M. Cerebellar development and autism spectrum disorder in tuberous sclerosis complex. *J Child Neurol*. 2015;30(14):1954–1962. https://doi.org/10.1177/0883073815600870.
61. Tsai PT, Hull C, Chu Y, et al. Autistic-like behaviour and cerebellar dysfunction in Purkinje cell Tsc1 mutant mice. *Nature*. 2012;488(7413):647–651. https://doi.org/10.1038/nature11310.
62. Tsai PT, Rudolph S, Guo C, et al. Sensitive periods for cerebellar-mediated autistic-like behaviors. *Cell Rep*. 2018;25(2):357–367. https://doi.org/10.1016/j.celrep.2018.09.039. e4.
63. Behesti H, Kocabas A, Buchholz DE, Carroll TS, Hatten ME. Altered temporal sequence of transcriptional regulators in the generation of human cerebellar granule cells. *Elife*. 2021;10. https://doi.org/10.7554/eLife.67074.
64. Tabar V, Studer L. Pluripotent stem cells in regenerative medicine: challenges and recent progress. *Nat Rev Genet*. 2014;15(2):82–92. https://doi.org/10.1038/nrg3563.
65. Zhou T, Tan L, Cederquist GY, et al. High-content screening in hPSC-neural progenitors identifies drug candidates that inhibit zika virus infection in fetal-like organoids and adult brain. *Cell Stem Cell*. 2017;21(2):274–283. https://doi.org/10.1016/j.stem.2017.06.017. e5.
66. Kondo T, Imamura K, Funayama M, et al. iPSC-based compound screening and in vitro trials identify a synergistic anti-amyloid β combination for Alzheimer's disease. *Cell Rep*. 2017;21(8):2304–2312. https://doi.org/10.1016/j.celrep.2017.10.109.
67. Wang C, Najm R, Xu Q, et al. Gain of toxic apolipoprotein E4 effects in human iPSC-derived neurons is ameliorated by a small-molecule structure corrector article. *Nat Med*. 2018;24(5):647–657. https://doi.org/10.1038/s41591-018-0004-z.
68. Burkhardt MF, Martinez FJ, Wright S, et al. A cellular model for sporadic ALS using patient-derived induced pluripotent stem cells. *Mol Cell Neurosci*. 2013;56:355–364. https://doi.org/10.1016/j.mcn.2013.07.007.
69. Behne R, Teinert J, Wimmer M, et al. Adaptor protein complex 4 deficiency: a paradigm of childhood-onset hereditary spastic paraplegia caused by defective protein trafficking. *Hum Mol Genet*. 2020;29(2):320–334. https://doi.org/10.1093/hmg/ddz310.

70. Di Nardo A, Lenoel I, Winden K, et al. Phenotypic screen with TSC-deficient neurons reveals Heat-shock machinery as a druggable pathway for mTORC1 and reduced cilia. *Cell Rep*. 2020;31(12).
71. Odawara A, Matsuda N, Ishibashi Y, Yokoi R, Suzuki I. Toxicological evaluation of convulsant and anticonvulsant drugs in human induced pluripotent stem cell-derived cortical neuronal networks using an MEA system. *Sci Rep*. 2018;8(1). https://doi.org/10.1038/s41598-018-28835-7.
72. Obergrussberger A, Friis S, Brüggemann A, Fertig N. Automated patch clamp in drug discovery: major breakthroughs and innovation in the last decade. *Expet Opin Drug Discov*. 2021;16(1):1–5. https://doi.org/10.1080/17460441.2020.1791079.
73. Cooper O, Seo H, Andrabi S, et al. Pharmacological rescue of mitochondrial deficits in iPSC-derived neural cells from patients with familial Parkinson's disease. *Sci Transl Med*. 2012;4(141). https://doi.org/10.1126/scitranslmed.3003985.

CHAPTER

6

Brain organoids: models of cell type diversity, connectivity, and disease phenotypes

Madison R. Glass[1], *Felix A. Kyere*[1], *Deborah L. French*[2,3], *Jason L. Stein*[1] and *Elisa A. Waxman*[3]

[1]Department of Genetics & UNC Neuroscience Center, University of North Carolina at Chapel Hill, Chapel Hill, NC, United States; [2]Department of Pathology and Laboratory Medicine, The Children's Hospital of Philadelphia, Philadelphia, PA, United States; [3]Center for Cellular and Molecular Therapeutics, The Children's Hospital of Philadelphia, Philadelphia, PA, United States

Introduction

An ongoing challenge in neuroscience research has been to create a representative model system that recapitulates the complexity of the human brain. The advent of the embryonic stem cell (ESC) and induced pluripotent stem cell (iPSC) technologies has revolutionized the neuroscience field by providing a consistent source of human cells that can self-replicate, be genetically manipulated, and be differentiated into neural cells (Table 6.1). Brain organoids, which are neural cells differentiated from pluripotent stem cells (PSCs) to form a three-dimensional (3D) structure, have recently been developed to provide the novel benefits of in vitro modeling of 3D cytoarchitecture for the study of additional phenotypes thought to underlie central nervous system (CNS) disorders. Brain organoids have the potential to overcome the limitations of other model systems, such as animal models, which lack the genetic and cytoarchitecture complexity of human brains, and primary fetal cell cultures, which for political reasons are difficult to obtain.

Rodent models have been the most commonly used system in neuroscience preclinical research, because they can be genetically or chemically manipulated and have representative cytoarchitecture, neuronal function, and quantifiable behaviors. However, the translation of

Phenotyping of Human iPSC-derived Neurons
https://doi.org/10.1016/B978-0-12-822277-5.00005-5

TABLE 6.1 Terminology for three-dimensional neuronal differentiation.

Term	Definition
Embryonic stem cells (ESC)	Pluripotent stem cells cultured from the inner cell mass of the blastocyst
Induced pluripotent stem cells (iPSC)	Pluripotent stem cells reprogrammed from somatic cells by overexpression of sets of transcription factors
Human pluripotent stem cells (hPSC)	Includes both ESC and iPSC
2D culture	Cells plated in a monolayer
Embryoid bodies (EBs)	Clustered cells grown in suspension
Spheroids[a]	3D spherical culture of cells that may display some structure
Organoids[a]	3D spherical culture of cells that displays some architecture of a particular brain region
Assembloids	The assembly or fusion of multiple organoid or spheroid models
Neurosphere[a]	3D spherical culture derived from a suspension of neural progenitor cells

[a]*These terms are sometimes used interchangeably in the literature, such that there is an undefined continuum between these definitions.*

these animal studies to human brain conditions has been limited.[1] Rodent models lack the ability to adequately model human-specific genetic variation, including that of non-coding regions of the genome that are important for CNS disorder risk,[2] human cellular complexity (e.g., minimal outer radial glia), and gross structure (e.g., lack of gyri).[3,4] The inability of animal models to fully recapitulate human brain structure, function, and development has led to the need for an accessible, high fidelity, and high throughput model system of the human brain.[5]

Primary neural cultures have been generated from human fetal tissue, providing a model system with a human genetic background.[6] However, political influences have limited access to this source of material, stunting research in human brain development.[7] Human pluripotent stem cell (hPSC)-derived neural cultures have the benefit of providing a human genetic background with greater accessibility. Additionally, these cells can be generated from individual patients with extensive phenotypic characterization of their disease and genetically modified to create isogenic lines. Great progress has been made in differentiating hPSC cultures to neural cell types; however, the majority of neural differentiation protocols use 2D adherent cultures that exhibit abnormal migratory dynamics, morphology, maturation, and limited connectivity.[8]

Organoids provide a 3D model of in vivo brain development, recapitulating aspects of neuronal subtype variability, interconnectivity, and cytoarchitecture. These models replicate facets of human brain development, including regional specificity, cell type heterogeneity, and structural organization. However, organoids are an imperfect representation of brain development with challenges in reproducibility and fidelity.[9,10] Nevertheless, organoids are a useful tool for investigations into developmental CNS disorders, such as microcephaly, autism, and epilepsy. As a rapidly developing technology, organoids provide potential for additional CNS disorder modeling in the future.

This chapter will discuss the development of brain organoids as a model system and specific brain organoid protocols generated to-date, focusing predominantly on human cortical or cerebral organoids (hCOs), which represent the majority of publications at the time of this writing. For the purposes of this chapter, our discussion is limited to hPSC-derived organoids and some spheroid models, which contain 3D architecture similar to fetal brain development, excluding the more simplistic embryoid body (EB) models (Table 6.1). We will further discuss neuronal activity and connectivity in organoid models, non-neuronal cell types that may be included in organoids to provide a more in vivo-like environment, and the use of organoids to model brain disorders. We will additionally present some mechanisms for evaluating organoid fidelity and reproducibility, which should be considered during project design and throughout project development and execution. We anticipate the future of brain organoids as playing a vital role in neuroscience and CNS disease research.

Cerebral organoids

Human cortical organoids (hCOs) represent the majority of 3D brain model systems published, to-date. These organoid models primarily produce neural progenitor cells (NPCs) and immature excitatory, glutamatergic neurons of the dorsal forebrain in an organization that mimics primary fetal cortex development.[11] Although organoid model systems exist for other brain regions (see the Section "Other brain region specific organoids"), cortical deficits are thought to underlie many CNS disorders, elevating the importance of modeling this region.[12,13] This section will describe human corticogenesis, hCO differentiation, and the fidelity of these in vitro cell types to their in vivo counterparts to better understand hCO models of CNS disorders.

Human corticogenesis overview

To create a high fidelity model system of human cortical development, it is first necessary to understand stereotypical corticogenesis. Corticogenesis occurs in a series of transient laminar structures that develop from neuroepithelial cells and lay the foundation for organizing cortical cell types and circuitry (Fig. 6.1).[3] Tightly packed radial glial cells form the ventricular zone (VZ) at the innermost ventricular surface and extend radial glia fibers to the pial surface. Radial glia proliferation generates additional progenitors or neurons. The first neurons migrate along radial glia fibers and form the preplate.[14] With radial glia proliferation and differentiation into intermediate progenitors and outer radial glial cells, the ventricular zone grows into the subventricular zone (SVZ) and eventually the outer subventricular zone (oSVZ).[15–17] The outer subventricular zone is greatly expanded in humans relative to lower species including rodents[3] and is hypothesized to underlie the expansion of human cortical surface area.[18,19] Simultaneously, the preplate splits to form the subplate and the marginal zone. The marginal zone contains Cajal–Retzius cells and borders the pial surface.[20] The subplate, containing migrating excitatory neurons, invading

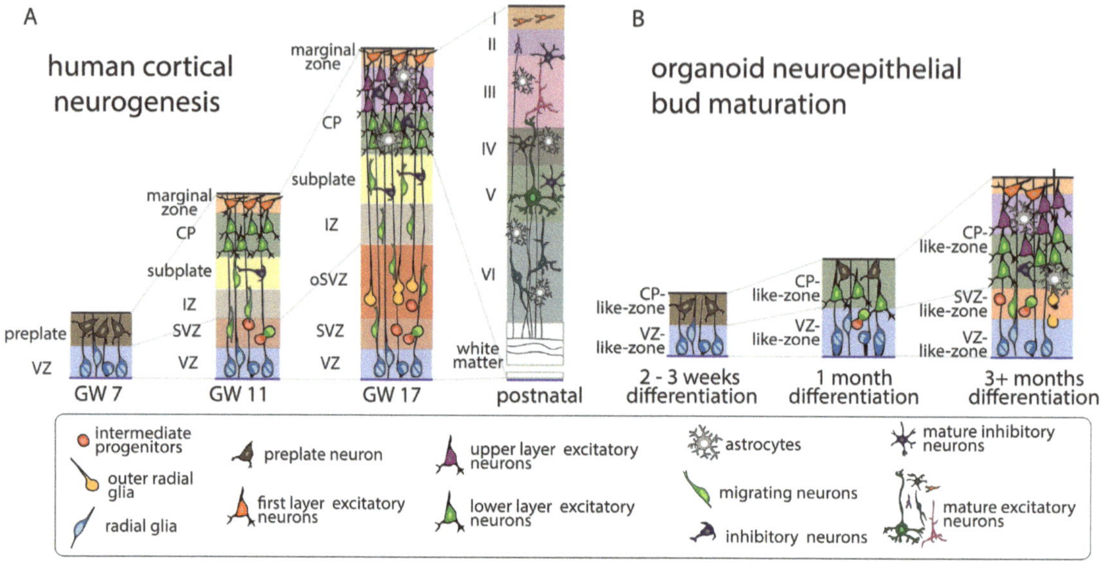

FIGURE 6.1 **Structural and Cell Type Comparison of Human Cortical Organoids to Human Fetal Cortex.** Representative timeline of (A) human fetal corticogenesis shown on the left, and (B) organoid maturation shown on the right. Approximate developmental time points are indicated. (A) In corticogenesis, early preplate formation splits into the marginal zone and subplate, with the cortical plate (CP) developing in between. The VZ expands into the SVZ and later the oSVZ. At the end of gestation, the CP contains the six layers of the cortex, and white matter areas develop. (B) hCO development imperfectly mimics in vivo corticogenesis. hCOs are organized into CP-like and VZ-like structures with simplified, less segregated laminae. Neural progenitors and both upper and lower layer excitatory neurons are present in hCOs. However, oSVZ and outer radial glia are underrepresented, and some transient structures are not reported in organoids. Mature hCOs best model mid-fetal development, but contain astrocytes after extended differentiations, consistent with later development. VZ (ventricular zone). SVZ (subventricular zone). oSVZ (outer subventricular zone). IZ (intermediate zone). CP (cortical plate). GW (gestational week). *No Permission Required.*

interneurons, and thalamocortical projections, influences early circuitry formation.[12,21–23] Newly born neurons migrate through the intermediate zone to the cortical plate, which develops between the marginal zone and subplate. The cortical plate forms in an inside-out fashion, such that the youngest neurons migrate to the outermost layers below the marginal zone. Late in gestation, radial glia terminally differentiate into astrocytes that are distributed throughout the cortex.[24,25] The goal of hCO modeling is to recapitulate with high fidelity each of these aspects of in vivo human cortical development.

Organoid differentiation overview

Shortly following the discovery of human iPSC technology,[26,27] methods were modified to efficiently differentiate hPSCs to neural lineages.[28,29] Early protocols differentiated ESCs as free floating clusters, termed EBs, followed by replating as monolayer neural cultures.[30–32] However, by maintaining EBs in suspension, 3D cellular structure and connectivity allowed for the formation of organoids.[11,29] The first hPSC organoid protocols were based on mouse

ESC 2D differentiations protocols to generate neural populations, but were optimized to create hCOs.[33,34] Further protocol improvements have provided reproducibility and laminar organization of hCOs (Fig. 6.2). Studies have also directly compared organoids to primary tissue and include more thorough characterizations of cell type composition.[35–38]

Cerebral organoid differentiations can generally be described in three steps: induction, differentiation, and maturation, each of which model important transitions that occur during in vivo brain development.[39] In the Induction step, neuroepithelial cells are generated in 3D EB structures by the addition of specific sets of morphogens, growth factors, and/or small molecules (designated as directed differentiation) or through spontaneous cell-to-cell interactions (designated as undirected differentiation). This step can take 5–18 days.[29,40,41] In the differentiation step, the neuroepithelial cells differentiate to NPCs that self-renew and potentially acquire regional identities through the addition of growth factors, embedding in extracellular matrix like Matrigel, or supplementing with serum.[11,40,41] Differentiation can take between 4 and 18 days and includes early neurogenesis.[10,42] The Maturation step is the last and longest step during which cortical wall-like structures form. The hCOs mature for at least 3 months to generate neurons with spontaneous action potentials, but cultures can be kept indefinitely, with the longest published hCO maturation lasting 2 years.[37,40,43] Protocols vary widely in their strategies to support maturation, ranging from specialized equipment like spinning bioreactors to aid with media diffusion,[10,11,41,44] growth factors to promote maturation,[40] extracellular matrix to support structural organization,[41,45] and hyperoxic conditions to reduce cell death in long-term cultures.[37]

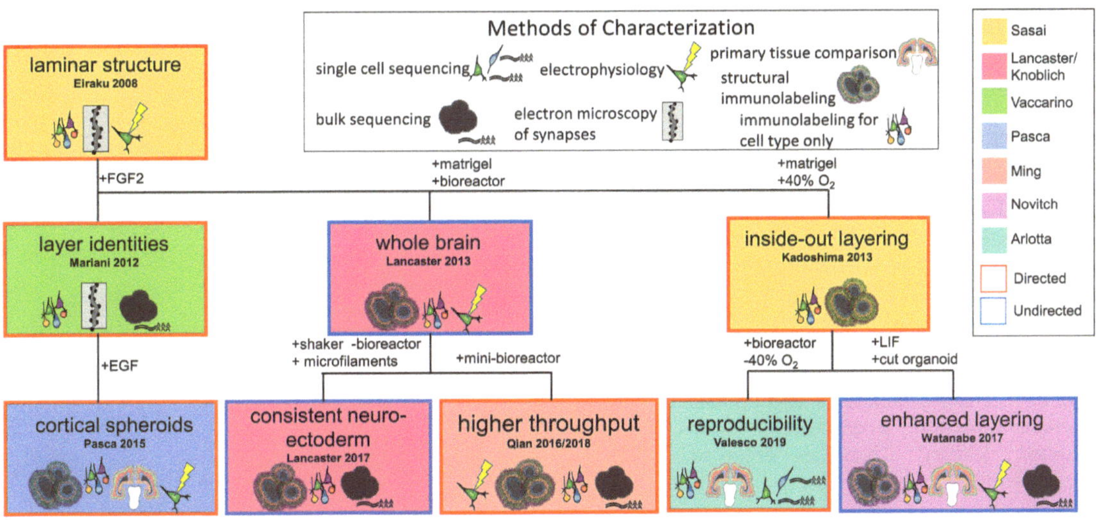

FIGURE 6.2 **Lineage Tree of Selected Cerebral Organoid Protocols.** Each box represents papers published of hCO protocols by different research groups (fill color) and are classified as using either directed or undirected differentiation (outline color). A major innovation of the publication is indicated as the header of each box and first author, year is indicated underneath. Methods of characterization used in each publication are provided in each of the boxes using the provided key. All protocols represented here used hPSCs except for Eiraku 2008, which used both mouse and human ESCs. Lineage of protocols was determined by the protocol cited as the preceding study in the indicated publication. Major changes are indicated with the lineage lines with "+" for adding a component or "−" for removing a component. *No Permission Required.*

Different induction steps can influence the cell types produced in hCOs, and can be broadly defined into two categories of protocols. Directed differentiation protocols use dual-SMAD inhibitors to prevent the generation of endoderm and mesoderm and enhance the generation of neuroectoderm by inhibiting bone morphogenic protein (BMP) and transforming growth factor (TGF) pathways.[28] Some protocols additionally add a WNT inhibitor to aid in dorsal specification.[36,38] Conversely, undirected protocols use serum-free conditions and no additional signaling factors, but rely on cell-to-cell signaling to induce a neuroectoderm fate (Fig. 6.3A and B).[10,11,37,45] Undirected protocols generally produce a more heterogeneous mix of cells, sporadically including regions of ventral forebrain, hindbrain, choroid plexus, and immature retina, as well as mesodermal cells.[10,45] Increased cell type diversity can be a strength, enabling the production of multiple cell types and connectivity within the same experimental structure. The Lancaster protocol[11,45] has been widely adapted for this reason to allow study of a range of cortical and non-cortical cell types.[11,46–50] On the other hand, increased cell type diversity can create high variability in cell type composition across organoids, making assessment and comparisons difficult between donors, operators, and/or laboratories.[10,36]

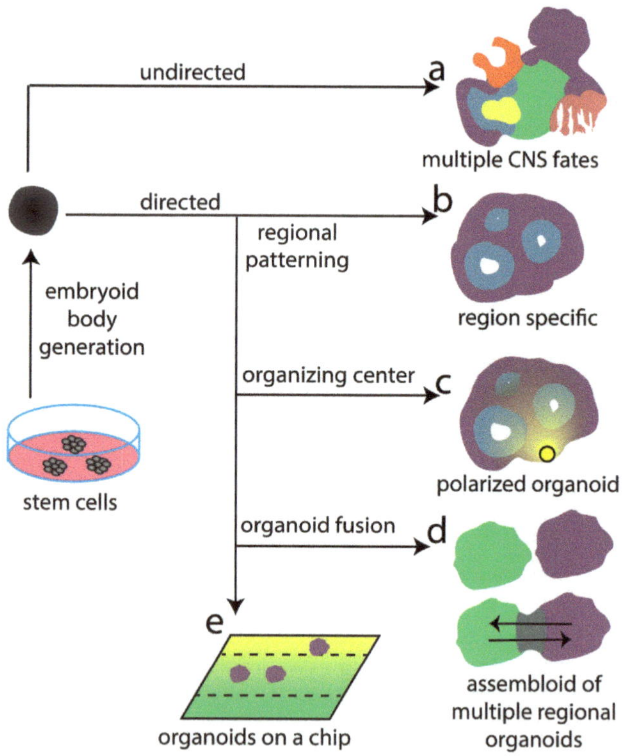

FIGURE 6.3 **Types of organoid models.** Organoids are made from stem cells that go through embryoid body generation and then one of the indicated broad protocols. (A) Undirected organoids do not use exogenous signaling molecules and instead rely on intrinsic cell-cell signaling to create CNS cell types and organization. (B) Directed organoids use exogenous signaling molecules and can produce more homogeneous cultures with regional identities. (C) Polarizing centers are created with the addition of a subpopulation of cells providing signaling molecules that diffuse through an organoid. (D) Multiple organoids generated as in (B) can be fused together to create an "assembloid" for more complex connectivity and modeling of multiple regions. (E) A microfluidic device can provide signaling molecules and diffusion of nutrients providing a mechanical means for directed delivery across an organoid. *No Permission Required.*

Fidelity of hCO cell types and organization

Prior to use in disease modeling, organoids must be evaluated for cellular, organizational, and functional fidelity relative to the developing in vivo brain. The expression of key transcription factors, intermediate filaments, and cell surface markers can be tested within hCOs to identify specific cell types (Table 6.2). Immunolabeling is commonly used for cell type identification, but antibody specificity, the lack of negative controls, and insufficient quantification can hinder experimental interpretation.[51,52] The analysis of multiple cell specific markers and inclusion of negative controls, such as undifferentiated iPSCs, provide increased confidence in immunolabeling-based cell type identification. More recently, single cell RNA sequencing (scRNAseq) studies classify cells based on the expression of many genes, which is useful for distinguishing cell types.[9,36,38,39,43] Publicly available data sets of bulk and single cell transcriptomes from the human brain are available to benchmark organoid fidelity.[9,53,54] However, the ability of single cell transcriptomic data to reflect the cell type composition is dependent on the number of genes detected per cell and the number of cells sequenced.[55]

Generally, cell type composition of hCO includes all three types of excitatory neural progenitors and immature excitatory upper- and lower-layer neurons, depending on the maturity of the hCOs.[56] A direct comparison of 37 organoids from three different hCO protocols to primary tissue using scRNAseq demonstrated overlapping gene expression profiles for excitatory neurons and radial glia, which has been corroborated by other studies.[57,58] However, organoids also contain small populations of "unknown" cell types without identifiable counterparts, in vivo, possibly due to non-specific differentiation,[9] and organoids vary in cell type composition due to technical differences between protocols. Less directed hCO protocols can also stochastically produce cells from many regions of the brain, making defined cell-cell interactions difficult to study reliably, and impacting the reproducibility of hCOs (see the section "Reproducibility").

Organoids can be further evaluated for fidelity by comparing cellular spatial organization and function. Progenitor cell types are generally found in ventricular-like zones within organoids.[37,48,59] The development of radial glia populations is critical in hCOs because they produce intermediate progenitors and outer radial glia, as well as neurons and astrocytes within the organoid, and the radial glia fibers guide the other cell types to create cortical wall-like organization.[40,41,60,61] Organoid radial glia have similar polarity, cell adhesion, and interkinetic nuclear migration to radial glia in primary tissue.[29,35,45,62] However, some radial glia in organoids are "pan radial glia" rather than of a specific subtype or topographical organization.[9] Some protocols with expanded outer radial glia populations also increase structural fidelity by including a distinct outer subventricular-like zone.[37,41] Like primary outer radial glia, hCO outer radial glia lack apical processes making them morphologically distinct from ventricular radial glia.[11] Overall, these progenitor cell types bear strong similarities to their in vivo counterparts.

In vivo and within the hCOs, neurons are generally located in the cortical plate with inside-out organization. However, structures resembling the subplate and intermediate zone may be small or not present,[56] and distinct layering, particularly in long-term cultures, may vary between protocols.[63] Individual hCO neurons can form synapses and generate

TABLE 6.2 **Cell populations and markers in the fetal cortex and cerebral organoid models**. Cell populations observed in human cortical development and cerebral organoids and the markers frequently used to assess them. The developmental timepoint is indicated in gestational weeks (GW), which is equivalent to post-conception week + 2. The developmental timepoint provided is the first timepoint that the marker was noted by immunostaining in the developing human fetal cortex with current literature. Organoid timeline provides a window of the timepoints across studies. Development of populations by specific protocols from the labs of Pasca, Novitch, and Ming are indicated (PMIDs for specific publications are provided underneath). The most frequently utilized marker(s) is indicated under the primary marker column. Specificity of the marker to detect the specific population is indicated under "Marker may also detect ...." Additional markers may allow for independent tools to identify the population of interest. PMIDs are provided for references.

Population	Developmental timepoint[a]	Organoid timeline*	Pasca* (26005811)	Novitch* (29020636)	Ming* (27118425)	Primary marker	Marker may also detect .... (Specificity)[b]	Additional markers to use	References (PMIDs)
Proliferative radial glia	GW4	~1 week	NT	day 7	day 14	SOX2	Undifferentiated iPSC; radial glia of other brain regions	PAX6, Ki67	24097039, 19796614, 29020636
Proliferative radial glia	GW4	~1 week	day 18	day 7	day 14	PAX6	Ventral forebrain (LGE); hippocampus; some dopaminergic neurons; some astrocytes	SOX2, Ki67, LHX2	24097039, 29020636, 26295830, 24636737, 27231702, 25383901, 25882041
Early forebrain progenitors	GW5.5	~2.5 weeks	day 18	day 17	day 14	FOXG1	Throughout forebrain		25383901
Layer I: Marginal zone/Cajal-Retzius cells	GW5-7	1.5 –4 weeks	day 76	day 9	day 28	Reelin	Areas of the hippocampus[c]	p73	12077194, 10684887, 30861578, 24917793
Layer VI: Lower layer neurons	GW8-9	4–8 weeks	day 52	day 35 –56	day 28	TBR1	Layer V; hippocampal glutamatergic neurons	NURR1, FOXP2	24917793, 25882041, 20727056
Layer V: Lower layer neurons	GW10	4–8 weeks	day 40	day 56	day 28	CTIP2	Layer VI; striatal neurons;	ROBO1, SRGAP1	25383901, 25882041,

							hippocampus; subplate		20624841, 21060114
Layer IV: Upper layer neurons	GW13-14	8–14 weeks	day 76	day 98	day 56	SATB2	Layers II-III; hippocampus	FOXP1	25882041, 20624841, 21060114, 32923614
Layer III: Upper layer neurons	GW15.5	8–14 weeks	day 52 –76	day 98	day 84	BRN2	Layer II		29020636
Layer II: Upper layer neurons	~GW20	11 –22 weeks	day 76	day 154	day 84	CUX1	Layer III-IV; hippocampus	CUX2	25882041, 20624841
Outer radial glia	GW13.5	~8 weeks	NT	day 56	day 56	HOPX	Quiescent NSCs; mature astrocytes	FAM107A	26406371
Intermediate progenitors	GW10	4–8 weeks	day 52	day 35 –56	day 28	EOMES	Developing hippocampus	PPP1R17	20154730, 30861584
Astrocytes	GW17	8–24 weeks	day 56	day 84	day 100	GFAP	Radial glia[d]	S100B, ALDH1L1, Glutamine synthetase, "star like" morphology	16243597, 10344592, 28700615
OPC/ Oligodendrocytes	GW15-20 (MBP ~14 weeks post-natal)	16 –36 weeks[e]	NT	None	NT	OLIG2	Ventral, inhibitory progenitor populations; motor neurons	MBP, CNPase, OLIG3	16267213, 15736232
Excitatory neurons	GW15	6–18 weeks	day 130	day 42 –56	day 84	VGLUT1	N/A	Glutaminase, VGLUT2[f]	22684983, 26438332
Inhibitory neurons	GW11-13.5	~12 weeks[g]	NT	NT	day 84	GAD67	N/A	GAT1, GAD65, GABA	24047602, 17965125

[a] *Developmental timeline based on first time primary marker was documented in the developing neocortex by IHC analysis. Timing may vary based on cortical area.*
[b] *this column is not all-inclusive.*
[c] *also contain Cajal-Rezius cells.*
[d] *Glial markers identified as early as GW6 as radial glia (PMID: 16158418).*
[e] *primarily observed in ventral forebrain and midbrain organoids.*
[f] *may identify separate populations of neurons.*
[g] *in primarily excitatory, hCO models, GABAergic neurons may not be detected outside of some reelin-positive cells.*
NA, not available; *NT*, not tested; *none*, markers were assessed and no population was identified. *Expression timeline provided is the first timepoint noted in published studies, but markers may appear prior to the time noted.
*Expression timeline was determined based on when the population was first detected in published works. This may not be representative of when the population first arises in the hCO model.

spontaneous action potentials, suggesting some functionality.[10,37,43,62] Additional details on neuronal activity of hCOs can be found in the section "Neuronal activity and connectivity."

In summary, hCO protocols can model excitatory neurogenesis, including appropriate structural organization, some progenitor functionality, and neuronal activity. Human organoids also produce outer radial glia, a critical component of hCO fidelity and the study of disorders related to cortical organization and size. A hCO protocol demonstrating high fidelity to all aspects of human corticogenesis has not yet been demonstrated, but many protocols capture individual aspects of development, including the breadth of cell types,[10,45] organization,[35,37] gliogenesis,[64] and/or electrophysiological properties of neurons.[43] Future improvements to organoid development will include the creation of structures with more consistent cell type composition (see the section "Reproducibility"), including non-neuronal cell types (see the section "Non-neuronal cell") and increased reproducibility (see the section "Reproducibility"). With current organoid methodology, investigators can choose a protocol that generates cell types, structural organization, and specific functions that model critical biological processes to address their research questions.

Other brain region specific organoids

To appropriately model the brain regions involved in CNS disorder pathology, regional identities beyond the cerebral cortex must be adequately recapitulated within organoid model systems. The study of developmental disorders of midbrain and hindbrain,[65–67] movement disorders, and neuromotor disorders[68–70] require modeling of these brain regions and non-brain tissues, including the cell types contained therein. Even for disorders predominantly involving the dorsal forebrain, inhibitory neuron populations are underrepresented in hCO models, and excitatory/inhibitory balance is necessary for appropriate neurodevelopment and function.[71–73] Combining multiple regional identities can also model connectivities between cell populations that otherwise would not be observed. This section will review differentiations into regionalized organoids, focusing on methods of directed patterning and specification.

While dual-SMAD inhibition treatment can direct differentiation toward hCOs,[28,29] non-cortical region-specific organoids can be generated when using dual-SMAD inhibition in combination with other signaling molecules. Many protocols that produce region-specific organoids were derived from 2D protocols, designed to direct patterning of specific brain regions/cell types. At the time of this writing, protocols have been generated for dorsal forebrain, ventral forebrain, midbrain, hypothalamic, thalamic, motor, retinal, cerebellar, and spinal cord organoids (Table 6.3). Directed differentiation protocols typically use small molecules or recombinant proteins to mimic developmental signals that facilitate patterning along the dorsal-ventral or rostral-caudal axes. For instance, sonic hedgehog (SHH) signaling during development is generated by the notochord and floor plate cells to establish ventral-dorsal polarity gradients.[74] Therefore, SHH activators are often used to generate ventral forebrain organoids[37,61,71,75–78] and other ventral regions, such as hypothalamic neurons and dopaminergic midbrain neurons.[41,79–83]

TABLE 6.3 **Directed patterning with signaling molecules**. Brain region-specific organoids can be produced with the use of signaling molecules for directed patterning specification. Signaling molecules mimic dorsal-ventral or rostral-caudal axes patterning signals and/or cell-cell interactions to create specific cell types. Only patterning signals are included in the table, omitting SMAD inhibitors, general growth factors, and specifics regarding timing of treatments. Many markers are used in these studies to identify a specific population of neurons, but only a subset are provided in this table. More detailed information can be found in the references indicated. WNT (Wingless-related integration site). SHH (Sonic hedgehog). FGF (Fibroblast growth factor). TGF (Transforming growth factor). ERK (Extracellular signal-regulated kinase). BMP (Bone morphogenic protein). RA (retinoic acid). VEGF (Vascular endothelial growth factor). IGF (Insulin like growth factor). HGF (Hepatocyte growth factor). SDF (Stromal cell-derived factor). DA (dopamine). SNpc (substantia nigra pars compacta). VTA (ventral tegmental area).

Brain region	Neuron type	General method of directed patterning[a]	Important markers identified	References (PMIDs)
Dorsal forebrain	Mostly glutamatergic, some GABAergic at later stage	WNT inhibitor	Reelin, SATB2, CTIP2, TBR1, TBR2	18983967, 24277810, 29058117, 28757360, 31168097
Dorsal forebrain	Pyramidal glutamatergic neurons	SHH inhibitor	TBR1, PAX6	28504681, 31091434
Ventral forebrain	GABAergic inhibitory neuron	WNT inhibitor + SHH activator	NKX2-1, GAD1, GAD2, VGAT, DLX1, DLX2, LHX6, NR2F2	28504681, 28445465, 30202107, 31363137, 31974223
Ventral forebrain	GABAergic inhibitory neuron	SHH activators[b]	NKX2-1, ASCL1, SST, DLX2, VGAT, GABA, GAD1, GAD2	31091434, 28757360, 29020636
Mixed forebrain	Range of neuron types in gradient	Induced-SHH in focal region/ organizer	FOXG1, PAX6, GSH2, NKX2-2, NKX2-1	30936566
Ventral hypothalamus (with pituitary)	Not defined	SHH activator	RX, NKX2-1	26762480
Hypothalamus	Peptidergic	SHH activators, WNT activator	NKX2-1, NKX2-2, RAX1, FOXA2, POMC, VIP, OXT, NPY, OTP	27118425
Midbrain	Dopaminergic (mostly A9 DA neurons)	SHH activators, WNT activator, FGF-8	FOXA2, TH, DAT, PITX3	27118425
Midbrain	Dopaminergic (A9 & A10)—SNpc & VTA	WNT activator, SHH activator, FGF-8, insulin	OTX2, LMX1A, CORIN, FOXA2, MASH1, NURR1, TH, DAT	27476966
Midbrain	Dopaminergic neurons (A9 & A10). Also, glutamatergic and GABAergic neurons (Berger)	WNT activator, SHH activator, TGF-B3	TH, LMX1A, FOXA2, NURR1, DAT, DDC	30204191, 28416282

(Continued)

TABLE 6.3 **Directed patterning with signaling molecules.** Brain region-specific organoids can be produced with the use of signaling molecules for directed patterning specification. Signaling molecules mimic dorsal-ventral or rostral-caudal axes patterning signals and/or cell-cell interactions to create specific cell types. Only patterning signals are included in the table, omitting SMAD inhibitors, general growth factors, and specifics regarding timing of treatments. Many markers are used in these studies to identify a specific population of neurons, but only a subset are provided in this table. More detailed information can be found in the references indicated. WNT (Wingless-related integration site). SHH (Sonic hedgehog). FGF (Fibroblast growth factor). TGF (Transforming growth factor). ERK (Extracellular signal-regulated kinase). BMP (Bone morphogenic protein). RA (retinoic acid). VEGF (Vascular endothelial growth factor). IGF (Insulin like growth factor). HGF (Hepatocyte growth factor). SDF (Stromal cell-derived factor). DA (dopamine). SNpc (substantia nigra pars compacta). VTA (ventral tegmental area).—cont'd

Brain region	Neuron type	General method of directed patterning[a]	Important markers identified	References (PMIDs)
Thalamus	Glutamatergic neurons (vGlut2, EAAT4)	Insulin + ERK inhibitor + BMP7	OTX2, DBX1, GBX2, TCF7L2	30799279
Motor neuron	Motor neuron	RA + SHH activator + VEGF inhibitor + gamma-secretase/notch inhibitor	HB9, SMI32	29107592
neuromuscular junction	Skeletal muscle, Schwann cells, spinal cord neurons	WNT activator + bFGF + IGF1 + HGF	Motor neuron—ISLET1, CHAT, TUJ1; myocyte side—DES,MYOD, myosin; Schwann cells—S100B; spinal cord neuron—SMI32, PAX2, LHX1	31956040
Retinal	Optical cup: retinal ganglion cells, photoreceptors, cones	WNT inhibitor early, WNT activator + SHH activator late	RX, PAX6, CHX10, MITF, Rhodopsin, Recoverin, NRL	22704518
Retinal	Optical cup: retinal ganglion cells, photoreceptors, cones, ciliary margin	BMP4[c], WNT activator + VEGF inhibitor late	RX, PAX6, CHX10, MITF, Rhodopsin, Recoverin, NRL	25695148
Cerebellum	Purkinje cell + other cell type precursors in the cerebellar plate	FGF2, FGF19, SDF1 (sequential)	KIRREL2, PFT1A, SKOR2, OLIG2, ATOH1, FOXA2, NKX6.1	25640179
Spinal cord	Primary populations – Vglut2+ neurons and GABA + neurons	WNT activator, bFGF, RA[d]	NKX6.1, OLIG2, HB9, ISLET1, LHX1A, OLIG3, PAX2, BRN3A, FOXA2, VGLUT2, GAD67	30061169
Hippocampus/choroid plexus	Choroid plexus- and medial pallium-like tissues	WNT inhibitor early, WNT activator + BMP4 late	FOXG1, LEF1, LHX2, NRP2, ZBTB20, PROX1, KA1, CaMKII	26573335

[a] *Use of SMAD inhibition, general growth factors, and specifics of timing are omitted from the table.*
[b] *SHH activator (SAG): lower concentrations produced more LGE/CGE, higher concentrations produced more Nkx2-1 (MGE).*
[c] *Step-wise decrease in BMP4 concentrations.*
[d] *Increased BMP4 dorsalized; increased SAG ventralized (concentration dependent).*

Regionalization may also be produced by creating a gradient, using inducible organizers or hydrofluidic devices (Fig. 6.3C,E). An SHH organizer was created within an EB structure by combining a doxycycline-inducible SHH-expressing EB with unmodified hPSC cells, resulting in an organoid with patterning of multiple brain regions.[84] The cells nearest to the SHH organizing center expressed markers of hypothalamic development, while cells most distal expressed dorsal forebrain markers. Markers of medial and lateral ganglionic eminence (MGE/LGE) regions were also observed, showing the production of multiple cell populations by the SHH organizer with higher fidelity to the developing forebrain. Hydrofluidic or microfluidic devices have similarly been shown to provide specific directional delivery of signaling molecules, forming a gradient across an organoid.[85,86] Creative facilitation of directed patterning using these methods may provide regionalization along a continuum for recapitulation of multiple brain regions within one organoid.

Regionalized organoids have an additional ability to create a more pure population of a specific brain region that can be fused with another specific region to create a functional circuit with distinct cell types (Fig. 6.3D). Additional details regarding this type of circuitry will be discussed in a subsequent section (see the section "Neuronal activity and connectivity"). Protocols have also been designed for the hippocampus with choroid plexus organoids and the neuromuscular junction organoids, both of which can model connectivity between diverse cell types.[87,88]

There are several challenges in developing region-specific organoid protocols, beyond that of identifying specific signaling molecules and timing to achieve the appropriate cell population(s). Developmental studies on human fetal tissue are limited for many brain regions and time periods, and animal studies may not be representative of human development, making a gold-standard assessment of fidelity difficult. In addition, many markers are not specific to a single region or cell type. For instance, OLIG2 expression is an early neural marker in both ventral telencephalon neurons and the spinal cord, and, later in development, oligodendrocytes.[89,90] Identification of cell populations can be supported through appropriate use of multiple markers and negative controls (see the subsection "Fidelity of hCO cell types"). Guidelines outlined above and those provided in subsequent sections (see the section "Reproducibility") will support the advancement of region-specific organoids and provide the validation necessary for future CNS disorder research.

Non-cortical region-specific organoids have a remarkable potential for modeling features of a number of CNS disorders. The ability to generate distinct cell types by mimicking developmental signaling cues provides an expanding toolkit for future research. Although these organoids are currently less utilized as compared to hCOs, they offer expanding possibilities to generate multiple brain regions and tissues to provide models of connectivity important for multiple CNS disorders.

Neuronal activity and connectivity

Neuronal activity and connectivity imbalances are observed in multiple CNS disorders, including autism, epilepsy, and schizophrenia,[91–93] and are therefore high priority phenotypes to model within hCO cultures. A main benefit of hCOs over 2D cultures is the potential to recapitulate regional connectivity.[94] The ability of organoids to be physiologically active and connect with other cell types, including connectivity between multiple organoids ("assembloids"), additionally allows for representative modeling of in vivo neuronal systems. Organoids are uniquely capable of modeling complex phenotypes and long-range connectivity, while containing the genetic background of the donor from which they were derived. Here, we describe current methods and studies on synaptic activity and connectivity of brain organoids.

Synaptic activity

Frequently used methods to measure activity in hCOs include patch-clamping, calcium imaging, and multi-electrode array (MEA) analyses. For patch-clamping studies, slice preparations of organoids are made, similar to methods used for recording of brain slice activity with electrodes used to record ionic currents. Calcium imaging, using fluorescent calcium indicators such as Fluo-4AM, is typically less sensitive in detecting neuronal activity compared to other methods, although additional sensitivity may be achieved through viral transduction with a genetically encoded calcium indicator (GCaMP).[95] MEA recordings allow for multiple electrophysiological recordings of cells in contact with an array. MEA works well with adherent cultures, since recordings depend on localized signals. However, hCOs require methods to facilitate contact, such as coating the array with laminin, followed by several weeks recovery to allow the organoid to become semi-adherent,[43] placing a coverslip or weight on top of the organoid,[81] or attaching the organoid to a small grid of spikes.[10]

Multiple studies have shown both spontaneous and evoked neuronal activity within brain organoids. Three to 4 months into hCO culture, ~50% of patch-clamped neurons showed spontaneous action potentials that were blocked with tetrodotoxin and largely inhibited by glutamate receptor antagonists, supporting spontaneous excitatory synaptic neurotransmission.[37,41] Responses were also evoked by exogenous glutamate, NMDA (*N*-methyl-D-aspartate), or GABA (γ-aminobutyric acid).[11,96] With increased organoid maturity, spontaneous synaptic activity increased,[43,96] evoked action potentials had higher amplitude and frequency,[48] and intracellular calcium transients had a greater dependence on AMPA (α-amino-3-hydroxy-5-methyl-4-isoxazolepropionic acid) receptor activity, consistent with synaptic maturation.[59]

Excitatory GABAergic responses were also noted in younger organoids, which diminished with age, consistent with the developmental switch of GABAergic neurons from excitatory to inhibitory.[41,59] In more mature organoids, GABA receptor agonists inhibit the number of spike events and GABA receptor antagonists increase synchronized events, showing late development of inhibitory neurons in predominantly dorsal forebrain organoids.[43] This GABAergic population is necessary for the maintenance of excitatory/inhibitory balance and mature synaptic connections, similar to that observed in vivo.[72,73]

Three-month old cortical organoids exhibited similar membrane potential, resistance, and capacitance to fetal tissue at gestational weeks 16–22. However, the peak amplitude sodium (Na^+) and potassium (K^+) currents differ between the two populations.[37] Using a regression model, where preterm EEG data were used to predict gestational age in vivo, MEA recordings of in vitro organoids were then classified along the developmental trajectory of functional age. Neuronal recordings of organoids older than 25 weeks exhibited higher correlations with preterm EEG data as compared to younger cultures, while recordings of human fetal brain primary cultures, iPSC monolayers, and mouse primary cultures did not correlate with EEG data.[43] More mature cortical organoids (35 weeks) resembled 25–38 week preterm EEG readings, such that the events per hour increased with age and the mean duration of activity decreased in both datasets.[43] Although these organoid cultures do not perfectly recapitulate brain development in vivo, organoids have more synchronous activity than monolayer culture systems.

Connectivity of neuronal organoids

Supporting neuronal connectivity, synchronous activity and paired firing between neurons have also been noted in multiple brain organoid models.[37,41,43,78,81,87,97] Synaptic connectivity in organoids can be visualized through neurons that are injected with neurobiotin, a tracer capable of anterograde, retrograde, and trans-neuronal transport,[98] and measured by dual patch clamp, where evoked currents are induced in 1 cell and activity is recorded in a separate cell.[59] Similarly, long-range connectivity has been observed where hCOs were placed on either side of a hydrogel micro-column, such that fluorescently labeled axons extended 1 cm across the column and synchronous calcium activity between the distanced hCOs was detected.[97] Further, when organoids were transplanted into mice, axons grew from the organoids into the host brain, and stimulation of the organoid produced local field potentials in the host brain, showing functional connectivity and signaling.[99]

Connectivity between brain regions can also be observed through "assembloids," which are the fusion of two brain organoids that were independently generated through directed differentiation (Fig. 6.3D). A benefit of assembloids is the creation of defined brain regions or cell types, prior to fusion, which reduces variability for disease modeling and analysis.[100] This method enables a more controlled study of connectivity and functionality, facilitated by diverse neuronal populations. Assembloids of dorsal and ventral forebrain organoids facilitate migration and interconnection of GABAergic inhibitory neurons from the ventral organoid side into the dorsal side containing excitatory neurons, similar to that observed in fetal gestational week 18–20 human forebrain.[71,75,78] Interestingly, minimal migration was observed if two organoids of the same type were fused together, and migration was predominantly from the ventral into the dorsal side,[71,75] supporting specific chemotaxic signaling. Inhibitory neurons that migrated into the dorsal side had two-times the maximum action potential generation rate and more branching of processes compared to those that did not migrate.[71] With the application of electrical stimulation to the dorsal side, excitatory postsynaptic currents were evoked from inhibitory cells, followed by inhibitory postsynaptic currents,[71] supporting an integrative network. Corticothalamic projections can also be modeled with assembloids, which result in bidirectional axon projections and increased firing frequency between cortical and thalamic organoids.[78] Coupling brain organoid models such as assembloids or an inducible organizer[84] with neural tracers, such as the rabies virus, which is a retrograde tracer requiring chemically active synapses,[101] could further support long-range connectivity between diverse neuronal populations.

Neurons in organoids are additionally capable of connectivity with non-neural tissues. Neuromuscular junction organoids, created with neuromesodermal progenitors, develop motor neurons, skeletal muscle, and terminal Schwann cells, resulting in electrophysiologically active neurons that release acetylcholine and facilitate muscle contractions.[87] In addition, air-liquid interface hCOs that are co-cultured with mouse spinal cord-muscle explants resulted in bundled axons that innervate the explant and create sporadic muscle contractions.[44] High-amplitude muscle contractions were abolished upon lesioning the neuronal axon tracts, or further evoked by stimulation of the axon tracts.[44] These organoids are therefore capable of producing functional output, such as muscle activity.

These studies support iPSC-derived neuronal organoid synaptic activity with connectivity of diverse populations and functional modeling. This ability to create fusions of region-specific organoids, with an interconnected morphology and migration mimicking that in development, is a major advance for the study of neurodevelopment. However, the generation of a valid assembloid model amplifies existing concerns of fidelity, reproducibility, analysis, and interpretation of multiple cell regions and cell-cell interactions that are compounded beyond that of a single organoid model. Nevertheless, future work creating more complex assembloid models and in combination with non-neuronal cells could provide more comprehensive models for CNS research.

Non-neuronal cells

The CNS consists of cell types other than neurons and progenitors, including astrocytes, oligodendrocytes, microglia, and vasculature, but these cell types are often missing from commonly used brain organoid cultures. The interplay between neurons and these other cell types facilitate synaptic signaling, neuronal maturity, and the health of the neuronal system, and can play a role in CNS disorders.[102–104] The inclusion of these non-neuronal cells into organoids increases complexity and creates a more realistic model of neurodevelopment. Below, we discuss current studies and mechanisms to include these additional cell types, which can develop spontaneously, be facilitated, or be supplemented into an organoid model. Through their inclusion, organoids can more precisely model in vivo development of the brain.

Astrocytes

Astrocytes play a vital role in synaptic development and neuronal maturity by facilitating both synapse formation, synaptic pruning, and modulation of neuronal excitation through glutamate uptake. Dysfunctional astrocytes have been shown to contribute to neurodevelopmental disorders, such as Fragile X and Down syndrome.[102] In addition, mature astrocytes uptake glutamate in order to help regulate neuronal excitability, and lack of mature astrocytes may result in excitotoxicity in certain cultures.[105] Astrocytes are therefore an important component of hCO models.

Astrocytes develop from radial glia later in gestation, after the formation of neurons, and continue to expand during postnatal stages. Glial fibrillary acidic protein (GFAP) is a typical marker to recognize astrocytes; however, radial glia also express GFAP and other astrocyte markers,[14,106,107] making it difficult to distinguish between the two populations. In addition, GFAP expression is higher in reactive astrocytes, promoting additional markers such as S100beta and ALDH1L1 to increase specificity of astrocyte labeling.[108–110] Therefore, morphological characteristics (i.e., star-like morphology) and immunolabeling (GFAP, S100beta, and/or ALDH1L1) should be assessed concurrently to adequately support the presence of astrocyte populations in an organoid model.

Astrocytes spontaneously develop in mature brain organoid models.[10,36,37,40,41,43,81,88,96,111] The production of astrocytes characteristically increases with longer culture periods, similar to the expansion of astrocytes observed in vivo.[58,111] Using scRNAseq analysis, astrocyte populations from 150 day old hCOs show higher similarity to those in the human postnatal brain as compared to fetal astrocytes.[35] This is consistent with organoid expression profiles acquired from 80 to 250 days in culture as correlating with the range of gestational week 10 to postnatal week 0, and organoids greater than 350 days old correlating with postnatal samples.[58]

The importance of astrocytes is highlighted in a model where astrocytes do not spontaneously develop, but can be supplemented. PSC-derived astrocytes co-cultured with NGN2-inducible neurons into a cortical-like spheroid model promoted neuronal synapse formation with increased density when compared to NGN2 cultures without astrocytes.[112] This supports the function of astrocytes as facilitating neuronal activity and synaptic maturity in organoids.

Oligodendrocytes

Oligodendrocytes myelinate axons in the brain and are therefore necessary for efficient conduction of neuronal electrical signals. While comparatively few organoid studies focus on oligodendrocytes, their presence in organoid models is essential to those investigating multiple sclerosis, schizophrenia, and Alzheimer's disease.[103] In vivo, oligodendrocytes are formed after astrocytes from oligodendrocyte progenitor cells (OPCs) that are created in waves in both cortical and spinal cord regions. They are considered the most proliferative population in the CNS.[113]

Only a minority of hCOs report spontaneous oligodendrocyte development.[35,111] However, several models of ventral forebrain and midbrain organoids spontaneously develop OPCs or oligodendrocytes, noted from 2 to 5 months into culture,[71,78,80,81] consistent with SHH acting as a morphogen to produce OPCs.[103]

Oligodendrocytes may also be facilitated within hCOs with signaling molecules.[42,64,76] These organoids develop oligodendrocyte markers after approximately 100 days in culture and likely oligodendrocyte functionality, given the observations of both myelin basic protein (MBP)-positive puncta surrounding axons and electron microscopy images of axon myelination. Interestingly, astrocyte populations also increased when oligodendrocytes were promoted.[42,64] Oligodendrocytes were also shown to be electrophysiologically active;[64] however, oligodendrocyte-induced changes to neuronal function or electrophysiology have yet to be documented in these models. Nevertheless, these studies show promising advances on the inclusion of oligodendrocytes in organoid models.

Microglia

Microglia are the innate immune cells of the CNS, where they provide functions of synaptic pruning, production of cytokines, and phagocytosis.[104,114] Microglia are derived from yolk-sac progenitors early in prenatal development, and migrate into the CNS prior to the development of the blood-brain barrier (BBB), after which local proliferation distributes microglial populations throughout the CNS.[104,114] Microglia can modulate differentiation and migration of NPCs, and dysregulation of their inflammatory responses in the CNS can contribute to diseases, such as autism and Alzheimer's disease.[115] The importance of microglia in neurological processes and neuroinflammation supports their inclusion in neural organoid models.

Since microglia do not originate from neural ectoderm or other glial precursors, they do not spontaneously develop in neural organoid models. Protocols to include microglia within brain organoid models include co-culturing,[59,77,116–118] modifying hCO differentiation to promote mesodermal cells,[119] or transplantation.[99]

iPSC-derived microglia-like cells release cytokines, respond to lipopolysaccharide (LPS), and perform phagocytosis.[116] Upon addition to the medium of cerebral organoids, microglia-like cells spontaneously migrate into the organoids, distribute through the organoid, extend processes within 1 week, and take on microglial morphology.[77,116,117] Further, these microglia-like cells respond to organoid damage by changing morphology to resemble activated microglia with phagocytotic properties.[116,117] Alternatively, by promoting mesodermal cell differentiation within hCOs, the mesoderm progenitors develop expression patterns and morphology similar to microglia concurrently with the development of cortical layering.[119] These microglia are in close proximity with neuronal processes, consistent with their role in synaptic pruning, and release cytokines in response to LPS.

These studies show that microglia in organoid models provide immune and synaptic functionality that is not present in standard hCO differentiation protocols, and may be essential to model some CNS disorders.

Vascularization/nutrient distribution

Oxygen and other nutrient distribution in organoids are limited by diffusion, which becomes less effective with organoid growth and maturity. This results in toxicity, cell stress pathways, and cell death, which often evolves into a necrotic core and limits the hCO's ability to generate cell type fidelity and regional specificity.[9,35,120] Nutrient diffusion can be promoted either by mechanical means that mimic vascularization or through inducing vascularization components within organoids.

Mechanical means to promote nutrient diffusion and oxygenation include high oxygen conditions, spinning bioreactors, and slicing of the organoids, each allowing for greater extension of the outer subventricular zone and development of cortical layering.[11,35,37,41,60] Microfluidic or hydrogel chips can create a homeostatic environment for oxygen and nutrients with similar permeability to that found in vivo. These models provide greater growth and differentiation and less cell death.[79,121] Air-liquid interface or organotypic-like cultures also improve survivability and growth of organoids, despite the flattening of tissue.[44,122] These mechanisms provide successful means to overcome limitations that come with organoid growth, even though they do not provide nutrient delivery through vascular networks, such as those that develop, in vivo.

Vascularization in vivo involves a network of brain endothelial cells and pericytes in close contact with astrocytes, microglia, and neurons. This network comprises the BBB, a selectively permeable barrier that allows for the diffusion of nutrients, but prevents neurotoxic substances from entering the brain.[123] Vascular organoids (vOrganoids) can be produced by co-culturing iPSC-derived endothelial cells with hCOs or signaling molecules to promote vascularization.[124,125] These methods result in tubular-like endothelial structures and expression of BBB markers. An inducible expression system initiated during hCO differentiation can also promote vascular-like endothelial cells that reduce cell death and inflammatory markers, and increase growth, expression of BBB markers, and neuronal maturity by both electrophysiological activity and gene expression.[126] Vascularization can be visualized with uptake of fluorescein isothiocyanate-dextran into the endothelial/vascular-like tubes, which intermingle with the hCO, demonstrating a perfusable vascular like-network. With transplantation, vOrganoids connect to the mouse vasculature, supporting functionality.[126] While the above described methods do not provide a true vascular system for brain organoids, and benefits of these methods with long-term culturing require further validation, they show encouraging advances and highlight important benefits that vOrganoids can provide.

Summary of non-neuronal cells

Significant advances have been made in producing non-neuronal cells that can be incorporated into organoid models. For CNS disorder research, some non-neuronal cells contribute to the disease phenotype and are therefore important to include. Furthermore, studies suggest these non-neuronal cells increase health and maturity of neurons in the organoid model. A multifaceted system with all of the above components would provide a more realistic environment, creating an ideal model system for neurodevelopmental and disease research.[11,37,41,48]

Use of models in disease

Recent advancements in 3D brain organoids provide promising approaches for modeling complex pathological disease processes leading to CNS disorders. Brain organoids have inherent human genetic variation and complexity of cell types present in the human cortex, neither of which are modeled well in animals.[2–4] Patient-derived iPSCs differentiated into brain organoids can be used to model phenotypes, investigate potential mechanisms underlying disease, and/or study a patient's disease in the absence of a known genetic cause. Phenotypic modeling of hCOs provides benefits over similar 2D systems, since they can model 3D cytoarchitectural deficits associated with a disorder. Brain organoids have additional advantages in that controls can be created either from unaffected family members or genome-edited isogenic cell lines, reducing genetic variability between experimental groups. hCOs have been used to model multiple CNS disorders, including microcephaly, Zika virus infection, autism spectrum disorder (ASD), schizophrenia, and Alzheimer's disease (Fig. 6.4).[11,37,41,48,127,128] This section will provide an overview of how hCOs are being used

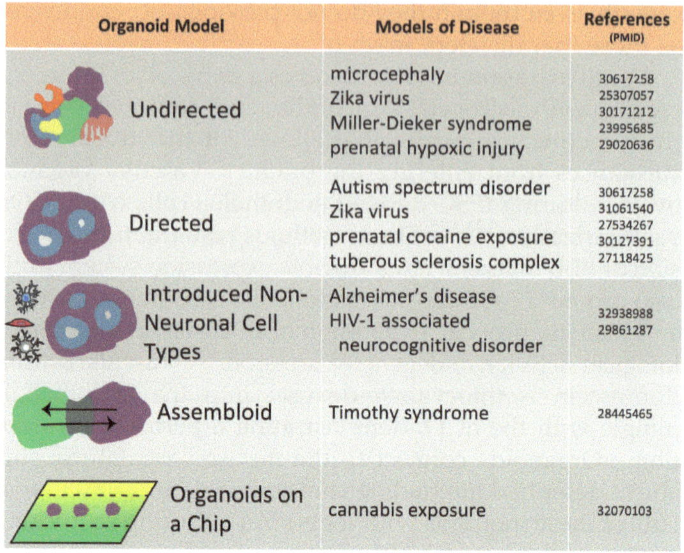

Organoid Model	Models of Disease	References (PMID)
Undirected	microcephaly Zika virus Miller-Dieker syndrome prenatal hypoxic injury	30617258 25307057 30171212 23995685 29020636
Directed	Autism spectrum disorder Zika virus prenatal cocaine exposure tuberous sclerosis complex	30617258 31061540 27534267 30127391 27118425
Introduced Non-Neuronal Cell Types	Alzheimer's disease HIV-1 associated neurocognitive disorder	32938988 29861287
Assembloid	Timothy syndrome	28445465
Organoids on a Chip	cannabis exposure	32070103

FIGURE 6.4 **Selected applications of organoids to model disease.** Recent work has utilized multiple organoid models to study CNS disorders, using the variety of methods discussed throughout this chapter. The type of organoid model is provided in the left column, diseases that have been modeled with the indicated model are provided in the center column, and select references (PMIDs) are provided in the right column. Non-neuronal cell types introduced to organoids are microglia, astrocytes, and endothelial cells. *No Permission Required.*

to study CNS disorders, focusing on microcephaly and ASD, as representative neurodevelopmental disorders for hCO disease modeling.

Microcephaly modeling with hCOs

A prototypical neurodevelopmental disease, where hCO modeling has revealed pathological phenotypes, is genetic or virally-caused microcephaly.[11,37,41] Both genetic microcephaly and Zika virus infection result in reduced brain volume and intellectual or neurological impairment, with Zika virus infection additionally causing cortical malformations and defective neuronal migration.[129,130] Using either directed or undirected hCOs, modeling of genetic microcephaly or Zika infection recapitulates the clinical phenotype by exhibiting smaller organoids, as compared to the unmanipulated controls. These studies additionally show that the organoids modeling genetic microcephaly have a decrease in progenitor populations,[11,48] while those modeling Zika virus infection have an increase in apoptosis.[37,41] This work highlights how hCOs can be used to model a phenotype indicative of the clinical disorder and understand mechanistic disease processes underlying brain structure changes.

ASD modeling with hCOs

Autism spectrum disorder (ASD) is a highly heritable and heterogenous neurodevelopmental disorder that is characterized by impairments in social communication, repetitive behaviors, highly restricted interests, and/or sensory behaviors with symptoms measured reliably in the second year of life.[131] While diagnosis is not based on brain pathology, functional and anatomical differences have been observed with post-mortem brain expression analyses, as well as in vivo brain imaging.[132] Further, ASD is highly heritable, supporting genetic risk factors contributing to the disorder.[133] Several studies have used iPSC-derived hCOs to understand disease mechanisms for molecularly-defined ASD mutations with high penetrance, as well as idiopathic ASD, where individual high penetrance mutations are not known, both of which will be discussed next.

Molecularly defined ASD

Recent genome-wide testing has identified genetic variation associated with high levels of risk for ASD. Associated variants range from single base pair changes to large deletions or duplications in the genome.[133] Although molecularly defined ASD accounts for only ~5% of cases, mechanistic studies on these forms of ASD can provide important insights into the risks and causes of this disorder, including cellular, molecular, and structural changes.

Deletions and duplications of large genomic regions in 16p11.2 are associated with increased risk for ASD and reliably exhibit clinical phenotypes of macrocephaly and microcephaly, respectively.[134] Directed hCO models, using iPSCs generated from patient fibroblasts with 16p11.2 deletions and duplications, were assessed for altered gene expression and structural changes, as compared to diploid controls.[135] hCOs with deletions of 16p11.2 exhibited an increase in diameter, and those with duplications had a decrease in diameter, consistent with the macrocephaly and microcephaly clinical phenotypes, respectively.[135,136] Transcriptomic analysis of the hCOs revealed that copy number changes to the 16p11.2 region altered neuronal differentiation pathways, with deletions resulting in more terminally differentiated neurons and less proliferative progenitors, and duplications showing the reverse. Deficits in neuronal migration were further observed in both the deletion and duplication mutant hCOs when compared to control hCOs. Together, this work supports the ability of hCOs to model ASD from patient-derived iPSCs with large copy number variants, which produce a similar phenotype to that of patients and provide insights into contributory pathways.

Additional mutations associated with ASD have been modeled with hCOs, some of which perturb the PI3K/AKT/mTORC1 signaling pathway, which has been associated with ASD risk and can lead to overgrowth and tumor formation when improperly regulated.[137,138] In these studies, hCOs were created with genetically modified iPSCs that have loss of function mutations in phosphatase and tensin homolog (PTEN) or tuberous sclerosis complex (TSC)1/2, and compared to isogenic (unmodified) lines.[46,139] Each of these loss of function mutations resulted in larger hCO size, similar to the macrocephaly observed in patients.[46,138,139] The PTEN mutation was associated with increased progenitor growth rate and delayed neural maturation, increasing overall progenitor production,[139] and the TSC1/2 mutation resulted in dysmorphic hCOs that underwent increased glial fate decisions.[46] Control phenotypes were restored via dominant negative AKT, AKT inhibition, or mTOR inhibitor treatment.[46,139] A genetic loss of function RAB39b mutation associated with ASD and macrocephaly[140] was similarly modeled with hCOs, where it was identified as part of the PI3K/AKT/mTORC1 pathway.[141] Together, this work highlights the usefulness of hCOs in examining molecular and cellular pathways leading to brain structure changes associated with ASD risk, like macrocephaly, and the advantage of using isogenic controls to study individual genetic differences.

Idiopathic ASD

The majority of individuals with ASD do not harbor a known variant associated with high risk for the disorder. Instead, these individuals have, as yet largely undefined, polygenic and environmental contributions leading to risk, and are referred to as idiopathic. A major benefit of iPSC-derived hCO cultures is their ability to study phenotype and cellular and molecular mechanisms underlying idiopathic ASD, since they can be derived from patient cells and compared to unaffected or familial controls.

To investigate cellular states and associated dysregulated molecular networks in idiopathic ASD development, time-dependent molecular and phenotypic developmental changes were examined in both a 2D model of excitatory neurons and 3D hCO differentiations.[49] ASD-derived neurons had significantly more complex neurite structures in both 2D and 3D cultures compared to unaffected controls. However, hCOs generated from cells of ASD individuals exhibited increased thickness of $TBR1^+$ cortical plate-like regions, whereas no difference was observed in the 2D culture.[49] These observations are novel findings that could contribute to the ASD clinical phenotype and promote further discoveries into contributory pathways of idiopathic ASD. Further, this study highlights the utility of 3D hCOs in modeling spatial developmental changes that are difficult to assess in 2D cultures and the identification of a phenotype associated with the disorder in the absence of a known genetic cause. Together, these studies provide new mechanistic pathways that can increase ASD risk, including alterations in cell morphology and gene expression.

Limitations of hCO modeling for CNS disorders

While hCOs model several aspects of CNS disorders, the approach also has limitations. hCO analyses are generally limited to measurements of morphology, expression analyses, and electrophysiological measures, but cannot assess behavioral outcomes, which often define CNS disorders. In addition, environmental factors that contribute to a disorder, including gene-environment interactions, would not be captured in an iPSC line and might otherwise be difficult to model with hCOs. Another major challenge is that the relative maturity of hCOs approximate that of mid-fetal development (see the section "Neuronal activity and connectivity"), when most CNS disorders are first observed during postnatal development or even in adulthood. In fact, extended differentiation times of up to 200 days are required to generate most neocortical cell types, including functional neurons and glia, substantially increasing experimental costs.[40] Thus, there are unmet needs to accelerate the differentiation and maturation speed of human neural stem cells for CNS disorder research.

In summary, hCOs have been used to model several aspects of human diseases, including structural changes, proliferation, and neurogenesis that can be used to examine mechanistic pathways. While hCOs may not be a perfect model system, hCOs can provide major contributions to our understanding of a wide range of CNS disorders and may direct the course of future therapeutic interventions for unmet clinical needs.

Reproducibility

Reproducibility of phenotypes associated with CNS disorders in human brain organoids is a consistent challenge. Although reliable generation of disease-associated phenotypes in organoid model systems is critical for mechanistic studies or drug screening, hCO studies of the same disorders often report non-overlapping observations. For example, direct comparisons of differentially expressed genes in hCOs derived from cases of idiopathic ASD and neurotypical controls reveal that there is little overlap across studies—only 71 differentially expressed genes were shared in three publications compared to 203 to 2478 unique differentially expressed genes identified in each publication.[142] Though low power (typically $N \leq 8$) and heterogeneity within disorder etiologies likely contribute to variability in the results, differences between organoid protocols and outcome measures can both generate false positive results and mask biological effects. This section will describe the known variability of hCOs, the attempts at improving culture homogeneity, and suggestions as to how research groups may develop and apply organoid methods to improve the reproducibility of hCO studies.

Sources of variability in organoid model systems

There is known variability in cell type and organizational outcomes in hCOs across different batches of differentiation, cell lines, and protocols.[143,144] Differences during PSC culture, such as colony size, morphology, and transcriptional states at the initiation of differentiation, can influence organoid outcomes.[145,146] 3D organoid cultures almost undoubtedly have laboratory- and researcher-dependent effects similar to those observed in 2D neuronal differentiations.[147] These potential technical artifacts are not quantitatively assessed in most research papers because technical differences, including individual handling or time of culture, are not reported and may contribute to discrepancies between studies.

In addition to technical artifacts, research groups improve their own protocols over time,[35,38,40,41,61,148] and new protocols are often generated by adapting previously described differentiation methods (see Fig. 6.2). Changes in base media, growth factors, timing, and machinery can alter cell types that are produced and cell-fate decisions.[149] To test the impact of protocol changes on differentiation capacity, Tanaka et al.[39] combined scRNAseq data from eight distinct protocols and predicted three separate developmental trajectories favored by different hCO protocols. All protocols produced neuroepithelial cells, cortical excitatory neurons, oligodendrocytes, and astrocytes in similar proportions. However, interneurons and cell types without a direct primary tissue correlation were produced in different quantities. Some protocols additionally generated mesodermal cells, while others did not. Consequently, molecular and cellular phenotypes of organoid disease models may not replicate between different protocols with small changes that are optimized for different uses or by different labs.[39] Uncorrected technical variability within culture systems will lead to false positives and negatives driven by technical artifacts rather than disease biology. Previous efforts to standardize media for other culture systems have greatly improved reproducibility.[150]

Addressing reproducibility

Directed hCO methods aim to generate more reproducible, homogenous cultures of excitatory cortical neurons and progenitors. Often reproducibility is tested by replicating the same differentiation on the same cell line,[10,44] across multiple clones from the same originating cells,[40,41,151] or across different neurotypical control lines.[11,17,35,37,62] Two research groups published directed protocols that produced uniform cell type composition and maturation across multiple cell lines as determined by scRNAseq, and each reported their protocol to be highly reproducible.[36,38] However, scRNAseq data from these protocols have not been directly compared, and this research and other organoid characterizations have not been independently replicated. Individual labs use their own standards of internal reproducibility and organoid success, but establishing universal standards for reproducibility could reduce variability across research groups. Standardized protocols would be a first step in enabling meta-analyses from multiple research groups, which would increase statistical power and thereby confidence in identified group or population differences. A series of best practices for establishing reproducibility and quantifying variability for technical factors has been proposed, including the use of "Rosetta" lines for developing protocols as part of a set of best practices.[152] PGP1 and H9 PSC lines have been used across several research groups and may be good candidate Rosetta lines.

Defining markers of successful patterning both early and late in differentiation would be useful for reducing variability in long term cultures.[10,146] Early markers of organoid success are promising because homogeneity of early progenitor cell types is critical for successful long-term culture and has been noted as a defining feature in highly reproducible protocols.[36,38] We recommend that standards of early organoid culture quality, such as 80% FOXG1$^+$ cells that form apical membranes, should be widely applied.[146] A standardized set of markers for validation and a definition of "success" would also allow for organoid models across independent studies to be more directly compared.

Despite the need for more regimented guidance to increase reproducibility, organoids remain an excellent technique to study otherwise inaccessible developmental processes within disease research. Even though less than 100 differentially expressed genes were shared across three studies of idiopathic ASD, all studies implicated synaptic dysfunction in ASD through various assays.[142] Through careful documentation of technical artifacts and intentional adaptation of organoid protocols, these studies provide useful observations for understanding CNS disorders. In the future, standardized protocols and well powered, multisite research efforts will surely advance the understanding of CNS disorders using hCOs.

Conclusions and future directions

The advent of brain organoid research has allowed novel in vitro modeling of the human brain that has already resulted in significant advancements in understanding the cellular and molecular basis of multiple CNS disorders. Over the past decade, great strides have been made in brain organoid technology such that organoids exhibit laminar structure, diverse cell type representation, connectivity, and the inclusion of non-neuronal cells, which together lead to the ability to model the brain with ever increasing fidelity. Nevertheless, differences exist between organoids and the in vivo brain. Further technological advancements would promote the capabilities of organoid research in the future, including enhanced laminar architecture and gyrification,[121] multicomponent assembloids to mimic complex circuitry systems, culture reagents optimized to minimize metabolic stress,[9] and acceleration of maturation.[153] Continued comparisons of organoids to the in vivo developing brain are especially important to ensure that biological processes in the dish correspond to the in vivo brain. Since iPSC-based research is still quite expensive and individual labs may not have the resources to embark on large scale, high-powered projects, increased collaborations will be necessary with multi-center replication studies. These efforts will support the generation of optimized protocols, enhancing fidelity and reproducibility, and lead to high throughput, cross-site meta-analyses. As a relatively new in vitro model of human brain development, and given the encouraging advancements, to-date, we expect brain organoids will gain increasing utility in the future for the understanding of CNS disorders and support the development of new therapeutics.

References

1. McGonigle P. Animal models of CNS disorders. *Biochem Pharmacol*. 2014;87(1):140–149. https://doi.org/10.1016/j.bcp.2013.06.016.
2. de la Torre-Ubieta L, Stein JL, Won H, et al. The dynamic landscape of open Chromatin during human cortical neurogenesis. *Cell*. 2018;172(1–2):289–304. https://doi.org/10.1016/j.cell.2017.12.014. e18.
3. Lui JH, Hansen DV, Kriegstein AR. Development and evolution of the human neocortex. *Cell*. 2011;146(1):18–36. https://doi.org/10.1016/j.cell.2011.06.030.
4. Rakic P. Evolution of the neocortex: a perspective from developmental biology. *Nat Rev Neurosci*. 2009;10(10):724–735. https://doi.org/10.1038/nrn2719.
5. Dolmetsch R, Geschwind DH. The human brain in a dish: the promise of iPSC-derived neurons. *Cell*. 2011;145(6):831–834. https://doi.org/10.1016/j.cell.2011.05.034.
6. Buc-Caron MH. Neuroepithelial progenitor cells explanted from human fetal brain proliferate and differentiate in vitro. *Neurobiol Dis*. 1995;2(1):37–47.
7. https://www.sciencemag.org/news/2020/08/new-us-ethics-board-rejects-most-human-fetal-tissue-research-proposals.
8. Marton RM, Paşca SP. Organoid and assembloid technologies for investigating cellular Crosstalk in human brain development and disease. *Trends Cell Biol*. 2020;30(2):133–143. https://doi.org/10.1016/j.tcb.2019.11.004.
9. Bhaduri A, Andrews MG, Mancia Leon W, et al. Cell stress in cortical organoids impairs molecular subtype specification. *Nature*. 2020;578(7793):142–148. https://doi.org/10.1038/s41586-020-1962-0.
10. Quadrato G, Nguyen T, Macosko EZ, et al. Cell diversity and network dynamics in photosensitive human brain organoids. *Nature*. 2017;545(7652):48–53. https://doi.org/10.1038/nature22047.
11. Lancaster MA, Renner M, Martin C-A, et al. Cerebral organoids model human brain development and microcephaly. *Nature*. 2013;501(7467):373–379. https://doi.org/10.1038/nature12517.
12. Hadders-Algra M. Early human brain development: Starring the subplate. *Neurosci Biobehav Rev*. 2018;92:276–290. https://doi.org/10.1016/j.neubiorev.2018.06.017.
13. Manzini MC, Walsh CA. What disorders of cortical development tell us about the cortex: one plus one does not always make two. *Curr Opin Genet Dev*. 2011;21(3):333–339. https://doi.org/10.1016/j.gde.2011.01.006.
14. Howard BM, Mo Z, Filipovic R, Moore AR, Antic SD, Zecevic N. Radial glia cells in the developing human brain. *Neuroscientist*. 2008;14(5):459–473. https://doi.org/10.1177/1073858407313512.
15. Betizeau M, Cortay V, Patti D, et al. Precursor diversity and complexity of lineage relationships in the outer subventricular zone of the primate. *Neuron*. 2013;80(2):442–457. https://doi.org/10.1016/j.neuron.2013.09.032.
16. Hansen DV, Lui JH, Parker PRL, Kriegstein AR. Neurogenic radial glia in the outer subventricular zone of human neocortex. *Nature*. 2010;464(7288):554–561. https://doi.org/10.1038/nature08845.

17. Pollen AA, Nowakowski TJ, Chen J, et al. Molecular identity of human outer radial glia during cortical development. *Cell*. 2015;163(1):55–67. https://doi.org/10.1016/j.cell.2015.09.004.
18. Andrews MG, Nowakowski TJ. Human brain development through the lens of cerebral organoid models. *Brain Res*. 2019;1725:146470. https://doi.org/10.1016/j.brainres.2019.146470.
19. Nowakowski TJ, Pollen AA, Sandoval-Espinosa C, Kriegstein AR. Transformation of the radial glia Scaffold Demarcates two Stages of human cerebral cortex development. *Neuron*. 2016;91(6):1219–1227. https://doi.org/10.1016/j.neuron.2016.09.005.
20. Bystron I, Blakemore C, Rakic P. Development of the human cerebral cortex: Boulder Committee revisited. *Nat Rev Neurosci*. 2008;9(2):110–122. https://doi.org/10.1038/nrn2252.
21. Duque A, Krsnik Z, Kostović I, Rakic P. Secondary expansion of the transient subplate zone in the developing cerebrum of human and nonhuman primates. *Proc Natl Acad Sci USA*. 2016;113(35):9892–9897. https://doi.org/10.1073/pnas.1610078113.
22. Hoerder-Suabedissen A, Molnár Z. Development, evolution and pathology of neocortical subplate neurons. *Nat Rev Neurosci*. 2015;16(3):133–146. https://doi.org/10.1038/nrn3915.
23. Ozair MZ, Kirst C, van den Berg BL, Ruzo A, Rito T, Brivanlou AH. hPSC modeling reveals that fate selection of cortical Deep projection neurons occurs in the subplate. *Cell Stem Cell*. 2018;23(1):60–73. https://doi.org/10.1016/j.stem.2018.05.024. e6.
24. Holst CB, Brøchner CB, Vitting-Seerup K, Møllgård K. Astrogliogenesis in human fetal brain: complex spatiotemporal immunoreactivity patterns of GFAP, S100, AQP4 and YKL-40. *J Anat*. 2019;235(3):590–615. https://doi.org/10.1111/joa.12948.
25. Miller FD, Gauthier AS. Timing is everything: making neurons versus glia in the developing cortex. *Neuron*. 2007;54(3):357–369.
26. Takahashi K, Tanabe K, Ohnuki M, et al. Induction of pluripotent stem cells from adult human fibroblasts by defined factors. *Cell*. 2007;131(5):861–872.
27. Yu J, Vodyanik MA, Smuga-Otto K, et al. Induced pluripotent stem cell lines derived from human somatic cells. *Science (New York, N.Y.)*. 2007;318(5858):1917–1920.
28. Chambers SM, Fasano CA, Papapetrou EP, Tomishima M, Sadelain M, Studer L. Highly efficient neural conversion of human ES and iPS cells by dual inhibition of SMAD signaling. *Nat Biotechnol*. 2009;27(3):275–280. https://doi.org/10.1038/nbt.1529.
29. Eiraku M, Watanabe K, Matsuo-Takasaki M, et al. Self-organized formation of polarized cortical tissues from ESCs and its active manipulation by extrinsic signals. *Cell Stem Cell*. 2008;3(5):519–532. https://doi.org/10.1016/j.stem.2008.09.002.
30. Bain G, Kitchens D, Yao M, Huettner JE, Gottlieb DI. Embryonic stem cells express neuronal properties in vitro. *Dev Biol*. 1995;168(2):342–357.
31. Shamblott MJ, Axelman J, Littlefield JW, et al. Human embryonic germ cell derivatives express a broad range of developmentally distinct markers and proliferate extensively in vitro. *Proc Natl Acad Sci USA*. 2001;98(1):113–118.
32. Zhang SC, Wernig M, Duncan ID, Brüstle O, Thomson JA. In vitro differentiation of transplantable neural precursors from human embryonic stem cells. *Nat Biotechnol*. 2001;19(12):1129–1133.
33. Barberi T, Klivenyi P, Calingasan NY, et al. Neural subtype specification of fertilization and nuclear transfer embryonic stem cells and application in parkinsonian mice. *Nat Biotechnol*. 2003;21(10):1200–1207.
34. Kawasaki H, Mizuseki K, Nishikawa S, et al. Induction of midbrain dopaminergic neurons from ES cells by stromal cell-derived inducing activity. *Neuron*. 2000;28(1):31–40.
35. Qian X, Su Y, Adam CD, et al. Sliced human cortical organoids for modeling distinct cortical Layer Formation. *Cell Stem Cell*. 2020;26(5):766–781. https://doi.org/10.1016/j.stem.2020.02.002. e9.
36. Velasco S, Kedaigle AJ, Simmons SK, et al. Individual brain organoids reproducibly form cell diversity of the human cerebral cortex. *Nature*. 2019;570(7762):523–527. https://doi.org/10.1038/s41586-019-1289-x.
37. Watanabe M, Buth JE, Vishlaghi N, et al. Self-organized cerebral organoids with human-specific features predict effective drugs to Combat Zika virus infection. *Cell Rep*. 2017;21(2):517–532. https://doi.org/10.1016/j.celrep.2017.09.047.
38. Yoon S-J, Elahi LS, Paşca AM, et al. Reliability of human cortical organoid generation. *Nat Methods*. 2019;16(1):75–78. https://doi.org/10.1038/s41592-018-0255-0.

39. Tanaka Y, Cakir B, Xiang Y, Sullivan GJ, Park I-H. Synthetic analyses of single-cell transcriptomes from multiple brain organoids and fetal brain. *Cell Rep*. 2020;30(6):1682–1689. https://doi.org/10.1016/j.celrep.2020.01.038. e3.
40. Pasca AM, Sloan SA, Clarke LE, et al. Functional cortical neurons and astrocytes from human pluripotent stem cells in 3D culture. *Nat Methods*. 2015;12(7):671–678. https://doi.org/10.1038/nmeth.3415.
41. Qian X, Nguyen HN, Song MM, et al. Brain-region-specific organoids using mini-bioreactors for modeling ZIKV exposure. *Cell*. 2016;165(5):1238–1254. https://doi.org/10.1016/j.cell.2016.04.032.
42. Madhavan M, Nevin ZS, Shick HE, et al. Induction of myelinating oligodendrocytes in human cortical spheroids. *Nat Methods*. 2018;15(9):700–706. https://doi.org/10.1038/s41592-018-0081-4.
43. Trujillo CA, Gao R, Negraes PD, et al. Complex oscillatory waves emerging from cortical organoids model early human brain network development. *Cell Stem Cell*. 2019;25(4):558–569. https://doi.org/10.1016/j.stem.2019.08.002. e7.
44. Giandomenico SL, Mierau SB, Gibbons GM, et al. Cerebral organoids at the air-liquid interface generate diverse nerve tracts with functional output. *Nat Neurosci*. 2019;22(4):669–679. https://doi.org/10.1038/s41593-019-0350-2.
45. Lancaster MA, Knoblich JA. Generation of cerebral organoids from human pluripotent stem cells. *Nat Protoc*. 2014;9(10):2329–2340. https://doi.org/10.1038/nprot.2014.158.
46. Blair JD, Hockemeyer D, Bateup HS. Genetically engineered human cortical spheroid models of tuberous sclerosis. *Nat Med*. 2018;24(10):1568–1578. https://doi.org/10.1038/s41591-018-0139-y.
47. Ghatak S, Dolatabadi N, Trudler D, et al. Mechanisms of hyperexcitability in Alzheimer's disease hiPSC-derived neurons and cerebral organoids vs isogenic controls. *Elife*. 2019;8. https://doi.org/10.7554/eLife.50333.
48. Li R, Sun L, Fang A, Li P, Wu Q, Wang X. Recapitulating cortical development with organoid culture in vitro and modeling abnormal spindle-like (ASPM related primary) microcephaly disease. *Protein & Cell*. 2017;8(11):823–833. https://doi.org/10.1007/s13238-017-0479-2.
49. Schafer ST, Paquola ACM, Stern S, et al. Pathological priming causes developmental gene network heterochronicity in autistic subject-derived neurons. *Nat Neurosci*. 2019;22(2):243–255. https://doi.org/10.1038/s41593-018-0295-x.
50. Daviaud N, Chevalier C, Friedel RH, Zou H. Distinct vulnerability and resilience of human neuroprogenitor subtypes in cerebral organoid model of prenatal hypoxic Injury. *Front Cell Neurosci*. 2019;13:336. https://doi.org/10.3389/fncel.2019.00336.
51. Uhlen M, Bandrowski A, Carr S, et al. A proposal for validation of antibodies. *Nat Methods*. 2016;13(10):823–827. https://doi.org/10.1038/nmeth.3995.
52. Weller MG. Ten basic Rules of antibody validation. *Anal Chem Insights*. 2018;13. https://doi.org/10.1177/1177390118757462, 1177390118757462.
53. Polioudakis D, de la Torre-Ubieta L, Langerman J, et al. A single-cell transcriptomic Atlas of human neocortical development during mid-gestation. *Neuron*. 2019;103(5):785–801. https://doi.org/10.1016/j.neuron.2019.06.011. e8.
54. Zhong S, Zhang S, Fan X, et al. A single-cell RNA-seq survey of the developmental landscape of the human prefrontal cortex. *Nature*. 2018;555(7697):524–528. https://doi.org/10.1038/nature25980.
55. Haque A, Engel J, Teichmann SA, Lönnberg T. A practical guide to single-cell RNA-sequencing for biomedical research and clinical applications. *Genome Med*. 2017;9(1):75. https://doi.org/10.1186/s13073-017-0467-4.
56. Di Lullo E, Kriegstein AR. The use of brain organoids to investigate neural development and disease. *Nat Rev Neurosci*. 2017;18(10):573–584. https://doi.org/10.1038/nrn.2017.107.
57. Amiri A, Coppola G, Scuderi S, et al. Transcriptome and epigenome landscape of human cortical development modeled in organoids. *Science (New York, N.Y.)*. 2018;362(6420). https://doi.org/10.1126/science.aat6720.
58. Trevino AE, Sinnott-Armstrong N, Andersen J, et al. Chromatin accessibility dynamics in a model of human forebrain development. *Science (New York, N.Y.)*. 2020;367(6476). https://doi.org/10.1126/science.aay1645.
59. Shi Y, Sun L, Wang M, et al. Vascularized human cortical organoids (vOrganoids) model cortical development in vivo. *PLoS Biol*. 2020;18(5):e3000705. https://doi.org/10.1371/journal.pbio.3000705.
60. Kadoshima T, Sakaguchi H, Nakano T, et al. Self-organization of axial polarity, inside-out layer pattern, and species-specific progenitor dynamics in human ES cell-derived neocortex. *Proc Natl Acad Sci USA*. 2013;110(50):20284–20289. https://doi.org/10.1073/pnas.1315710110.

61. Sloan SA, Darmanis S, Huber N, et al. Human astrocyte maturation captured in 3D cerebral cortical spheroids derived from pluripotent stem cells. *Neuron*. 2017;95(4):779–790. https://doi.org/10.1016/j.neuron.2017.07.035. e6.
62. Mariani J, Simonini MV, Palejev D, et al. Modeling human cortical development in vitro using induced pluripotent stem cells. *Proc Natl Acad Sci USA*. 2012;109(31):12770–12775. https://doi.org/10.1073/pnas.1202944109.
63. Tambalo M, Lodato S. Brain organoids: human 3D models to investigate neuronal circuits assembly, function and dysfunction. *Brain Res*. 2020;1746:147028. https://doi.org/10.1016/j.brainres.2020.147028.
64. Marton RM, Miura Y, Sloan SA, et al. Differentiation and maturation of oligodendrocytes in human three-dimensional neural cultures. *Nat Neurosci*. 2019;22(3):484–491. https://doi.org/10.1038/s41593-018-0316-9.
65. Barkovich AJ. Developmental disorders of the midbrain and hindbrain. *Front Neuroanat*. 2012;6:7. https://doi.org/10.3389/fnana.2012.00007.
66. Barkovich AJ, Millen KJ, Dobyns WB. A developmental and genetic classification for midbrain-hindbrain malformations. *Brain*. 2009;132(Pt 12):3199–3230. https://doi.org/10.1093/brain/awp247.
67. Matilla-Dueñas A, Corral-Juan M, Rodríguez-Palmero Seuma A, et al. Rare neurodegenerative diseases: clinical and genetic update. *Adv Exp Med Biol*. 2017;1031:443–496. https://doi.org/10.1007/978-3-319-67144-4_25.
68. Ebrahimi-Fakhari D, Van Karnebeek C, Münchau A. Movement disorders in treatable Inborn Errors of metabolism. *Mov Disord*. 2019;34(5):598–613. https://doi.org/10.1002/mds.27568.
69. Bamford NS, McVicar K. Localising movement disorders in childhood. *Lancet ChildAdoles Health*. 2019;3(12):917–928. https://doi.org/10.1016/S2352-4642(19)30330-X.
70. Zwicker JG, Missiuna C, Harris SR, Boyd LA. Developmental coordination disorder: a review and update. *Eur J Paediatr Neurol*. 2012;16(6):573–581. https://doi.org/10.1016/j.ejpn.2012.05.005.
71. Birey F, Andersen J, Makinson CD, et al. Assembly of functionally integrated human forebrain spheroids. *Nature*. 2017;545(7652):54–59. https://doi.org/10.1038/nature22330.
72. Landau ID, Egger R, Dercksen VJ, Oberlaender M, Sompolinsky H. The impact of structural heterogeneity on excitation-inhibition balance in cortical networks. *Neuron*. 2016;92(5):1106–1121. https://doi.org/10.1016/j.neuron.2016.10.027.
73. Wang DD, Kriegstein AR. Defining the role of GABA in cortical development. *J Physiol*. 2009;587(Pt 9):1873–1879. https://doi.org/10.1113/jphysiol.2008.167635.
74. Ribes V, Briscoe J. Establishing and interpreting graded Sonic Hedgehog signaling during vertebrate neural tube patterning: the role of negative feedback. *Cold Spring Harbor Perspect Biol*. 2009;1(2):a002014. https://doi.org/10.1101/cshperspect.a002014.
75. Bagley JA, Reumann D, Bian S, Lévi-Strauss J, Knoblich JA. Fused cerebral organoids model interactions between brain regions. *Nat Methods*. 2017;14(7):743–751. https://doi.org/10.1038/nmeth.4304.
76. Kim H, Xu R, Padmashri R, et al. Pluripotent stem cell-derived cerebral organoids reveal human Oligodendrogenesis with dorsal and ventral Origins. *Stem Cell Rep*. 2019;12(5):890–905. https://doi.org/10.1016/j.stemcr.2019.04.011.
77. Song L, Yuan X, Jones Z, et al. Functionalization of brain region-specific spheroids with isogenic microglia-like cells. *Sci Rep*. 2019;9(1):11055. https://doi.org/10.1038/s41598-019-47444-6.
78. Xiang Y, Tanaka Y, Patterson B, et al. Fusion of regionally specified hPSC-derived organoids models human brain development and interneuron migration. *Cell Stem Cell*. 2017;21(3):383–398. https://doi.org/10.1016/j.stem.2017.07.007. e7.
79. Berger E, Magliaro C, Paczia N, et al. Millifluidic culture improves human midbrain organoid vitality and differentiation. *Lab Chip*. 2018;18(20):3172–3183. https://doi.org/10.1039/c8lc00206a.
80. Jo J, Xiao Y, Sun AX, et al. Midbrain-like organoids from human pluripotent stem cells contain functional dopaminergic and neuromelanin-producing neurons. *Cell Stem Cell*. 2016;19(2):248–257. https://doi.org/10.1016/j.stem.2016.07.005.
81. Monzel AS, Smits LM, Hemmer K, et al. Derivation of human midbrain-specific organoids from neuroepithelial stem cells. *Stem Cell Rep*. 2017;8(5):1144–1154. https://doi.org/10.1016/j.stemcr.2017.03.010.
82. Ogura T, Sakaguchi H, Miyamoto S, Takahashi J. Three-dimensional induction of dorsal, intermediate and ventral spinal cord tissues from human pluripotent stem cells. *Development*. 2018;145(16). https://doi.org/10.1242/dev.162214.
83. Ozone C, Suga H, Eiraku M, et al. Functional anterior pituitary generated in self-organizing culture of human embryonic stem cells. *Nat Commun*. 2016;7:10351. https://doi.org/10.1038/ncomms10351.

84. Cederquist GY, Asciolla JJ, Tchieu J, et al. Specification of positional identity in forebrain organoids. *Nat Biotechnol*. 2019;37(4):436–444. https://doi.org/10.1038/s41587-019-0085-3.
85. Demers CJ, Soundararajan P, Chennampally P, et al. Development-on-chip: in vitro neural tube patterning with a microfluidic device. *Development*. 2016;143(11):1884–1892. https://doi.org/10.1242/dev.126847.
86. O'Grady B, Balikov DA, Wang JX, et al. Spatiotemporal control and modeling of morphogen delivery to induce gradient patterning of stem cell differentiation using fluidic channels. *Biomaterials science*. 2019;7(4):1358–1371. https://doi.org/10.1039/c8bm01199k.
87. Faustino Martins J-M, Fischer C, Urzi A, et al. Self-organizing 3D human Trunk neuromuscular organoids. *Cell Stem Cell*. 2020;26(2):172–186. https://doi.org/10.1016/j.stem.2019.12.007. e6.
88. Sakaguchi H, Kadoshima T, Soen M, et al. Generation of functional hippocampal neurons from self-organizing human embryonic stem cell-derived dorsomedial telencephalic tissue. *Nat Commun*. 2015;6:8896. https://doi.org/10.1038/ncomms9896.
89. Jakovcevski I, Zecevic N. Olig transcription factors are expressed in oligodendrocyte and neuronal cells in human fetal CNS. *J Neurosci*. 2005;25(44):10064–10073.
90. Ligon KL, Alberta JA, Kho AT, et al. The oligodendroglial lineage marker OLIG2 is universally expressed in diffuse gliomas. *J Neuropathol Exp Neurol*. 2004;63(5):499–509.
91. Blanco-Suárez E, Caldwell ALM, Allen NJ. Role of astrocyte-synapse interactions in CNS disorders. *J Physiol*. 2017;595(6):1903–1916. https://doi.org/10.1113/JP270988.
92. Chaudhury S, Sharma V, Kumar V, Nag TC, Wadhwa S. Activity-dependent synaptic plasticity modulates the critical phase of brain development. *Brain & Dev*. 2016;38(4):355–363. https://doi.org/10.1016/j.braindev.2015.10.008.
93. John D, Berg DK. Long-lasting changes in neural networks to compensate for altered nicotinic input. *Biochem Pharmacol*. 2015;97(4):418–424. https://doi.org/10.1016/j.bcp.2015.07.020.
94. Bitar M, Barry G. Building a human brain for research. *Front Mol Neurosci*. 2020;13:22. https://doi.org/10.3389/fnmol.2020.00022.
95. Chen T-W, Wardill TJ, Sun Y, et al. Ultrasensitive fluorescent proteins for imaging neuronal activity. *Nature*. 2013;499(7458):295–300. https://doi.org/10.1038/nature12354.
96. Logan S, Arzua T, Yan Y, et al. Dynamic characterization of structural, molecular, and electrophysiological phenotypes of human-induced pluripotent stem cell-derived cerebral organoids, and comparison with fetal and adult gene profiles. *Cells*. 2020;9(5). https://doi.org/10.3390/cells9051301.
97. Cullen DK, Gordián-Vélez WJ, Struzyna LA, et al. Bundled three-dimensional human axon tracts derived from brain organoids. *iScience*. 2019;21:57–67. https://doi.org/10.1016/j.isci.2019.10.004.
98. Huang Q, Zhou D, DiFiglia M. Neurobiotin, a useful neuroanatomical tracer for in vivo anterograde, retrograde and transneuronal tract-tracing and for in vitro labeling of neurons. *J Neurosci Methods*. 1992;41(1):31–43.
99. Mansour AA, Gonçalves JT, Bloyd CW, et al. An in vivo model of functional and vascularized human brain organoids. *Nat Biotechnol*. 2018;36(5):432–441. https://doi.org/10.1038/nbt.4127.
100. Sloan SA, Andersen J, Paşca AM, Birey F, Paşca SP. Generation and assembly of human brain region-specific three-dimensional cultures. *Nat Protoc*. 2018;13(9):2062–2085. https://doi.org/10.1038/s41596-018-0032-7.
101. Ugolini G. Use of rabies virus as a transneuronal tracer of neuronal connections: implications for the understanding of rabies pathogenesis. *Dev Biologicals*. 2008;131:493–506.
102. Clarke LE, Barres BA. Emerging roles of astrocytes in neural circuit development. *Nat Rev Neurosci*. 2013;14(5):311–321. https://doi.org/10.1038/nrn3484.
103. Kuhn S, Gritti L, Crooks D, Dombrowski Y. Oligodendrocytes in development, myelin generation and beyond. *Cells*. 2019;8(11). https://doi.org/10.3390/cells8111424.
104. Salter MW, Stevens B. Microglia emerge as central players in brain disease. *Nat Med*. 2017;23(9):1018–1027. https://doi.org/10.1038/nm.4397.
105. Rothstein JD, Dykes-Hoberg M, Pardo CA, et al. Knockout of glutamate transporters reveals a major role for astroglial transport in excitotoxicity and clearance of glutamate. *Neuron*. 1996;16(3):675–686.
106. Middeldorp J, Boer K, Sluijs JA, et al. GFAPdelta in radial glia and subventricular zone progenitors in the developing human cortex. *Development*. 2010;137(2):313–321. https://doi.org/10.1242/dev.041632.
107. Mori T, Buffo A, Götz M. The novel roles of glial cells revisited: the contribution of radial glia and astrocytes to neurogenesis. *Curr Top Dev Biol*. 2005;69:67–99.

108. Bramanti V, Tomassoni D, Avitabile M, Amenta F, Avola R. Biomarkers of glial cell proliferation and differentiation in culture. *Front Biosci*. 2010;2:558–570.
109. Yang Y, Vidensky S, Jin L, et al. Molecular comparison of GLT1+ and ALDH1L1+ astrocytes in vivo in astroglial reporter mice. *Glia*. 2011;59(2):200–207. https://doi.org/10.1002/glia.21089.
110. Yoon H, Walters G, Paulsen AR, Scarisbrick IA. Astrocyte heterogeneity across the brain and spinal cord occurs developmentally, in adulthood and in response to demyelination. *PLoS One*. 2017;12(7):e0180697. https://doi.org/10.1371/journal.pone.0180697.
111. Renner M, Lancaster MA, Bian S, et al. Self-organized developmental patterning and differentiation in cerebral organoids. *EMBO J*. 2017;36(10):1316–1329. https://doi.org/10.15252/embj.201694700.
112. Krencik R, Seo K, van Asperen JV, et al. Systematic three-dimensional coculture rapidly recapitulates interactions between human neurons and astrocytes. *Stem Cell Rep*. 2017;9(6):1745–1753. https://doi.org/10.1016/j.stemcr.2017.10.026.
113. Craig GA, Yoo S, Du TY, Xiao J. Plasticity in oligodendrocyte lineage progression: an OPC puzzle on our nerves. *Eur J Neurosci*. June 2020. https://doi.org/10.1111/ejn.14845.
114. Sominsky L, De Luca S, Spencer SJ. Microglia: key players in neurodevelopment and neuronal plasticity. *Int J Biochem Cell Biol*. 2018;94:56–60. https://doi.org/10.1016/j.biocel.2017.11.012.
115. Hammond TR, Robinton D, Stevens B. Microglia and the brain: complementary Partners in development and disease. *Annu Rev Cell Dev Biol*. 2018;34:523–544. https://doi.org/10.1146/annurev-cellbio-100616-060509.
116. Abud EM, Ramirez RN, Martinez ES, et al. iPSC-derived human microglia-like cells to study neurological diseases. *Neuron*. 2017;94(2):278–293. https://doi.org/10.1016/j.neuron.2017.03.042. e9.
117. Lin Y-T, Seo J, Gao F, et al. APOE4 causes Widespread molecular and cellular alterations associated with Alzheimer's disease phenotypes in human iPSC-derived brain cell types. *Neuron*. 2018;98(6):1141–1154. https://doi.org/10.1016/j.neuron.2018.05.008. e7.
118. Nzou G, Wicks RT, Wicks EE, et al. Human cortex spheroid with a functional blood brain barrier for high-throughput neurotoxicity screening and disease modeling. *Sci Rep*. 2018;8(1):7413. https://doi.org/10.1038/s41598-018-25603-5.
119. Ormel PR, Vieira de Sá R, van Bodegraven EJ, et al. Microglia innately develop within cerebral organoids. *Nat Commun*. 2018;9(1):4167. https://doi.org/10.1038/s41467-018-06684-2.
120. Nasu M, Takata N, Danjo T, et al. Robust formation and maintenance of continuous stratified cortical neuroepithelium by laminin-containing matrix in mouse ES cell culture. *PLoS One*. 2012;7(12):e53024. https://doi.org/10.1371/journal.pone.0053024.
121. Karzbrun E, Kshirsagar A, Cohen SR, Hanna JH, Reiner O. Human brain organoids on a chip reveal the physics of folding. *Nat Phys*. 2018;14(5):515–522. https://doi.org/10.1038/s41567-018-0046-7.
122. Nestor MW, Paull D, Jacob S, et al. Differentiation of serum-free embryoid bodies from human induced pluripotent stem cells into networks. *Stem Cell Res*. 2013;10(3):454–463. https://doi.org/10.1016/j.scr.2013.02.001.
123. Lauschke K, Frederiksen L, Hall VJ. Paving the Way toward complex blood-brain barrier models using pluripotent stem cells. *Stem Cell Dev*. 2017;26(12):857–874. https://doi.org/10.1089/scd.2017.0003.
124. Ham O, Jin YB, Kim J, Lee M-O. Blood vessel formation in cerebral organoids formed from human embryonic stem cells. *Biochem Biophys Res Commun*. 2020;521(1):84–90. https://doi.org/10.1016/j.bbrc.2019.10.079.
125. Pham MT, Pollock KM, Rose MD, et al. Generation of human vascularized brain organoids. *Neuroreport*. 2018;29(7):588–593. https://doi.org/10.1097/WNR.0000000000001014.
126. Cakir B, Xiang Y, Tanaka Y, et al. Engineering of human brain organoids with a functional vascular-like system. *Nat Methods*. 2019;16(11):1169–1175. https://doi.org/10.1038/s41592-019-0586-5.
127. Raja WK, Mungenast AE, Lin Y-T, et al. Self-organizing 3D human neural tissue derived from induced pluripotent stem cells recapitulate Alzheimer's disease phenotypes. *PLoS One*. 2016;11(9):e0161969. https://doi.org/10.1371/journal.pone.0161969.
128. Ye F, Kang E, Yu C, et al. DISC1 regulates neurogenesis via modulating Kinetochore attachment of ndel1/nde1 during Mitosis. *Neuron*. 2017;96(5):1041–1054. https://doi.org/10.1016/j.neuron.2017.10.010. e5.
129. Faheem M, Naseer MI, Rasool M, et al. Molecular genetics of human primary microcephaly: an overview. *BMC Med Genom*. 2015;8(Suppl 1):S4. https://doi.org/10.1186/1755-8794-8-S1-S4.
130. Vhp L, Aragão MM, Pinho RS, et al. Congenital Zika virus infection: a review with Emphasis on the spectrum of brain Abnormalities. *Curr Neurol Neurosci Rep*. 2020;20(11):49. https://doi.org/10.1007/s11910-020-01072-0.
131. Ozonoff S, Heung K, Byrd R, Hansen R, Hertz-Picciotto I. The onset of autism: patterns of symptom emergence in the first years of life. *Autism Res*. 2008;1(6):320–328. https://doi.org/10.1002/aur.53.
132. Lord C, Brugha TS, Charman T, et al. Autism spectrum disorder. *Nat Rev Dis Prim*. 2020;6(1):5. https://doi.org/10.1038/s41572-019-0138-4.

133. de la Torre-Ubieta L, Won H, Stein JL, Geschwind DH. Advancing the understanding of autism disease mechanisms through genetics. *Nat Med*. 2016;22(4):345–361. https://doi.org/10.1038/nm.4071.
134. Sønderby IE, Gústafsson Ó, Doan NT, et al. Dose response of the 16p11.2 distal copy number variant on intracranial volume and basal ganglia. *Mol Psychiatr*. 2020;25(3):584–602. https://doi.org/10.1038/s41380-018-0118-1.
135. Urresti J, Zhang P, Moran-Losada P, et al. Cortical organoids model early brain development disrupted by 16p11.2 copy number variants in autism. *bioRxiv*. 2020. https://doi.org/10.1101/2020.06.25.172262, 2020.06.25.172262.
136. Qureshi AY, Mueller S, Snyder AZ, et al. Opposing brain differences in 16p11.2 deletion and duplication carriers. *J Neurosci*. 2014;34(34):11199–11211. https://doi.org/10.1523/JNEUROSCI.1366-14.2014.
137. Nathan N, Keppler-Noreuil KM, Biesecker LG, Moss J, Darling TN. Mosaic disorders of the PI3K/PTEN/AKT/TSC/mTORC1 signaling pathway. *Dermatol Clin*. 2017;35(1):51–60. https://doi.org/10.1016/j.det.2016.07.001.
138. Sato A. mTOR, a potential target to Treat autism spectrum disorder. *CNS Neurol Disord - Drug Targets*. 2016;15(5):533–543.
139. Li Y, Muffat J, Omer A, et al. Induction of expansion and folding in human cerebral organoids. *Cell Stem Cell*. 2017;20(3):385–396. https://doi.org/10.1016/j.stem.2016.11.017. e3.
140. Woodbury-Smith M, Deneault E, Yuen RKC, et al. In individuals with intellectual disability, autism spectrum disorder, and macrocephaly. *Mol Autism*. 2017;8:59. https://doi.org/10.1186/s13229-017-0175-3.
141. Zhang W, Ma L, Yang M, et al. Cerebral organoid and mouse models reveal a RAB39b-PI3K-mTOR pathway-dependent dysregulation of cortical development leading to macrocephaly/autism phenotypes. *Genes Dev*. 2020;34(7–8):580–597. https://doi.org/10.1101/gad.332494.119.
142. Griesi-Oliveira K, Fogo MS, Pinto BGG, et al. Transcriptome of iPSC-derived neuronal cells reveals a module of co-expressed genes consistently associated with autism spectrum disorder. *Mol Psychiatr*. February 2020. https://doi.org/10.1038/s41380-020-0669-9.
143. Kelava I, Lancaster MA. Stem cell models of human brain development. *Cell Stem Cell*. 2016;18(6):736–748. https://doi.org/10.1016/j.stem.2016.05.022.
144. Qian X, Song H, Ming G-L. Brain organoids: advances, applications and challenges. *Development*. 2019;146(8). https://doi.org/10.1242/dev.166074.
145. Bauwens CL, Peerani R, Niebruegge S, et al. Control of human embryonic stem cell colony and aggregate size heterogeneity influences differentiation trajectories. *Stem Cell*. 2008;26(9):2300–2310. https://doi.org/10.1634/stemcells.2008-0183.
146. Watanabe M, Haney JR, Vishlaghi N, et al. TGFβ superfamily signaling regulates the state of human stem cell pluripotency and competency to create telencephalic organoids. *bioRxiv*. 2019. https://doi.org/10.1101/2019.12.13.875773, 2019.12.13.875773.
147. Volpato V, Smith J, Sandor C, et al. Reproducibility of molecular phenotypes after long-term differentiation to human iPSC-derived neurons: a multi-site Omics study. *Stem Cell Rep*. 2018;11(4):897–911. https://doi.org/10.1016/j.stemcr.2018.08.013.
148. Qian X, Jacob F, Song MM, Nguyen HN, Song H, Ming G-L. Generation of human brain region-specific organoids using a miniaturized spinning bioreactor. *Nat Protoc*. 2018;13(3):565–580. https://doi.org/10.1038/nprot.2017.152.
149. Khakipoor S, Crouch EE, Mayer S. Human organoids to model the developing human neocortex in health and disease. *Brain Res*. 2020;1742:146803. https://doi.org/10.1016/j.brainres.2020.146803.
150. Hayflick L, Jacobs P, Perkins F. A procedure for the standardization of tissue culture media. *Nature*. 1964;204:146–147.
151. Bershteyn M, Nowakowski TJ, Pollen AA, et al. Human iPSC-derived cerebral organoids model cellular features of lissencephaly and reveal prolonged Mitosis of outer radial glia. *Cell Stem Cell*. 2017;20(4):435–449. https://doi.org/10.1016/j.stem.2016.12.007. e4.
152. Volpato V, Webber C. Addressing variability in iPSC-derived models of human disease: guidelines to promote reproducibility. *Dis Models Mecha*. 2020;13(1). https://doi.org/10.1242/dmm.042317.
153. Sood D, Cairns DM, Dabbi JM, et al. Functional maturation of human neural stem cells in a 3D bioengineered brain model enriched with fetal brain-derived matrix. *Sci Rep*. 2019;9(1):17874. https://doi.org/10.1038/s41598-019-54248-1.

SECTION II

The use of iPSC-derived neurons to study neurological disorders

CHAPTER

7

Human models as new tools for drug development and precision medicine

Alfred K. Dei-Ampeh[1,2], *Manuj Shah*[2,3], *Gerarda Cappuccio*[1,2], *Damian W. Young*[1,2,4,5] and *Mirjana Maletic-Savatic*[1,2,5]

[1]Department of Pediatrics-Neurology, Baylor College of Medicine, Houston, TX, United States; [2]Jan and Dan Duncan Neurological Research Institute, Texas Children's Hospital, Houston, TX, United States; [3]Department of Neuroscience, Rice University, Houston, TX, United States; [4]Department of Pharmacology and Chemical Biology, Baylor College of Medicine, Houston, TX, United States; [5]Center for Drug Discovery, Department of Pathology and Immunology, Baylor College of Medicine, Houston, TX, United States

Introduction

In the past 15 years, even as the number of new drug candidates nearly doubled, 90% of those that succeeded in preclinical animal studies failed in human trials. According to recent estimates, this high attrition rate, due to lack of efficacy approximately 50% of the time and toxicity 30% of the time, raises the cost per new drug to up to $2.8 billion. To assess the efficacy, safety, pharmacokinetics, and pharmacodynamics of new drug entities before investing more resources in clinical trials, the Federal Drug Administration (FDA) relies on data generated from animal models.[1] However, the ability to predict the effects of drug candidates in humans using these models is limited and "first-in-human" tests present substantial risks for participants who can experience serious, even lethal, harm.

While animal models remain invaluable for the discovery and development of new therapeutics, particularly when combined with recent advances in genome sequencing (homologous recombination, CRISPR/Cas9, and random transgenesis), they are burdened with critical pitfalls. Rodent models—the most frequently used models for deciphering the causes and pathology of disease that are used in preclinical drug studies—cannot replicate certain aspects of disease that are uniquely human. First, they cannot replicate human genetic

Phenotyping of Human iPSC-derived Neurons
https://doi.org/10.1016/B978-0-12-822277-5.00004-3

heterogeneity found across the global population.[2] Unlike mice, humans are not inbred. And while we understand more and more about risk loci in humans, the genetic causes of all diseases are not well understood and therefore cannot be modeled in rodents with gene manipulation. Furthermore, individual responses to drugs cannot be anticipated without taking human genetic diversity into account. Second, the human brain is far more complex than the mouse brain. Human neuronal and glial populations are more diverse than those in the rodent brain, and the magnitude of gene expression in human neurons and glia compared to mouse neurons and glia further underscores human brain complexity with respect to function.[3,4] Finally, human metabolism, including drug metabolism, is unique. For example, several commonly used drugs like ibuprofen and warfarin are toxic to rodents, underscoring how a drug's availability, distribution, efficiency, and excretion can vary widely between species. Thus, even though rodents are indispensable for drug development and approval, rodent animal models do not predict subjective drug effects or idiosyncratic activities in humans.

These factors contribute to the high failure rate of promising new drugs during the approval process for clinical studies and the high attrition rate of clinical trials.[5,6] Such widespread failure of preclinical models to adequately predict efficacy in humans has led drug developers to look for other models to better inform decisions about which targets to pursue and for which indications.[7,8]

Drug development pipeline

The discovery and development of any drug is invariably a lengthy, complex, and expensive process. It starts with the identification and validation of the putative molecular/cellular drug target, followed by extensive preclinical and clinical investigations, and ends with a series of regulatory approvals. The process involves comprehensive testing in in vitro, ex vivo, and in vivo preclinical studies, and clinical examination in human subjects (phase 1, 2, and 3 trials) (Fig. 7.1).[1,9] Preclinical drug development can be divided into three principal stages: (1) drug discovery stage: target identification, validation with disease modeling, and identification of lead compounds; (2) preclinical screening and testing stage: pharmacokinetic and pharmacodynamic profiles, critical to determining initial drug regimens for clinical trials; and (3) preclinical trial and translational stage: examination of drug toxicity and efficacy. At each stage, ineligible drug candidates, which are either ineffective or toxic, are eliminated. The process has been estimated to take 10–15 years and cost over $2 billion US dollars per approved drug.

Only approximately 10% of drugs tested are approved by regulatory agencies and brought to market for human use.[10–12] Multiple factors contribute to the high rate of attrition in drug discovery, including the inefficacy of new compounds, in vivo and/or in vitro models that do not mirror the complexity of human biology, and variations in patient responses to drugs. An FDA review of drug development reported that 80%–90% of drug candidates entering phase 1 clinical trials fail. The most common causes of failure were lack of efficacy (accounting for 57% of failure) and toxicity. For instance, in a study involving Fragile X Mental Retardation Protein (FMR1) levels in Fragile X patients, out of the 5000 compounds screened, 4000 were

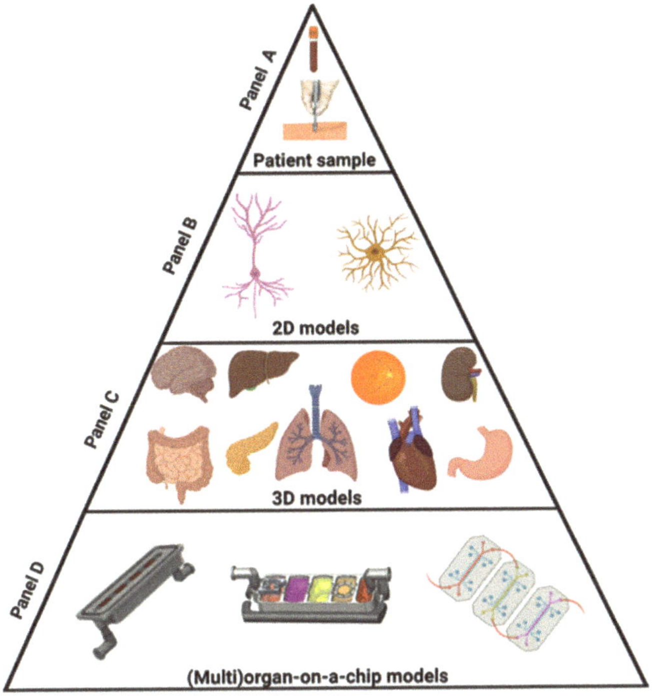

FIGURE 7.1 Human induced pluripotent stem cell (iPSC) derived models. Panel A: iPSCs involves the reprogramming of human peripheral blood mononuclear cells (PBMCs) or fibroblasts from skin biopsies. Panel B: Neuronal/glial progenitor cells are derived in monolayer, two-dimensional (2D) cultures from iPSCs. Panel C: three-dimensional organoids can be derived from iPSCs to generate diverse tissues. Panel D: Organ-on-chip platforms contain organoids interconnected by microfluid systems. *Created with BioRender.com.*

FDA-approved and six were identified as promising based on preclinical results. However, none of the selected compounds corrected FMR1 levels[13,14] and the trial was closed. Toxicities—most commonly involving the cardiovascular system, liver, gastrointestinal system, and brain—accounted for drug failure in 35% and 28% of phase 1 and phase 2 clinical trials, respectively. Toxicity remains one of the most critical problems for clinical trials as many reactions are not predicted in animal models: one study reported that 57% of 221 human toxicity events caused by 150 compounds were not predicted by preclinical rodent studies.[15]

Given the complexities of drug development, which translates into billions of dollars in costs and is accompanied by high attrition rates, it is in the best interests of the pharmaceutical industry, health agencies, and the public to improve the efficiency and success rates of this process. One of the emerging solutions is the identification of highly specific molecular targets, defined as molecules with (1) a specific role in pathology, (2) an available three-dimensional structure to assess drug binding and activity, and (3) chemical evidence of potential activity with the least amount of toxicity. Once such molecular targets are identified, deployment of specific compounds developed to modify them should improve drug-target kinetics, time-dependent drug activity, and treatment efficacy of a given disease. Another solution is development of highly specific drugs. Over the past decade, combinatorial chemistry paired with novel high-throughput functional assays, fragment screening,[16–18] and physiological screening have substantially improved the pipeline, accelerating the identification of putative drug candidates.[9,19–22] The compound libraries used for screenings are designed with different technologies, such as diversity-oriented synthesis aiming at generating small molecules with diverse appendages, functional groups, stereochemistry, and skeletons, which yield diverse biological activities capable of modulating a wide variety of biological processes.[23–25] Once the biophysical and pharmacokinetics studies are done and best hits are selected, the next step is to determine their efficacy and toxicity in the preclinical realm. However, it is important to remember that improvements in compound libraries and chemistry do not improve the chances for success or translate to increased clinical trial efficacy if the right targets are not identified in the first place. Finally, additional preclinical models that allow personalized medicine approach need to be developed to allow precise testing of the identified targets and drug compounds.

Human models as a screening tool for personalized precision medicine

The best models for drug development are those that account for the effects that genetic and environmental diversity from the patient population may have on basic tissue biology, are reproducible, and accurately reflect drug responses. Emerging technologies utilizing human pluripotent stem cells show great promise for developing such reliable, rapid, and cost-effective models to precisely reflect the diversity of human disease and responses to therapeutic agents that cannot be recapitulated in animals (Figs. 7.2 and 7.3).[26]

Monolayer models

To date, the majority of cell-based, high-throughput screens that use small compound libraries for lead identification utilize monolayers of immortalized cell lines grown on planar, rigid plastic surfaces optimized for cell attachment and growth. The homogeneity of these cultures allows for reproducible results across wells, which is critical for a successful screen. However, these cell lines fall short of modeling the molecular mechanisms found in human neurons and the utility of human neuronal cell culture models as preclinical testing platforms for drug development has only recently been appreciated. New modeling methods involve the use of human induced pluripotent stem cells (iPSCs), which are derived from patient skin or blood cells that have been reprogrammed into an embryonic-like pluripotent state through the introduction of key reprogramming factors.[27] This seminal discovery revolutionized the study of human disease through the ability to create a robust in vitro platform for patient-specific disease modeling, development of cell therapy, drug discovery, and preclinical validation, and provide an opportunity for personalized precision medicine.[28] Reprogramming somatic cells into iPSCs allows researchers to generate cells from patients with different disease severities, drug metabolism responses, and genetic backgrounds.[29,30] These patient-derived iPSCs can be expanded indefinitely and can differentiate into any cell type of the body, including neurons and glia, while retaining the same genetic background as the originating cells. Therefore, iPSC-derived models overcome many of the limitations recognized in animal models of disease and other popular in vitro models, such as immortalized cell lines and primary mouse neural cell types, which can be plagued by a lack of biological relevance and ethical issues with regard to animal models.

	iPSC-derived 2D models	iPSC-derived 3D models	Organ-on-a chip models	Mouse models
Physiological relevance	Limited	Better than 2D models	Better than 3D model	Most physiological environment
Cost	Low	Moderate to high	High	High
Ease of setup	Easy to moderately challenging	Very challenging	Challenging and special expertise required	Expertise required
Production	Fast	Slow	Slow	Time consuming
Cell types	Usually monotype; co-culture is possible	Highly diverse cell types similar to in vivo tissue	As in 3D models	Closest to the human brain but with intrinsic differences
Tissue architecture	Absent	Complex architecture reminiscent of organ development	Complex based on the design	Complex
Signal factors and nutrients diffusion	Short diffusion distance exchanged through cell membrane	Inefficient transport to the interior cells leading to cell death	Precisely controlled spatial and diffusive gradient	Physiological
Vascularization/ perfusion	Absent	Absent	Present	Present
Genome-editing capability	Easy	Hard	Hard	Hard
Efficient high-throughput screening	Yes	Possible	Possible	No

FIGURE 7.2 Comparisons among animal and different human models used for drugs studies.

In addition to the important discovery of transcription factors that reprogram somatic cells into pluripotent cells, subsequent studies have illuminated methods for differentiating pluripotent cells into various subtypes of neuronal and glial cells. With respect to neural cell types, iPSCs can differentiate into specified and highly homogeneous populations of neuroprogenitor cells, excitatory and inhibitory cortical neurons,[31–33] midbrain dopaminergic neurons,[34] motor neurons,[35] astrocytes,[36–38] oligodendrocytes,[39,40] and microglia.[41–45] The level of homogeneity and maturity of these cultures often depends on the method used, from transcription factors to patterning-based methods, as described in other chapters in this book. Typically, neurons differentiated using these methods are most similar to early embryonic neurons in terms of maturation level; however, these neurons can also be modified to acquire

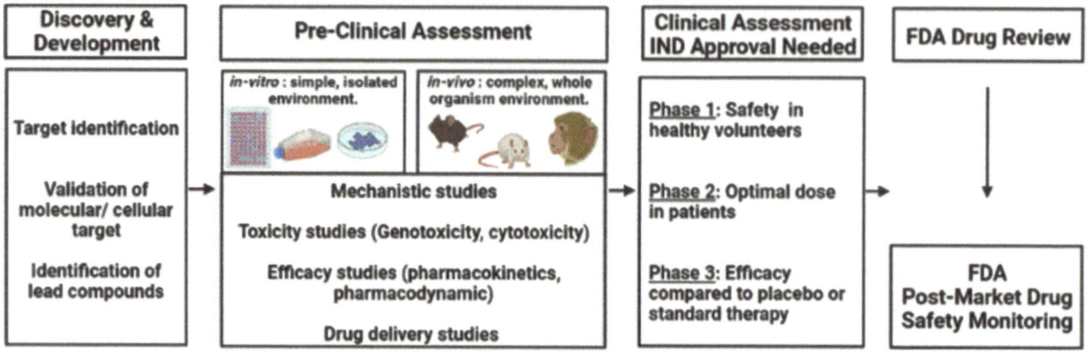

FIGURE 7.3 Drug discovery pipeline. Preclinical development begins with the discovery phase that involves in-silico, in-vitro, and in-vivo studies. A select number of hit compounds with desired therapeutic effect on biological targets are further developed and subjected to preclinical assessment. Filing and approval of the Investigational New Drug (IND) assesses all preclinical findings before clinical assessments commence. Increasing number of participants take part in clinical trials. Eventually, FDA Drug Review determines whether the drug can be approved for marketing. If approved, patients can start to be treated while FDA continues to monitor drug's safety. *Created with BioRender.com.*

age-specific features like senescence markers and decreased telomere length,[46] permitting studies of age-related neurological diseases.[47–49] In addition, these models have the added advantage of facilitating studies of personalized treatments using genetic engineering strategies like the CRISPR/Cas9 system. Gene editing can be used to introduce specific disease-causing mutations into normal donor iPSCs or to correct patient-derived mutant iPSCs at disease-causing loci to see if specific genetic changes are causative of disease phenotypes while maintaining the patient's genetic background.[50–53] Beyond gene correction, iPSC-derived neurons in monolayers can be used for small molecule screens to reverse phenotypes found in patient-derived neurons that are missing in healthy control neurons. These studies are critical for improving target identification that will hopefully translate to improved efficacy in clinical trials.

Numerous studies have underscored the utility of these patient-specific models to glean insights into disease pathology. Here, we highlight a few that used patient-derived two-dimensional (2D) monolayers of neurons. In Phelan-McDermid Syndrome, for example, studies of patient-derived neurons revealed that the attenuated SHANK3 expression caused altered excitatory synaptic transmission,[54] reported for the first time in human cells. Recently, seminal studies of the 16p copy number variation effects on patient-derived dopaminergic neurons have pointed to altered neuronal soma size, synaptic marker expression, neuronal differentiation and hyperactivity compared to isogenic control lines, which could be corrected with Rhosin, an inhibitor of the RhoA pathway.[55] In Fragile X, the first studies of iPSC-derived neuroprogenitors have observed an abnormal cell cycle and aberrant proliferation,[56] confirming findings from murine neural stem cells[57] and other reports on Fragile X patient-derived neuroprogenitors.[58] Most recently, Bhattacharyya and collaborators have generated new gene edited reporter cell lines allowing high throughput small molecule screens for untargeted discovery of those molecules that activate human FMR1 gene.[59]

In addition to neurological disorders, drug screenings for certain neuropsychiatric conditions have taken advantage of neuronal 2D models. For example, in the past, we learned the most about human brain pathology in schizophrenia from postmortem studies.[60] Thus, human models that allow mechanistic investigations of this disease have been highly welcomed. For example, using schizophrenia patients-derived iPSCs, Brennand and collaborators (2011) studied the effects of five antipsychotic drugs on neuronal connectivity.[61] To specifically evaluate synaptic connectivity, the authors used rabies virus and fluorescent reporters to label synapses and sorted them to quantitatively compare drugs' impact on neuronal connectivity. These neuronal 2D studies clarified the underlying molecular mechanisms in schizophrenia and underscored the aberrant synaptic plasticity as one of the key drivers of pathology.[61]

Finally, human 2D neuroprogenitor models provided key knowledge about the pathophysiological mechanisms of Zika virus infection and potential drug treatments for the disease. For example, a 6000-compound screen examining Zika infection of iPSC-derived neuroprogenitors identified Emricasan and PHA-690509 as some of the hit compounds that prevented Zika infection.[62] A second screen utilizing 1000 FDA-approved drugs on brain organoids identified another set of hit compounds that prevented infection, including Hippeastrine and Amodiaquine.[63] Based on all these studies, it has become apparent that drugs that show high efficacy in monolayer in vitro models do not always have the same level of efficacy in vivo. This is partially explained by the fact that cells grown in monolayer cultures do not have a complex three-dimensional tissue architecture and do not exactly reflect the complex interactions between the extracellular matrix and other microenvironmental factors influencing cellular phenotype and drug responses.[64,65] Thus, the inability of these 2D models to recapitulate in vivo-like cytoarchitectural organization, the diversity of synaptic connections in the brain, and cell morphology and density in native microenvironments[66] has impeded their use in precise disease modeling and drug screening applications.

Organoids

More recently, in an attempt to better reproduce three-dimensional (3D) organization of tissues, a new model system utilizing human iPSCs has emerged. Organoids are self-organizing, stem-cell-derived structures that resemble in vivo tissue counterparts in both cell composition and organ architecture more successfully than 2D models. Because the nomenclature of these models is still under debate, the terms "organoids" and "spheroids" are used interchangeably. However, in general, spheroids are cell aggregates made from immortalized cell lines, primary cell cultures, or primary human tissues such as tumors. They were first observed in the 1970s, when hamster lung cells grown in suspension arranged themselves into a nearly perfect spherical form. Organoids are organ-specific structures generated from stem/progenitor cells that self-assemble. Their formation is driven by internal developmental programs while spheroids develop primarily through cell-to-cell adhesion. Organoids can be derived from primary tissue stem cells, embryonic stem cells, or iPSCs.[67–71] Multiple organoid differentiation methods have been developed to generate organoids for different tissues, including the generation of retinal,[72–74] cerebral,[75–80] hippocampal,[81] cerebellar,[82] and mid-brain organoids,[83,84] esophageal,[85] gut,[86–88] liver,[89] pancreas organoids,[90–92] heart,[93] lung,[94–98] and kidney.[99,100] Organoids have quickly become an important tool for basic and translational research with wide applications for disease modeling, drug screening, drug optimization, and personalized and regenerative medicine.

Because organoids are produced from patient samples, the patient-specific genetic background is preserved, making them an ideal model system to study genetic disorders and develop gene or other therapies to counteract them. For example, gene-editing (CRISPR, CRISPR/Cas9) is now used to incorporate specific genetic modifications to study the effects of specific genes on organ pathology or to counteract existing mutations and rescue cellular functions. The complex responses of the variety of cell types in the organoid can then be studied using multi-omic proteome, transcriptome, and metabolome profiling for molecular characterization,[101,102] calcium imaging and neuronal transmission for electrophysiology characterization, and immunohistochemistry for morphological evaluation,[103] all leading to better understanding of the mechanism of disease.[104–106] In addition, patient-specific organoid models are particularly useful for studies of patients harboring rare mutations that require development of personalized drug regimens. Unlike 2D models, 3D models replicate treatment responses as in vivo studies.[107] In a striking example, Dekkers and colleagues pioneered the use of intestinal organoids as a valuable model of cystic fibrosis for drug testing.[108] They used patient-specific intestinal organoids to identify the drug VX-770 and successfully treated two patients with the same rare mutations causing cystic fibrosis,[109] pioneering future research toward personalized drug discovery for therapy of genetic syndromes.

Numerous studies have demonstrated the value of brain organoids for mechanistic and drug development studies.[52,110–112] For example, Raj and collaborators (2021) modeled neurogenesis in Fragile X syndrome using iPSC-derived cortical organoids and showed that overactive PI3K signaling leads to over-proliferation of neuroprogenitors and less differentiation into mature neurons. Observed defects in cell fate could be rescued by inhibiting PI3K signaling using p110β and/or S6K1 inhibitors. This corrected both excess protein synthesis and cell proliferation in a subset of patient neural cells.[56] Thus, the organoid model system can be used to screen for compounds that reverse such molecular and cellular changes. In addition, progress has been made in studies of Tuberous Sclerosis, a neurocutaneous syndrome characterized by intellectual disability and autism in about half of the patients. Blair and collaborators (2018) established human cellular models of Tuberous Sclerosis using CRISPR/Cas9-mediated gene editing to establish loss of function of Tuberous Sclerosis Complex 1 (TSC1) or TSC2 genes (causative of Tuberous Sclerosis) in iPSCs used to generate brain organoids. Many of the developmental differences seen in 2D neuronal cultures were also observed in brain organoids, including a strong bias toward an astroglial cell fate, altered cell morphology, cytomegaly, and hyperactivation of mTORC1 signaling. Using heterozygous TSC2 iPSCs with a conditional mutation in the functional allele, the authors investigated the "two-hit" hypothesis of cortical tuber development in brain organoids and demonstrated causal evidence that second-hit mutations are necessary and sufficient to generate dysplastic cells in the developing human cortex. Interestingly, rapamycin analogues and other mTOR inhibitors such as AZD-8055, which have been used in rodent and human studies for treatment of Tuberous Sclerosis, rescued neuronal differentiation and morphology.[113] These results show that brain organoids can be used to identify additional compounds that reverse phenotypes found in TSC2-deficient neurons that may not have the side effects found in rapamycin. In another example, midbrain organoids derived from iPSCs with a Leucine-Rich Repeat Kinase 2 (LRRK2) G2019S mutation associated with Parkinson's disease showed pathological hallmarks of the disease, such as decreased expression of dopaminergic and mature neuronal markers, and reduced neurite length. Additional studies using this model uncovered a new target, the Thioredoxin Interacting Protein (TXNIP) (a core cellular redox signaling regulator), which modulates α-synuclein levels and could potentially be targeted with novel drug compounds.[114]

Further, patient-derived forebrain organoids with DISC1 mutations have been used to study phenotypes and affected pathways in schizophrenia. Altered cell cycle progression proliferation, decreased neuronal development, an overactivated WNT signaling pathway, and reduced expression of Fibroblast Growth Factor Receptor 1 (FGFR1) protein were all uncovered as potential novel drug targets in schizophrenia patients bearing DISC1 mutations using brain organoids.[115–117]

Finally, organoids have proven useful for studies of glioblastoma. The first 3D glioblastoma model was reported by Galli et al. in 2004, consisting of 3D neurospheres that displayed some degree of polarization and spatial organization.[118] Recently, organoids have been generated from human glioblastoma cells which remarkably reproduced in vivo stem cell density, gradient, and morphology,[119,120] as well as responses to drug treatments, such as temozolomide and radiation therapy.[120] These organoids offer the potential to conduct genotype-drug association analyses and personalized drug screening studies.[120,121] These are just some examples demonstrating the power of organoid technology to pave the way for future studies of personalized precision therapies for a wide variety of diseases.

While human brain organoid models provide better representation of human tissues compared to cell lines, they are still not ideal. Some limitations include the lack of a native microenvironment (e.g., extra cellular matrix composition and growth factor gradients), vascularization, and the presence of immune cells such as microglia, which restricts the delivery of nutrients and growth factors to the organoid core and modeling immune responses and immune cell interactions. As penetration of the blood-brain barrier is critical for successful delivery of drug compounds to the brain, lack of it in organoid models poses a major pitfall for drug screens. Organoids containing blood-brain barrier are now being generated to carry out permeability studies on drug candidates. These novel strategies to vascularize brain organoids and to model the blood-brain-barrier[122,123] promise to accelerate the use of organoids for drug development and screening.

Organ-on-chip platforms

Another major challenge for the use of organoid models for drug screening is the difficulty of preserving individual tissue phenotypes over long periods while allowing communication between different tissues. Although tissues cultured in isolation can recapitulate some aspects of physiological functions, drug screens and pharmacokinetic studies require multiorgan interactions. To this end, different groups have developed Micro Physiological Systems (MPS), in vitro platforms that contain bioengineered human tissues connected into physiologically relevant integration units (microfluidic cell culture chips), called "organ-on-chip" platforms.[124,125] These platforms are designed to maintain the tissues and provide vascular niches, thereby preserving mature tissue fidelity, facilitating tissue communication by vascular perfusion, and upholding the separation of interstitial and intravascular compartments by selectively permeable endothelial barriers.[126] These complex model systems support drug testing and toxicity studies by mimicking more physiologically relevant microanatomies and processes while offering fine control over cellular composition, function, and outputs: linking the tissues by interconnecting fluid enables crosstalk between tissues and facilitates studies of drug delivery, distribution, and uptake.[127–130] Different organs-on-a-chip can be connected according to physiological hierarchy, creating a "body-on-chip" to better mimic the in vivo environment.[127,128,131] For example, Chramiec and collaborators (2020) reported the generation of an integrated three-organs-on-chip (heart-liver-vascular) for in vitro studies. Similarly, the generation of a four-organs-on-chip (intestine-liver-skin-kidney) was recently reported.[128] MPS models are particularly useful for high-throughput drug screening as they require only nanograms to micrograms of a given compound for testing. Recognizing the huge potential of these models to improve drug screens and preclinical trials and thereby reduce drug failure rates, the National Institute of Health and Defense Advanced Research Projects Agency have worked together with the FDA and other agencies to launch a series of Organ-On-a-Chip programs to foster the use of this technology for drug development pipelines.[132] Given that toxicology studies are critical for passing a drug candidate to next stage of testing, these multiorgan platforms could quickly inform which key organs are particularly prone to toxicity and whether the drug or its products accumulate somewhere. Moreover, because the critical window for therapy could be different in different patients, the organ-on-chip platforms could assess this question more effectively and on a more personalized level compared to traditional animal models.

However, as all models do, the MPS suffers from limitations, such as standardization of the design and quality controls. These are important considerations as numerous organs require different support systems, media, and timeframe for development, which can influence their longevity in the MPS platform and data interpretation. To support the development of the next generation of these models, the National Center for Advancing Translational Sciences initiated a multicenter study that involves independent quality control studies of a diverse range of models by The Tissue Chip Testing Centers at Massachusetts Institute of Technology and Texas A&M University; data from the studies are deposited in a database developed by the University of Pittsburgh. Similarly, the European Organ-on-Chip Society supervises the implementation of the European Organ-on-Chip Roadmap.

Finally, 3D bioprinting has emerged as a new tool that leverages the strengths of organoid and microfluidic systems to produce living tissue, bone, blood vessels, and, potentially, whole organs.[133] 3D organ bioprinting allows for the assembly of multiple cell types or stem cells/growth factors along with other biomaterials in a layer-by-layer fashion to produce bioartificial organs that maximally imitate their natural counterparts. A critical component of successful bioprinting of clinically useful tissue is the selection of optimal biomaterials, since many could be biologically active and cause unwanted cellular interactions and premature or undesired stem cell differentiation. Additional limitation of bioprinting is the absence of vascular networks, which can lead to poor nutrition of engineered tissue and incomplete tissue formation or necrosis.[134] Nevertheless, these models herald a new generation of tools capable of developing personalized precision therapy for a variety of human diseases.

Conclusion

Recent advances in human models of disease have paved the way for a paradigm shift in pharmaceutical drug discovery and development. These models offer several advantages over traditional cell lines as they provide not only human cells for toxicity testing, but also an avenue for personalized precision medicine. Human models are particularly useful for studies of pathology, drug efficacy, and safety for those neurodevelopmental disorders that start to develop in utero. The human cerebral cortex expands substantially during prenatal development, which presents a significant challenge for the discovery of critical processes that are not easily replicated in animal models and cannot be studied in live human fetuses. As early events during gestation may underlie many neurodevelopmental and neuropsychiatric disorders that manifest later in life, human brain organoids represent an invaluable model for advancing our understanding of early disease development and pathology, accelerating the development of better therapies for these disorders. In addition, human models are useful for orphan diseases, as these disorders are rare and often do not attract the substantial investments required for drug development. Testing gene therapies in patient-specific models is also more precise than testing them in mouse models where the genetic background is far less complex than in humans. Finally, toxicology studies in human models should be considered as an add-on to preclinical studies, given that certain metabolic processes in humans differ from those in other mammals.

Although the generation of patient-specific models is time-consuming and costly, the benefits for not only the pharmaceutical industry, but also for public health and individual patients are enormous. The direct involvement of a patient in a clinical trial that tests his/her own cells/tissues in a human model platform will ensure better insight, engagement, and ownership of the drug development process and, ultimately, better therapies for all patients.

References

1. Singh VK, Seed TM. How necessary are animal models for modern drug discovery? *Expert Opin Drug Discov.* 2021;16(12):1391–1397. https://doi.org/10.1080/17460441.2021.1972255.
2. Zuberi A, Lutz C. Mouse models for drug discovery. Can new tools and technology improve translational power? *ILAR J.* 2016;57(2):178–185. https://doi.org/10.1093/ilar/ilw021.
3. Hodge RD, Bakken TE, Miller JA, et al. Conserved cell types with divergent features in human versus mouse cortex. *Nature.* 2019;573(7772):61–68. https://doi.org/10.1038/s41586-019-1506-7.
4. Zecevic N, Chen Y, Filipovic R. Contributions of cortical subventricular zone to the development of the human cerebral cortex. *J Comp Neurol.* 2005;491(2):109–122. https://doi.org/10.1002/cne.20714.
5. Kitaeva KV, Rutland CS, Rizvanov AA, Solovyeva VV. Cell culture based in vitro test systems for anticancer drug screening. *Front Bioeng Biotechnol.* 2020;8. https://doi.org/10.3389/fbioe.2020.00322.
6. Stevens JL, Baker TK. The future of drug safety testing: expanding the view and narrowing the focus. *Drug Discov Today.* 2009;14(3–4):162–167. https://doi.org/10.1016/j.drudis.2008.11.009.
7. Kim J, Koo BK, Knoblich JA. Human organoids: model systems for human biology and medicine. *Nat Rev Mol Cell Biol.* 2020;21(10):571–584. https://doi.org/10.1038/s41580-020-0259-3.
8. Pound P, Ritskes-Hoitinga M. Is it possible to overcome issues of external validity in preclinical animal research? Why most animal models are bound to fail. *J Transl Med.* 2018;16(1). https://doi.org/10.1186/s12967-018-1678-1.
9. Hughes JP, Rees SS, Kalindjian SB, Philpott KL. Principles of early drug discovery. *Br J Pharmacol.* 2011;162(6):1239–1249. https://doi.org/10.1111/j.1476-5381.2010.01127.x.

10. Hay M, Thomas DW, Craighead JL, Economides C, Rosenthal J. Clinical development success rates for investigational drugs. *Nat Biotechnol*. 2014;32(1):40–51. https://doi.org/10.1038/nbt.2786.
11. Van Norman GA. Drugs, Devices, and the FDA: Part 2: an Overview of approval processes: FDA approval of medical Devices. *J Am Coll Cardiol Basic Transl Sci*. 2016;1(4):277–287. https://doi.org/10.1016/j.jacbts.2016.03.009.
12. Wong CH, Siah KW, Lo AW. Estimation of clinical trial success rates and related parameters. *Biostatistics*. 2019;20(2):273–286. https://doi.org/10.1093/biostatistics/kxx069.
13. Kaufmann M, Schuffenhauer A, Fruh I, et al. High-throughput screening using iPSC-derived neuronal progenitors to identify compounds counteracting epigenetic gene silencing in fragile X syndrome. *J Biomol Screen*. 2015;20(9):1101–1111. https://doi.org/10.1177/1087057115588287.
14. Kumari D, Swaroop M, Southall N, Huang W, Zheng W, Usdin K. High-throughput screening to identify compounds that increase fragile x mental retardation protein expression in neural stem cells differentiated from fragile x syndrome patient-derived induced pluripotent stem cells. *Stem Cells Transl Med*. 2015;4(7):800–808. https://doi.org/10.5966/sctm.2014-0278.
15. Shanks N, Greek R, Greek J. Are animal models predictive for humans? *Philos Ethics Humanit Med*. 2009;4(1). https://doi.org/10.1186/1747-5341-4-2.
16. Chamakuri S, Chung M-K, Samuel ELG, et al. Design and construction of a stereochemically diverse piperazine-based DNA-encoded chemical library. *Bioorg Med Chem*. 2021;48:116387. https://doi.org/10.1016/j.bmc.2021.116387.
17. Chamakuri S, Shah MM, Yang DCH, Santini C, Young DW. Practical and scalable synthesis of orthogonally protected-2-substituted chiral piperazines. *Org Biomol Chem*. 2020;18(43):8844–8849. https://doi.org/10.1039/d0ob01713b.
18. Chang L, Campbell J, Raji IO, et al. Discovery of small molecules targeting the tandem tudor domain of the epigenetic factor UHRF1 using fragment-based ligand discovery. *Sci Rep*. 2021;11(1). https://doi.org/10.1038/s41598-020-80588-4.
19. Lage OM, Ramos MC, Calisto R, Almeida E, Vasconcelos V, Vicente F. Current screening methodologies in drug discovery for selected human diseases. *Mar Drugs*. 2018;16(8). https://doi.org/10.3390/md16080279.
20. Law R, Barker O, Barker JJ, et al. The multiple roles of computational chemistry in fragment-based drug design. *J Comput Aided Mol Des*. 2009;23(8):459–473. https://doi.org/10.1007/s10822-009-9284-1.
21. Lee JA, Uhlik MT, Moxham CM, Tomandl D, Sall DJ. Modern phenotypic drug discovery is a viable, neoclassic pharma strategy. *J Med Chem*. 2012;55(10):4527–4538. https://doi.org/10.1021/jm201649s.
22. Phatak SS, Stephan CC, Cavasotto CN. High-throughput and in silico screenings in drug discovery. *Expert Opin Drug Discov*. 2009;4(9):947–959. https://doi.org/10.1517/17460440903190961.
23. Chauhan J, Luthra T, Gundla R, Ferraro A, Holzgrabe U, Sen S. A diversity oriented synthesis of natural product inspired molecular libraries. *Org Biomol Chem*. 2017;15(43):9108–9120. https://doi.org/10.1039/c7ob02230a.
24. Horman SR. *Complex High-Content Phenotypic Screening*. InTech; 2016. https://doi.org/10.5772/65387.
25. Monty OBC, Simmons N, Chamakuri S, Matzuk MM, Young DW. Solution-phase Fmoc-based Peptide synthesis for DNA-encoded chemical libraries: reaction conditions, protecting group strategies, and pitfalls. *ACS Comb Sci*. 2020;22(12):833–843. https://doi.org/10.1021/acscombsci.0c00144.
26. Rossetti AC, Koch P, Ladewig J. Drug discovery in psychopharmacology: from 2D models to cerebral organoids. *Dialogues Clin Neurosci*. 2019;21(2):203–210. https://doi.org/10.31887/DCNS.2019.21.2/jladewig.
27. Takahashi K, Tanabe K, Ohnuki M, et al. Induction of pluripotent stem cells from adult human fibroblasts by defined factors. *Cell*. 2007;131(5):861–872. https://doi.org/10.1016/j.cell.2007.11.019.
28. Soldner F, Jaenisch R. iPSC disease modeling. *Science*. 2012;338(6111):1155–1156. https://doi.org/10.1126/science.1227682.
29. Chang CY, Ting HC, Liu CA, et al. Induced pluripotent stem cells: a powerful neurodegenerative disease modeling tool for mechanism study and drug discovery. *Cell Transplant*. 2018;27(11):1588–1602. https://doi.org/10.1177/0963689718775406.
30. Hung SSC, Khan S, Lo CY, Hewitt AW, Wong RCB. Drug discovery using induced pluripotent stem cell models of neurodegenerative and ocular diseases. *Pharmacol Ther*. 2017;177:32–43. https://doi.org/10.1016/j.pharmthera.2017.02.026.

31. Pang ZP, Yang N, Vierbuchen T, et al. Induction of human neuronal cells by defined transcription factors. *Nature*. 2011;476(7359):220–223. https://doi.org/10.1038/nature10202.
32. Yang N, Chanda S, Marro S, et al. Generation of pure GABAergic neurons by transcription factor programming. *Nat Methods*. 2017;14(6):621–628. https://doi.org/10.1038/nmeth.4291.
33. Zhang Y, Pak CH, Han Y, et al. Rapid single-step induction of functional neurons from human pluripotent stem cells. *Neuron*. 2013;78(5):785–798. https://doi.org/10.1016/j.neuron.2013.05.029.
34. Stathakos P, Jiménez-Moreno N, Crompton LA, et al. A monolayer hiPSC culture system for autophagy/mitophagy studies in human dopaminergic neurons. *Autophagy*. 2021;17(4):855–871. https://doi.org/10.1080/15548627.2020.1739441.
35. Sepehrimanesh M, Ding B. Generation and optimization of highly pure motor neurons from human induced pluripotent stem cells via lentiviral delivery of transcription factors. *Am J Physiol Cell Physiol*. 2020;319(4):C771–C780. https://doi.org/10.1152/ajpcell.00279.2020.
36. Byun JS, Lee CO, Oh M, et al. Rapid differentiation of astrocytes from human embryonic stem cells. *Neurosci Lett*. 2020;716:134681. https://doi.org/10.1016/j.neulet.2019.134681.
37. Raman S, Srinivasan G, Brookhouser N, et al. A defined and scalable peptide-based platform for the generation of human pluripotent stem cell-derived astrocytes. *ACS Biomater Sci Eng*. 2020;6(6):3477–3490. https://doi.org/10.1021/acsbiomaterials.0c00067.
38. TCW J, Wang M, Pimenova AA, et al. An efficient platform for astrocyte differentiation from human induced pluripotent stem cells. *Stem Cell Rep*. 2017;9(2):600–614. https://doi.org/10.1016/j.stemcr.2017.06.018.
39. García-León JA, García-Díaz B, Eggermont K, et al. Generation of oligodendrocytes and establishment of an all-human myelinating platform from human pluripotent stem cells. *Nat Protoc*. 2020;15(11):3716–3744. https://doi.org/10.1038/s41596-020-0395-4.
40. Wang C, Luan Z, Yang Y, et al. High purity of human oligodendrocyte progenitor cells obtained from neural stem cells: Suitable for clinical application. *J Neurosci Methods*. 2015;240:61–66. https://doi.org/10.1016/j.jneumeth.2014.10.017.
41. Banerjee P, Paza E, Perkins EM, et al. Generation of pure monocultures of human microglia-like cells from induced pluripotent stem cells. *Stem Cell Res*. 2020;49:102046. https://doi.org/10.1016/j.scr.2020.102046.
42. Douvaras P, Sun B, Wang M, et al. Directed differentiation of human pluripotent stem cells to microglia. *Stem Cell Rep*. 2017;8(6):1516–1524. https://doi.org/10.1016/j.stemcr.2017.04.023.
43. McQuade A, Coburn M, Tu CH, Hasselmann J, Davtyan H, Blurton-Jones M. Development and validation of a simplified method to generate human microglia from pluripotent stem cells. *Mol Neurodegener*. 2018;13(1). https://doi.org/10.1186/s13024-018-0297-x.
44. Muffat J, Li Y, Yuan B, et al. Efficient derivation of microglia-like cells from human pluripotent stem cells. *Nat Med*. 2016;22(11):1358–1367. https://doi.org/10.1038/nm.4189.
45. Popova G, Soliman SS, Kim CN, et al. Human microglia states are conserved across experimental models and regulate neural stem cell responses in chimeric organoids. *Cell Stem Cell*. 2021;28(12):2153–2166. https://doi.org/10.1016/j.stem.2021.08.015. e6.
46. Liu GH, Ding Z, Izpisua Belmonte JC. IPSC technology to study human aging and aging-related disorders. *Curr Opin Cell Biol*. 2012;24(6):765–774. https://doi.org/10.1016/j.ceb.2012.08.014.
47. Ashapkin VV, Kutueva LI, Kurchashova SY, Kireev II. Are there common mechanisms between the Hutchinson-Gilford progeria syndrome and natural aging? *Front Genet*. 2019;10(MAY). https://doi.org/10.3389/fgene.2019.00455.
48. Mertens J, Herdy JR, Traxler L, et al. Age-dependent instability of mature neuronal fate in induced neurons from Alzheimer's patients. *Cell Stem Cell*. 2021;28(9):1533–1548. https://doi.org/10.1016/j.stem.2021.04.004. e6.
49. Miller JD, Ganat YM, Kishinevsky S, et al. Human iPSC-based modeling of late-onset disease via progerin-induced aging. *Cell Stem Cell*. 2013;13(6):691–705. https://doi.org/10.1016/j.stem.2013.11.006.
50. Cong L, Ran FA, Cox D, et al. Multiplex genome engineering using CRISPR/Cas systems. *Science*. 2013;339(6121):819–823. https://doi.org/10.1126/science.1231143.
51. González F, Zhu Z, Shi ZD, et al. An iCRISPR platform for rapid, multiplexable, and inducible genome editing in human pluripotent stem cells. *Cell Stem Cell*. 2014;15(2):215–226. https://doi.org/10.1016/j.stem.2014.05.018.
52. Hoffman GE, Schrode N, Flaherty E, Brennand KJ. New considerations for hiPSC-based models of neuropsychiatric disorders. *Mol Psychiatr*. 2019;24(1):49–66. https://doi.org/10.1038/s41380-018-0029-1.

53. Musunuru K. Genome editing of human pluripotent stem cells to generate human cellular disease models. *Dis Models Mech*. 2013;6(4). https://doi.org/10.1242/dmm.012054.
54. Shcheglovitov A, Shcheglovitova O, Yazawa M, et al. SHANK3 and IGF1 restore synaptic deficits in neurons from 22q13 deletion syndrome patients. *Nature*. 2013;503(7475):267–271. https://doi.org/10.1038/nature12618.
55. Sundberg M, Pinson H, Smith RS, et al. 16p11.2 deletion is associated with hyperactivation of human iPSC-derived dopaminergic neuron networks and is rescued by RHOA inhibition in vitro. *Nat Commun*. 2021;12(1). https://doi.org/10.1038/s41467-021-23113-z.
56. Raj N, McEachin ZT, Harousseau W, et al. Cell-type-specific profiling of human cellular models of fragile X syndrome reveal PI3K-dependent defects in translation and neurogenesis. *Cell Rep*. 2021;35(2):108991. https://doi.org/10.1016/j.celrep.2021.108991.
57. Liu B, Li Y, Stackpole EE, et al. Regulatory discrimination of mRNAs by FMRP controls mouse adult neural stem cell differentiation. *Proc Natl Acad Sci USA*. 2018;115(48):E11397–E11405. https://doi.org/10.1073/pnas.1809588115.
58. Sunamura N, Iwashita S, Enomoto K, Kadoshima T, Isono F. Loss of the fragile X mental retardation protein causes aberrant differentiation in human neural progenitor cells. *Sci Rep*. 2018;8(1). https://doi.org/10.1038/s41598-018-30025-4.
59. Hunt JFV, Li M, Risgaard R, et al. High throughput small molecule screen for reactivation of fmr1 in fragile x syndrome human neural cells. *Cells*. 2022;11(1). https://doi.org/10.3390/cells11010069.
60. Eastwood SL, Cairns NJ, Harrison PJ. Synaptophysin gene expression in schizophrenia: Investigation of synaptic pathology in the cerebral cortex. *Br J Psychiatry*. 2000;176:236–242. https://doi.org/10.1192/bjp.176.3.236.
61. Brennand KJ, Simone A, Jou J, et al. Modelling schizophrenia using human induced pluripotent stem cells. *Nature*. 2011;473(7346):221–225. https://doi.org/10.1038/nature09915.
62. Xu M, Lee EM, Wen Z, et al. Identification of small-molecule inhibitors of Zika virus infection and induced neural cell death via a drug repurposing screen. *Nat Med*. 2016;22(10):1101–1107. https://doi.org/10.1038/nm.4184.
63. Zhou T, Tan L, Cederquist GY, et al. High-content screening in hPSC-neural progenitors identifies drug candidates that Inhibit Zika virus infection in fetal-like organoids and adult brain. *Cell Stem Cell*. 2017;21(2):274–283. https://doi.org/10.1016/j.stem.2017.06.017. e5.
64. Langhans SA. Three-dimensional in vitro cell culture models in drug discovery and drug repositioning. *Front Pharmacol*. 2018;9. https://doi.org/10.3389/fphar.2018.00006.
65. Rizvanov AA, Yalvaç ME, Shafigullina AK, et al. Interaction and self-organization of human mesenchymal stem cells and neuro-blastoma SH-SY5Y cells under co-culture conditions: a novel system for modeling cancer cell micro-environment. *Eur J Pharm Biopharm*. 2010;76(2):253–259. https://doi.org/10.1016/j.ejpb.2010.05.012.
66. Miranda C, Fernandes T, Diogo M, Cabral J. Towards multi-organoid systems for drug screening applications. *Bioengineering*. 2018;5(3):49. https://doi.org/10.3390/bioengineering5030049.
67. Artegiani B, Hendriks D, Beumer J, et al. Fast and efficient generation of knock-in human organoids using homology-independent CRISPR–Cas9 precision genome editing. *Nat Cell Biol*. 2020;22(3):321–331. https://doi.org/10.1038/s41556-020-0472-5.
68. Clevers H. Modeling development and disease with organoids. *Cell*. 2016;165(7):1586–1597. https://doi.org/10.1016/j.cell.2016.05.082.
69. Sasai Y. Next-generation regenerative medicine: Organogenesis from stem cells in 3D culture. *Cell Stem Cell*. 2013;12(5):520–530. https://doi.org/10.1016/j.stem.2013.04.009.
70. Takebe T, Wells JM. Organoids by design. *Science*. 2019;364(6444):956–959. https://doi.org/10.1126/science.aaw7567.
71. Yoon SJ, Elahi LS, Pasca AM, et al. Reliability of human cortical organoid generation. *Nat Methods*. 2019;16(1):75–78. https://doi.org/10.1038/s41592-018-0255-0.
72. Eiraku M, Takata N, Ishibashi H, et al. Self-organizing optic-cup morphogenesis in three-dimensional culture. *Nature*. 2011;472(7341):51–56. https://doi.org/10.1038/nature09941.
73. Nakano T, Ando S, Takata N, et al. Self-formation of optic cups and storable stratified neural retina from human ESCs. *Cell Stem Cell*. 2012;10(6):771–785. https://doi.org/10.1016/j.stem.2012.05.009.
74. Völkner M, Zschätzsch M, Rostovskaya M, et al. Retinal organoids from pluripotent stem cells efficiently recapitulate Retinogenesis. *Stem Cell Rep*. 2016;6(4):525–538. https://doi.org/10.1016/j.stemcr.2016.03.001.

75. Giandomenico SL, Sutcliffe M, Lancaster MA. Generation and long-term culture of advanced cerebral organoids for studying later stages of neural development. *Nat Protoc*. 2021;16(2):579–602. https://doi.org/10.1038/s41596-020-00433-w.
76. Lancaster MA, Knoblich JA. Generation of cerebral organoids from human pluripotent stem cells. *Nat Protoc*. 2014;9(10):2329–2340. https://doi.org/10.1038/nprot.2014.158.
77. Lancaster MA, Renner M, Martin CA, et al. Cerebral organoids model human brain development and microcephaly. *Nature*. 2013;501(7467):373–379. https://doi.org/10.1038/nature12517.
78. Pasca AM, Sloan SA, Clarke LE, et al. Functional cortical neurons and astrocytes from human pluripotent stem cells in 3D culture. *Nat Methods*. 2015;12(7):671–678. https://doi.org/10.1038/nmeth.3415.
79. Pollen AA, Bhaduri A, Andrews MG, et al. Establishing cerebral organoids as models of human-specific brain evolution. *Cell*. 2019;176(4):743–756. https://doi.org/10.1016/j.cell.2019.01.017. e17.
80. Qian X, Jacob F, Song MM, Nguyen HN, Song H, Ming GL. Generation of human brain region–specific organoids using a miniaturized spinning bioreactor. *Nat Protoc*. 2018;13(3):565–580. https://doi.org/10.1038/nprot.2017.152.
81. Sakaguchi H, Kadoshima T, Soen M, et al. Generation of functional hippocampal neurons from self-organizing human embryonic stem cell-derived dorsomedial telencephalic tissue. *Nat Commun*. 2015;6(1). https://doi.org/10.1038/ncomms9896.
82. Muguruma K, Nishiyama A, Kawakami H, Hashimoto K, Sasai Y. Self-organization of polarized cerebellar tissue in 3D culture of human pluripotent stem cells. *Cell Rep*. 2015;10(4):537–550. https://doi.org/10.1016/j.celrep.2014.12.051.
83. Jo J, Xiao Y, Sun AX, et al. Midbrain-like organoids from human pluripotent stem cells contain functional dopaminergic and neuromelanin-producing neurons. *Cell Stem Cell*. 2016;19(2):248–257. https://doi.org/10.1016/j.stem.2016.07.005.
84. Monzel AS, Smits LM, Hemmer K, et al. Derivation of human midbrain-specific organoids from neuroepithelial stem cells. *Stem Cell Rep*. 2017;8(5):1144–1154. https://doi.org/10.1016/j.stemcr.2017.03.010.
85. Kasagi Y, Chandramouleeswaran PM, Whelan KA, et al. The esophageal organoid system reveals functional Interplay between Notch and Cytokines in reactive epithelial changes. *Cell Mol Gastroenterol Hepatol*. 2018;5(3):333–352. https://doi.org/10.1016/j.jcmgh.2017.12.013.
86. Fujii M, Matano M, Nanki K, Sato T. Efficient genetic engineering of human intestinal organoids using electroporation. *Nat Protoc*. 2015;10(10):1474–1485. https://doi.org/10.1038/nprot.2015.088.
87. Seidlitz T, Merker SR, Rothe A, et al. Human gastric cancer modelling using organoids. *Gut*. 2019;68(2):207–217. https://doi.org/10.1136/gutjnl-2017-314549.
88. Wosen JE, Ilstad-Minnihan A, Co JY, et al. Human intestinal enteroids model MHC-II in the gut epithelium. *Front Immunol*. 2019;10. https://doi.org/10.3389/fimmu.2019.01970.
89. Takebe T, Sekine K, Enomura M, et al. Vascularized and functional human liver from an iPSC-derived organ bud transplant. *Nature*. 2013;499(7459):481–484. https://doi.org/10.1038/nature12271.
90. Boj SF, Hwang CI, Baker LA, et al. Organoid models of human and mouse ductal pancreatic cancer. *Cell*. 2015;160(1–2):324–338. https://doi.org/10.1016/j.cell.2014.12.021.
91. Broutier L, Andersson-Rolf A, Hindley CJ, et al. Culture and establishment of self-renewing human and mouse adult liver and pancreas 3D organoids and their genetic manipulation. *Nat Protoc*. 2016;11(9):1724–1743. https://doi.org/10.1038/nprot.2016.097.
92. Wang D, Wang J, Bai L, et al. Long-term expansion of pancreatic islet organoids from resident Procr+ progenitors. *Cell*. 2020;180(6):1198–1211. https://doi.org/10.1016/j.cell.2020.02.048. e19.
93. Rossi G, Broguiere N, Miyamoto M, et al. Capturing cardiogenesis in gastruloids. *Cell Stem Cell*. 2021;28(2):230–240. https://doi.org/10.1016/j.stem.2020.10.013. e6.
94. Duan X, Tang X, Nair MS, et al. An airway organoid-based screen identifies a role for the HIF1α-glycolysis axis in SARS-CoV-2 infection. *Cell Rep*. 2021;37(6):109920. https://doi.org/10.1016/j.celrep.2021.109920.
95. Kim M, Mun H, Sung CO, et al. Patient-derived lung cancer organoids as in vitro cancer models for therapeutic screening. *Nat Commun*. 2019;10(1). https://doi.org/10.1038/s41467-019-11867-6.
96. Miller AJ, Dye BR, Ferrer-Torres D, et al. Generation of lung organoids from human pluripotent stem cells in vitro. *Nat Protoc*. 2019;14(2):518–540. https://doi.org/10.1038/s41596-018-0104-8.
97. Sachs N, Papaspyropoulos A, Zomer-van Ommen DD, et al. Long-term expanding human airway organoids for disease modeling. *EMBO J*. 2019;38(4). https://doi.org/10.15252/embj.2018100300.

98. Salahudeen AA, Choi SS, Rustagi A, et al. Progenitor identification and SARS-CoV-2 infection in long-term human distal lung organoid cultures. *bioRxiv*. July 2020. https://doi.org/10.1101/2020.07.27.212076.
99. Taguchi A, Kaku Y, Ohmori T, et al. Redefining the in vivo origin of metanephric nephron progenitors enables generation of complex kidney structures from pluripotent stem cells. *Cell Stem Cell*. 2014;14(1):53–67. https://doi.org/10.1016/j.stem.2013.11.010.
100. Takasato M, Er PX, Chiu HS, et al. Kidney organoids from human iPS cells contain multiple lineages and model human nephrogenesis. *Nature*. 2015;526(7574):564–568. https://doi.org/10.1038/nature15695.
101. Chen K, Baluya D, Tosun M, Li F, Maletic-Savatic M. Imaging Mass Spectrometry: a new tool to assess molecular Underpinnings of neurodegeneration. *Metabolites*. 2019;9(7):135. https://doi.org/10.3390/metabo9070135.
102. Tang C, Chen K, Bajic A, Choi WT, Baluya DL, Maletic-Savatic M. Analytical platforms and techniques to study stem cell metabolism. In: *Methods in Molecular Biology*. Vol. 1842. United States: Humana Press Inc.; 2018:265–281. https://doi.org/10.1007/978-1-4939-8697-2_20.
103. Manganas LN, Durá I, Osenberg S, et al. BASP1 labels neural stem cells in the neurogenic niches of mammalian brain. *Sci Rep*. 2021;11(1). https://doi.org/10.1038/s41598-021-85129-1.
104. Fair SR, Julian D, Hartlaub AM, et al. Electrophysiological maturation of cerebral organoids Correlates with dynamic morphological and cellular development. *Stem Cell Rep*. 2020;15(4):855–868. https://doi.org/10.1016/j.stemcr.2020.08.017.
105. Kathuria A, Lopez-Lengowski K, Jagtap SS, et al. Transcriptomic landscape and functional characterization of induced pluripotent stem cell-derived cerebral organoids in schizophrenia. *JAMA Psychiatr*. 2020;77(7):745–754. https://doi.org/10.1001/jamapsychiatry.2020.0196.
106. Shou Y, Liang F, Xu S, Li X. The application of brain organoids: from neuronal development to neurological diseases. *Front Cell Dev Biol*. 2020;8. https://doi.org/10.3389/fcell.2020.579659.
107. Riedl A, Schlederer M, Pudelko K, et al. Comparison of cancer cells in 2D vs 3D culture reveals differences in AKT-mTOR-S6K signaling and drug responses. *J Cell Sci*. 2017;130(1):203–218. https://doi.org/10.1242/jcs.188102.
108. Dekkers JF, Wiegerinck CL, De Jonge HR, et al. A functional CFTR assay using primary cystic fibrosis intestinal organoids. *Nat Med*. 2013;19(7):939–945. https://doi.org/10.1038/nm.3201.
109. Dekkers JF, Berkers G, Kruisselbrink E, et al. Characterizing responses to CFTR-modulating drugs using rectal organoids derived from subjects with cystic fibrosis. *Sci Transl Med*. 2016;8(344). https://doi.org/10.1126/scitranslmed.aad8278.
110. Mead BE, Karp JM. All models are wrong, but some organoids may be useful. *Genome Biol*. 2019;20(1). https://doi.org/10.1186/s13059-019-1677-4.
111. Qian X, Song H, Ming GL. Brain organoids: advances, applications and challenges. *Development (Camb)*. 2019;146(8). https://doi.org/10.1242/dev.166074.
112. Quadrato G, Brown J, Arlotta P. The promises and challenges of human brain organoids as models of neuropsychiatric disease. *Nat Med*. 2016;22(11):1220–1228. https://doi.org/10.1038/nm.4214.
113. Blair JD, Hockemeyer D, Bateup HS. Genetically engineered human cortical spheroid models of tuberous sclerosis. *Nat Med*. 2018;24(10):1568–1578. https://doi.org/10.1038/s41591-018-0139-y.
114. Kim H, Park HJ, Choi H, et al. Modeling G2019S-LRRK2 sporadic Parkinson's disease in 3D midbrain organoids. *Stem Cell Rep*. 2019;12(3):518–531. https://doi.org/10.1016/j.stemcr.2019.01.020.
115. Srikanth P, Lagomarsino VN, Muratore CR, et al. Shared effects of DISC1 disruption and elevated WNT signaling in human cerebral organoids. *Transl Psychiatry*. 2018;8(1). https://doi.org/10.1038/s41398-018-0122-x.
116. Stachowiak EK, Benson CA, Narla ST, et al. Cerebral organoids reveal early cortical maldevelopment in schizophrenia—computational anatomy and genomics, role of FGFR1. *Transl Psychiatry*. 2017;7(11). https://doi.org/10.1038/s41398-017-0054-x.
117. Ye F, Kang E, Yu C, et al. DISC1 regulates neurogenesis via modulating Kinetochore attachment of ndel1/nde1 during Mitosis. *Neuron*. 2017;96(5):1041–1054. https://doi.org/10.1016/j.neuron.2017.10.010. e5.
118. Galli R, Binda E, Orfanelli U, et al. Isolation and characterization of tumorigenic, stem-like neural precursors from human glioblastoma. *Cancer Res*. 2004;64(19):7011–7021. https://doi.org/10.1158/0008-5472.CAN-04-1364.

119. Hubert CG, Rivera M, Spangler LC, et al. A three-dimensional organoid culture system derived from human glioblastomas recapitulates the hypoxic gradients and cancer stem cell heterogeneity of tumors found in vivo. *Cancer Res.* 2016;76(8):2465–2477. https://doi.org/10.1158/0008-5472.CAN-15-2402.
120. Jacob F, Salinas RD, Zhang DY, et al. A patient-derived glioblastoma organoid model and Biobank recapitulates inter- and intra-tumoral heterogeneity. *Cell.* 2020;180(1):188–204. https://doi.org/10.1016/j.cell.2019.11.036. e22.
121. Azzarelli R. Organoid models of glioblastoma to study brain tumor stem cells. *Front Cell Dev Biol.* 2020;8. https://doi.org/10.3389/fcell.2020.00220.
122. Bergmann S, Lawler SE, Qu Y, et al. Blood–brain-barrier organoids for investigating the permeability of CNS therapeutics. *Nat Protoc.* 2018;13(12):2827–2843. https://doi.org/10.1038/s41596-018-0066-x.
123. Cakir B, Xiang Y, Tanaka Y, et al. Engineering of human brain organoids with a functional vascular-like system. *Nat Methods.* 2019;16(11):1169–1175. https://doi.org/10.1038/s41592-019-0586-5.
124. Tavakol DN, Fleischer S, Vunjak-Novakovic G. Harnessing organs-on-a-chip to model tissue regeneration. *Cell Stem Cell.* 2021;28(6):993–1015. https://doi.org/10.1016/j.stem.2021.05.008.
125. Vunjak-Novakovic G, Ronaldson-Bouchard K, Radisic M. Organs-on-a-chip models for biological research. *Cell.* 2021;184(18):4597–4611. https://doi.org/10.1016/j.cell.2021.08.005.
126. Chramiec A, Teles D, Yeager K, et al. Integrated human organ-on-a-chip model for predictive studies of anti-tumor drug efficacy and cardiac safety. *Lab Chip.* 2020;20(23):4357–4372. https://doi.org/10.1039/d0lc00424c.
127. Edington CD, Chen WLK, Geishecker E, et al. Interconnected microphysiological systems for quantitative biology and Pharmacology studies. *Sci Rep.* 2018;8(1). https://doi.org/10.1038/s41598-018-22749-0.
128. Maschmeyer I, Lorenz AK, Schimek K, et al. A four-organ-chip for interconnected long-term co-culture of human intestine, liver, skin and kidney equivalents. *Lab Chip.* 2015;15(12):2688–2699. https://doi.org/10.1039/c5lc00392j.
129. Raasch M, Rennert K, Jahn T, et al. Microfluidically supported biochip design for culture of endothelial cell layers with improved perfusion conditions. *Biofabrication.* 2015;7(1):015013. https://doi.org/10.1088/1758-5090/7/1/015013.
130. Vila OF, Qu Y, Vunjak-Novakovic G. In vitro models of neuromuscular junctions and their potential for novel drug discovery and development. *Expert Opin Drug Discov.* 2020;15(3):307–317. https://doi.org/10.1080/17460441.2020.1700225.
131. Skardal A, Murphy SV, Devarasetty M, et al. Multi-tissue interactions in an integrated three-tissue organ-on-a-chip platform. *Sci Rep.* 2017;7(1). https://doi.org/10.1038/s41598-017-08879-x.
132. Ma C, Peng Y, Li H, Chen W. Organ-on-a-Chip: a new paradigm for drug development. *Trends Pharmacol Sci.* 2021;42(2):119–133. https://doi.org/10.1016/j.tips.2020.11.009.
133. Kingsley DM, Roberge CL, Rudkouskaya A, et al. Laser-based 3D bioprinting for spatial and size control of tumor spheroids and embryoid bodies. *Acta Biomater.* 2019;95:357–370. https://doi.org/10.1016/j.actbio.2019.02.014.
134. Ren Y, Yang X, Ma Z, et al. Developments and Opportunities for 3D bioprinted organoids. *Int J Bioprinting.* 2021;7(3):364. https://doi.org/10.18063/ijb.v7i3.364.

CHAPTER

8

Use of cerebral organoids to model environmental and gene x environment interactions in the developing fetus and neurodegenerative disorders

Herbert M. Lachman[1,2,3,4]

[1]Department of Psychiatry and Behavioral Sciences, Albert Einstein College of Medicine, Bronx, NY, United States; [2]Department of Medicine, Albert Einstein College of Medicine, Bronx, NY, United States; [3]Department of Genetics, Albert Einstein College of Medicine, Bronx, NY, United States; [4]Dominick P. Purpura Department of Neuroscience, Albert Einstein College of Medicine, Bronx, NY, United States

Introduction

Neurodevelopmental disorders have a strong underlying genetic component with varying heritability estimates depending on the condition and clinical ascertainment.[1–7] Variation in the severity of autism spectrum disorders (ASD), intellectual and developmental disabilities (IDD), as well as schizophrenia (SCZ) is substantially influenced by both shared and non-shared environmental factors, as well as "second hit" genetic factors.[8–16] The two-hit model is supported by the finding that seemingly powerful risk variants are sometimes found in typically developing individuals.[17] In a study by Rosenfeld et al., for example, the penetrance of 15q11.2 del and 16p11.2 del was 10.4% and 62.4%, respectively.[18,19] In addition, loss of function variants are found in IDD candidate genes in the general population.[20] Finally, the identification of genetic hits in genome wide association studies (GWAS) in ASD, IDD, and SCZ implies a multi-hit model since such studies detect biologically weak, common

https://doi.org/10.1016/B978-0-12-822277-5.00006-7

genetic variants that require multiple genetic/environmental factors in order for a behavioral phenotype to emerge.[21–25]

Prenatal maternal stress, obstetric complications, infections, and nutritional deprivation can adversely affect the developing brain and increase the risk of neurodevelopmental and neuropsychiatric disorders[26,27] since the fetal brain is vulnerable to changes in the maternal/fetal environment.[26–30] For obvious ethical reasons, the experimental manipulations investigators use in animal models cannot be carried out on pregnant women, so the effects of environmental stressors on human fetal brain development are limited to extrapolating from studies carried out in lower animals or from clinical observations. Consequently, the underlying molecular processes by which environmental factors can adversely affect human fetal brain development and the subsequent effects they have on cognitive function and behavior, are not well-understood. Although studies carried out using animal models, primarily mice, have been enlightening, data from the mouse and human ENCODE (Encyclopedia of DNA Elements) projects, as well as studies on Repressor Element 1 (RE1) Silencing Transcription Factor (REST/NRSF) binding sites, show that substantial differences in gene regulatory networks exist between rodents and humans.[31–34] This suggests that unique pathophysiological features may exist in humans, and that novel pharmacotherapies, and the toxic effects of some environmental stressors, may not be evident in both species. The interspecies divergence in intergenic regulatory domains and the finding that a substantial number of GWAS signals are in intergenic DNA, near or within enhancer elements,[22,25] highlights the importance of using human cells. In addition, in vitro model systems using human cells are ideal for high throughput drug screening.

A deeper understanding of human brain development has been hampered by restrictions on the use of fetal brains obtained from spontaneous or induced abortions. Thus, the discovery of induced pluripotent stem cells (iPSCs) in 2006 by Shinya Yamanaka, from which models of the fetal brain were eventually developed, has forever revolutionized the field of human brain development and the disorders that occur upon its disruption.[35] The extraordinarily short period of time between the first peer-reviewed paper on iPSCs in 2006 to the 2012 Nobel Prize in Physiology or Medicine to Yamanaka (shared by John Gurdon for his pioneering work on somatic nuclear transfer) attests to the importance of the discovery of iPSCs. The most relevant model system using iPSCs for understanding human brain development and the effects disease-causing mutations and environmental factors have on human brain development are cerebral organoids. The different methods for growing cerebral organoids are discussed in greater detail in an earlier chapter. This chapter will focus on the use of these structures to study the impact of environmental factors, and gene x environment interactions on brain development, and how experimental findings could have an impact on preventing and treating their adverse effects on the fetal brain. Although the technology is new and the methods for identifying the most robust and reproducible model for the fetal brain are still works in progress, the 3-D organoid system is already paying dividends in understanding the pathogenesis of new viral scourges that adversely affect fetal brain development, such as Zika virus. In an ever-increasingly toxic world to which the developing fetus is exposed, cerebral organoids will play an important role in screening chemicals and drugs for their adverse effect on the developing human brain.

In addition to using cerebral organoids to model fetal brain development, researchers have expanded their usefulness to study the impact of environmental factors that might play a role in conditions that affect the adult brain, such as HIV-associated neurocognitive disease (HAND) and neurodevelopmental disorders. These important topics will also be discussed in this chapter.

Maternal immune activation

Arguably the most important environmental stressor for the developing fetus is maternal immune activation (MIA), a proinflammatory state in pregnancy mediated by proinflammatory cytokines that include, among others, TNF-α, IL-1β, and IL-6.[36–42] MIA can be induced by maternal infection caused by a variety of infectious microbes.[27,43–50] Although exposure of the developing fetus to certain infectious agents often leads to specific abnormalities, as described in detail below, the wide range of infectious agents that can cause MIA suggests that *immune activation* per se is an important common denominator. Similarly, a variety of non-infectious inducers of MIA can adversely affect fetal brain development, one of which is maternal autoimmune disorders. For example, an increased risk of ASD is seen in families with a maternal history of autoimmune disorders, such as rheumatoid arthritis and celiac disease, and a family history of type 1 diabetes.[51,52] Other stressors in pregnancy associated with a proinflammatory state and implicated in the risk of neurodevelopmental disorders in the fetus include metabolic syndrome (MS), obesity, and pre-eclampsia. Several groups have shown that MS and maternal obesity increase ASD risk in their offspring with odds ratios (ORs) between 1.5 and 2.5.[11,53] The combination of maternal obesity and prematurity increased the OR to 9.88, demonstrating the potential risk of exposure to multiple environmental risk factors.[54] And, in a large review by Han et al., maternal states associated with the future development of ADHD and Tourette Syndrome in exposed fetuses include obesity, pre-eclampsia, smoking, low socioeconomic status (SES), emotional stress, and autoimmune disorders.[27] Maternal thyroid disease is also associated with an increased risk for the development of ADHD, epilepsy (hyperthyroidism), and ASD (hypothyroidism).[55,56]

Since experimental manipulation of pregnant women cannot be carried out and only correlative studies are possible (e.g., measuring inflammatory cytokines in maternal blood followed by longitudinal assessment of exposed fetuses in the postnatal period), most of what is known about the molecular and developmental underpinnings of MIA have come from animal models. Mice, rats, and Rhesus monkeys exposed prenatally to proinflammatory cytokines, or the innate immune system activators/viral mimetics lipopolysaccharide (LPS) and poly(I:C), exhibit abnormal behaviors and have molecular and neuroanatomical changes similar to those seen in patients with ASD and SCZ.[41,57–64] Additional validity of these models was provided by a recent study in which MIA in pregnant rat dams was found to induce changes in gene expression in fetal brains that overlapped with targets of the ASD and IDD candidate genes, *FMR1* and *CHD8*.[29] These adverse effects on brain development, which have been observed after only a single injection of LPS or poly(I:C), are mediated by the brain's innate immune system, primarily fetal microglia.[65–71]

Cerebral organoids as a model system to study MIA and neuroinflammation

Cerebral organoids represent an ideal model system for studying the effects of immune activation on the developing fetal brain. Although novel insights into the pathogenesis of

ASD and SCZ have been gleaned from the mouse MIA model system, human cells and tissue are needed as well, as noted earlier. The ability to use cerebral organoids to study MIA and neuroinflammation most effectively hinges on the requirement that these structures contain each of the components that contribute to the pathophysiology of these conditions. On the cellular level, neurons, astrocytes, oligodendrocytes, and microglia are all critical for regulating neuronal network formation, synapse formation, synapse pruning, and responses to inflammation. However, the early methods used to develop cerebral organoids did not result in the formation of microglia. Thus, the use of human cerebral organoids to model MIA has been given a boost by the ability to grow microglia from iPSCs that resemble isolated human microglia at the gene expression and functional levels.[72–75] and the finding that both immortalized microglia cell lines and iPSC-generated microglia are capable of migrating into cerebral organoids.[76] These embedded microglia respond to LPS by increasing the expression of several proinflammatory cytokines. In addition, the investigators showed that organoids infected with Zika and Dengue Virus generated a stronger inflammatory response when microglia were present than when absent.[76] In another technological development, Ormel, et al. showed that microglia can develop innately from cerebral organoids using a modification of the Lancaster protocol.[77] These microglia also respond to LPS. An approach in which microglia and organoids develop separately and then are fused has the advantage of providing researchers with the opportunity to mix and match mutant and control preps to study the relative roles of neurons, astrocytes, and microglia on neuroinflammation.

Aside from studying the developing brain, model systems containing microglia are also important from a broader perspective in examining the role neuroinflammation plays in the pathogenesis of a subgroup of neuropsychiatric and neurodegenerative disorders. For example, transcriptome studies carried out on ASD and SCZ postmortem samples show that microglia activation is one of the top gene ontology (GO) terms among differentially expressed genes (DEGs).[78–80] Microglia activation is most likely induced by synaptic abnormalities caused by underlying risk alleles that disrupt neuronal function and synaptic architecture. This is supported by genetic findings showing associations of risk variants to neuronal genes, rather than genes that directly affect microglia function. On the other hand, several ASD candidate genes are expressed predominantly, or exclusively, in microglia, suggesting that *intrinsic* microglia deficits have a direct role in the pathogenesis of ASD in a subgroup of cases. An intriguing example is Rett Syndrome (RTT), which is caused by loss of function mutations in *MeCP2*. Although *MeCP2* is expressed at a higher level in neurons than microglia, bone marrow transplantation in a mouse model rescues aberrant behaviors following engraftment of phagocytic myeloid cells in the brain that can mimic the phagocytic function of microglia.[81] However, this finding remains controversial.[82] Recently, genetic variants in *CX3CR1*, a G protein-coupled receptor expressed exclusively in microglia have been identified as risk variants in ASD and SCZ,[83] showing that primary defects in microglia can cause neurodevelopmental problems. Parenthetically, one of the best illustrations of a direct effect of genetic variation on microglia function related to disease is in Alzheimer's disease, where several microglia-expressed genes have been identified by GWAS (e.g., *PLCG2, ABI3,* and *TREM2*).[84,85]

Genetic studies also implicate innate immune mechanisms in SCZ, supporting the use of model systems containing microglia. The most robust finding in SCZ GWAS is an association to the Human Leukocyte Antigen (HLA) locus.[25,86–88] Genetic associations between the HLA locus and ASD, and IDD have also been found by a number of investigators.[89–95]These findings suggest an autoimmune or autoinflammatory etiology in a subgroup of patients,[37,52,96–98] which is consistent with the finding that maternal autoimmune disorders are risk factors for ASD in offspring.[96,99] Alternatively, the association to the HLA locus could be due to nonimmune effects of HLA proteins on synaptic plasticity, which is mediated by microglia.[37,100–105]

Another component of the innate immune system is the complement pathway, components of which are found in developing synapses and are required for normal brain wiring.[106–110] Complement system dysfunction has been implicated in ASD and MIA,[111] and structural variants in the complement component 4 (C4) gene that increase C4A and C4B expression in the brain were recently identified.[112] Excessive complement activity resulting in a reduction in synaptic complexity could underlie the effect of C4 structural variants on the development of SZ. Such an effect is consistent with the findings that mutations in *CSMD1*, which codes for a negative regulator of complement activation, have been found in SZ and ASD.[2,113,114] In addition, a number of studies show an increase in various proinflammatory cytokines in patients with SCZ and ASD in the blood and brain.[115–120] The use of cerebral organoids to model neuroinflammation provides a platform for screening drugs that modulate cytokine responses.

Finally, cerebral organoids containing microglia will be very useful in studying the enigmatic conditions known as Pediatric Acute-Onset Neuropsychiatric Syndrome (PANS) and Pediatric Autoimmune Neuropsychiatric Disorders Associated with Streptococcal Infections (PANDAS). In its original description, PANDAS was characterized by acute onset obsessive-compulsive disorder (OCD) and tics following infection with group A beta-hemolytic streptococcus (GABHS).[121,122] The concept of PANS was introduced a decade later to expand the diagnosis to include a multitude of acute-onset neuropsychiatric problems and exclude the requirement of GABHS infection and tics. PANDAS is now viewed as a subgroup of PANS. Supporting an inflammatory or autoimmune etiology is the finding that statistically significant MRI differences suggestive of local inflammation were found in the thalamus, basal ganglia, and amygdala of PANS patients.[123] In addition, many patients respond to immunomodulators, such as nonsteroidal anti-inflammatory drugs (NSAIDs), intravenous immunoglobulin (IVIg), corticosteroids, and more recently, the B-cell inhibitor Rituximab.[124,125] The ability to understand how acute infections affect neuronal activity and molecular mechanisms in cerebral organoid cultures will allow investigators to better understand why these infections cause behavioral deficits.

Thus, cerebral organoid model systems, in particular those that contain microglia, can be used to examine MIA, as well as a range of clinical disorders likely initiated by neuroinflammation that go beyond the developing fetus.

Cerebral organoids as a model system to study infectious diseases that cause neurodevelopmental disorders

Prior to the modern era, the infectious disease most feared by women of child-bearing age was rubella (German measles), which can lead to congenital rubella syndrome (CRS) in exposed fetuses. Discovered in 1941 by Norman McAlister Gregg, CRS is characterized by the triad of sensorineural deafness, congenital cataracts, and congenital heart disease. However, a host of other developmental and neurodevelopmental problems can occur as well, including microcephaly, IDD, ASD, and SCZ. The risk is highest with first trimester exposure (~80%). However, the risk is also substantial if women acquire the infection a few weeks before conception. Gregg, an ophthalmologist, discovered the connection between maternal rubella and congenital cataracts in 1941 after a severe outbreak in Australia.[126] The discovery was initially met with skepticism because it was generally believed that congenital disorders were genetic, even though very little was known about how genetic information was transmitted (the Avery–MacLeod–McCarty experiment showing that DNA and not protein was the agent of genetic transmission, was published in 1944). Prior to the development of a vaccine, either as a stand-alone preventative or in combination with measles and mumps vaccines (MMR) in the late 1960s, tens of thousands of children were born with CRS in the U.S. alone. After mandatory vaccination was initiated (a quaint notion in an age of political divide and the anti-vaccine movement), rubella cases dramatically decreased, and by 1994, the disease was largely eliminated in the United States. Today, only a handful of rubella infections occur in the United States, primarily in travelers from developing nations where vaccination rates are low. Indeed, approximately 100,000 CRS cases still occur every year in the developing world, because of suboptimal MMR vaccination. One point that seems to be lost to the vaccine skeptics and those who falsely claim an association between the vaccine and ASD is that more than a million cases of CRS have been prevented by MMR (as well as the millions of lives saved from measles using the same vaccine; there are still >100,000 deaths a year from measles in nations with low vaccination rates).

The discovery of CRS was only made after thousands of clinical cases were observed and an association was established. Modern-day viruses that periodically emerge and damage the developing brain can now be confronted experimentally using 21st century molecular tools and cell culture techniques, which can help establish cause and effect and the underlying pathogenesis more rapidly. Although developing vaccines against emerging viruses that damage fetal brains is still the mainstay of the public health response, understanding the pathogenesis could lead to therapeutic interventions. Although there are no published studies examining the effect of rubella on the developing brain using an organoid model, the following are infectious diseases that adversely affect the brains of developing fetuses and the post-development period that are being addressed experimentally using these technologies.

Zika virus

The most recent and severe infectious disease affecting the fetal brain is Zika virus (ZIKV), which leads to the development of crippling microcephaly and other neurodevelopmental disorders, collectively referred to as Congenital Zika Syndrome (CZS). In addition to affecting brain development, developmental abnormalities in the eyes, and craniofacial and musculoskeletal abnormalities occur.[127] ZIKV first appeared in South and Central America, and the Caribbean in 2015. It is an insect borne illness (usually mosquitoes) that can also be sexually transmitted. In children and adults, ZIKV causes relatively mild to moderate symptoms characterized by fever, rash, headache, and joint pain—a nondescript viral illness in many ways. It is rarely fatal and is often asymptomatic. However, pregnant women can transmit the virus to the fetus. Overall, according to the CDC, approximately 5%–10% of exposed fetuses will develop a ZIKV-associated birth defect. Effects on the developing brain are due to the virus' neuronal cell tropism.[128] Forebrain organoids infected with ZIKV show a preference for neural progenitor cells (NPCs), which leads to increased cell death, premature neuronal differentiation, and reduced proliferation, resulting in decreased neuronal cell-layer volume, a finding compatible with the development of microcephaly. The original finding has been replicated and extended to show that ZIKV infection directly activates a subset of interferon (IFN)-stimulated genes in NPCs derived from cerebral organoids.[129]

In a recent study, ZIKV was found to trigger the production of virus-derived small interfering RNAs in NPCs, and that enoxacin, a broad-spectrum antibiotic that is also a nonspecific RNAi enhancer, has potent anti-ZIKV activity that prevents in vitro ZIKV infection.[129] In addition, methylene blue, a broad-spectrum antiviral agent that inhibits viral protease activity, was found to limit Zika virus infection in cerebral organoids and reduce viremia in a mouse model.[130] Finally, the antiviral drug sofosbuvir, which is used to treat hepatitis C, has been found to inhibit ZIKV replication in a variety of neuronal cell lines and cerebral organoids.[131]

Thus, only a few years after the emergence of ZIKV as an important public health problem, cerebral organoids were successfully employed to provide a valid model for the pathogenesis of microcephaly and novel therapies for infected pregnant women. However, moving from in vitro studies and mouse models to clinical translation will be challenging, and it is likely that a vaccine administered prior to pregnancy will emerge as the best option, rather than carrying out clinical trials in pregnant women to test novel pharmacotherapies.

SARS-CoV-2

As this chapter is being written, the worldwide pandemic caused by SARS-CoV-2 is still vexing the world, especially with the emergence of the highly infectious Omicron variant, although the vaccine roll-out is having a significant impact on mortality. The primary cause of death in patients with COVID-19, the disease caused by SARS-CoV-2, is due to respiratory complications. However, many organ systems are affected, including the CNS. Symptoms caused by CNS involvement include headache, persistent fatigue, confusion, psychiatric symptoms, and seizures. Although nasal neuroepithelium is not part of the CNS, anosmia is among the more common manifestations (but less so with Omicron variants compared to earlier strains). Early in the pandemic, it was not clear whether the neurological complications were due to the indirect effects of systemic inflammation that is especially pronounced in COVID-19 (the notorious cytokine storm) or vascular involvement. However, recent work using organoid models suggests that SARS-CoV-2 has tropism for some cell types in the brain. For example, Pellegrini et al., showed, in an organoid model of the choroid plexus, that SARS-CoV-2 pseudovirus expressing the critical spike protein, and live virus, attack choroid plexus epithelial cells, which express ACE2, the spike protein receptor the virus uses to gain entry into cells.[132] Little to no direct infection of neurons or glia was detected. Involvement of the choroid plexus leads to leakage across the brain blood barrier (BBB) that normally functions to block entry of infectious microbes, immune cells, and many cytokines into the brain.

Similarly, tropism for choroid plexus epithelium was detected in organoids by Jacob et al., who also found modest numbers of infected neurons and astrocytes.[133] Infected cells showed increased cell death, induction of an inflammatory response, and increased expression of CCL7, IL-32, CCL2, IL-18, and IL-8. Gene ontology (GO) analysis of upregulated genes showed enrichment for genes related to viral responses, RNA processing, response to cytokines, cytoskeletal rearrangement, and cell death. Among the down-regulated genes, enrichment was found for genes related to ion transport, transmembrane transport, cilium, and cell junction, which could explain the breakdown of the BBB.

On the other hand, Song, et al. used single cell sequencing of SARS-CoV-2 infected cerebral organoids to show that the virus can enter neurons, radial glia, and NPCs.[134] Interestingly, although cell death was detected in several neuronal clusters, there was only minimal overlap with SARS-CoV-2, suggesting that infected cells were inducing cell loss in neighboring cells through a bystander effect. Infected cells showed an enrichment for genes involved in metabolic processes and response to hyperoxia, while uninfected cells showed a mitochondrial catabolic state and response to hypoxia, suggesting that SARS-CoV-2 infected cells may create a resource-restricted environment for neighboring cells. They also detected SARS-CoV-2 in cortical neurons in autopsy samples.

Similarly, Ramani et al. showed that SARS-CoV-2 targets neurons in cerebral organoids.[135] Interestingly, they also showed that infected cells caused an altered distribution of Tau from axons to soma, which could potentially lead to long-term neurotoxic effects.

Involvement of the choroid plexus and proinflammatory cytokines is similar to that seen in neuroinflammatory and autoimmune disorders that affect the brain, such as anti-NMDA encephalitis and PANS/PANDAS.[136–140] This suggests that neuromodulators that have a therapeutic effect on these conditions, such as IVIg, nonsteroidal anti-inflammatory drugs NSAIDs, and glucocorticoids, could be effective in treating the cognitive and neurological aspects of SARS-CoV-2. Immunotherapies, such as glucocorticoids, are currently being used to treat severely affected hospitalized patients to prevent respiratory failure that would require intubation and mechanical ventilation (WHO/2019-nCoV/Corticosteroids/2020.1). In addition, there have been several studies in which IVIg was found to be effective in hospitalized patients with COVID-19 who developed encephalopathy.[141,142] However, the sample sizes were very small and there was no control group. With the development of cerebral organoid/brain blood barrier model systems derived from iPSCs, in vitro research could be used as an adjunct to clinical trials to assess immunomodulators and other novel treatments.[135]

Human immunodeficiency virus (HIV)

HIV is arguably the most important pathogen that directly and indirectly affects the adult brain. The indirect effects are due to the increased risk patients with AIDS have for opportunistic infections and neoplasms that attack the brain. Among the pathogens that act in this manner are Cytomegalovirus (CMV), *Treponema* pallidum, *Cryptococcus* neoformans, *Herpes Simplex Virus, Toxoplasmosis* gondii, and JC virus (JCV), the latter of which causes progressive multifocal leukoencephalopathy. Among the brain tumors associated with AIDS, CNS lymphoma is the most common. Opportunistic infections affecting the brain have decreased dramatically in recent years with the widespread use of antiretroviral therapy (ART). However, a large subgroup of AIDS patients who have been successfully treated with ART and have achieved an immunological remission, have persistent brain involvement due to latent reservoirs of genome-integrated HIV in microglia, astrocytes, and perivascular macrophages.[143,144] The HIV pandemic affects more than 30 million individuals worldwide. Consequently, despite the effectiveness of ART, the cognitive problems that persist even after successful suppression of the virus in the periphery, represents a substantial public health problem.[144–146] These reservoirs of infected cells are relatively resistant to ART because the majority of antiretroviral drugs are unable to effectively penetrate the BBB.[147]

The cognitive problems associated with latent HIV in the brain are referred to as HIV-1-associated neurocognitive diseases (HAND), which affects approximately 50% of HIV-infected individuals irrespective of HIV-1 clades or geography.[148–150] The neuropathogenesis of HAND is mediated, in part, by the response of the non-neuronal cells (perivascular macrophages, microglia and astrocytes) that are infected with HIV, which initiates an inflammatory response. In addition, HIV proteins have a direct toxic effect on neurons. Neurons themselves do not support HIV-1-infection or replication, but they express several cell-surface receptors (e.g., CCR5, CXCR4, NMDAR) that render them sensitive to insults by viral proteins (e.g., Tat, gp120), cytokines (e.g., TNF-α, IL1β) and metabolites (e.g., nitric oxide and arachidonic acid) secreted by immune cells in the brain.[151] In addition, it has been shown that HIV proteins inhibit adult neurogenesis, thereby impairing neuronal recovery. Other key events that contribute to cognitive impairment are dysregulation of neuronal support cells, and synapto-dendritic injury.

Investigators are using cerebral organoids to dig deeper into the pathogenesis HIV-related cognitive dysfunction and the problem of latent infection.[152] It is important to point out that such studies are not designed to determine the effect of HIV-infection per se on fetal brain development. With current treatment regimens to reduce the risk of maternal to fetal HIV transmission, the risk of an adverse effect on fetal brain development is low. Even if fetuses are exposed, there are apparently no significant cognitive impairments in childhood.[153] Rather, the cerebral organoid system is viewed as a proxy for the adult brain containing, with some coaxing, all of the brain cells needed to understand HAND, in particular microglia, a major reservoir of latent virus. Recently, Dos Reis et al., optimized culture conditions to introduce primary HIV-infected microglia into cerebral organoids.[152] They showed that HIV-infected microglia release TNF-α and IL-1β, mimicking the chronic neuroinflammatory environment observed in patients. They also found decreased neuronal viability following HIV-1 infection, but have not yet established the underlying mechanism. Nevertheless, their model system can be used to screen drugs that might limit neuronal damage.

Another novel approach to HAND that could be tested in vitro is gene therapy. There is a history of gene therapy in HIV using ex vivo manipulated hematopoietic stem cells targeting the *CCR5* gene, which codes for the HIV co-receptor used to gain entry into cells. Individuals homozygous for a polymorphic 32 base pair (bp) deletion in the *CCR5* gene are resistant to HIV progression.[154] An AIDS patient with acute myeloid leukemia treated by hematopoietic stem cell transplant using a donor homozygous for the *CCR5* 32 bp deletion was the first patient cured of HIV, since he remained HIV negative for more than a decade without ART, before succumbing to a leukemia relapse.[155] Similarly, AIDS patients treated with autologous CD34+ hematopoietic stem cells in which a null *CCR5* variant was generated using a ZNF nuclease gene editing approach, had restoration of CD4+ T-cells.[156] Kunze et al. described a gene therapy approach in cerebral organoids that directly targets HIV rather than CCR5. They transduced human neurons and cerebral organoids with a novel Adeno-associated virus-based vector (AAV9P1) that is tropic for astrocytes.[157] The virus was used to deliver a CRISPR/Cas9 tool kit to incapacitate latent HIV-1. Although gene modification approaches would be cost prohibitive as a standard of care for HAND, which affects such a large number of individuals, with time and experimental fine-tuning perhaps the cost will decrease, and the efficiency will improve to a level that will make gene therapy for refractory HAND more feasible.

Toxoplasmosis

A discussion about HIV and the brain is a good segue into reviewing the effects of infectious diseases seen in AIDS patients, some of which have public health consequences beyond HIV, namely toxoplasmosis and CMV. Maternal transmission of toxoplasmosis, which is caused by the protozoan *Toxoplasma gondii* (*T. gondii*), is ubiquitous, affecting more than one-third of the world's population. Infected cats shed infective oocytes in their feces, which are ingested by rodents and other mammals—intermediate hosts—that develop into tissue cysts. The life cycle closes when infected rodents are eaten by cats. The parasite picked the ideal definitive and intermediate host to complete its life cycle. Cats were domesticated nearly ten-thousand years ago, prized by most humans for their affable behavior and rodent-hunting prowess, the latter of which was an asset with advent of the agricultural revolution that provided an opportunity to store surplus food, a smorgasbord for vermin. According to some estimates, domestic cats kill between 1.3 and 4.0 billion birds and 6.3–22.3 billion small animals a year in the United States, creating a never-ending toxoplasmosis surplus.[158] Cats are also the major carrier of rabies among domesticated animals in the United States; not dogs, as most people assume. Humans are infected with toxoplasmosis when they come in contact with microscopic oocytes shed in cat feces when cat owners clean litter boxes. Soil and water supplies can also be contaminated by free-roaming outdoor cats (the major offenders of feline-causing bird and small mammal depopulations).

The proclivity of *T. gondii* for brain tissue is an important component of its life cycle because infected rodents lose their natural fear and avoidance of cats. This extraordinary life cycle could be admired for its Darwinian perfection—approximately, three-quarters of domesticated cats are infected with *T. gondii*. The evolutionary success comes with a price—an adverse effect on human health, especially the developing fetus.

In adults, cerebral involvement is very rare except in immunosuppressed populations, as highlighted by the HIV pandemic. With the advent of ART, toxoplasmosis in HIV-infected patients is declining. On the other hand, the prevalence is increasing in patients who are immunocompromised because of potent immunosuppressants taken by transplant recipients and patients with severe autoimmune disorders. In these patients, toxoplasmosis can emerge from latent infections, as well as from infected donor tissue. In contrast to the benign course in immunocompetent individuals, severe illness characterized by encephalitis, chorioretinitis, meningitis, and involvement of other tissues can occur in transplant patients.[159] In addition, it has been suggested that chronic latent toxoplasmosis plays a substantial role in the development of SCZ, bipolar disorder and other psychiatric disorders. However, most of the findings are based on serological correlations.[160–163] There is a lack of direct experimental evidence.

Unequivocal, however, are the adverse effects that occur upon the maternal transfer of *T. gondii* to the developing fetus, which can lead to preterm births, stillbirths, and damage to the fetus' brain and eyes. Women who are infected months before pregnancy can also transmit the parasite. Pregnant women with toxoplasmosis have a 30% chance of passing the infection to their fetus, with the risk of transmission increasing during gestation. The adverse effects are greater with first trimester exposure. This can lead to microcephaly, chorioretinitis, hydrocephalus, and cerebral calcifications.[164,165] According to a recent meta-analysis involving more than 900,000 pregnant women, the worldwide prevalence of toxoplasmosis infection in this population is approximately 1%.[166]

There is only one report in which the effect of *T. gondii* on cerebral organoids was investigated.[167]The authors showed that tachyzoites can infect human cerebral organoids, are transformed to bradyzoites, and can replicate in parasitophorous vacuoles to form cysts. This indicates that the *T. gondii* asexual life cycle is efficiently modeled in organoids. Transcriptomic analysis of *T. gondii*-infected organoids revealed activation of the type I interferon immune response against infection pathway, and transcriptome changes related to protozoan invasion and replication. Thus, cerebral organoids are a physiologically relevant in vitro model system for understanding *T. gondii* infections and host interactions. Although existing treatments are effective, cerebral organoids can be used to test new treatments that can serve as a backup to existing therapies in the event of drug resistance or severe side effects.

In addition to organoids, studies show that monolayer neurons derived from iPSCs and direct differentiation from fibroblasts can also be infected *T. gondii* and form the characteristic cysts.[168,169]

Cytomegalovirus (CMV)

CMV is a ubiquitous viral pathogen that is usually benign. However, maternal to fetal transmission can result in hearing loss, IDD, microcephaly, cerebral palsy, prematurity, and seizures.[170] It is the most common congenital infection and the main non-genetic cause of sensorineural hearing loss.[171] Congenital CMV affects approximately 1/200 live births. There has only been one report at the time of this writing in which cerebral organoids were used to model CMV infections in the CNS.[172] The organoids were reduced in size and showed impaired formation of cortical layers, abnormal calcium signaling and neural network activity. These phenomena were prevented by neutralizing antibodies.

Herpes simplex virus (HSV)

Neonatal HSV infections are usually transmitted from an infected mother during delivery, although some cases occur during the intrapartum period. Despite the ubiquity of genital herpes, the frequency of neonatal transmission is relatively low, occurring in approximately one in 3000–20,000 live births. Transmission from asymptomatic carriers, which is more common than transmission from symptomatic carriers, is an especially serious problem. Most cases, ~70%–85%, are caused by HSV type 2. Fetuses infected in utero can be born with low birth weight, microcephaly, hydrocephalus, chorioretinitis, and vesicular skin lesions.[173,174] The more common HSV infection in pregnancy is acquired during delivery, which leads to irritability, vesicular lesions, breathing difficulties, jaundice, and encephalitis.

There have been several recent reports in which cerebral organoids were used to model HSV infection in the CNS. Qiao et al. showed that HSV-1-infected organoids exhibited impaired neuronal differentiation, and dysregulated cortical layer and brain regionalization.[175] In addition, there was abnormal activation of microglia leading to the induction of inflammatory cytokines, including TNF-α, IL-6, IL-10, and IL-4, several of which have been implicated as mediators of MIA (e.g., TNF-α, IL-6).

Cerebral organoids and cellular stress

Heat shock

In addition to the direct effects of infectious diseases on the developing brain, and the influence of proinflammatory cytokines, a febrile episode per se can have an independent adverse effect on the fetus, in particular, fever persisting for several days.[44,50,96,176,177] Confounding the effect of fever are other studies suggesting that prenatal exposure to acetaminophen, which is commonly used by pregnant women to treat febrile episodes, may increase ASD risk.[178] This could be a concern considering the fact that approximately 20% of pregnant women experience one or more febrile episodes,[179] and the use of acetaminophen is ubiquitous. The cerebral organoid system is an ideal model for studying the effect of heat shock (HS) and other stressors. Several years ago, we analyzed the effect of HS (39°C for 24 h) on 50-day old cerebral organoids.[180] Control organoids were maintained at 37°C. Transcriptome profiling using bulk RNA-seq resulted in the identification of 186 differentially expressed genes (DEGs) that were nominally significant ($p < .05$: 105 increased in expression; 81 decreased). Of these, 12 achieved genome wide significance (FDR < 0.05), the majority of which were, as expected, HS-inducible genes (*HSPA1A, HSPA1B, HSP90AA1, DNAJB1, HSP90AB1, CRYAB, HSPH1, HSPA6, HSPA4L*). Interestingly, the other three genes, *SLC5A3, NGRF*, and *FKBP4*, have been implicated in subgroups of SCZ and ASD based on genetic and molecular studies.[181–184] In addition, a number of DEGs that were nominally significant are also ASD and SZ candidates, including *SMARCA2, PRODH, ADNP, KAT2A*, and *ZNF804A*. The changes in gene expression in response to HS were the same degree (~2-fold increase or 50% decrease) that would be expected from a copy gain or copy loss CNV (or null variant) respectively. Among the top GO terms for up-regulated DEGs were response to unfolded protein (UPR), and negative regulation of apoptosis, consistent with the expected response to HS stress. Although the sample size was small and bulk sequencing rather than single cell RNA sequencing (scRNA-seq) was used, the findings suggested a gene x environment (G x E) interaction in which an environmental stressor could have an impact on the expression of key developmentally regulated genes that act in concert with genetic variants implicated in neurodevelopmental disorders. However, it should be pointed out that the induction of HS and other genes in our experiment was transient, so the long-term impact on gene expression and brain function is not known at this time. However, several nominally significant DEGs code for chromatin modifiers (e.g., *KAT2A, SMARCA2*), which could potentially induce more long-term effects on gene expression.

Fetal alcohol syndrome

Exposure to alcohol during fetal development can result in physical and behavioral problems, collectively called fetal alcohol spectrum disorders (FASD).[185–187] Physical anomalies include short height, small head size, and craniofacial abnormalities. Behavioral, neurological, and cognitive problems include poor coordination, epilepsy, impaired fine motor skills, neurosensory hearing loss, poor impulse control, memory and attention problems, and learning disabilities. Although there is no recommendation for "safe" alcohol consumption during pregnancy, binge drinking (four drinks within 2 h) is the most dangerous for the developing fetus. Public health pronouncements have had an impact on reducing alcohol consumption during pregnancy, yet, about 20%–30% of pregnant women consume alcohol during pregnancy, and many women maintain regulator drinking habits before realizing they are pregnant, a situation reported by nearly 20% of Australian women.[188] Consequently, FASD remains a public health concern, and is among the most important preventable causes of behavioral problems in the world. According to a recent study in JAMA by May et al., the prevalence of FASD among first graders in the U.S. in four communities ranged between 1% and 5%.[189]

Although prevention is the key factor in reducing the impact of FASD, basic research is still needed to see if pharmacological intervention can mitigate some of the clinical consequences. This is especially important for prospective mothers who might have inadvertently consumed an excess of alcohol during early gestation before realizing they were pregnant. The most common basic research approach is the use of animal models, which provide researchers with the opportunity to control for confounding factors that affect fetal outcomes upon prenatal alcohol exposure, such as dosage, timing, duration, stress, diet, other drugs, and genetic background.[190] Studies show that exposing mice to ethanol prenatally leads to neuronal apoptosis[191] and functional abnormalities in microglia and astrocytes.[192] In addition, postnatal day 6–8 exposure to ethanol in rats, equivalent to a third trimester human fetus, leads to long-term disruption of the neuroimmune response by microglia and an increase in the expression of proinflammatory cytokines (e.g., TNF-α and IL-6) in the hypothalamus.

There have been a few reports on the use of cerebral organoids to model FASD. Arzua et al. found that alcohol exposure resulted in defects in a multitude of cellular and molecular processes, such as bioenergetic metabolism, gene expression, neuronal apoptosis, cytoskeleton disorganization, mitochondrial dysfunction, and metabolic stress.[193]Gene expression profiling carried out using microarrays showed that DEGs were enriched for those involved in neurodevelopmental and neurodegenerative disorders, very similar to the heat shock findings described earlier. In another study by Zhu et al., alcohol-exposed organoids displayed attenuated neurite outgrowth, skewed neural maturation, and transcriptome analysis showed enrichment for genes involved in the Hippo signaling pathway.[194] The use of cerebral organoids to study FASD and the development of novel treatments are in their infancy. However, given the interesting findings in rodent models, and the early findings with cerebral organoids, iPSC researchers have some guidance to test promising drugs for at-risk fetuses.

Cerebral organoids to model neurodegenerative disorders

Alzheimer's disease (AD)

Although the focus of this chapter has been on the use of cerebral organoids to model neurodevelopmental disorders, this section will cover their application to neurodegenerative disorders. Cerebral organoids contain each of the cellular elements that play a role in neurodegeneration (e.g., neurons, astrocytes, microglia). In addition, organoids contain an extracellular space matrix into which some neurodegeneration-associated protein aggregates are secreted. Consequently, 3-D organoids are better suited for modeling neurodegenerative disorders than 2-D monolayer cultures.

With the aging of the population, neurodegenerative disorders are increasing in prevalence. The most common is Alzheimer's disease (AD), which, according to the CDC affects nearly six million Americans. With current population trends, that number is projected to increase to 14 million by 2060. AD is genetically heterogeneous. Despite many years of research, which has resulted in a fairly comprehensive understanding of disease pathogenesis, treatments based on these findings, such as monoclonal antibodies targeting amyloid-β (Aβ) peptide, have been less than stellar. One problem is that such treatments need to be initiated prior to the onset of severe symptoms, which are associated with extensive neuronal degeneration. Recently, the monoclonal antibody Donanemab showed some promising findings in a small phase II clinical trial in which treatment was initiated early in the illness, with improved AD rating scale (ADRS) baseline scores at 76 weeks and reductions in amyloid plaque levels and global tau load.[195] However, there were no significant improvements in secondary outcomes. Longer and larger clinical trials beginning even earlier in the disease will be needed. Recently, the FDA approved the monoclonal antibody aducanumab after viewing the results of a phase III trial. However, the very modest, debatable gains and very high price tag, made the decision a controversial one, leading to the resignation of several members of the FDA expert advisory panel.

Because of these somewhat tepid findings, *in vitro* studies using human cerebral organoids will be useful in developing a deeper understanding of disease pathogenesis that could aid in earlier diagnosis. In addition, *in vitro* models can be used for drug testing, in particular pharmacological modalities other than monoclonal antibodies, which are expensive and cannot be taken orally.

Early-onset familial AD is usually caused by dominantly acting genetic variants in the amyloid precursor protein (APP) gene on chromosome 21, presenilin 1 (*PSEN1; PS1*) and presenilin 2 (*PSEN2; PS1*).[196–198] Late onset AD, and sporadic AD are generally viewed as combinations of genetic and environment factors, with *APOE4* being the strongest genetic risk allele.[199] *APOE* codes for apolipoproteins, which are fat-binding proteins. In the brain, APOE is expressed in astrocytes, which transport cholesterol to neurons expressing APOE receptors. APOE connects neuroinflammation to neurodegenerative disorders through its capacity to act as a checkpoint inhibitor of complement component C1q.[200]

The majority of AD is late-onset and sporadic. The heritability of early onset AD is very high, estimated to be >90%. Heritability estimates for late onset AD is also considerable (58%–79%), with GWAS revealing more than 50 loci containing genes involved in immunity, endocytosis, cholesterol transport, ubiquitination, microglia function, and Aβ and tau processing.[201]

For the purpose of disease modeling neurodegenerative disorders using iPSCs, there are two fundamental approaches. One is to generate lines from patients with risk variants that have the strongest impact on function so that cellular and molecular phenotypes can more readily be detected *in vitro*. These would include genetic variants in *APP*, *PSEN1*, and *PSEN2* that cause early-onset, familial AD. High risk variants could also be generated from control iPSC lines by introducing disease-causing variants using CRISPR-Cas9 gene editing. CRISPR edited isogenic lines are useful for transcriptomics and proteomics because confounding factors related to the development of iPSCs from patients and unaffected controls are essentially eliminated. Isogenic lines can also be generated by replacing mutant alleles in patient-derived iPSCs using CRISPR and homology directed repair. This has the advantage of preserving genetic factors other than the primary gene of interest that could be playing a role in the emergence of disease.

Cerebral organoids in familial AD

Monolayer neuronal cells, while able to produce elevated levels of Aβ and phosphorylated tau (p-tau), fail to produce Aβ plaques or neurofibrillary tangles.[202] However, cerebral organoids offer a solution to this problem. In the first organoid model system using patients with familial AD, Choi et al. showed that organoids made from patients with familial AD mutations in the *APP* and *PSEN1* genes are able to induce extracellular deposition of Aβ and produce plaques, as well as aggregates of p-tau.[202]

In a study by Zhao et al., cerebral organoids from AD patients homozygous for the APOE4 allele showed greater apoptosis and decreased synaptic integrity, and an increase in p-tau.[203] RNA-seq revealed an enrichment of genes involved in the formation of stress granules and disrupted RNA metabolism. Remarkably, these molecular phenotypes were corrected by the isogenic conversion of APOE4 to APOE3.

In another organoid study investigating the effects of APOE4, Lin et al. found hundreds of DEGs with enrichment of genes involved in synaptic function, lipid metabolism, and immune response (the latter of which were found in microglia-like cells).[204] In addition, APOE4 neurons exhibited increased synaptic APOE4 with impaired Aβ uptake and cholesterol accumulation in astrocytes, while reduced Aβ phagocytosis was observed in the microglia-like cells. Like the Zhao findings, isogenic conversion of APOE4 to APOE3 using CRIPSPR attenuated AD-related pathology, such as Aβ accumulation and p-tau in 6-month old organoids.

APOE4 has also been investigated as a risk factor for the effect of SARS-CoV-2 infection on the brain.[205] Low grade infection in neurons and astrocytes was observed in both neuron-astrocyte co-cultures and cerebral organoids, accompanied by an increase in astrocyte size and nuclear fragmentation occurring in APOE4 homozygotes. Infection was blocked by the antiviral drug remdesivir, which is used in clinical practice in hospitalized COVID-19 patients because it reduces the number of days spent in hospital.[206] Whether the in vitro findings translate to clinical outcome with respect to neurological, psychiatric, and cognitive problems in the acute setting and in long-COVID-19 syndrome remains to be seen.

Additionally, in an analysis of cerebral organoids harboring familial AD-causing *PSEN2* variants, Arber et al. found deficits in Notch signaling and neurogenesis.[207] In another study using *PSEN2* mutant cerebral organoids, a higher Aβ42/Aβ40 ratio, asynchronous calcium transients, and enhanced neuronal hyperactivity were found.[208] In addition, another study evaluating the effect of APP gene duplication and *PSEN1* variants on cerebral organoids found amyloid aggregation, hyperphosphorylated tau protein, and endosome abnormalities, similar to that seen in the brains of patients. These changes were reduced upon treatment with β- and γ-secretase inhibitors.[209] Similarly, Aβ and tau pathology were observed in a study using cerebral organoids derived from patients with familial AD[210]. These studies attest to the usefulness of the organoid model system in studying familial AD.

Modeling sporadic AD

Although most of the work carried out using iPSCs to model AD has focused on patients with familial AD and APOE4 homozygotes, sporadic AD is far more prevalent, accounting for over 90% of the AD cases in the population.[211]Sporadic AD is most likely caused by a combination of weak genetic variants interacting with environmental factors. While studying organoids and other neuronal cell types containing strong genetic determinants is better for understanding disease pathogenesis, modeling sporadic AD is arguably a better system for studying the impact of environmental factors.

Proof of concept that sporadic AD can be modeled in the absence of a strong genetic determinant was demonstrated by Parvoni et al., who showed that cerebral organoids derived from controls developed time-dependent increases in Aβ40 and Aβ42 following treatment with Aftin-5, an Aβ-42 inducer, leading to the formation of Aβ oligomers.[212]

However, suspected environmental factors implicated in sporadic AD, such as head injury, cardiovascular disease, and diabetes, and protective factors (diet and exercise) are difficult to model in vitro. On the other hand, stressors associated with AD that can be modeled in vitro are oxidative stress, pollutants, and heavy metals. However, these factors have not yet been applied to cerebral organoids.

Cerebral organoids for drug development in AD

Since monolayer neurons do not recapitulate key molecular features of AD, cerebral organoids are preferred for drug development and testing. Proof of principle for this approach was demonstrated by Choi et al., who showed that β- and γ-secretase inhibitors decreased Aβ pathology and tauopathy in cerebral organoids.[202] Importantly, Park et al. established a high-throughput, drug-screening platform to test FDA approved drugs that can penetrate the blood-brain barrier.[213] The group then evaluated the effect of these drugs on various AD molecular and cellular phenotypes, such as Aβ, p-tau, synapse loss, apoptosis, and autophagy. They showed that in APOE4 organoids, treatment with Flibanserin, an inhibitor of PTEN, and Ripasudil, an inhibitor of Dkk1, decreased the production of Aβ by increasing α-secretase activity and decreasing BACE1 activity. In addition, suppression of PTEN inhibited the production of p-tau. Together, these studies show that cerebral organoids are a good model system for identifying novel therapeutic options for AD.

Modeling Parkinson Disease using organoid cultures

The other major neurodegenerative disorder to which organoid technology has been applied is Parkinson disease (PD), which is the second most common in the United States after AD. PD affects more than 1% of the population over the age of 65. It is characterized by the accumulation of abnormal intracellular protein aggregates called Lewy bodies, which are composed of α-synuclein (α-syn).[214,215] α-syn is a neuronal protein encoded by the *SNCA* gene that regulates synaptic vesicle trafficking. Like AD, PD is caused by high impact genetic variants that are familial and monogenic (autosomal dominant and recessive forms exist), leading to early onset disease, representing approximately 10% of cases. Some of the genes that cause familial PD include *PINK1, LRRK2, SNCA, GBA, PARK7, GCH1, PLA2G6,* and *ATP13A2,* among others.[216] However, the majority of PD cases are sporadic and late onset, which are caused by common variants probably in combination with environmental factors. In a recent meta-analysis of 17 PD GWAS datasets, it was found that 90 common variants explained 16%–36% of the heritable risk.[217] Significant genetic correlations were found for brain volumes, intracranial volume, smoking status, and educational attainment. In addition, Mendelian randomization between cognitive performance and PD risk showed a robust association.

Organoids derived from iPSCs are increasingly being used to model PD. However, unlike the telencephalic organoids used for studying neurodevelopmental and neurodegenerative disorders, midbrain organoids are generally used for PD since the disorder is caused by the selective loss of dopaminergic neurons in the substantia nigra pars compacta originating in the midbrain ventral tegmental area (although neurodegeneration is more widespread in later stages).[214,215]

Using midbrain organoids, Wulansari et al., found that loss-of-function mutations in the early onset PD candidate gene *DNAJC6* resulted in pathologic α-synuclein aggregation and mitochondrial and lysosomal abnormalities.[218] In another midbrain organoid study using iPSCs containing a mutation in *PRKN,* a decrease in glial fibrillary acidic protein (GFAP)-positive astrocytes was found, suggesting that non-neuronal cells could be playing a role in the neuronal cell death associated with PD caused by this particular gene.[219] Additionally, Shin et al. found that isogenic midbrain organoids with a PD variant, LRRK2 G2019S, recapitulated pathological hallmarks observed in patients, including the abnormal localization of serine 129 phosphorylated α-synuclein in vesicular endosomal compartments.[220] Interestingly, in a G x E experiment, the investigators found that *LRRK1* mutated samples showed an enhancement of PD-associated phenotypes when treated with the dopaminergic neurotoxin, MPTP. Similarly, Kwak et al. showed that treating control dopamine organoids with MPTP resulted in dopaminergic neuron-specific cell death.[221]

Midbrain organoids are also being used to test the effects of novel drugs, and as a potential source of progenitors for cell replacement therapy. Boussaad et al., for example, developed automated protocols for neuronal differentiation of neural progenitor cells in both 2D culture and 3D midbrain organoid cultures for high-throughput drug screening.[222] And, neural stem precursor cells isolated from human midbrain organoids were successfully transplanted into an animal PD model, after which they developed into dopamine neurons and integrated into the midbrain circuit.[223] These studies show promise for the identification of novel therapeutic options for PD using cerebral organoids.

Conclusion

Since their discovery nearly a decade ago, cerebral organoids continue to fascinate the scientific and lay public. Although much work still needs to be carried out to find the optimal growth conditions and improve reproducibility, there have been significant discoveries using these structures, as described throughout this chapter. I have focused primarily on the use of organoid cultures in the context of environmental factors that can affect the developing brain, such as infectious diseases and maternal immune activation. I also described how cerebral organoids are being used to study neurodegenerative disorders and infectious diseases that affect adults. The rapid incorporation of 3-D organoid cultures by investigators to study newly emerging infectious diseases, such as Zika virus and SARS-CoV-2, and finding plausible disease pathogenesis mechanisms attests to the utility of this model system.

Future studies incorporating new technological and methodological advances will add substantially to findings described in this chapter. One is scRNA-seq, which provides single cell discrimination of transcriptomes in samples that are heterogeneous, an especially useful tool for complex structures like cerebral organoids. Another is the development of cerebral organoids that incorporate epithelial and vascular endothelial cells into the organoid preps to mimic the BBB. These will be useful in understanding the role of the BBB in the movement of inflammatory cytokines and toxins from the periphery to the brain parenchyma, especially in conditions associated with loss of BBB integrity.[224]

References

1. Ben-David E, Shifman S. Combined analysis of exome sequencing points toward a major role for transcription regulation during brain development in autism. *Mol Psychiatr*. 2013;18(10):1054–1056. https://doi.org/10.1038/mp.2012.148.
2. Cukier HN, Dueker ND, Slifer SH, et al. Exome sequencing of extended families with autism reveals genes shared across neurodevelopmental and neuropsychiatric disorders. *Mol Autism*. 2014;5(1). https://doi.org/10.1186/2040-2392-5-1.
3. Glessner JT, Wang K, Cai G, et al. Autism genome-wide copy number variation reveals ubiquitin and neuronal genes. *Nature*. 2009;459(7246):569–572. https://doi.org/10.1038/nature07953.
4. Marshall CR, Noor A, Vincent JB, et al. Structural variation of chromosomes in autism spectrum disorder. *Am J Hum Genet*. 2008;82(2):477–488. https://doi.org/10.1016/j.ajhg.2007.12.009.
5. Pinto D, Delaby E, Merico D, et al. Convergence of genes and cellular pathways dysregulated in autism spectrum disorders. *Am J Hum Genet*. 2014;94(5):677–694. https://doi.org/10.1016/j.ajhg.2014.03.018.
6. Sanders SJ, Murtha MT, Gupta AR, et al. De novo mutations revealed by whole-exome sequencing are strongly associated with autism. *Nature*. 2012;485(7397):237–241. https://doi.org/10.1038/nature10945.
7. Stefansson H, Meyer-Lindenberg A, Steinberg S, et al. CNVs conferring risk of autism or schizophrenia affect cognition in controls. *Nature*. 2014;505(7483):361–366. https://doi.org/10.1038/nature12818.
8. Chun H, Leung C, Wen SW, McDonald J, Shin HH. Maternal exposure to air pollution and risk of autism in children: a systematic review and meta-analysis. *Environ Pollut*. 2020;256:113307. https://doi.org/10.1016/j.envpol.2019.113307.
9. Grabrucker AM. Environmental factors in autism. *Front Psychiatr*. 2013;3. https://doi.org/10.3389/fpsyt.2012.00118.
10. Jung Y, Lee AM, McKee SA, Picciotto MR. Maternal smoking and autism spectrum disorder: meta-analysis with population smoking metrics as moderators. *Sci Rep*. 2017;7(1). https://doi.org/10.1038/s41598-017-04413-1.
11. Krakowiak P, Walker CK, Bremer AA, et al. Maternal metabolic conditions and risk for autism and other neurodevelopmental disorders. *Pediatrics*. 2012;129(5):e1121–e1128. https://doi.org/10.1542/peds.2011-2583.

12. Rossignol DA, Genuis SJ, Frye RE. Environmental toxicants and autism spectrum disorders: a systematic review. *Transl Psychiatry*. 2014;4(2). https://doi.org/10.1038/tp.2014.4.
13. Sandin S, Lichtenstein P, Kuja-Halkola R, Larsson H, Hultman CM, Reichenberg A. The familial risk of autism. *JAMA*. 2014;311(17):1770–1777. https://doi.org/10.1001/jama.2014.4144.
14. Saunders A, Woodland J, Gander S. A Comparison of prenatal exposures in children with and without a diagnosis of autism spectrum disorder. *Cureus*. 2019. https://doi.org/10.7759/cureus.5223.
15. Wang M, Hossain F, Sulaiman R, Ren X. Exposure to inorganic arsenic and lead and autism spectrum disorder in children: a systematic review and meta-analysis. *Chem Res Toxicol*. 2019;32(10):1904–1919. https://doi.org/10.1021/acs.chemrestox.9b00134.
16. Castelbaum L, Sylvester CM, Zhang Y, Constantino JN. On the nature of monozygotic Twin concordance and discordance for autistic Trait severity: a quantitative analysis. *Behav Genet*. 2019;50(4):263–272.
17. Bucan M, Abrahams BS, Wang K, et al. Genome-wide analyses of exonic copy number variants in a family-based study point to novel autism susceptibility genes. *PLoS Genet*. 2009;5(6):e1000536. https://doi.org/10.1371/journal.pgen.1000536.
18. Kirov G, Rees E, Walters JTR, et al. The penetrance of copy number variations for schizophrenia and developmental delay. *Biol Psychiatr*. 2014;75(5):378–385. https://doi.org/10.1016/j.biopsych.2013.07.022.
19. Rosenfeld JA, Coe BP, Eichler EE, Cuckle H, Shaffer LG. Estimates of penetrance for recurrent pathogenic copy-number variations. *Genet Med*. 2013;15(6):478–481. https://doi.org/10.1038/gim.2012.164.
20. Ropers HH, Wienker T. Penetrance of pathogenic mutations in haploinsufficient genes for intellectual disability and related disorders. *Eur J Med Genet*. 2015;58(12):715–718. https://doi.org/10.1016/j.ejmg.2015.10.007.
21. Coleman JRI, Bryois J, Gaspar HA, et al. Biological annotation of genetic loci associated with intelligence in a meta-analysis of 87,740 individuals. *Mol Psychiatr*. 2019;24(2):182–197. https://doi.org/10.1038/s41380-018-0040-6.
22. Gudmundsson OO, Walters GB, Ingason A, et al. Attention-deficit hyperactivity disorder shares copy number variant risk with schizophrenia and autism spectrum disorder. *Transl Psychiatry*. 2019;9(1). https://doi.org/10.1038/s41398-019-0599-y.
23. Hussman JP, Chung RH, Griswold AJ, et al. A noise-reduction GWAS analysis implicates altered regulation of neurite outgrowth and guidance in autism. *Mol Autism*. 2011;2(1). https://doi.org/10.1186/2040-2392-2-1.
24. Nudel R, Simpson NH, Baird G, et al. Genome-wide association analyses of child genotype effects and parent-of-origin effects in specific language impairment. *Gene Brain Behav*. 2014;13(4):418–429. https://doi.org/10.1111/gbb.12127.
25. Schizophrenia Working Group of the Psychiatric Genomics Consortium. Biological insights from 108 schizophrenia-associated genetic loci. *Nature*. 2014;511(7510):421–427. https://doi.org/10.1038/nature13595.
26. Paquin V, Lapierre M, Veru F, King S. Early environmental upheaval and the risk for schizophrenia. *Annu Rev Clin Psychol*. 2021;17:285–311. https://doi.org/10.1146/annurev-clinpsy-081219-103805.
27. Han VX, Patel S, Jones HF, et al. Maternal acute and chronic inflammation in pregnancy is associated with common neurodevelopmental disorders: a systematic review. *Transl Psychiatry*. 2021;11(1). https://doi.org/10.1038/s41398-021-01198-w.
28. Goasdoué K, Miller SM, Colditz PB, Björkman ST. Review: the blood-brain barrier; protecting the developing fetal brain. *Placenta*. 2017;54:111–116. https://doi.org/10.1016/j.placenta.2016.12.005.
29. Lombardo MV, Moon HM, Su J, Palmer TD, Courchesne E, Pramparo T. Maternal immune activation dysregulation of the fetal brain transcriptome and relevance to the pathophysiology of autism spectrum disorder. *Mol Psychiatr*. 2018:1001–1013. https://doi.org/10.1038/mp.2017.15.
30. Spann MN, Monk C, Scheinost D, Peterson BS. Maternal immune activation during the third trimester is associated with neonatal functional connectivity of the salience network and fetal to toddler behavior. *J Neurosci*. 2018;38(11):2877–2886. https://doi.org/10.1523/JNEUROSCI.2272-17.2018.
31. Cracking ENCODE. *Lancet*. 2012;380(9846):61520–61527. https://doi.org/10.1016/S0140-6736(12)61520-7.
32. Kavanagh DH, Dwyer S, O'Donovan MC, Owen MJ. The ENCODE project: implications for psychiatric genetics. *Mol Psychiatr*. 2013;18(5):540–542. https://doi.org/10.1038/mp.2013.13.
33. Rockowitz S, Zheng D. Significant expansion of the REST/NRSF cistrome in human versus mouse embryonic stem cells: potential implications for neural development. *Nucleic Acids Res*. 2015;43(12):5730–5743. https://doi.org/10.1093/nar/gkv514.

34. The ENCODE Project Consortium. An integrated encyclopedia of DNA elements in the human genome. *Nature*. 2012;489(7414):57–74. https://doi.org/10.1038/nature11247.
35. Takahashi K, Yamanaka S. Induction of pluripotent stem cells from mouse embryonic and adult fibroblast cultures by defined factors. *Cell*. 2006;126(4):663–676. https://doi.org/10.1016/j.cell.2006.07.024.
36. Amodeo DA, Lai CY, Hassan O, Mukamel EA, Behrens MM, Powell SB. Maternal immune activation impairs cognitive flexibility and alters transcription in frontal cortex. *Neurobiol Dis*. 2019;125:211–218. https://doi.org/10.1016/j.nbd.2019.01.025.
37. Boulanger-Bertolus J, Pancaro C, Mashour GA. Increasing role of maternal immune activation in neurodevelopmental disorders. *Front Behav Neurosci*. 2018;12. https://doi.org/10.3389/fnbeh.2018.00230.
38. Dozmorov MG, Bilbo SD, Kollins SH, et al. Associations between maternal cytokine levels during gestation and measures of child cognitive abilities and executive functioning. *Brain Behav Immun*. 2018;70:390–397. https://doi.org/10.1016/j.bbi.2018.03.029.
39. Guma E, Plitman E, Chakravarty MM. The role of maternal immune activation in altering the neurodevelopmental trajectories of offspring: a translational review of neuroimaging studies with implications for autism spectrum disorder and schizophrenia. *Neurosci Biobehav Rev*. 2019;104:141–157. https://doi.org/10.1016/j.neubiorev.2019.06.020.
40. Singer AB, Windham GC, Croen LA, et al. Maternal exposure to occupational Asthmagens during pregnancy and autism spectrum disorder in the study to explore early development. *J Autism Dev Disord*. 2016;46(11):3458–3468. https://doi.org/10.1007/s10803-016-2882-6.
41. Smith SEP, Li J, Garbett K, Mirnics K, Patterson PH. Maternal immune activation alters fetal brain development through interleukin-6. *J Neurosci*. 2007;27(40):10695–10702. https://doi.org/10.1523/JNEUROSCI.2178-07.2007.
42. Xu, S., Hanganu-Opatz IL. Knock-Down of Hippocampal DISC1 in Immune-Challenged Mice Impairs the Prefrontal-Hippocampal Coupling and the Cognitive Performance throughout Development.
43. Abdallah MW, Larsen N, Grove J, et al. Amniotic fluid inflammatory cytokines: potential markers of immunologic dysfunction in autism spectrum disorders. *World J Biol Psychiatr*. 2013;14(7):528–538. https://doi.org/10.3109/15622975.2011.639803.
44. Atladottir HO, Henriksen TB, Schendel DE, Parner ET. Autism after infection, febrile episodes, and antibiotic use during pregnancy: an exploratory study. *Pediatrics*. 2012;130(6):e1447–e1454. https://doi.org/10.1542/peds.2012-1107.
45. Brown AS. Epidemiologic studies of exposure to prenatal infection and risk of schizophrenia and autism. *Dev Neurobiol*. 2012;72(10):1272–1276. https://doi.org/10.1002/dneu.22024.
46. Fatemi SH, Folsom TD, Rooney RJ, et al. The viral theory of schizophrenia revisited: abnormal placental gene expression and structural changes with lack of evidence for H1N1 viral presence in placentae of infected mice or brains of exposed offspring. *Neuropharmacology*. 2012;62(3):1290–1298. https://doi.org/10.1016/j.neuropharm.2011.01.011.
47. Paquette SG, Banner D, Zhao Z, et al. Interleukin-6 is a potential biomarker for severe pandemic H1N1 Influenza A infection. *PLoS One*. 2012;7(6):e38214. https://doi.org/10.1371/journal.pone.0038214.
48. Parker-Athill EC, Tan J. Maternal immune activation and autism spectrum disorder: interleukin-6 signaling as a key mechanistic pathway. *Neurosignals*. 2010:113–128. https://doi.org/10.1159/000319828.
49. Richetto J, Calabrese F, Riva MA, Meyer U. Prenatal immune activation induces maturation-dependent alterations in the prefrontal GABAergic transcriptome. *Schizophr Bull*. 2014:351–361. https://doi.org/10.1093/schbul/sbs195.
50. Zerbo O, Iosif AM, Walker C, Ozonoff S, Hansen RL, Hertz-Picciotto I. Is maternal Influenza or fever during pregnancy associated with autism or developmental delays? Results from the CHARGE (childhood autism risks from genetics and environment) study. *J Autism Dev Disord*. 2013;43(1):25–33. https://doi.org/10.1007/s10803-012-1540-x.
51. Atladóttir HO, Pedersen MG, Thorsen P, et al. Association of family history of autoimmune diseases and autism spectrum disorders. *Pediatrics*. 2009;124(2):687–694. https://doi.org/10.1542/peds.2008-2445.
52. Hafizi S, Tabatabaei D, Lai MC. Review of clinical studies targeting inflammatory pathways for individuals with autism. *Front Psychiatr*. 2019;10. https://doi.org/10.3389/fpsyt.2019.00849.
53. Sullivan EL, Nousen EK, Chamlou KA. Maternal high fat diet consumption during the perinatal period programs offspring behavior. *Physiol Behav*. 2014;123:236–242. https://doi.org/10.1016/j.physbeh.2012.07.014.

54. Reynolds LC, Inder TE, Neil JJ, Pineda RG, Rogers CE. Maternal obesity and increased risk for autism and developmental delay among very preterm infants. *J Perinatol*. 2014;34(9):688–692. https://doi.org/10.1038/jp.2014.80.
55. Ge GM, Leung MTY, Man KKC, et al. Maternal thyroid dysfunction during pregnancy and the risk of adverse outcomes in the offspring: a systematic review and meta-analysis. *J Clin Endocrinol Metabol*. 2020;105(12). https://doi.org/10.1210/clinem/dgaa555.
56. Jones HF, Ho ACC, Sharma S, et al. Maternal thyroid autoimmunity associated with acute-onset neuropsychiatric disorders and global regression in offspring. *Dev Med Child Neurol*. 2019;61(8):984–988. https://doi.org/10.1111/dmcn.14167.
57. Arnold SE, Franz BR, Gur RC, et al. Smaller neuron size in schizophrenia in hippocampal subfields that mediate cortical-hippocampal interactions. *Am J Psychiatr*. 1995;152(5):738–748. https://doi.org/10.1176/ajp.152.5.738.
58. Bauman MD, Iosif AM, Smith SEP, Bregere C, Amaral DG, Patterson PH. Activation of the maternal immune system during pregnancy alters behavioral development of rhesus monkey offspring. *Biol Psychiatr*. 2014;75(4):332–341. https://doi.org/10.1016/j.biopsych.2013.06.025.
59. Benes FM. Regulation of cell cycle and DNA repair in post-mitotic GABA neurons in psychotic disorders. *Neuropharmacology*. 2011;60(7–8):1232–1242. https://doi.org/10.1016/j.neuropharm.2010.12.011.
60. Benes FM, Lim B, Matzilevich D, Subburaju S, Walsh JP. Circuitry-based gene expression profiles in GABA cells of the trisynaptic pathway in schizophrenics versus bipolars. *Proc Natl Acad Sci USA*. 2008;105(52):20935–20940. https://doi.org/10.1073/pnas.0810153105.
61. Benes FM, Lim B, Matzilevich D, Walsh JP, Subburaju S, Minns M. Regulation of the GABA cell phenotype in hippocampus of schizophrenics and bipolars. *Proc Natl Acad Sci USA*. 2007;104(24):10164–10169. https://doi.org/10.1073/pnas.0703806104.
62. Ducharme G, Lowe GC, Goutagny R, Williams S. Early alterations in hippocampal circuitry and theta rhythm generation in a mouse model of prenatal infection: implications for schizophrenia. *PLoS One*. 2012;7(1). https://doi.org/10.1371/journal.pone.0029754.
63. Meyer U, Feldon J, Fatemi SH. In-vivo rodent models for the experimental investigation of prenatal immune activation effects in neurodevelopmental brain disorders. *Neurosci Biobehav Rev*. 2009;33(7):1061–1079. https://doi.org/10.1016/j.neubiorev.2009.05.001.
64. Oskvig DB, Elkahloun AG, Johnson KR, Phillips TM, Herkenham M. Maternal immune activation by LPS selectively alters specific gene expression profiles of interneuron migration and oxidative stress in the fetus without triggering a fetal immune response. *Brain Behav Immun*. 2012;26(4):623–634. https://doi.org/10.1016/j.bbi.2012.01.015.
65. Blank T, Prinz M. Microglia as modulators of cognition and neuropsychiatric disorders. *Glia*. 2013;61(1):62–70. https://doi.org/10.1002/glia.22372.
66. Ikezu S, Yeh H, Delpech JC, et al. Inhibition of colony stimulating factor 1 receptor corrects maternal inflammation-induced microglial and synaptic dysfunction and behavioral abnormalities. *Mol Psychiatr*. 2020. https://doi.org/10.1038/s41380-020-0671-2.
67. Kerns D, Vong GS, Barley K, et al. Gene expression abnormalities and oligodendrocyte deficits in the internal capsule in schizophrenia. *Schizophr Res*. 2010;120(1–3):150–158. https://doi.org/10.1016/j.schres.2010.04.012.
68. Maezawa I, Calafiore M, Wulff H, Jin LW. Does microglial dysfunction play a role in autism and Rett syndrome? *Neuron Glia Biol*. 2012;7(1):85–97. https://doi.org/10.1017/S1740925X1200004X.
69. Prins JR, Eskandar S, Eggen BJL, Scherjon SA. Microglia, the missing link in maternal immune activation and fetal neurodevelopment; and a possible link in preeclampsia and disturbed neurodevelopment? *J Reprod Immunol*. 2018;126:18–22. https://doi.org/10.1016/j.jri.2018.01.004.
70. Rodriguez JI, Kern JK. Evidence of microglial activation in autism and its possible role in brain underconnectivity. *Neuron Glia Biol*. 2011;7(2–4):205–213. https://doi.org/10.1017/S1740925X12000142.
71. Zhan Y, Paolicelli RC, Sforazzini F, et al. Deficient neuron-microglia signaling results in impaired functional brain connectivity and social behavior. *Nat Neurosci*. 2014;17(3):400–406. https://doi.org/10.1038/nn.3641.
72. Abud EM, Ramirez RN, Martinez ES, et al. iPSC-derived human microglia-like cells to study neurological diseases. *Neuron*. 2017;94(2):278–293. https://doi.org/10.1016/j.neuron.2017.03.042. e9.
73. Douvaras P, Sun B, Wang M, et al. Directed differentiation of human pluripotent stem cells to microglia. *Stem Cell Rep*. 2017;8(6):1516–1524. https://doi.org/10.1016/j.stemcr.2017.04.023.

74. Haenseler W, Sansom SN, Buchrieser J, et al. A highly efficient human pluripotent stem cell microglia model Displays a neuronal-Co-culture-Specific expression profile and inflammatory response. *Stem Cell Rep.* 2017;8(6):1727–1742. https://doi.org/10.1016/j.stemcr.2017.05.017.
75. Pandya H, Shen MJ, Ichikawa DM, et al. Differentiation of human and murine induced pluripotent stem cells to microglia-like cells. *Nat Neurosci.* 2017;20(5):753–759. https://doi.org/10.1038/nn.4534.
76. Abreu CM, Gama L, Krasemann S, et al. Microglia increase inflammatory responses in iPSC-derived human BrainSpheres. *Front Microbiol.* 2018. https://doi.org/10.3389/fmicb.2018.02766.
77. Ormel PR, Vieira de Sá R, van Bodegraven EJ, et al. Microglia innately develop within cerebral organoids. *Nat Commun.* 2018;9(1). https://doi.org/10.1038/s41467-018-06684-2.
78. Gupta S, Ellis SE, Ashar FN, et al. Transcriptome analysis reveals dysregulation of innate immune response genes and neuronal activity-dependent genes in autism. *Nat Commun.* 2014;5. https://doi.org/10.1038/ncomms6748.
79. Voineagu I, Eapen V. Converging pathways in autism spectrum disorders: Interplay between synaptic dysfunction and immune responses. *Front Hum Neurosci.* 2013. https://doi.org/10.3389/fnhum.2013.00738.
80. Voineagu I, Wang X, Johnston P, et al. Transcriptomic analysis of autistic brain reveals convergent molecular pathology. *Nature.* 2011;474(7351):380–386. https://doi.org/10.1038/nature10110.
81. Derecki NC, Katzmarski N, Kipnis J, Meyer-Luehmann M. Microglia as a critical player in both developmental and late-life CNS pathologies. *Acta Neuropathol.* 2014;128(3):333–345. https://doi.org/10.1007/s00401-014-1321-z.
82. Wang J, Wegener JE, Huang TW, et al. Wild-type microglia do not reverse pathology in mouse models of Rett syndrome. *Nature.* 2015;521(7552):E1–E4. https://doi.org/10.1038/nature14444.
83. Ishizuka K, Fujita Y, Kawabata T, et al. Rare genetic variants in CX3CR1 and their contribution to the increased risk of schizophrenia and autism spectrum disorders. *Transl Psychiatry.* 2017;7(8). https://doi.org/10.1038/tp.2017.173.
84. Colonna M, Wang Y. TREM2 variants: new keys to decipher Alzheimer disease pathogenesis. *Nat Rev Neurosci.* 2016;17(4):201–207. https://doi.org/10.1038/nrn.2016.7.
85. Sims R, Van Der Lee SJ, Naj AC, et al. Rare coding variants in PLCG2, ABI3, and TREM2 implicate microglial-mediated innate immunity in Alzheimer's disease. *Nat Genet.* 2017;49(9):1373–1384. https://doi.org/10.1038/ng.3916.
86. Amare AT, Schubert KO, Hou L, et al. Association of polygenic score for schizophrenia and HLA antigen and inflammation genes with response to lithium in bipolar affective disorder: a genome-wide association study. *JAMA Psychiatr.* 2018;75(1):65–74. https://doi.org/10.1001/jamapsychiatry.2017.3433.
87. Cross-Disorder Group of the Psychiatric Genomics Consortium. Genomic relationships, novel loci, and pleiotropic mechanisms across Eight psychiatric disorders. In: *Group of the Psychiatric Genomics Consortium. Electronic Address: Plee0@mgh.Harvard.Edu.* Vol. 179. 2019.
88. Irish Schizophrenia Genomics Consortium and the Wellcome Trust Case Control Consortium 2. Genome-wide association study implicates HLA-C*01:02 as a risk factor at the major histocompatibility complex locus in schizophrenia. *Biol Psychiatr.* 2012;72(8):620–628. https://doi.org/10.1016/j.biopsych.2012.05.035.
89. Bennabi M, Gaman A, Delorme R, et al. HLA-class II haplotypes and autism spectrum disorders. *Sci Rep.* 2018;8(1). https://doi.org/10.1038/s41598-018-25974-9.
90. Guerini F, Bolognesi E, Chiappedi M. HLA-G *14bp insertion and the KIR2DS1-HLAC2 complex impact on behavioral impairment in children with. *Autism Spectr Disord Neurosci.* 2018;370(17):30410–30414.
91. Johnson WG, Buyske S, Mars AE, et al. HLA-DR4 as a risk allele for autism acting in mothers of probands possibly during pregnancy. *Arch Pediatr Adolesc Med.* 2009;163(6):542–546. https://doi.org/10.1001/archpediatrics.2009.74.
92. Lee LC, Zachary AA, Leffell MS, et al. HLA-DR4 in families with autism. *Pediatr Neurol.* 2006;35(5):303–307. https://doi.org/10.1016/j.pediatrneurol.2006.06.006.
93. Nudel R, Benros ME, Krebs MD, et al. Immunity and mental illness: findings from a Danish population-based immunogenetic study of seven psychiatric and neurodevelopmental disorders. *Eur J Hum Genet.* 2019;27(9):1445–1455. https://doi.org/10.1038/s41431-019-0402-9.
94. Tamouza R, Fernell E, Eriksson MA, et al. HLA polymorphism in regressive and non-regressive autism: a preliminary study. *Autism Res.* 2020;13(2):182–186. https://doi.org/10.1002/aur.2217.

95. Torres AR, Sweeten TL, Cutler A, et al. The association and linkage of the HLA-A2 class I allele with autism. *Hum Immunol.* 2006;67(4–5):346–351. https://doi.org/10.1016/j.humimm.2006.01.001.
96. Croen LA, Qian Y, Ashwood P, et al. Family history of immune conditions and autism spectrum and developmental disorders: findings from the study to explore early development. *Autism Res.* 2019;12(1):123–135. https://doi.org/10.1002/aur.1979.
97. Gesundheit B, Rosenzweig JP, Naor D, et al. Immunological and autoimmune considerations of autism spectrum disorders. *J Autoimmun.* 2013;44:1–7. https://doi.org/10.1016/j.jaut.2013.05.005.
98. Vogel Ciernia A, Careaga M, LaSalle JM, Ashwood P. Microglia from offspring of dams with allergic asthma exhibit epigenomic alterations in genes dysregulated in autism. *Glia.* 2018;66(3):505–521. https://doi.org/10.1002/glia.23261.
99. Gata-Garcia A, Diamond B. Maternal antibody and ASD: clinical data and animal models. *Front Immunol.* 2019;10(MAY). https://doi.org/10.3389/fimmu.2019.01129.
100. Boulanger LM. MHC class I in activity-dependent structural and functional plasticity. *Neuron Glia Biol.* 2004;1(3):283–289. https://doi.org/10.1017/S1740925X05000128.
101. Corriveau RA, Huh GS, Shatz CJ. Regulation of class I MHC gene expression in the developing and mature CNS by neural activity. *Neuron.* 1998;21(3):505–520. https://doi.org/10.1016/S0896-6273(00)80562-0.
102. Huh GS, Boulanger LM, Du H, Riquelme PA, Brotz TM, Shatz CJ. Functional requirement for class I MHC in CNS development and plasticity. *Science.* 2000;290(5499):2155–2159. https://doi.org/10.1126/science.290.5499.2155.
103. Lee H, Brott BK, Kirkby LA, et al. Synapse elimination and learning rules co-regulated by MHC class i H2-Db. *Nature.* 2014;509(7499):195–200. https://doi.org/10.1038/nature13154.
104. McAllister AK. Major histocompatibility complex i in brain development and schizophrenia. *Biol Psychiatr.* 2014;75(4):262–268. https://doi.org/10.1016/j.biopsych.2013.10.003.
105. Shatz CJ. MHC class I: an unexpected role in neuronal plasticity. *Neuron.* 2009;64(1):40–45. https://doi.org/10.1016/j.neuron.2009.09.044.
106. Allen NJ, Barres BA. Signaling between glia and neurons: focus on synaptic plasticity. *Curr Opin Neurobiol.* 2005;15(5):542–548. https://doi.org/10.1016/j.conb.2005.08.006.
107. Bolton MML, Eroglu C. Look who is weaving the neural web: glial control of synapse formation. *Curr Opin Neurobiol.* 2009;19(5):491–497. https://doi.org/10.1016/j.conb.2009.09.007.
108. Fourgeaud L, Boulanger LM. Synapse remodeling, Compliments of the complement system. *Cell.* 2007;131(6):1034–1036. https://doi.org/10.1016/j.cell.2007.11.031.
109. Presumey J, Bialas AR, Carroll MC. Complement system in neural synapse elimination in development and disease. In: *Advances in Immunology.* Vol. 135. United States: Academic Press Inc.; 2017:53–79. https://doi.org/10.1016/bs.ai.2017.06.004.
110. Stephan AH, Barres BA, Stevens B. The complement system: an unexpected role in synaptic pruning during development and disease. *Annu Rev Neurosci.* 2012;35:369–389. https://doi.org/10.1146/annurev-neuro-061010-113810.
111. Magdalon J, Mansur F, Teles e Silva AL, de Goes VA, Reiner O, Sertié AL. Complement system in brain architecture and neurodevelopmental disorders. *Front Neurosci.* 2020;14. https://doi.org/10.3389/fnins.2020.00023.
112. Sekar A, Bialas AR, De Rivera H, et al. Schizophrenia risk from complex variation of complement component 4. *Nature.* 2016;530(7589):177–183. https://doi.org/10.1038/nature16549.
113. Guo H, Peng Y, Hu Z, et al. Genome-wide copy number variation analysis in a Chinese autism spectrum disorder cohort. *Sci Rep.* 2017;7. https://doi.org/10.1038/srep44155.
114. Ortega-Alonso A, Ekelund J, Sarin AP, et al. Genome-wide association study of psychosis proneness in the Finnish population. *Schizophr Bull.* 2017;43(6):1304–1314. https://doi.org/10.1093/schbul/sbx006.
115. Bergink V, Gibney SM, Drexhage HA. Autoimmunity, inflammation, and psychosis: a search for peripheral markers. *Biol Psychiatr.* 2014;75(4):324–331. https://doi.org/10.1016/j.biopsych.2013.09.037.
116. Careaga M, Rogers S, Hansen RL, Amaral DG, Van de Water J, Ashwood P. Immune Endophenotypes in children with autism spectrum disorder. *Biol Psychiatr.* 2017;81(5):434–441. https://doi.org/10.1016/j.biopsych.2015.08.036.
117. Goldsmith DR, Rapaport MH, Miller BJ. A meta-analysis of blood cytokine network alterations in psychiatric patients: Comparisons between schizophrenia, bipolar disorder and depression. *Mol Psychiatr.* 2016;21(12):1696–1709. https://doi.org/10.1038/mp.2016.3.

118. Hoseth E, Ueland T, Dieset I. A study of TNF pathway activation in schizophrenia and bipolar disorder in plasma and brain tissue.
119. Pandey GN, Ren X, Rizavi HS, Zhang H. Proinflammatory cytokines and their membrane-bound receptors are altered in the lymphocytes of schizophrenia patients. *Schizophr Res*. 2015;164(1–3):193–198. https://doi.org/10.1016/j.schres.2015.02.004.
120. Prata J, Santos SG, Almeida MI, Coelho R, Barbosa MA. Bridging Autism Spectrum Disorders and Schizophrenia through inflammation and biomarkers—pre-clinical and clinical investigations. *J Neuroinflammation*. 2017;14(1). https://doi.org/10.1186/s12974-017-0938-y.
121. Calaprice D, Tona J, Parker-Athill EC, Murphy TK. A Survey of pediatric acute-onset neuropsychiatric syndrome characteristics and course. *J Child Adolesc Psychopharmacol*. 2017;27(7):607–618. https://doi.org/10.1089/cap.2016.0105.
122. Swedo SE, Leonard HL, Mittleman BB, et al. Identification of children with pediatric autoimmune neuropsychiatric disorders associated with streptococcal infections by a marker associated with rheumatic fever. *Am J Psychiatr*. 1997;154(1):110–112. https://doi.org/10.1176/ajp.154.1.110.
123. Zheng J, Frankovich J, McKenna ES, et al. Association of pediatric acute-onset neuropsychiatric syndrome with microstructural differences in brain Regions detected via diffusion-weighted magnetic resonance imaging. *JAMA Netw Open*. 2020;3(5):e204063. https://doi.org/10.1001/jamanetworkopen.2020.4063.
124. Vitaliti G, Tabatabaie O, Matin N, et al. The usefulness of immunotherapy in pediatric neurodegenerative disorders: a systematic review of literature data. *Hum Vaccines Immunother*. 2015;11(12):2749–2763. https://doi.org/10.1080/21645515.2015.1061161.
125. Wilbur C, Bitnun A, Kronenberg S, et al. PANDAS/PANS in childhood: Controversies and evidence. *Paediatr Child Health*. 2019;24(2):85–91. https://doi.org/10.1093/pch/pxy145.
126. Dunn PM. Perinatal lessons from the past: Sir Norman Gregg, ChM, MC, of Sydney (1892–1966). *Arch Dis Child Fetal Neonatal Ed*. 2007;92(6):513–514.
127. Freitas BC, Beltrão-Braga PCB, Marchetto MC. Modeling inflammation on neurodevelopmental disorders using pluripotent stem cells. In: *Advances in Neurobiology*. Vol. 25. Brazil: Springer; 2020:207–218. https://doi.org/10.1007/978-3-030-45493-7_7.
128. Boppana SB, Britt WJ, Fowler K, et al. Pathogenesis of non-zika congenital viral infections. *J Infect Dis*. 2017;216:S912–S918. https://doi.org/10.1093/infdis/jix431.
129. Liu L, Chen Z, Zhang X, et al. Protection of ZIKV infection-induced neuropathy by abrogation of acute antiviral response in human neural progenitors. *Cell Death Differ*. 2019;26(12):2607–2621. https://doi.org/10.1038/s41418-019-0324-7.
130. Li Z, Lang Y, Sakamuru S, et al. Methylene blue is a potent and broad-spectrum inhibitor against Zika virus in vitro and in vivo. *Emerg Microb Infect*. 2020;9(1):2404–2416. https://doi.org/10.1080/22221751.2020.1838954.
131. Sacramento CQ, De Melo GR, De Freitas CS, et al. The clinically approved antiviral drug sofosbuvir inhibits Zika virus replication. *Sci Rep*. 2017;7. https://doi.org/10.1038/srep40920.
132. Pellegrini L, Albecka A, Mallery DL, et al. SARS-CoV-2 infects the brain choroid plexus and disrupts the blood-CSF barrier in human brain organoids. *Cell Stem Cell*. 2020;27(6):951–961. https://doi.org/10.1016/j.stem.2020.10.001. e5.
133. Jacob F, Pather SR, Huang W-K, et al. Human pluripotent stem cell-derived neural cells and brain organoids reveal SARS-CoV-2 neurotropism Predominates in choroid plexus epithelium. *Cell Stem Cell*. 2020:937–950. https://doi.org/10.1016/j.stem.2020.09.016. e9.
134. Song E, Zhang C, Israelow B, et al. Neuroinvasion of SARS-CoV-2 in human and mouse brain. *J Exp Med*. 2021;218(3). https://doi.org/10.1084/jem.20202135.
135. Li Y, Terstappen GC, Zhang W. *Differentiation of Human Induced Pluripotent Stem Cells (hiPSC) into Endothelial-type Cells and Establishment of an in Vitro Blood-Brain Barrier Model*. Springer Science and Business Media LLC; 2021. https://doi.org/10.1007/7651_2021_363.
136. Frankovich J, Leibold CMC, Farmer C, et al. The burden of caring for a child or adolescent with pediatric acute-onset neuropsychiatric syndrome (PANS): an observational longitudinal study. *J Clin Psychiatr*. 2019;80(1):E1–E8. https://doi.org/10.4088/JCP.17m12091.
137. Hyman SE. PANDAS: Too narrow a view of the neuroimmune landscape. *Am J Psychiatr*. 2021;178(1):5–7. https://doi.org/10.1176/appi.ajp.2020.20111598.

138. Melamed I, Kobayashi RH, O'Connor M, et al. Evaluation of intravenous immunoglobulin in pediatric acute-onset neuropsychiatric syndrome. *J Child Adolesc Psychopharmacol.* 2021;31(2):118–128. https://doi.org/10.1089/cap.2020.0100.
139. Miya K, Takahashi Y, Mori H. Anti-NMDAR autoimmune encephalitis. *Brain Dev.* 2014:645–652. https://doi.org/10.1016/j.braindev.2013.10.005.
140. Sørensen CB, Skov L, Lundby L, Grejsen J, Aaslet L, Debes NM. PANDAS and PANS in children and adolescents are still controversial diagnoses. *Ugeskr Laeger.* 2018;180(48).
141. Muccioli L, Pensato U, Bernabè G, et al. Intravenous immunoglobulin therapy in COVID-19-related encephalopathy. *J Neurol.* 2020. https://doi.org/10.1007/s00415-020-10248-0.
142. Perrin P, Collongues N, Baloglu S, et al. Cytokine release syndrome-associated encephalopathy in patients with COVID-19. *Eur J Neurol.* 2021;28(1):248–258. https://doi.org/10.1111/ene.14491.
143. Valdebenito S, Castellano P, Ajasin D, Eugenin EA. Astrocytes are HIV reservoirs in the brain: a cell type with poor HIV infectivity and replication but efficient cell-to-cell viral transfer. *J Neurochem.* 2021. https://doi.org/10.1111/jnc.15336.
144. Wallet C, De Rovere M, Van Assche J, et al. Microglial cells: the main HIV-1 reservoir in the brain. *Front Cell Infect Microbiol.* 2019;9. https://doi.org/10.3389/fcimb.2019.00362.
145. Simioni S, Cavassini M, Annoni JM, et al. Cognitive dysfunction in HIV patients despite long-standing suppression of viremia. *AIDS.* 2010;24(9):1243–1250. https://doi.org/10.1097/QAD.0b013e3283354a7b.
146. Winston A, Spudich S. Cognitive disorders in people living with HIV. *The Lancet HIV.* 2020;7(7):e504–e513. https://doi.org/10.1016/S2352-3018(20)30107-7.
147. Osborne PN, Nair M, Daunert T. The paradox of HIV blood-brain barrier penetrance and antiretroviral drug delivery Deficiencies. *Trends Neurosci.* 2020;43(9): 30149-1.
148. Gannon P, Khan MZ, Kolson DL. Current understanding of HIV-associated neurocognitive disorders pathogenesis. *Curr Opin Neurol.* 2011;24(3):275–283. https://doi.org/10.1097/WCO.0b013e32834695fb.
149. Saylor D, Dickens AM, Sacktor N, et al. HIV-associated neurocognitive disorder - pathogenesis and prospects for treatment. *Nat Rev Neurol.* 2016;12(4):234–248. https://doi.org/10.1038/nrneurol.2016.27.
150. Tyor W, Fritz-French C, Nath A. Effect of HIV clade differences on the onset and severity of HIV-associated neurocognitive disorders. *J Neurovirol.* 2013;19(6):515–522. https://doi.org/10.1007/s13365-013-0206-6.
151. Rao VR, Ruiz AP, Prasad VR. Viral and cellular factors underlying neuropathogenesis in HIV associated neurocognitive disorders (HAND). *AIDS Res Ther.* 2014;11(1). https://doi.org/10.1186/1742-6405-11-13.
152. Dos Reis RS, Sant S, Keeney H, Wagner MCE, Ayyavoo V. Modeling HIV-1 neuropathogenesis using three-dimensional human brain organoids (hBORGs) with HIV-1 infected microglia. *Sci Rep.* 2020;10(1):15209. https://doi.org/10.1038/s41598-020-72214-0.
153. Chaudhury S, Williams PL, Mayondi GK, et al. Neurodevelopment of HIV-exposed and HIV-unexposed uninfected children at 24 months. *Pediatrics.* 2017;140(4):e20170988. https://doi.org/10.1542/peds.2017-0988.
154. Becker Y. The molecular mechanism of human resistance to HIV-1 infection in persistently infected individuals—a review, hypothesis and implications. *Virus Gene.* 2005;31(1):113–119.
155. Brown TR. I am the Berlin patient: a personal reflection. *AIDS Res Hum Retrovir.* 2015;31(1):2–3. https://doi.org/10.1089/AID.2014.0224.
156. Tebas P, Stein D, Tang WW, et al. Gene editing of CCR5 in autologous CD4 T cells of persons infected with HIV. *N Engl J Med.* 2014;370(10):901–910. https://doi.org/10.1056/NEJMoa1300662.
157. Kunze C, Börner K, Kienle E, et al. Synthetic AAV/CRISPR vectors for blocking HIV-1 expression in persistently infected astrocytes. *Glia.* 2018;66(2):413–427. https://doi.org/10.1002/glia.23254.
158. Loss SR, Will T, Marra PP. The impact of free-ranging domestic cats on wildlife of the United States. *Nat Commun.* 2013;4. https://doi.org/10.1038/ncomms2380.
159. Khurana S, Batra N. Toxoplasmosis in organ transplant recipients: Evaluation, implication, and prevention. *Trop Parasitol.* 2016;6(2):123–128. https://doi.org/10.4103/2229-5070.190814.
160. Chaudhury A, Ramana BV. Schizophrenia and bipolar disorders: the Toxoplasma connection. *Trop Parasitol.* 2019;9(2):71–76. https://doi.org/10.4103/tp.TP_28_19.
161. Lin HA, Chien WC, Huang KY, et al. Infection with Toxoplasma gondii increases the risk of psychiatric disorders in Taiwan: a nationwide population-based cohort study. *Parasitology.* 2020;147(13):1577–1586. https://doi.org/10.1017/S0031182020001183.

162. Torrey EF, Yolken RH. Schizophrenia as a pseudogenetic disease: a call for more gene-environmental studies. *Psychiatr Res*. 2019;278:146–150. https://doi.org/10.1016/j.psychres.2019.06.006.
163. Xiao J, Prandovszky E, Kannan G, et al. Toxoplasma gondii: biological Parameters of the connection to schizophrenia. *Schizophr Bull*. 2018;44(5):983–992. https://doi.org/10.1093/schbul/sby082.
164. Elsheikha HM, Marra CM, Zhu XQ. Epidemiology, pathophysiology, diagnosis, and management of cerebral toxoplasmosis. *Clin Microbiol Rev*. 2020;34(1):1–28. https://doi.org/10.1128/CMR.00115-19.
165. Karshima SN, Karshima MN. Human Toxoplasma gondii infection in Nigeria: a systematic review and meta-analysis of data published between 1960 and 2019. *BMC Publ Health*. 2020;20(1). https://doi.org/10.1186/s12889-020-09015-7.
166. Rostami A, Riahi SM, Contopoulos-Ioannidis DG, et al. Acute Toxoplasma infection in pregnant women worldwide: a systematic review and meta-analysis. *PLoS Negl Trop Dis*. 2019;13(10). https://doi.org/10.1371/journal.pntd.0007807.
167. Seo HH, Han HW, Lee SE, et al. Modelling Toxoplasma gondii infection in human cerebral organoids. *Emerg Microb Infect*. 2020;9(1):1943–1954. https://doi.org/10.1080/22221751.2020.1812435.
168. Passeri E, Jones-Brando L, Bordón C, et al. Infection and characterization of Toxoplasma gondii in human induced neurons from patients with brain disorders and healthy controls. *Microb Infect*. 2016;18(2):153–158. https://doi.org/10.1016/j.micinf.2015.09.023.
169. Tanaka N, Ashour D, Dratz E, Halonen S. Use of human induced pluripotent stem cell-derived neurons as a model for Cerebral Toxoplasmosis. *Microb Infect*. 2016;18(7–8):496–504. https://doi.org/10.1016/j.micinf.2016.03.012.
170. Kagan KO, Sonek J, Hamprecht K. Antenatal treatment options for primary cytomegalovirus infections. *Curr Opin Obstet Gynecol*. 2018;30(6):355–360. https://doi.org/10.1097/GCO.0000000000000486.
171. Leruez-Ville M, Ville Y. Is it time for routine prenatal serological screening for congenital cytomegalovirus? *Prenat Diagn*. 2020;40(13):1671–1680. https://doi.org/10.1002/pd.5757.
172. Sun G, Chiuppesi F, Chen X, et al. Modeling human cytomegalovirus-induced microcephaly in human iPSC-derived brain organoids. *Cell Rep Med*. 2020;1(1):100002. https://doi.org/10.1016/j.xcrm.2020.100002.
173. elDirini AA, Ogden TE, Ryan SJ. Subretinal endophotocoagulation. A new model of subretinal neovascularization in the rabbit. *Retina*. 1991;11(2):244–249.
174. Pinninti SG, Kimberlin DW. Neonatal herpes simplex virus infections. *Semin Perinatol*. 2018;42(3):168–175. https://doi.org/10.1053/j.semperi.2018.02.004.
175. Qiao H, Guo M, Shang J, et al. Herpes simplex virus type 1 infection leads to neurodevelopmental disorder-associated neuropathological changes. *PLoS Pathog*. 2020;16(10):e1008899. https://doi.org/10.1371/journal.ppat.1008899.
176. Hornig M, Bresnahan MA, Che X, et al. Prenatal fever and autism risk. *Mol Psychiatr*. 2018;23(3):759–766. https://doi.org/10.1038/mp.2017.119.
177. Mazina V, Gerdts J, Trinh S, et al. Epigenetics of autism-related impairment: copy number variation and maternal infection. *J Dev Behav Pediatr*. 2015;36(2):61–67. https://doi.org/10.1097/DBP.0000000000000126.
178. Bauer AZ, Kriebel D, Herbert MR, Bornehag C-G, Swan SH. Prenatal paracetamol exposure and child neurodevelopment: a review. *Horm Behav*. 2018;101:125–147. https://doi.org/10.1016/j.yhbeh.2018.01.003.
179. Collier SA, Rasmussen SA, Feldkamp ML, Honein MA. Prevalence of self-reported infection during pregnancy among control mothers in the National Birth Defects Prevention Study. *Birth Defects Res Part A Clin Mol Teratol*. 2009;85(3):193–201. https://doi.org/10.1002/bdra.20540.
180. Lin M, Zhao D, Hrabovsky A, Pedrosa E, Zheng D, Lachman HM. Heat shock alters the expression of schizophrenia and autism candidate genes in an induced pluripotent stem cell model of the human telencephalon. *PLoS One*. 2014;9(4). https://doi.org/10.1371/journal.pone.0094968.
181. Liu L, Lei J, Sanders SJ, et al. DAWN: a framework to identify autism genes and subnetworks using gene expression and genetics. *Mol Autism*. 2014;5(1):22. https://doi.org/10.1186/2040-2392-5-22.
182. Pirooznia M, Niranjan T, Chen Y-C, et al. Affected sib-pair analyses identify signaling networks associated with social behavioral deficits in autism. *Front Genet*. 2019;10:1186. https://doi.org/10.3389/fgene.2019.01186.
183. Vawter MP, Hamzeh AR, Muradyan E, Civelli O, Abbott GW, Alachkar A. Association of Myoinositol Transporters with schizophrenia and bipolar disorder: evidence from human and animal studies. *Mol Neuropsychiatry*. 2019;5(4):200–211. https://doi.org/10.1159/000501125.

184. Zakharyan R, Atshemyan S, Gevorgyan A, Boyajyan A. Nerve growth factor and its receptor in schizophrenia. *BBA Clin*. 2014;1:24–29. https://doi.org/10.1016/j.bbacli.2014.05.001.
185. Klug MG, O'Connell AM, Palme A, Kobrinsky N, Burd L. A validation study of the alcohol related neurodevelopmental disorders behavioral checklist. *Alcohol Clin Exp Res*. 2021;45(4):765–772. https://doi.org/10.1111/acer.14570.
186. Mattson SN, Bernes GA, Doyle LR. Fetal alcohol spectrum disorders: a review of the neurobehavioral deficits associated with prenatal alcohol exposure. *Alcohol Clin Exp Res*. 2019;43(6):1046–1062. https://doi.org/10.1111/acer.14040.
187. Schölin L, Mukherjee RAS, Aiton N, et al. Fetal alcohol spectrum disorders: an overview of current evidence and activities in the UK. *Arch Dis Child*. 2021;106(7):636–640. https://doi.org/10.1136/archdischild-2020-320435.
188. Muggli E, O'Leary C, Donath S, et al. Did you ever drink more?" A detailed description of pregnant women's drinking patterns. *BMC Publ Health*. 2016;16:683. https://doi.org/10.1186/s12889-016-3354-9.
189. May PA, Chambers CD, Kalberg WO, et al. Prevalence of fetal alcohol spectrum disorders in 4 US communities. *JAMA*. 2018;319(5):474–482. https://doi.org/10.1001/jama.2017.21896.
190. Petrelli B, Weinberg J, Hicks GG. Effects of prenatal alcohol exposure (PAE): insights into FASD using mouse models of PAE. *Biochem Cell Biol*. 2018;96(2):131–147. https://doi.org/10.1139/bcb-2017-0280.
191. Li H, Wen W, Xu H, et al. 4-Phenylbutyric acid Protects against ethanol-induced damage in the developing mouse brain. *Alcohol Clin Exp Res*. 2019;43(1):69–78. https://doi.org/10.1111/acer.13918.
192. Kane CJM, Drew PD. Neuroinflammatory contribution of microglia and astrocytes in fetal alcohol spectrum disorders. *J Neurosci Res*. 2021;99(8):1973–1985. https://doi.org/10.1002/jnr.24735.
193. Arzua T, Yan Y, Jiang C, et al. Modeling alcohol-induced neurotoxicity using human induced pluripotent stem cell-derived three-dimensional cerebral organoids. *Transl Psychiatry*. 2020;10(1):347. https://doi.org/10.1038/s41398-020-01029-4.
194. Zhu Y, Wang L, Yin F, et al. Probing impaired neurogenesis in human brain organoids exposed to alcohol. *Integr Biol*. 2017;9(12):968–978. https://doi.org/10.1039/c7ib00105c.
195. Mintun MA, Lo AC, Duggan Evans C, et al. Donanemab in early Alzheimer's disease. *N Engl J Med*. 2021;384(18):1691–1704. https://doi.org/10.1056/NEJMoa2100708.
196. Hsu S, Gordon BA, Hornbeck R, et al. Discovery and validation of autosomal dominant Alzheimer's disease mutations. *Alzheimer's Res Ther*. 2018;10(1):67. https://doi.org/10.1186/s13195-018-0392-9.
197. Hunter S, Brayne C. Understanding the roles of mutations in the amyloid precursor protein in Alzheimer disease. *Mol Psychiatr*. 2018;23(1):81–93. https://doi.org/10.1038/mp.2017.218.
198. Williams A. Drug administration. Breaking the chain reaction. *Nurs Times*. 1990;86(44):39–41.
199. Belloy ME, Napolioni V, Greicius MD. A quarter century of APOE and Alzheimer's disease: progress to date and the Path forward. *Neuron*. 2019;101(5):820–838. https://doi.org/10.1016/j.neuron.2019.01.056.
200. Yin C, Ackermann S, Ma Z, et al. ApoE attenuates unresolvable inflammation by complex formation with activated C1q. *Nat Med*. 2019;25(3):496–506. https://doi.org/10.1038/s41591-018-0336-8.
201. Sims R, Hill M, Williams J. The multiplex model of the genetics of Alzheimer's disease. *Nat Neurosci*. 2020;23(3):311–322. https://doi.org/10.1038/s41593-020-0599-5.
202. Choi SH, Kim YH, Hebisch M, et al. A three-dimensional human neural cell culture model of Alzheimer's disease. *Nature*. 2014;515(7526):274–278. https://doi.org/10.1038/nature13800.
203. Zhao J, Fu Y, Yamazaki Y, et al. APOE4 exacerbates synapse loss and neurodegeneration in Alzheimer's disease patient iPSC-derived cerebral organoids. *Nat Commun*. 2020;11(1):5540. https://doi.org/10.1038/s41467-020-19264-0.
204. Lin Y-T, Seo J, Gao F, et al. APOE4 causes widespread molecular and cellular alterations associated with Alzheimer's disease phenotypes in human iPSC-derived brain cell types. *Neuron*. 2018;98(6):1141–1154. https://doi.org/10.1016/j.neuron.2018.05.008. e7.
205. Wang C, Zhang M, Garcia G, et al. ApoE-isoform-dependent SARS-CoV-2 neurotropism and cellular response. *Cell Stem Cell*. 2021;28(2):331–342. https://doi.org/10.1016/j.stem.2020.12.018. e5.
206. Beigel JH, Tomashek KM, Dodd LE, et al. Remdesivir for the treatment of covid-19—final report. *N Engl J Med*. 2020;383(19):1813–1826. https://doi.org/10.1056/NEJMoa2007764.
207. Arber C, Lovejoy C, Harris L, et al. Familial Alzheimer's disease mutations in PSEN1 lead to premature human stem cell neurogenesis. *Cell Rep*. 2021;34(2):108615. https://doi.org/10.1016/j.celrep.2020.108615.

208. Yin J, VanDongen AM. Enhanced neuronal activity and asynchronous calcium transients revealed in a 3D organoid model of Alzheimer's disease. *ACS Biomater Sci Eng*. 2021;7(1):254–264. https://doi.org/10.1021/acsbiomaterials.0c01583.
209. Raja WK, Mungenast AE, Lin Y-T, et al. Self-organizing 3D human neural tissue derived from induced pluripotent stem cells recapitulate Alzheimer's disease phenotypes. *PLoS One*. 2016;11(9):e0161969. https://doi.org/10.1371/journal.pone.0161969.
210. Gonzalez C, Armijo E, Bravo-Alegria J, Becerra-Calixto A, Mays CE, Soto C. Modeling amyloid beta and tau pathology in human cerebral organoids. *Mol Psychiatr*. 2018;23(12):2363–2374. https://doi.org/10.1038/s41380-018-0229-8.
211. Riemens RJM, Kenis G, van den Beucken T. Human-induced pluripotent stem cells as a model for studying sporadic Alzheimer's disease. *Neurobiol Learn Mem*. 2020;175:107318. https://doi.org/10.1016/j.nlm.2020.107318.
212. Pavoni S, Jarray R, Nassor F, et al. Small-molecule induction of Aβ-42 peptide production in human cerebral organoids to model Alzheimer's disease associated phenotypes. *PLoS One*. 2018;13(12):e0209150. https://doi.org/10.1371/journal.pone.0209150.
213. Park J-C, Jang S-Y, Lee D, et al. A logical network-based drug-screening platform for Alzheimer's disease representing pathological features of human brain organoids. *Nat Commun*. 2021;12(1):280. https://doi.org/10.1038/s41467-020-20440-5.
214. Coccia E, Ahfeldt T. Towards physiologically relevant human pluripotent stem cell (hPSC) models of Parkinson's disease. *Stem Cell Res Ther*. 2021;12(1):253. https://doi.org/10.1186/s13287-021-02326-5.
215. Zagare A, Gobin M, Monzel AS, Schwamborn JC. A robust protocol for the generation of human midbrain organoids. *STAR Protocols*. 2021;2(2):100524. https://doi.org/10.1016/j.xpro.2021.100524.
216. Li B, Zhao G, Zhou Q, et al. Gene4PD: a comprehensive genetic database of Parkinson's disease. *Front Neurosci*. 2021;15:679568. https://doi.org/10.3389/fnins.2021.679568.
217. Nalls MA, Blauwendraat C, Vallerga CL, et al. Identification of novel risk loci, causal insights, and heritable risk for Parkinson's disease: a meta-analysis of genome-wide association studies. *Lancet Neurol*. 2019;18(12):1091–1102. https://doi.org/10.1016/S1474-4422(19)30320-5.
218. Wulansari N, Darsono WHW, Woo H-J, et al. mutations. *Sci Adv*. 2021;7(8). https://doi.org/10.1126/sciadv.abb1540.
219. Kano M, Takanashi M, Oyama G, et al. Reduced astrocytic reactivity in human brains and midbrain organoids with PRKN mutations. *NPJ Parkinsons Dis*. 2020;6(1):33. https://doi.org/10.1038/s41531-020-00137-8.
220. Kim H, Park HJ, Choi H, et al. Modeling G2019S-LRRK2 sporadic Parkinson's disease in 3D midbrain organoids. *Stem Cell Rep*. 2019;12(3):518–531. https://doi.org/10.1016/j.stemcr.2019.01.020.
221. Kwak TH, Kang JH, Hali S, et al. Generation of homogeneous midbrain organoids with in vivo-like cellular composition facilitates neurotoxin-based Parkinson's disease modeling. *Stem Cells (Dayton)*. 2020;38(6):727–740. https://doi.org/10.1002/stem.3163.
222. Boussaad I, Obermaier CD, Hanss Z, et al. A patient-based model of RNA mis-splicing uncovers treatment targets in Parkinson's disease. *Sci Transl Med*. 2020;12(560). https://doi.org/10.1126/scitranslmed.aau3960.
223. Kim S, et al. Neural stem cells derived from human midbrain organoids as a stable source for treating Parkinson's disease: Midbrain organoid-NSCs (Og-NSC) as a stable source for PD treatment. *Prog Neurobiol*. 2021;204:102086. https://doi.org/10.1016/j.pneurobio.2021.102086. In this issue.
224. Bose R, Banerjee S, Dunbar GL. Modeling neurological disorders in 3D organoids using human-derived pluripotent stem cells. *Front Cell Dev Biol*. 2021;9:640212. https://doi.org/10.3389/fcell.2021.640212.

CHAPTER

9

iPSC-derived models of autism: Tools for patient phenotyping and assay-based drug discovery

Gilles Maussion[1,2,], Cecilia Rocha[1,*], Luisa Pimentel[1,*] and Thomas M. Durcan[1]*

[1]Early Drug Discovery Unit (EDDU), Montreal Neurological Institute and Hospital, McGill University, Montreal, QC, Canada; [2]Montreal Neurological Institute, Neurology and Neurosurgery, Montreal, QC, Canada

Introduction

Syndromic autisms have generated particular attention for research focused on understanding the molecular mechanisms responsible for autism spectrum disorder (ASD). Genetically well characterized and often caused by a single gene mutation, the studies of these syndromes have allowed researchers to determine how a known genetic defect can be responsible for cellular and/or molecular dysregulation which, among others, lead to behavioral features observed in patients diagnosed with ASDs.[1,2] The term "spectrum" reflects the genetic and phenotypic heterogeneities of this group of diseases,[3] which can include common behavioral characteristics, such as (i) deficits in verbal and nonverbal communication, (ii) impairment in social interaction, and (iii) the presence of restricted and stereotyped behaviors[4]; the autistic triad defined by Leo Kanner.

During the past 25–30 years, a wealth of research has been conducted on postmortem brain tissue from ASD patients and on mouse models of ASD to identify potentially relevant phenotypes at the cellular and molecular levels that may underlie behavioral features observed in ASDs. Postmortem brain studies have given descriptive information related to the ASDs at the end of patients' lives. However, they do not elucidate the trajectories of ASDs and pathophysiological mechanisms leading to the observed phenotypes. Mouse

* These authors have equally contributed.

Phenotyping of Human iPSC-derived Neurons
https://doi.org/10.1016/B978-0-12-822277-5.00001-8

models of ASD are able to follow develop of ASD-like behaviors and how molecular mechanisms change over time, but present obvious differences compared to human patients, due to inter-species barriers. Furthermore, mouse models neither fully recapitulate the multigenic risk factors encompassed in the human genome nor the full spectrum of phenotypes observed in patients.[5]

As mentioned, ASDs are heterogeneous in genetic and phenotypic presentation and can be classified into three distinct categories. The first is idiopathic autism, the cause of which remains unknown, and that represents approximately 80% of ASDs cases. The second group includes syndromic autisms, which are defined by specific symptoms and known genetic causes account for 5%−15% of the cases.[6] Finally, sporadic nonsyndromic autisms include patients in which the genetic basis, such as gene deletion, duplication, or disruption, have been identified by high-throughput sequencing and which present clinical profiles with autistic features, but the symptomatology does not fit within the syndromic criteria. This last category represents 1%−5% of the ASDs.[7,8]

Taking into consideration all the studies performed on human cohorts, which includes the three categories of ASDs, more than 1100 genes have been implicated in its pathophysiology.[9] This further illustrates the limitations encountered by research projects conducted on mouse models and post-mortem brain tissue from patients. While these model systems provide some information about genetic and molecular mechanisms responsible for ASDs, other model systems are required for further investigation into how these genetic changes alter the function of human neurons or glia.

Since 2006 (in mice) and 2007 (in humans), somatic cells have been reprogrammed into induced pluripotent stem cells (iPSCs) by overexpression of four transcription factors (Nanog, C-Myc, Klf, and Oct3/4).[10,11] iPSCs constitute a cell type that can generate, under the appropriate conditions, any other cell type from the three germinal layers: the endoderm, the mesoderm, and the ectoderm.[12] When treated with neural induction factors, iPSCs can generate in vitro cellular models of patient-derived cells without requiring a brain sample biopsy, which apart from specific medical conditions requiring brain surgery, cannot be performed before the patient's death.

In recent years, iPSC technology has generated a lot of interest from the scientific community, representing an approach to overcome the gaps left from neurological studies utilizing postmortem brain tissue and mouse models.[5] Looking to the future, as the number of research studies using iPSC-derived cell models continues to increase, it will start to become possible for iPSCs to be developed as a tool for personalized medicine, where treatments may be developed for specific patients from research that is conducted on iPSC-derived neurons carrying their own genetic background.[13]

While most of the studies to date have been performed to identify phenotypes and functional deficits relevant for disease initiation and progression,[14] characterizing robust phenotypes that can be repeatedly observed and reversed through assay development and improved through drug screening[15] is crucial to cover the genetic and phenotypic heterogeneity of the ASDs. Such characterization of recovery phenotypes, using IPSC-derived patient line constitute a paradigm shift[16] and the missing link toward a personalized medicine.[5]

This chapter aims to elucidate how iPSC-derived neuronal and glial cell models from patients diagnosed with syndromic autism can complement the information collected from previous approaches. We foremost want to emphasize the potential of the iPSC-derived model to develop robust assay and drug screening platforms that will further allow researchers to investigate potentially all types of autisms based on observations obtained through studies of syndromic cases. We finally want to provide insight on the requirements that must be fulfilled to perform properly developed assays to study ASDs using iPSC-derived cells.

Syndromic autisms

Unlike idiopathic autism, whose diagnoses are only based on interviews and questionnaires, syndromic autism constitutes a group of diseases characterized by symptoms that include the autistic triad and by known genetic causes. The known function of mutated or deleted genes in syndromic ASD provides indications on the potential mechanisms leading to a given patient phenotype. Consequently, data collected from investigations on syndromic autisms have allowed researchers to develop hypotheses and infer potential dysregulations that are likely to occur in cases diagnosed with idiopathic autism, whose genetic causes remain undetermined. In this section, we review the genetic causes and symptomatology of the main syndromic autisms.

Fragile X syndrome

Fragile X Syndrome is the most prevalent syndromic autism, mostly affecting male individuals with a sex ratio estimated at 2:1[17] and mainly caused by a CGG expansion in the 5′UTR sequence of the FMRP translational regulator 1 (*FMR1*) gene located on Xq27.[18] *FMR1* encodes an RNA binding protein that modulates the translation of specific transcripts, with many of these encoding synaptic proteins. While sequences containing less than 55 CGG repeats lead to physiological *FMR1* expression, expansions that increase in size from 55 to 200 repeats are considered as premutation repeat levels and lead to a partial repression of *FMR1* expression. Expansions above 200 repeats completely abolish *FMR1* expression. Patients diagnosed with Fragile X Syndrome exhibit symptoms including social impairment, developmental delay, macroorchidism, hyperactivity, and intellectual disability.

Rett syndrome

Rett syndrome is caused, in most cases, by mutations of the Methyl-CpG-Binding Protein 2 (*MECP2*) gene located on the Xq28.[19] *MECP2* codes for a Methyl CpG DNA binding protein that acts as a transcriptional regulator included in the SWItch/Sucrose Non-Fermenting (SWI/SNF) complex, which is an important regulator of neuronal development.[20] Actually, many other mutations have been identified in genes coding proteins that belong to the SWI/SNF complex leading to developmental disorders.[21] Rett syndrome, with an incidence estimated at 1/9000[22] mostly affects women (in cases of X-related Rett syndrome). Patients with Rett syndrome present symptoms including repetitive behaviors, intellectual disabilities, epilepsies, and regression of learned skills is observed during childhood.

Other mutations have also been described in the following genes -Cyclin-Dependent Kinase Like 5 (*CDKL5*),[23] Netrin G1 (*NTNG*1),[24] and Myocyte Enhancer Factor 2C (*MEF2C*)[25] in patients with Rett or Rett-like syndromes. Interestingly, patients carrying mutation in the *CDKL5* gene were initially diagnosed as Atypical Rett syndrome or Early Onset Rett syndrome. With the increasing patient recruitment, a deeper investigation of the occurrence of the symptoms in Rett patients with *MECP2* mutations and patients with *CDKL5* mutations has led to defining for those, the CDKL5 Deficiency disorder.[26]

FOXG1 deletion syndrome

The *FOXG1* deletion syndrome, as its name suggests, is caused by a deletion or loss of function mutation in the Forkhead Box G1 (*FOXG1*)gene, located on 14p12 and encoding the transcription factor that regulates genes involved in brain development.[27] *FOXG1* deletion syndrome is also considered an atypical Rett syndrome. Patients diagnosed with the disease present common traits with Rett syndrome, such as intellectual disability, autistic features, and epilepsy. Microcephaly and partial or complete agenesis of the corpus callosum have been observed in patients with *FOXG1* deletion syndrome.

Tuberous sclerosis

Gene mutations in patients diagnosed with tuberous sclerosis were found in the TSC Complex Subunit 1 (*TSC1*)[28] and TSC Complex Subunit 2 (*TSC2*)[29] genes which are respectively located in 9q34.13 and 16p13.3. The *TSC1* gene codes for the hamartin protein, which interacts with tuberin, a GTPase translated from the *TSC2* gene. The interaction between these two proteins negatively regulates the (FK506-binding protein)-rapamycin-associated protein/mammalian Target Of Rapamycin (FRAP/mTOR) pathway. As a consequence, mutations in either one of these two genes lead to de-repression of the mTOR pathway. Tuberous sclerosis is considered as a multi-systemic disease, as it affects several organs. Regarding the neurological phenotype, patients with tuberous sclerosis commonly present with symptoms including learning deficits, intellectual, disability, and epilepsy, as well as structural brain abnormalities call tubers. Other organs including the skin, kidney, heart, and lungs are also affected.

Pheland McDermid syndrome

The Pheland McDermid syndrome is characterized by a global developmental delay with an absence of speech or a speech delay, hypotonia and dysmorphic feature. This syndrome is caused by mutations in the SH3 And Multiple Ankyrin Repeat Domains 3 (*SHANK3*) gene located on Chr22q13.33 and that encodes for a post-synaptic scaffolding protein.[30] Interestingly, in mouse models, mutations or disruptions in Shank3 gene were shown to leading changes in post-translational modifications of other synaptic interacting molecules which affect synapse functioning.[31,32] Those studies highlight the importance of the scaffolding protein SHANK3 for accurate synapse formation and function.

Prader-Willi and Angelman syndromes

Prader-Willi and Angelman syndromes are caused by deletion of the 15q11-15q13 region from the paternal and maternal[33] chromosome, respectively. Genes located in this region have been associated with ASDs[34] and are related to parental imprinting.[35] The prevalence of Prader-Willi syndrome (PWS) has been estimated between 1/10,000 to 1/30,000.[36] The deletion of the paternal copy is expected to generate an haploinsufficiency of many known genes, including Makorin Ring Finger Protein 3 (*MKRN3*), MAGE Family Member L2 (*MAGEL2*), Necdin (*NDN*), and, *SNURF-SNRPN*. Interestingly, the SNRPN Upstream Open Reading Frame-Small Nuclear Ribonucleoprotein Polypeptide N (*SNURF-SNRPN*) is a polycistronic gene that encodes for SHARPIN, a ribonucleoprotein that interacts with SHANK3, whose defect is expected to be responsible for neurological phenotypes observed in PWS. Patients with Prader-Willi syndrome present with obsessive-compulsive behavior, facial dysmorphic features, muscular hypotonia, and hypogonadism.[37] The deletion of the maternal copy of the 15q11-q13 region generates haploinsufficiency of Ubiquitin Protein Ligase E3A (*UBE3A*) and ATPase Phospholipid Transporting 10A (Putative) (*ATP10C*) genes. The UBE3A protein ubiquitinates substrate proteins (including synaptic proteins such as Arc)[38] that are then going to be degraded by the proteasome. Targeted degradation of proteins through the proteasome pathway is required to modulate cell cycle, synapse formation as well as cell integrity by getting rid of misfolded proteins. Interestingly, dysfunction of the proteasome complex has been associated with motor dysfunction due to an excess of misfolded protein complexes which compromise cell integrity. ATP10C encodes a component of an ATP-dependent cation transport. It contributes to regulating the transport of phospatidylcholine which is a key component of the cell membrane. Then, haploinsufficiency of both genes UBE3A and ATP10C are likely to be responsible, in the developing brain of patients with Angelman syndrome, for changes in the balance between proliferation and differentiation as well as motor deficits. Patients with Angelman syndrome whose incidence has been estimated at 1 for 15,000 birth[39] present with microcephaly, ataxia, intellectual disability, severe speech impairment, abnormal EEG patterns, and epilepsy.[40]

Timothy syndrome

Timothy syndrome is a multi-systemic disease caused by mutations in the Calcium Voltage-Gated Channel Subunit Alpha1 C (*CACNA1C*) gene located on 12q13.33 and encoding for a Calcium Voltage-Gated Channel Subunit.[41] Timothy syndrome is characterized by syndactyly, cognitive impairments, and autistic features. Furthermore, mutations in that calcium channel subunit, expressed in the cardiomyocytes induce changes in the cardiac action potential; a severe prolongation of the QT interval and cardiac arrhythmia[42] that leads to premature death around 2.5 years of age, cognitive impairments, and autistic features.

The syndromic autisms described here illustrate the variety of genes that can be causal of autistic features in patients. Studying the mechanisms through which gene-specific mutations are responsible for neuronal alterations observed in patients constitutes a unique way of further understanding how the phenotypes observed in ASD patients without any known gene variants associated with ASD can lead to autistic features. A better understanding of the etiology of the ASDs will be helpful for establishing personalized therapeutic strategies.

iPSC studies to model ASDs in vitro

iPSC technology has allowed scientists to obtain cells from patients and to generate neurons and other specialized cell types in vitro with the aim of investigating neurodevelopmental disorders. While iPSCs were originally generated from skin fibroblasts, which require a biopsy, procedures have been adapted to less invasive techniques that collect somatic cells from peripheral blood, hair, urine, and dental stem cells.[43] These less invasive methods facilitate the collection of samples from younger subjects under the age of 18 in which a noninvasive procedure would be preferred. Patient-derived iPSCs can be then used to identify disease-specific phenotypes, helping us uncover why a given patient has progressed to develop a given form of autism.

iPSC studies focused on syndromic and sporadic autisms

During the development of the central nervous system (CNS), neuronal activity and intracellular trafficking play significant roles in regulating the development of neuronal subtypes and brain cytoarchitecture.[44,45] Intracellular mitochondrial distribution and dendritic spine formation on developing neurons were shown to be modulated by cellular activity.[46] Interestingly, subcellular and morphological changes were observed on post-mortem brain samples from patients diagnosed with syndromic autisms. An increased dendritic spine density was observed in layer V pyramidal neurons of the temporal cortex from fragile X patients.[47] The examination of cortical areas of postmortem brain samples from patients diagnosed with Rett syndrome has shown decreased numbers of dendrites and dendritic spines.[48] Finally, immunohistochemical analyses of postmortem brain samples from patients with tuberous sclerosis revealed accumulations of markers of autophagy proteins DOR and P62 as well as accumulation of Lysosomal Associated Membrane Proteins LAMP1 and LAMP2.[49] Considering subcellular and morphological changes observed in neurons from postmortem brain samples of patients with Fragile X,[47] Rett syndrome,[48] and tuberous sclerosis,[49] activity-dependent developmental processes were hypothesized to be impaired in the context of these ASDs. These findings illustrate the importance of understanding how genetic variants associated with these syndromes lead to changes in neuronal activity and iPSC-derived neurons provide a useful model system for this work.

A study utilizing iPSC cells from Timothy syndrome (TS) patients showed that mutations in L-type voltage-gated calcium channel $Ca_V1.2$ correlated with increased amplitude of calcium signaling events in neural progenitor cells (NPCs) and activity-dependent development defects.[50] Increased intracellular calcium in NPCs and disruption of the expression of deep layer neuronal marker SATB Homeobox 2 (SATB2) that could lead to a reduction in projecting neurons were detected in TS cells. These findings link, for the first time, neuronal development alterations with ASD features, as they might emerge from cortical connectivity defects resulting from the reduced projecting neurons in TS patients.[50] Interestingly, an increased calcium response to calcium-permeable α-Amino-3-hydroxy-5-Methyl-4-isoxazole Propionic Acid (AMPA) and N-Methyl-D-Aspartate (NMDA) receptors was shown in iPSC-derived neurons from Fragile X patients.[51] Fragile X patient-derived neurons also revealed an increase in dendrite length. This data corroborates the increased size and number of dendritic spines observed in neurons of the layer V temporal cortex from Fragile X patients.[52] The application of Naspm trihydrochloride; an antagonist of the Calcium Permeable AMPA receptor lacking GluA2 subunit (GluA2-lacking/CP-AMPARs) reversed the morphological and activity phenotypes seen in these Fragile X-derived neurons.[51] Aside from ion channel-dependent conductance, signaling mediated by metabotropic receptors has also been investigated in the context of Fragile X syndrome. Interestingly, mGluR5 dependent signaling is known to de-represses FMRP-dependent inhibition and contributes to the local translation of synaptic transcripts. An increased mGLUR1 response has been observed in iPSC-derived neurons from Fragile X lines. These iPSC-derived neurons were also more prone to differentiate toward a glutamatergic phenotype, which could contribute to phenotypes in patients. Application of the GluR1 antagonist promotes the differentiation of neural progenitors toward a glial fate and reduces the mGLRU5 response.[53] These studies demonstrate that ion conductance and cell signaling impairments in the contexts of Fragile X and Timothy syndromes strongly affect the capability of neural progenitors cells to differentiate properly into neurons.

Impaired maturation and signaling pathways were also observed in iPSC-derived cells from patients diagnosed with Tuberous Sclerosis. Decreased expression of SKI Family Transcriptional Corepressor 2 (SKOR2) and increased expression of the marker of proliferation KI67 were reported in iPSC-derived Purkinje cells from patients with TSC2 mutations.[54] Those data suggest that TSC2 mutations are responsible for a delayed differentiation of IPSC-derived Purkinje cells. A decrease in FMRP expression, as well as increased dendrite length, implies that Fragile X syndrome and Tuberous Sclerosis may have common pathways affected by both disorders. These phenotypes were partially rescued by the application of rapamycin, an allosteric mTOR inhibitor. A second study performed on directly induced neurons from patients with *TSC2* mutations identified mitochondrial defects, including reduced mitochondrial membrane potential, impaired lysosomal activity, and accumulation of mitochondria in dendrites. These observations from induced neurons further highlight the importance of dendritic trafficking and mitochondria for proper neuronal function and show how these mechanisms may be involved in disease progression for Tuberous Sclerosis. Importantly, treatment with rapamycin was able to restore mitochondrial membrane potential and lysosomal activity to healthy control levels.[55] These in vitro iPSC-derived Purkinje cell phenotypes also corroborate the cerebellar abnormalities observed in patients with ASDs. Independent studies have reported decreased number[56,57] and size[58] of Purkinje cells in the cerebellum from patients with ASDs. These morphometric changes are likely to contribute to motor dysfunction in ASDs because the cerebellum is critical for motor coordination. Furthermore, mitochondrial dysfunction had previously been suggested from primary patient tissue studies that showed the accumulation of autophagic proteins, P62 and DOR, observed in the temporal lobe patients with TSC.[49]

Mitochondrial defects have also been implied in research using iPSC-derived neurons from patients with Rett syndrome. A high throughput experiment with a transcriptome profiling readout was performed using iPSC-derived cells from patients diagnosed with Rett syndrome.[59] This study revealed an increase in the expression of genes related to cell cycle and mitochondrial function, while genes involved in neuronal development and cell signaling were significantly decreased. This study also demonstrated that iPSCs from patients with Rett syndrome have an impaired balance between proliferation and differentiation, with delayed maturation of neurons,[59] as well as deficits in mitochondrial function. Furthermore, morphological analysis of Rett patient-derived neurons revealed reductions in the number of dendritic spines and synapses, as well as a decrease in cell soma size. The Rett patient-derived neurons also presented with reduced frequency of calcium transient signaling compared to healthy control neurons.[60] The disruption of proper neuronal activity, as well as disrupted differentiation, lead to conclusions that *MECP2* mutations induce gene dysregulations that are likely to impair activity-dependent developmental mechanisms leading to morphological changes and functional impairment of developing neurons in patients with Rett syndrome.

Intriguingly, impaired mitochondrial trafficking was shown to alter synapse and dendritic spine stability.[46] The altered expression of mitochondrial genes has been demonstrated in postmortem brain tissue from ASD patients. An increase in expression of Solute Carrier Family 25 Member 12 (*SLC25A12*) that codes an aspartate glutamate transporter of the inner mitochondrial membrane, associated with Autism[61] was found in the Brodmann Area 46 (BA46) from autistic patients.[62] Similarly, increased expression of mitochondrial proteins, including Aspartate/Glutamate Carrier and Cytochrome C Oxidase were observed in brain homogenate from ASD patients.[63] An expression study performed on anterior cingulate gyrus, motor cortex, and thalamus from autistic patients has identified a decrease in transcripts of the electron transport chain (ETC) in those brain areas.[64] Furthermore, altered levels of electron transport chain complexes and elevated reactive oxygen species (ROS) resulting from mitochondrial dysfunction are linked with impaired dendritic trafficking.[65] Decreased expression of the electron transport chain correlated with increased ROS in the temporal cortex and cerebellum of autistic children.[66] Combined together, such findings imply that mitochondrial dysfunction is likely a contributing factor to the morphological and functional changes observed with the iPSC model of Rett syndrome. In support of this idea, in postmortem brain tissue from Rett patients, neurons located in cortical layer III and V harbor fewer and shorter dendrites, as well as a decreased number of dendritic spines.[48,67] Similarly, a reduced dendritic spine density was also observed in CA1 neurons.[68] These studies highlight the utility of patient-specific iPSC-derived neurons for identifying and validating phenotypes observe in rodent or postmortem models and allow for the possibility of drug screening to reverse these phenotypes in vitro.

iPSC studies focusing on sporadic non-syndromic autism

Several other iPSC-based studies were performed based on gene mutations or chromosomal changes identified in patients with ASDs. High throughput sequencing performed on autistic patients identified genes whose disruption are likely to contribute to ASD phenotype development. These genes include *MBD5* (Methyl-CpG Binding Domain Protein 5), *SATB2* (SATB homeobox 2), *TCF4* (Transcription Factor 4), which encode for DNA binding proteins, *EHMT1* (Euchromatic Histone Lysine Methyltransferase 1), and *GRIN2B* (Glutamate Ionotropic Receptor NMDA Type Subunit 2B).[7,69,70] From these findings, neural progenitors were developed expressing stable Short Hairpin RNA (shRNAs) against *MBD5, SATB2, TCF4, EHMT1* transcripts.[71,72] A description of the results from these efforts and how they have transformed the field's understanding of ASD is below.

Combining whole-genome transcription profiles, DNA methylation, and DNA protein interaction profiles of lines with shRNA targeted knockdown of ASD-associated genes, common molecular signatures were identified between *EHMT1*-and *TCF4*-silenced NPCs,[71] as well as between *SATB2* and *MBD5*-silenced NPCs.[72] The transcriptional profiles from knockdown lines compared to healthy control lines and in differentiating NPCs suggest that those four silenced genes lead to premature differentiation of NPCs. Common differentially methylated regions emerged from two independent comparisons of the DNA methylation profiles, including the comparison of control NPCs and differentiated cells, as well as control NPCs and gene-specific silenced NPCs. These DNA methylation changes provide evidence for the role of the silenced genes in NPC differentiation. Finally, common binding sites for TCF4 and EHMT1 proteins were found by chromatin immunoprecipitation. Similar changes in DNA methylation and gene expression occurring during the differentiation of NPCs and in silenced NPCs suggest that impaired dosage of genes found mutated or disrupted in ASD patients are responsible for a premature differentiation of neural progenitor cells through common molecular pathways, notably the WNT pathway. These disruptions in differentiation affect the developmental balance between cell proliferation and cell differentiation.

Impaired cell differentiation has also been observed in iPSC-derived neurons with a deletion or mutation in the *GRIN2B* gene.[73] Whole transcriptome profiles of *GRIN2B*-deleted or mutated iPSC-derived neurons have shown an increased expression of cell cycle genes and a decreased expression of genes involved in synapse function and ion transport. NMDA signaling blockade by APV application in differentiating control iPSC-derived neurons recapitulates dysregulations in KI67 and Synapsin2 (SYN2) expression observed in *GRIN2B*-deleted or mutated iPSC-derived neurons. Finally, impaired NMDA response in *GRIN2B*-mutated or -deleted iPSC-derived neurons were found. Combined together, this data implies that NMDA receptors, which are expressed in extrasynaptic regions during early developmental stages, are required for appropriate neuronal differentiation and dysfunction could be responsible for the development of ASD in some patients.

Further studies have revealed that different mechanisms affecting neuronal activity can lead to disruption of neuronal differentiation and neuronal morphology.Transient Receptor Potential Cation Channel Subfamily Member C6 (*TRPC6*) gene, whose expression is controlled by *MECP2* gene was found disrupted in cases with ASDs. The *TRPC6* gene, located on the chromosome 11q22.1, encodes for a transient calcium-activated channel receptor. iPSC-derived neurons with TRPC6 disruption present with reduced expression of this synaptic gene, a reduced dendritic spine density, reduced branch length, and synaptic contact. These observations are comparable to the neuronal morphologies reported in the studies using postmortem brain tissues from Rett patients. Interestingly, these phenotypes were restored by Hyperforin application or *TRPC6* over-expression.[43] Both studies demonstrate that impaired calcium conductance in developing neurons disrupts the neuronal differentiation process, as well as neuronal morphological features that are important for neuronal network formation.

Another gene, *SHANK3*, encoding for a postsynaptic scaffolding protein, was found disrupted in two ASD patients.[74] iPSC-derived placodal neurons with *SHANK3* microdeletion at 30 days in vitro have an increased number of primary neurites, but smaller cell body sizes, suggesting those iPSC-derived cells are directed toward premature differentiation. These phenotypes were rescued with *SHANK3* overexpression showing that SHANK3 function was responsible for initiating the morphological changes. Interestingly, the number of synaptic contacts made by *SHANK3* deletion iPSC-derived neurons was reduced compared to healthy control cultures. Morphological changes observed with 2D iPSC-derived neurons reflect observations from post-mortem brain tissue of Fragile X syndrome patients.[47] They also suggest that premature differentiation of *SHANK3* mutated iPSC-derived placodal developing neurons does not permit the formation of mature synapses.[74] Together, these studies implicate the disruption in synapse formation on neuronal network deficits in *SHANK3*-deficient patients.

Studies performed using iPSCs derived from patients carrying ASD-associated mutations or deletions with syndromic autism or in genome-edited iPSC-derived cells, have provided the tools and the cells needed to identify transcriptional dysregulation, morphological abnormalities, and changes in cellular electrophysiological properties, all of which are identified phenotypes that underlie the autistic behavioral features of patients. These phenotypes merit further investigation in patients with idiopathic autism. Taken together, the studies described here have demonstrated how investigating gene-specific mutations that lead to different syndromes with common autistic features opens up pathways to be explored in idiopathic autism and considerations for assay development and drug screening at a larger scale.

Data collected by studies focused on iPSCs from idiopathic autism

Several studies were performed on iPSC-derived cells from patients diagnosed with idiopathic autism. Whole transcription profiles, cell morphology, and electrophysiological properties were analyzed to determine if any common phenotypes could be observed between genetically defined and undefined ASDs.

Gene expression profiling

Transcriptomic analysis performed on iPSC-derived neurons from three autistic patients and their siblings led to the identification of 161 differentially expressed genes (DEGs) that are related to the development of the nervous system, synapse and ion channels, as well as neurotransmitter receptor activity. Treatment of those DEGs using Ingenuity Pathway Analyses software highlighted a network of genes controlled by REST/NRSF factors, known to repress neuronal genes in non-neuronal cells, and involved in neuronal development.[75] Transcriptional deregulation of NRSF-dependent genes is suggested to affect the neuronal differentiation process in iPSC-derived cells from ASD patients. In a separate study, whole transcriptome profiling of iPSC-derived neurons from six patients and five controls at the days in vitro (DIV)35 and DIV135 corroborate the results from Liu et al. After correction for multiple testing, 531 and 257 DEGs were found respectively at DIV35 and DIV135. Differentially expressed genes at DIV35 were primarily related to anatomical structure morphogenesis and at DIV135 were related to brain development. Weighted correlation network analysis (WGCNA) at both stages identified networks of co-expressed genes. At DIV35, genes upregulated in ASD iPSC-derived cells were involved in the regulation of transcription. At DIV135, upregulated genes were found in control iPSC-derived neurons that are involved in neuronal commitment and differentiation.[76] Transcriptomic data collected from both studies strongly suggest that changes in the transcriptional regulation occurring at early stages of development are likely to affect neuronal differentiation in the developing brain from patients with ASDs.

Another study utilized cells from six autistic patients (five with high functioning autism and one with more severe ASD) and six non-ASD controls. This study analyzed the transcriptome of neural progenitors and 4 week-old iPSC-derived neurons purified by FACS in and between the two groups. After correction for multiple testing, no DEGs were identified between the two groups at the NPC stage and only 20 DEGs were found between the two groups in 4-week-old iPSC-derived neurons. From this study, a WGCNA analysis identified a gene network involved in protein biosynthesis, RNA processing, and splicing, implicating mitochondrial function, which positively correlates with the disease. A module of genes involved in translational processes and regulation of gene expression was downregulated in the 4 week-old iPSC-derived neurons from ASDs patients, whereas a module of genes involved in genes related to the synapse, ion channels, neurotransmission were upregulated.[77] Data collected from that study led to identifying modules of genes consistently co-expressed, many of them being associated with ASDs suggesting that the variety of genetic causes or vulnerability factors in ASDs led to changes in common molecular pathways.

Several studies were conducted on iPSC-derived NPCs and neurons from ASD patients with macrocephaly. iPSC-derived neurons from ASD patients were compared with healthy controls and showed an increase of KI67 expression and decreased expression of neuronal markers such as VGLUT and GABA. Furthermore, a WGCNA analysis, performed on IPSC-derived neurons from ASD patients identified a module of down-regulated genes coding for voltage-gated cation channels.[78] Those data suggest that IPSC-derived cells from ASD patients harbor an impaired differentiation process combined increase in proliferating cells which may reflect the macrocephaly observed in patients.[78]

To further study differential time-dependent transcriptional profiles and regulation in iPSC-derived differentiating neurons, RNA sequencing was performed on the same cohort at 2, 4, 7, and 14 days of differentiation. A heterochronicity in the expression of time-dependent gene modules was observed between ASD and control iPSC-derived neurons. Data collected in this study implied that the entire transcriptomic profile throughout the neuronal development is altered starting from the very earliest stages and should be considered as a continuum of dynamic dysfunctions that disturbs the neuronal development and function in ASDs.[79]

Concordances in gene expression profiles obtained from studies on iPSC-derived cells and post-mortem brain tissue from idiopathic autism

Gene expression profiles of iPSC-derived neurons from idiopathic autism are consistent with findings from post-mortem brain tissue from patients with idiopathic autism. Dysregulation of synaptic genes in iPSC-derived neurons from patients with idiopathic autism corroborates the functional annotation performed on transcriptome profiles from BA9; BA41/BA42 and the cerebellar vermis of autistic patients. In 2011, Voineagu et al., showed a downregulation of synaptic genes combined with increased expression of genes involved in immune response and inflammation in brains of patients with ASD.[80] Similar expression changes were observed in the prefrontal cortex of autistic patients compared to controls.[81] A list of genes coding for transcription factors able of controlling synaptic genes was also identified. Furthermore, a meta-analysis of several gene expression studies performed on postmortem brain tissue from ASDs patients[80,82–84] showed the correlation of two gene networks downregulated in the samples from ASDs patients. The downregulated gene networks included genes involved in synapses and mitochondrial function.[85] As we have discussed earlier, synapse gene networks are enriched in autism risk genes. Finally, Mahfouz and colleagues used the Brainspan database to investigate the longitudinal expression profiles of 455 autism-related genes in 16 brain areas. Results emerging from functional annotation and gene co-expression analyses identified enrichment of genes involved in mitochondrial and synaptic functions, in GABAergic neuronal transmission, and the regulation of translation.[86]

Combined together, gene expression profiling performed with iPSC-derived neurons and post mortem brain tissue from idiopathic autism have pointed to a convergence of functional dysregulations that are likely to underlie genetic vulnerability factors or gene-specific mutations in the cases of syndromic autism.

Although studies using iPSC-derived neurons suggest that the neuronal differentiation process is altered in the context of idiopathic autism, the heterogeneity of gene expression profiles between studies may depend on various phenotypes of ASD patients, as well as experimental design regarding the cell types and the timepoints that are analyzed. Despite these variations, the studies described here, with the exception of Liu et al. from 2017, reported, either in the list of DEGs or in the list of co-expressed genes correlated with the disease, a significant representation of ASD risk genes based on comparison to the master list of ASD-associated genes established by the Simons foundation.[87] That list collects genetic data from all types of ASDs. The significant representation of ASD-risk genes in the transcriptome analyses of idiopathic autism partially validates the contribution of the genetics in the etiology of ASD and the subsequent mechanistic changes primarily under investigation in syndromic autisms.

Morphological and electrophysiological properties in iPSC-derived neurons from patients with idiopathic autism

Abnormalities in morphological and electrophysiological properties have also been observed with iPSC-derived neurons from patients with idiopathic autism. Compared to healthy control sibling iPSC-derived neurons, neurons derived from patients with ASD showed decreased frequency and kinetics of spontaneous post-synaptic excitatory currents.[75] Similar observations were made when comparing iPSC-derived neurons from ASD patients to unpaired controls. In this study, a decrease in spontaneous firing and calcium transients was observed on DIV30 in iPSC-derived neurons from patients with ASD.[76]

These studies imply that a decrease in the electrophysiological activity of iPSC-derived neurons from patients with idiopathic ASD is occurring. A decrease in the number of spikes per burst was recorded from cultures of iPSC-derived neurons at DIV50 from patients with ASD who had macrocephaly, compared to control cultures.[78] On the same cohort of iPSC lines, a study has correlated the electrophysiological pattern of neurons with RNA sequencing analyses by using principal component analysis.[88] For each cell line, a total of 15 recordings of extracellular field potentials were performed using a multielectrode array across 47 days in vitro. The authors used a mathematical algorithm (Minimum Embedded Dimension-MED) to assess and compare the nonlinear dynamics of IPSC-derived neurons in differentiation (which reflects the dynamic complexity) from ASD patients and controls. A decreased dynamic complexity was identified during the differentiation of iPSC-derived neurons from patients with ASD, which reflects the decreases in the number of bursting electrodes and in the coefficient of variation of spike interval measures.[88] The changes in dynamic complexity in ASD-derived neurons were correlated to sequencing from iPSC-derived neurons with a cluster of 53 genes differentially expressed and involved in neurodevelopment, cell morphology, and cell migration. Furthermore, ASD-risk genes were shown to be overrepresented in that network. That study strongly suggests that changes in the electrophysiological profiles in idiopathic ASD iPSC-derived neurons are underlined by changes in gene expression profiles.

Similar phenotypes between iPSC-derived neurons from patients with sporadic or syndromic autisms and idiopathic autism

Data collected from cultures of iPSC-derived neurons from patients with sporadic or syndromic autisms have shown that a mutation or a deletion of specific genes, such as *FMR1*[51,53] *GRIN2B*[73] and *SHANK3*[74] can lead to impaired gene dosage, transcriptomic dysregulation, altered neuronal differentiation and activity, notably those mediated through calcium conductance. Whereas no genetic causes have been established for idiopathic autisms, the repeated observation of decreased neuronal activity combined with gene dysregulation related to cell differentiation and synapse functioning, along with a significant overlap of ASD risk genes in the list of DEGs, suggests that dysfunction of iPSC-derived neurons from ASD patients is underlined by common developmental-dependent molecular mechanisms also observed in syndromic autisms.

Studies performed with iPSC-derived cells from patients with syndromic, sporadic or idiopathic autism have highlighted molecular pathways and cellular processes potentially altered in the developing brain, in the context of ASDs. They have also shown that ASDs, which can be considered as a group of diseases with heterogeneous potential genetic cause, are underlined by common mechanisms that can be characterized by common molecular and cellular phenotypes and functional convergences supported by findings using postmortem brain tissue. However, the cellular systems that have been presented here so far do not fully recapitulate the heterogeneity and complexity of the developing human brain. Thus, moving from 2D iPSC-derived neuronal cultures, other more physiological models have been developed to take the next steps in advancing our understanding into ASDs.

3D models of ASDs—a focus on organoids, spheroids, and assembloids

Brain organoids are 3D cultures generated from iPSCs that are cultivated and maintained under low attachment conditions, creating structures that recapitulate brain development, architecture, and function. Brain organoids are long-term cultures that can be grown over the course of several months[89–91] and potentially years.[92,93] As discussed in an earlier chapter, different areas of the brain can be obtained with the addition specific neurodevelopmental factors to drive stem cell fate and neuronal differentiation.[94] In this section, the importance of using brain organoids to model neurodevelopmental diseases, including ASDs will be presented, highlighting the different approaches to generate and analyze these 3D cultures.

Studying neurodevelopmental diseases is challenging, as the key events of brain formation, such as neuronal differentiation, migration, and synaptogenesis, occur during gestational phases. Brain organoids can recapitulate these developmental events in a dish and cerebral organoids were shown to present epigenomic signatures of fetal brain, which makes them a fundamental tool to investigate neurodevelopment in health and disease.[95] Cerebral organoids can be formed using pluripotent stem cells from patients with different neurological diseases, providing researchers with the tools to study brain development in a dish. These model systems are being used to investigate the mechanisms underlying neurological alterations in lines with specific patient mutations and for testing potential therapeutics in patient-derived cells. This is an unprecedented approach connecting research and clinical practice to understand the biology behind the clinical presentation of patients with neurological disorders.

Velasco et al. described a method to generate dorsal forebrain organoids that ensures the generation of diverse progenitors and neuronal cell types while maintaining high reproducibility and a degree of variation similar to what is observed naturally in endogenous brains.[91] Although many groups have developed different protocols to generate and study brain organoids, constant work is in place to refine the methods to generate 3D neural tissue in ASD research.[91,96]

Using dorsal forebrain organoids, Amiri et al. modeled early human brain development[97] by looking at gene expression and regulation of gene expression changes in these organoids. By comparing enhancer elements, which are regions of the genome that are responsible for regulating and often increasing gene expression of particular gene targets, and corresponding transcripts of organoids and cortical brain tissue, this study revealed that brain organoids patterned to the dorsal forebrain mimic the longitudinal development of embryonic and early fetal cortical primordium.[97] This is of great importance for neurodevelopmental diseases research, especially for ASDs, as several genes listed on the SFARI database were differentially expressed and associated with enhancer elements. Thus, this work validated the use of organoids in early human brain development research and highlighted its potential in neurodevelopmental disorders such as ASDs.

Interestingly, Mariani et al. showed that circuit alterations can be modeled in brain organoids, highlighting how different genetic alterations and risk factors can converge on similar cellular mechanisms contributing to the pathogenesis of ASDs. In this study, idiopathic ASD was modeled using telencephalic organoids, and alterations in neuronal proliferation and differentiation were observed in organoids from ASD patients with different mutations.[98] ASD-derived organoids displayed upregulation of genes involved in proliferation and accelerated cell cycle that correlated with increased organoid size. Additionally, increased numbers of synapses were detected and overexpression of the factor FOXG1 was linked to overproduction of GABAergic inhibitory neurons and correlated with symptom severity.[98] This work showed that ASD-derived organoids presented with an overproduction of GABAergic progenitors and mature neurons that contributed to the imbalance between excitatory and inhibitory signals.

Changes in excitation and inhibition have been associated with ASD and observed in studies utilizing iPSC-derived neurons.[99] However, modeling brain circuits in different genetic backgrounds is challenging, as several areas of the brain that are interconnected to form a complex functional neural network that controls behaviors and cognitive function might be affected in ASD. To overcome this challenge, and to study the connections between different brain areas, 3D models called "assembloids" were developed. Assembloids are formed when organoids from different cell lineages are combined. This model facilitates the study of interactions between distinct cell types and areas of the brain to provide a valuable tool for examining neuronal networks during development and, importantly, to investigate alterations in neuronal circuits in ASDs.

Recent studies have utilized assembloids to understand the processes driving the formation of cortical circuits. In one study, Birey et al. merged glutamatergic organoids with GABAergic organoids to form assembloids containing both subtypes.[100] During brain development, the migration of GABAergic interneurons from ventral to dorsal forebrain, where glutamatergic neurons are generated, permits the integration of cortical circuits containing both subtypes. Interestingly, Birey et al. showed for the first time that assembloids recapitulate the fetal forebrain saltatory migration and functional cortical circuit integration using human spheroids resembling both pallium and subpallium regions.[100] Interestingly, this study showed that the fusion of pallium spheroids containing VGLUT1$^+$ glutamatergic neurons and subpallium spheroids containing in VGAT$^+$ GABAergic neurons was driven by unidirectional neuronal migration from subpallium into pallium spheroids by live imaging of spheroids that were virally labeled. This pattern of interneuronal migration was observed in rodent cultures,[101] however, this approach was able to model for the first time human neuronal migration during cortical circuit development.

Assembloids were then used to model migration defects during corticogenesis that are observed in neurodevelopmental diseases such as the Timothy Syndrome (TS). Assembloids generated from TS patient-derived iPSCs displayed increased neuronal saltatory frequency from subpallium spheroids into pallium. Observed migration defects were associated with reduced length and speed on conduction in TS patients.[100] The saltatory migration of interneurons is an important step for cortical circuit formation, and alterations in brain circuit are correlated with excitatory and inhibitory imbalance in schizophrenia, autism, and intellectual disabilities.[102]

In another study, Bagley et al. described the migration of neurons from ventral into dorsal forebrain, resembling typical interneuron migration.[103] Xiang et al. used a similar approach to study interneuron migration and described a system to generate functional human medial ganglionic eminence (MGE) and cortical organoids. In this model, progenitors from medial ganglionic eminence organoids were able to migrate into cortical organoids three days after fusion, mimicking tangential migration of interneuron progenitors. Importantly, neuronal morphology, electrical activity, and GABAergic identity were maintained during migration.[104]

Another important aspect of brain complexity that can be studied in organoids is the presence of different cell types that are crucial for proper neuronal development and activity. Sloan et al. described a method to pattern forebrain spheroids into dorsal and ventral 3D structures that are composed of glutamatergic and GABAergic neurons, modeling forebrain neuronal interaction.[93] Different from other protocols, this method enables the formation of glial cell types, such as astrocytes, that are crucial for neuronal development and for modulating circuit and synapse formation. Thus, this method recapitulates migration and integration of GABAergic neurons into the cerebral cortex and interactions of neuronal and glial populations.[93]

Connectivity between different areas of the brain has also been modeled using assembloids. Xiang et al. described a method to obtain iPSC-derived thalamic organoids, which were formed with a majority of vesicular glutamate transporter 2 (vGLUT2)-positive glutamatergic neurons, whereas cortical organoids contained both glutamatergic and

GABAergic neurons.[105] Thalamic organoids were fused with cortical structures to model the formation of thalamocortical axon connections observed during brain development.[105] After six days of fusion, axon projections exiting thalamic organoids were observed infiltrating cortical organoids and displayed typical growth cone morphology with expression of the pre-synaptic marker synaptophysin (SYP) and post-synaptic marker PSD95 (post-synaptic density protein 95). This implies that these processes were active and established functional thalamocortical connections. Importantly, increased firing frequencies in neurons from thalamic organoids were observed after fusion with cortical projections were detected,[105] confirming that assembloids can recapitulate structural and functional connections between different brain regions during brain development.

The next step in ASD research using 3D models is to identify cellular phenotypes that are amenable to screening and identifying biomarkers to guide the research for therapies. Such phenotypes may include, but are not limited to, morphological (axonal length and dendritic spine development), cell fate commitment (cell fate and differentiation), and functional biomarkers (spontaneous activity, hyperactivity, and hypoactivity). The fast-paced development of iPSC-derived assays is undoubtedly boosting neurodevelopmental disorders research.

The use of iPSCs to develop assays and novel therapies that can be translated to the clinic for ASD

CNS drug discovery is still a costly and challenging process. A 2018 analysis by the Tufts Center for the Study of Drug Development looked at the United States drug market from 2000 through 2017 and found that CNS drugs require on average 20% more time than non-CNS drugs to be developed. It also took 38% longer for these drugs to win market approval.[106] One contributing factor to the increased time for CNS drug discovery is the cell models historically used for CNS drug discovery.

Before the advent of patient-derived iPSCs, the cell lines used for CNS drug discovery included immortalized fibroblasts or neuronal progenitor cell lines and cell lines from nervous system tumors.[107–109] These lines are easy to culture and handle and they can be easily expanded. However, these cell lines often carry karyotypic abnormalities and they fail to reproduce signaling mechanisms present in endogenous neural cells, which are required for effective drug screening.[110,111] iPSC technology helps bridge the gap from preclinical to clinical studies by providing physiologically and therapeutically relevant cell models. More specifically, iPSC technology provides access to virtually unlimited human material, a great variety of tissue- and disease-specific cells,[112] and cell lines with a more stable genotype, when compared to transformed cell lines.[110]

More specifically to neuropsychiatric disorders-directed assays, the availability of patient-derived lines that can be differentiated into functional neural cells and neuronal networks is a big advantage.[113] One of the fields where the use of iPSC is growing is the field of ASD, as reviewed in the previous sections of this chapter.

The genetic causes of ASD are highly heterogeneous, with more than 1000 genes implicated in the disorder. Since iPSCs can be derived from any patient,[114] the use of this technology allows us to explore the genetic variability in ASD and identify patient-specific therapeutic approaches. In other words, this technology can be applied to understanding the functional effects of genetic variants on complex diseases, which contributes to both drug and disease model development.[113]

As it has been described previously, patient-specific iPSC-derived neural cells can exhibit disease-relevant cellular phenotypes. These cells and associated phenotypes can then be used to develop drug screening platforms directed toward the identification of novel therapeutic targets that reverse the patient-specific phenotypes. These screening efforts are called primary screens because they identify initial targets and compounds of interest for a particular phenotype. More complex phenotypes that are difficult to scale up, such as functional neuronal responses, or 3D organoid development, can be used for the validation of existing preclinical drug candidates, in an effort known as secondary screens.[115] In the review by Christensen et al.[115] iPSCs are proposed to be especially beneficial in secondary screening campaigns, that is, for hit validation, qualification, and profiling of drugs, among others. It is also suggested that iPSC models are especially important where a human target is essential, as is the case of antibodies or RNA therapeutics, where non-human drug discovery models have little to no use due to inter-species differences.

iPSCs are an important tool for drug screening platforms and can be used for both target-based and phenotypic screenings. Target-based screenings using iPSCs depend on the availability of protocols to generate specific subtypes of neurons or glia that express targets in the proper physiologically relevant location and pathway.[112] High-content screens, also called phenotypic screens, monitor changes in phenotypes of cells, including subcellular changes, changes in cell morphology, and in gene expression. Phenotypic screens are more challenging to perform as they are highly dependent on the nature of the targeted phenotypes and also its reproducibility[112] across different cell lines, for example. Phenotypic screens can also be used for target identification, where compounds with known mechanisms of action are screened to correct a particular cellular phenotype and reveal which mechanisms were involved in the correction.

For phenotypic assay development, isogenic iPSC control lines are important tools to determine whether a specific phenotype is indeed a direct result of a patient-specific mutation.[115] Isogenic lines are achieved through genome editing techniques to correct a variant of investigation, as discussed in a previous chapter. Gene editing techniques can also be applied to make targeted modifications in the genome of iPSCs and thus generate either loss-of-function alleles or to recreate monogenic mutations that are associated with a known disease.[116] Another important tool for the development of phenotypic assays are cell biobanks. Biobanks are often the best resource for quality human material and come with a wealth of clinical information regarding the donor. However, there are still few biobanks offering iPSC collections. This creates a barrier to the use of iPSC-derived neurons.[117]

In addition to iPSC cell banks, neural progenitor cell (NPC) banks are desired for the study of ASD and other neurodevelopmental conditions. Neural progenitors cell banks are important for scaling-up cell production for the numbers required for high-throughput screening campaigns.[115] Repetitive passaging of iPSC can lead to genomic alterations. To avoid repetitive passaging, iPSC lines and NPC lines can be banked in sufficient quantities to allow for single-use, as proposed by Christensen et al.[115] This group also claims that the banking of NPCs reduces well-to-well variability in an iPSC-based drug discovery platform.

Limitations for using iPSC-derived neurons in drug screening platforms

Variability and genomic alterations are some of the issues that can arise from using iPSC-derived models for drug screening campaigns. In addition to these, several other points need to be considered when applying iPSC and iPSC-derived neurons to phenotypic screenings. These considerations are mentioned below.

Quality control testing

The incorporation of iPSC-derived neurons in assay development requires constant quality control testing. To reduce potential artifacts and increase reproducibility, iPSCs need to be routinely assessed for pluripotency, genomic stability, differentiation capability, microbial contamination, among others.[112,115] These topics were discussed in greater detail in an earlier chapter. In addition, quality control metrics of the neurons produced need to be routinely assayed, including gene expression, morphology, and function of neuronal networks.

Automation challenges

The application of iPSCs into high-throughput screening faces many challenges, such as automated cell handling and miniaturization[115] and scaling up.[113] iPSCs are difficult to adapt to automated cell handling processes because they are not supposed to be routinely split up into single cells for passaging, and rather should be clump-passaged, as discussed in an earlier chapter. Repeated disruption of iPSC colony organization may introduce karyotype changes. In addition, neuronal culture can be difficult to withstand automated plating and feeding since neurons are more fragile than other cell types in culture and may be prone to lift off the plate or undergo neurite damage as a result of harsh manipulations.

Cost

The preparation and maintenance of individual lines of iPSC-derived neurons from large cohorts of patients and matched controls are associated with high costs[117] due to surface treatment, reagents, banking, handling, among others.

Small "n"

Since the generation of multiples iPSC cell lines can be time-consuming, costly, and labor-intensive,[118,119] most studies are done with low biological replicate numbers. Since each line represents a single individual (or sometimes even cells from the same individual it is used to create different lines), it is certainly not representative of large cohort population studies.

Epigenetic memory

iPSCs are derived from somatic cells and it has been suggested that the persistence of residual epigenetic memory from those cells can introduce a confounding phenotypic response to candidate therapies in drug screening.[120,121] Conversely, others have found that epigenetic signatures are erased during reprogramming of somatic cells to iPSCs and these epigenetic signatures are often important for patient-relevant phenotypes.[122]

Well-to-well variability

Well-to-well variability is important to control for any screening campaign. For example, it has been shown that there can be considerable differences in neurite outgrowth quantification between wells, which can be introduced by differences in initial cell seeding density of iPSCs or iPSC-derived neurons.[115] Ensuring that each well is coated and plated exactly in the same way is critical for screening success. In addition, media can evaporate from cells within the plate unevenly, leading to edge effects on 96-well or 384-well plates, where the outer wells evaporate more quickly than inner wells. Therefore, it's important to use incubators with strict humidity control and to ensure that all wells remain leveled with frequent half-media changes.

Variability within cell lines

Differences between donor individuals can impact many iPSC cellular traits, from DNA methylation and protein abundance to pluripotency, differentiation, and cell morphology.[123] Since residual DNA-methylation signatures of the somatic tissues of origin may affect transcriptomic profiles of cells,[121] different iPSC clones prepared from the same individual may harbor different somatic mutations[117] and it may be useful to study multiple clones from the same patient to visualize this variability. These differences in somatic mutations can yield different phenotypes. This can be even more impactful for the study of ASD because variability is already an enormous factor due to multiple genetic loci implicated in ASD.

Variability across differentiation batches

In addition to variability across wells or lines, a significant source of variability in a screening effort is likely to come from differentiation batches. Therefore, it is recommended to use one large batch of neurons for primary screening efforts, to control for variation across batches. Secondary and confirmatory assays can then use multiple batches to be sure that the batch was not responsible for the finding of the screening.

Disease modeling

iPSC disease models are frequently designed to study rare, monogenic, hereditary forms of the disease, rather than spontaneous forms that are characterized by complex genetic traits involving multiple unknown genetic and epigenetic factors.[114] However, multigene editing and synthetic biology approaches in patient-derived iPSC models could potentially address these challenges.[124,125]

Screening of simple phenotypes

iPSC-derived neuronal modeling technology is often used for the identification of simple phenotypes in screening efforts, such as neurite morphology or neuronal activity. For more complex assays, looking into circuit connectivity and other developmental processes, optimization is still needed before assays are ready to be adapted for screening.

Despite these limitations, iPSC-based assays have been developed to further the understanding of ASD and to investigate potential drug treatments.

The use of iPSC-derived neurons for personalized medicine

In the study by Marchetto et al.,[60] neuronal phenotypes of Rett syndrome were identified in iPSC-derived neurons from patients, including fewer synapses, reduced spine density, and smaller soma size, compared with iPSC-derived neurons from healthy controls. In addition, the iPSC-derived neurons from Rett patients showed signs of altered calcium signaling and electrophysiological abnormalities when compared to neurons derived from control iPSCs. In the same study, it was found that gentamycin treatment rescued some of the synaptic defects seen in the Rett neurons.

Another study showed the beneficial effects of IGF-1 to neurons derived from iPSCs of Rett patients. Treatment of these neurons with IGF-1 for 14 days increased the total neurite length in cultures compared to untreated Rett neurons.[126] Williams et al.[127] also investigated cells derived from iPSCs from Rett patients, but in their study, they focused on glial cells. This study revealed that mutant astrocytes differentiated from iPSCs derived from Rett patients elicited adverse effects on the survival and normal function of wild-type neurons and that both insulin growth factor 1 (IGF-1) and GPE (a peptide containing the first three amino acids of IGF-1) can partially rescue the neuronal defects caused by these mutant Rett astrocytes.

More relevant studies showed structural, molecular, and physiological Rett-related phenotypes, as reviewed by Ip et al.[128] From these combined studies, it has been suggested that drugs could potentially be tested on neuronal cultures derived from patients with Rett syndrome before deciding whether the drug should be administered to a particular individual,[129] which highlights the potential of iPSC-derived cells to be used for personalized medicine.

Personalized medicine tailors medical treatment for an individual patient based on genetic and genomic information coming from the biological samples of the patient itself.[117] Personalized medicine using iPSCs can help to tease out responders and nonresponders, as in the study of Schöndorf et al.[130] where they showed that lithium nonresponse in a subset of bipolar disease patients was recapitulated by in vitro lithium response in the iPSC-derived neurons of the same patients. Similar studies could be of great value for ASD drug development to address the complexity and heterogeneity that characterize these disorders.

Here, we have summarized how iPSC-based models hold promise, both as a patient phenotyping tool, as well as for drug discovery. While the development of new therapies for ASD remains an unmet need, the advent of iPSC technology brings hope for the identification of novel therapeutics that were not able to be identified with preclinical models that were historically used for ASD research.

Conclusions

The advent of human iPSC-derived neuron technology has opened new avenues for studying neurodevelopmental disorders that for so long had to rely on findings from postmortem and animal studies. Although those models have respectively allowed us to observe molecular and cellular changes in patients' brains and to partially recapitulate dysregulations observed in ASDs, they do not fully model the temporal and phenotypic complexities occurring in ASDs. Studies in iPSC-derived cells provide functional insights to observations made on other models and, therefore, constitute a powerful tool to answer questions left unanswered by postmortem and animal studies. This technology also makes patient-driven research possible.

iPSC-based studies of syndromic autisms, which are well-characterized in terms of genetic etiology and phenotypes, have allowed the research community to identify changes in gene expression profiles, cellular morphology, and cell activity, and to infer the functional effects of a given mutation on the dysregulation observed in vitro. Interestingly, iPSC-based studies

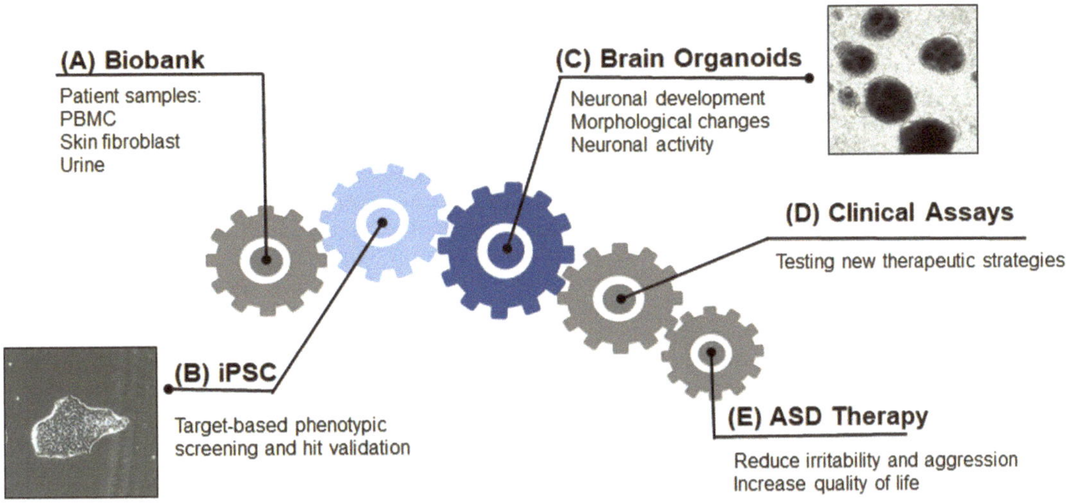

FIGURE 9.1 hiPSCs and brain organoids as important tools for assay development.

of idiopathic autisms have further highlighted the phenotypic heterogeneity of ASDs. These investigations have also provided further information regarding the longitudinal changes that potentially occur in the developing brain. Furthermore, iPSC-based studies facilitate the identification of molecular and cellular processes conserved between syndromic and idiopathic autisms, which are valuable findings for the drug development process.

However, 2D-iPSC models do present limitations regarding the recapitulation of brain complexity. In recent years, the development of the 3D-iPSC models in the form of spheroids and organoids has addressed the complexity missing from 2D models, through the presence of cell-cell interaction, neuronal maturation, and neuronal network formation.

Despite its limitations, both 2D and 3D-iPSC models are now being applied for the development of phenotypic assays in drug discovery (see Fig. 9.1). Since many phenotypes are shared not only among different syndromic autisms but also with idiopathic autism, phenotypic assay-based drug discovery has the potential to address unmet therapeutic needs in ASD despite a diverse genetic background. The avenue for the application of iPSC technology in ASD is a promising one. However, to be useful for drug discovery, rigorous quality control of iPSC-derived cells needs to be implemented to allow for proper biological readouts and to meet industry standards.

References

1. Fernandez BA, Scherer SW. Syndromic autism spectrum disorders: moving from a clinically defined to a molecularly defined approach. *Dialogues Clin Neurosci*. 2017;19(4):353–371. https://www.ncbi.nlm.nih.gov/pmc/articles/PMC5789213/pdf/DialoguesClinNeurosci-19-353.pdf.
2. Sztainberg Y, Zoghbi HY. Lessons learned from studying syndromic autism spectrum disorders. *Nat Neurosci*. 2016;19(11):1408–1418. https://doi.org/10.1038/nn.4420.
3. Geschwind DH, Levitt P. Autism spectrum disorders: developmental disconnection syndromes. *Curr Opin Neurobiol*. 2007;17(1):103–111. https://doi.org/10.1016/j.conb.2007.01.009.

4. Kanner L. Autistic disturbances of affective contact. *Nervous Child*. 1943;2:217–250.
5. Maussion G, Rocha C, Bernard G, Beitel LK, Durcan TM. Patient-derived stem cells, another in vitro model, or the missing link toward novel therapies for autism spectrum disorders? *Front Pediatr*. 2019;7. https://doi.org/10.3389/fped.2019.00225.
6. Abrahams BS, Geschwind DH. Advances in autism genetics: on the threshold of a new neurobiology. *Nat Rev Genet*. 2008;9(5):341–355. https://doi.org/10.1038/nrg2346.
7. O ' Roak BJ, Vives L, Girirajan S, et al. Sporadic autism exomes reveal a highly interconnected protein network of de novo mutations. *Nature*. 2012;485(7397):246–250. https://doi.org/10.1038/nature10989.
8. O'Roak BJ, Deriziotis P, Lee C, et al. Exome sequencing in sporadic autism spectrum disorders identifies severe de novo mutations. *Nat Genet*. 2011;43(6):585–589. https://doi.org/10.1038/ng.835.
9. BS N, Ravi K, Sharmila B-B. AutDB: a gene reference resource for autism research. *Nucleic Acids Res*. 2009:D832–D836. https://doi.org/10.1093/nar/gkn835.
10. Takahashi K, Yamanaka S. Induction of pluripotent stem cells from mouse embryonic and adult fibroblast cultures by defined factors. *Cell*. 2006;126(4):663–676. https://doi.org/10.1016/j.cell.2006.07.024.
11. Takahashi K, Tanabe K, Ohnuki M, et al. Induction of pluripotent stem cells from adult human fibroblasts by defined factors. *Cell*. 2007;131(5):861–872. https://doi.org/10.1016/j.cell.2007.11.019.
12. Baghbaderani BA, Syama A, Sivapatham R, et al. Detailed characterization of human induced pluripotent stem cells manufactured for therapeutic applications. *Stem Cell Rev Rep*. 2016;12(4):394–420. https://doi.org/10.1007/s12015-016-9662-8.
13. Shi Y, Inoue H, Wu JC, Yamanaka S. Induced pluripotent stem cell technology: a decade of progress. *Nat Rev Drug Discov*. 2017;16(2):115–130. https://doi.org/10.1038/nrd.2016.245.
14. Hollingsworth EW, Vaughn JE, Orack JC, et al. iPhemap: an atlas of phenotype to genotype relationships of human iPSC models of neurological diseases. *EMBO Mol Med*. 2017;9(12):1742–1762. https://doi.org/10.15252/emmm.201708191.
15. Elitt MS, Barbar L, Tesar PJ. Drug screening for human genetic diseases using iPSC models. *Hum Mol Genet*. 2018;27(2):R89–R98. https://doi.org/10.1093/hmg/ddy186.
16. Nair M, Sandhu SS, Sharma AK. Induced pluripotent stem cell technology: a paradigm shift in medical science for drug screening and disease modeling. *Curr Med Chem*. 2017;24(39):4368–4398. https://doi.org/10.2174/0929867324666170727100508.
17. Hatton DD, Sideris J, Skinner M, et al. Autistic behavior in children with fragile X syndrome: prevalence, stability, and the impact of FMRP. *Am J Med Genet*. 2006;140(17):1804–1813. https://doi.org/10.1002/ajmg.a.31286.
18. Pieretti M, Zhang F, Fu YH, et al. Absence of expression of the FMR-1 gene in fragile X syndrome. *Cell*. 1991;66(4):817–822. https://doi.org/10.1016/0092-8674(91)90125-I.
19. Amir RE, Van Den Veyver IB, Wan M, Tran CQ, Francke U, Zoghbi HY. Rett syndrome is caused by mutations in X-linked MECP2, encoding methyl- CpG-binding protein 2. *Nat Genet*. 1999;23(2):185–188. https://doi.org/10.1038/13810.
20. Hsieh J, Gage FH. Chromatin remodeling in neural development and plasticity. *Curr Opin Cell Biol*. 2005;17(6):664–671. https://doi.org/10.1016/j.ceb.2005.09.002.
21. Sokpor G, Xie Y, Rosenbusch J, Tuoc T. Chromatin remodeling BAF (SWI/SNF) complexes in neural development and disorders. *Front Mol Neurosci*. 2017;10:243. https://doi.org/10.3389/fnmol.2017.00243.
22. Fehr S, Bebbington A, Nassar N, et al. Trends in the diagnosis of rett syndrome in Australia. *Pediatr Res*. 2011;70(3):313–319. https://doi.org/10.1203/PDR.0b013e3182242461.
23. Lin C, Franco B, Rosner MR. CDKL5/Stk9 kinase inactivation is associated with neuronal developmental disorders. *Hum Mol Genet*. 2005;14(24):3775–3786. https://doi.org/10.1093/hmg/ddi391.
24. Lin L, Lesnick TG, Maraganore DM, Isacson O. Axon guidance and synaptic maintenance: preclinical markers for neurodegenerative disease and therapeutics. *Trends Neurosci*. 2009;32(3):142–149. https://doi.org/10.1016/j.tins.2008.11.006.
25. Lambert L, Bienvenu T, Allou L, et al. MEF2C mutations are a rare cause of Rett or severe Rett-like encephalopathies. *Clin Genet*. 2012;82(5):499–501. https://doi.org/10.1111/j.1399-0004.2012.01861.x.
26. Fehr S, Wilson M, Downs J, et al. The CDKL5 disorder is an independent clinical entity associated with early-onset encephalopathy. *Eur J Hum Genet: EJHG (Eur J Hum Genet)*. 2013;21(3):266–273. https://doi.org/10.1038/ejhg.2012.156.

27. Jacob FD, Ramaswamy V, Andersen J, Bolduc FV. Atypical Rett syndrome with selective FOXG1 deletion detected by comparative genomic hybridization: case report and review of literature. *Eur J Hum Genet.* 2009;17(12):1577–1581. https://doi.org/10.1038/ejhg.2009.95.
28. Van Slegtenhorst M, De Hoogt R, Hermans C, et al. Identification of the tuberous sclerosis gene TSC1 on chromosome 9q34. *Science.* 1997;277(5327):805–808. https://doi.org/10.1126/science.277.5327.805.
29. Carsillo T, Astrinidis A, Henske EP. Mutations in the tuberous sclerosis complex gene TSC2 are a cause of sporadic pulmonary lymphangioleiomyomatosis. *Proc Natl Acad Sci USA.* 2000;97(11):6085–6090. https://doi.org/10.1073/pnas.97.11.6085.
30. Phelan K, Rogers RH, Boccuto L. In: Adam MP. Ardinger HH, Pagon RA, eds, et al. *GeneReviews((R)).* Seattle (WA; 1993).
31. Amal H, Barak B, Bhat V, et al. Shank3 mutation in a mouse model of autism leads to changes in the S-nitroso-proteome and affects key proteins involved in vesicle release and synaptic function. *Mol Psychiatr.* 2020;25(8):1835–1848. https://doi.org/10.1038/s41380-018-0113-6.
32. Reim D, Distler U, Halbedl S, et al. Mutant mice reveals brain region specific changes relevant to autism spectrum disorder. *Front Mol Neurosci.* 2017;10:26. https://doi.org/10.3389/fnmol.2017.00026.
33. Kishino T, Lalande M, Wagstaff J. UBE3A/E6-AP mutations cause Angelman syndrome. *Nat Genet.* 1997;15(1):70–73. https://doi.org/10.1038/ng0197-70.
34. Hogart A, Wu D, LaSalle JM, Schanen NC. The comorbidity of autism with the genomic disorders of chromosome 15q11.2-q13. *Neurobiol Dis.* 2010;38(2):181–191. https://doi.org/10.1016/j.nbd.2008.08.011.
35. Lalande M, Calciano MA. Molecular epigenetics of Angelman syndrome. *Cell Mol Life Sci.* 2007;64(7–8):947–960. https://doi.org/10.1007/s00018-007-6460-0.
36. Cassidy SB, Schwartz S, Miller JL, Driscoll DJ. Prader-Willi syndrome. *Genet Med.* 2012;14(1):10–26. https://doi.org/10.1038/gim.0b013e31822bead0.
37. Cassidy SB, Dykens E, Williams CA. Prader-Willi and Angelman syndromes: sister imprinted disorders. *Am J Med Genet Semin Med Genet.* 2000;97(2):136–146. https://doi.org/10.1002/1096-8628(200022)97:2<136::AID-AJMG5>3.0.CO;2-V.
38. Greer PL, Hanayama R, Bloodgood BL, et al. The Angelman syndrome protein Ube3A regulates synapse development by ubiquitinating arc. *Cell.* 2010;140(5):704–716. https://doi.org/10.1016/j.cell.2010.01.026.
39. Ludwig M, Katalinic A, Groß S, Sutcliffe A, Varon R, Horsthemke B. Increased prevalence of imprinting defects in patients with Angelman syndrome born to subfertile couples. *J Med Genet.* 2005;42(4):289–291. https://doi.org/10.1136/jmg.2004.026930.
40. Laan LAEM, Vein AA. Angelman syndrome: is there a characteristic EEG? *Brain Dev.* 2005;27(2):80–87. https://doi.org/10.1016/j.braindev.2003.09.013.
41. Splawski I, Timothy KW, Sharpe LM, et al. CaV1.2 calcium channel dysfunction causes a multisystem disorder including arrhythmia and autism. *Cell.* 2004;119(1):19–31. https://doi.org/10.1016/j.cell.2004.09.011.
42. Splawski I, Timothy KW, Decher N, et al. Severe arrhythmia disorder caused by cardiac L-type calcium channel mutations. *Proc Natl Acad Sci USA.* 2005;102(23):8089–8096. https://doi.org/10.1073/pnas.0502506102.
43. Griesi-Oliveira K, Acab A, Gupta AR, et al. Modeling non-syndromic autism and the impact of TRPC6 disruption in human neurons. *Mol Psychiatr.* 2015;20(11):1350–1365. https://doi.org/10.1038/mp.2014.141.
44. LeMasson G, Marder E, Abbott LF. Activity-dependent regulation of conductances in model neurons. *Science.* 1993;259(5103):1915–1917. https://doi.org/10.1126/science.8456317.
45. Spitzer NC. Activity-dependent neurotransmitter respecification. *Nat Rev Neurosci.* 2012;13(2):94–106. https://doi.org/10.1038/nrn3154.
46. Li Z, Okamoto KI, Hayashi Y, Sheng M. The importance of dendritic mitochondria in the morphogenesis and plasticity of spines and synapses. *Cell.* 2004;119(6):873–887. https://doi.org/10.1016/j.cell.2004.11.003.
47. Irwin SA, Galvez R, Greenough WT. Dendritic spine structural anomalies in fragile-X mental retardation syndrome. *Cerebr Cortex.* 2000;10(10):1038–1044. https://doi.org/10.1093/cercor/10.10.1038.
48. Belichenko PV, Belichenko PV, Oldfors A, Hagberg B, Dahlström A. Rett syndrome: 3-D confocal microscopy of cortical pyramidal dendrites and afferents. *Neuroreport.* 1994;5(12):1509–1513. https://doi.org/10.1097/00001756-199407000-00025.
49. Yasin SA, Ali AM, Tata M, et al. MTOR-dependent abnormalities in autophagy characterize human malformations of cortical development: evidence from focal cortical dysplasia and tuberous sclerosis. *Acta Neuropathol.* 2013;126(2):207–218. https://doi.org/10.1007/s00401-013-1135-4.

50. Pasca SP, Portmann T, Voineagu I, et al. Using iPSC-derived neurons to uncover cellular phenotypes associated with Timothy syndrome. *Nat Med*. 2011;17(12):1657–1662. https://doi.org/10.1038/nm.2576.
51. Achuta VS, Möykkynen T, Peteri UK, et al. Functional changes of AMPA responses in human induced pluripotent stem cell–derived neural progenitors in fragile x syndrome. *Sci Signal*. 2018;11(513). https://doi.org/10.1126/scisignal.aan8784.
52. Irwin SA, Patel B, Idupulapati M, et al. Abnormal dendritic spine characteristics in the temporal and visual cortices of patients with fragile-X syndrome: a quantitative examination. *Am J Med Genet*. 2001;98(2):161–167. https://doi.org/10.1002/1096-8628(20010115)98:2<161::AID-AJMG1025>3.0.CO;2-B.
53. Achuta VS, Grym H, Putkonen N, et al. Metabotropic glutamate receptor 5 responses dictate differentiation of neural progenitors to NMDA-responsive cells in fragile X syndrome. *Dev Neurobiol*. 2017;77(4):438–453. https://doi.org/10.1002/dneu.22419.
54. Sundberg M, Tochitsky I, Buchholz DE, et al. Purkinje cells derived from TSC patients display hypoexcitability and synaptic deficits associated with reduced FMRP levels and reversed by rapamycin. *Mol Psychiatr*. 2018;23(11):2167–2183. https://doi.org/10.1038/s41380-018-0018-4.
55. Ebrahimi-Fakhari D, Saffari A, Wahlster L, et al. Impaired mitochondrial dynamics and mitophagy in neuronal models of tuberous sclerosis complex. *Cell Rep*. 2016;17(4):1053–1070. https://doi.org/10.1016/j.celrep.2016.09.054.
56. Bailey A, Luthert P, Dean A, et al. A clinicopathological study of autism. *Brain*. 1998;121(5):889–905. https://doi.org/10.1093/brain/121.5.889.
57. Ritvo ER, Freeman BJ, Scheibel AB, Duong T, Robinson H, Guthrie D. Lower Purkinje cell counts in the cerebella of four autistic subjects: initial findings of the UCLA-NSAC autopsy research report. *Am J Psychiatr*. 1986;143(7):862–866. https://doi.org/10.1176/ajp.143.7.862.
58. Fatemi SH, Halt AR, Realmuto G, et al. Purkinje cell size is reduced in cerebellum of patients with autism. *Cell Mol Neurobiol*. 2002;22(2):171–175. https://doi.org/10.1023/A:1019861721160.
59. Tanaka Y, Kim KY, Zhong M, Pan X, Weissman SM, Park IH. Transcriptional regulation in pluripotent stem cells by methyl CpG-binding protein 2 (MeCP2). *Hum Mol Genet*. 2014;23(4):1045–1055. https://doi.org/10.1093/hmg/ddt500.
60. Marchetto MCN, Carromeu C, Acab A, et al. A model for neural development and treatment of rett syndrome using human induced pluripotent stem cells. *Cell*. 2010;143(4):527–539. https://doi.org/10.1016/j.cell.2010.10.016.
61. Ramoz N, Reichert JG, Smith CJ, et al. Linkage and association of the mitochondrial aspartate/glutamate carrier SLC25A12 gene with autism. *Am J Psychiatr*. 2004;161(4):662–669. https://doi.org/10.1176/appi.ajp.161.4.662.
62. Lepagnol-Bestel AM, Maussion G, Boda B, et al. SLC25A12 expression is associated with neurite outgrowth and is upregulated in the prefrontal cortex of autistic subjects. *Mol Psychiatr*. 2008;13(4):385–397. https://doi.org/10.1038/sj.mp.4002120.
63. Palmieri L, Papaleo V, Porcelli V, et al. Altered calcium homeostasis in autism-spectrum disorders: evidence from biochemical and genetic studies of the mitochondrial aspartate/glutamate carrier AGC1. *Mol Psychiatr*. 2010;15(1):38–52. https://doi.org/10.1038/mp.2008.63.
64. Anitha A, Nakamura K, Thanseem I, et al. Downregulation of the expression of mitochondrial electron transport complex genes in autism brains. *Brain Pathol*. 2013;23(3):294–302. https://doi.org/10.1111/bpa.12002.
65. Siddiqui MF, Elwell C, Johnson MH. Mitochondrial dysfunction in autism spectrum disorders. *Autism Open Access*. 2016;6(5):1000190. https://doi.org/10.4172/2165-7890.1000190.
66. Chauhan A, Gu F, Essa MM, et al. Brain region-specific deficit in mitochondrial electron transport chain complexes in children with autism. *J Neurochem*. 2011;117(2):209–220. https://doi.org/10.1111/j.1471-4159.2011.07189.x.
67. Armstrong D, Dunn JK, Antalffy B, Trivedi R. Selective dendritic alterations in the cortex of rett syndrome. *JNEN (J Neuropathol Exp Neurol)*. 1995;54(2):195–201. https://doi.org/10.1097/00005072-199503000-00006.
68. Chapleau CA, Calfa GD, Lane MC, et al. Dendritic spine pathologies in hippocampal pyramidal neurons from Rett syndrome brain and after expression of Rett-associated MECP2 mutations. *Neurobiol Dis*. 2009;35(2):219–233. https://doi.org/10.1016/j.nbd.2009.05.001.
69. O'Roak BJ, Vives L, Fu W, et al. Multiplex targeted sequencing identifies recurrently mutated genes in autism spectrum disorders. *Science*. 2012;338(6114):1619–1622. https://doi.org/10.1126/science.1227764.

70. Talkowski ME, Rosenfeld JA, Blumenthal I, et al. Sequencing chromosomal abnormalities reveals neurodevelopmental loci that confer risk across diagnostic boundaries. *Cell*. 2012;149(3):525–537. https://doi.org/10.1016/j.cell.2012.03.028.
71. Chen ES, Gigek CO, Rosenfeld JA, et al. Molecular convergence of neurodevelopmental disorders. *Am J Hum Genet*. 2014;95(5):490–508. https://doi.org/10.1016/j.ajhg.2014.09.013.
72. Gigek C, Chen E, Ota V, et al. A molecular model for neurodevelopmental disorders. *Transl Psychiatr*. 2015;5.
73. Bell S, Maussion G, Jefri M, et al. Disruption of GRIN2B impairs differentiation in human neurons. *Stem Cell Rep*. 2018;11(1):183–196. https://doi.org/10.1016/j.stemcr.2018.05.018.
74. Kathuria A, Nowosiad P, Jagasia R, et al. Stem cell-derived neurons from autistic individuals with SHANK3 mutation show morphogenetic abnormalities during early development. *Mol Psychiatr*. 2018;23(3):735–746. https://doi.org/10.1038/mp.2017.185.
75. Liu X, Campanac E, Cheung HH, et al. Idiopathic autism: cellular and molecular phenotypes in pluripotent stem cell-derived neurons. *Mol Neurobiol*. 2017;54(6):4507–4523. https://doi.org/10.1007/s12035-016-9961-8.
76. Derosa BA, El Hokayem J, Artimovich E, et al. Convergent pathways in idiopathic autism revealed by time course transcriptomic analysis of patient-derived neurons. *Sci Rep*. 2018;8(1). https://doi.org/10.1038/s41598-018-26495-1.
77. Griesi-Oliveira K, Fogo MS, Pinto BGG, et al. Transcriptome of iPSC-derived neuronal cells reveals a module of co-expressed genes consistently associated with autism spectrum disorder. *Mol Psychiatr*. 2020. https://doi.org/10.1038/s41380-020-0669-9.
78. Marchetto MC, Belinson H, Tian Y, et al. Altered proliferation and networks in neural cells derived from idiopathic autistic individuals. *Mol Psychiatr*. 2017;22(6):820–835. https://doi.org/10.1038/mp.2016.95.
79. Schafer ST, Paquola ACM, Stern S, et al. Pathological priming causes developmental gene network heterochronicity in autistic subject-derived neurons. *Nat Neurosci*. 2019;22(2):243–255. https://doi.org/10.1038/s41593-018-0295-x.
80. Voineagu I, Wang X, Johnston P, et al. Transcriptomic analysis of autistic brain reveals convergent molecular pathology. *Nature*. 2011;474(7351):380–384. https://doi.org/10.1038/nature10110.
81. Liu X, Han D, Somel M, et al. Disruption of an evolutionarily novel synaptic expression pattern in autism. *PLoS Biol*. 2016;14(9). https://doi.org/10.1371/journal.pbio.1002558.
82. Chow ML, Pramparo T, Winn ME, et al. Age-dependent brain gene expression and copy number anomalies in autism suggest distinct pathological processes at young versus mature ages. *PLoS Genet*. 2012;8(3). https://doi.org/10.1371/journal.pgen.1002592.
83. Garbett K, Ebert PJ, Mitchell A, et al. Immune transcriptome alterations in the temporal cortex of subjects with autism. *Neurobiol Dis*. 2008;30(3):303–311. https://doi.org/10.1016/j.nbd.2008.01.012.
84. Parikshak NN, Swarup V, Belgard TG, et al. Genome-wide changes in lncRNA, splicing, and regional gene expression patterns in autism. *Nature*. 2016;540(7633):423–427. https://doi.org/10.1038/nature20612.
85. Schwede M, Nagpal S, Gandal MJ, et al. Strong correlation of downregulated genes related to synaptic transmission and mitochondria in post-mortem autism cerebral cortex. *J Neurodev Disord*. 2018;10(1). https://doi.org/10.1186/s11689-018-9237-x.
86. Mahfouz A, Ziats MN, Rennert OM, Lelieveldt BPF, MJT R. Shared pathways among autism candidate genes determined by Co-expression network analysis of the developing human brain transcriptome. *J Mol Neurosci*. 2015;57(4):580–594. https://doi.org/10.1007/s12031-015-0641-3.
87. Simons Foundation Autism Research Initiative. *SFARI Gene Database*; 2014. https://gene.sfari.org/.
88. Amatya DN, Linker SB, Mendes APD, et al. Dynamical electrical complexity is reduced during neuronal differentiation in autism spectrum disorder. *Stem Cell Rep*. 2019;13(3):474–484. https://doi.org/10.1016/j.stemcr.2019.08.001.
89. Lancaster MA, Renner M, Martin CA, et al. Cerebral organoids model human brain development and microcephaly. *Nature*. 2013;501(7467):373–379. https://doi.org/10.1038/nature12517.
90. Mansour AA, Gonçalves JT, Bloyd CW, et al. An in vivo model of functional and vascularized human brain organoids. *Nat Biotechnol*. 2018;36(5):432–441. https://doi.org/10.1038/nbt.4127.
91. Velasco S, Kedaigle AJ, Simmons SK, et al. Individual brain organoids reproducibly form cell diversity of the human cerebral cortex. *Nature*. 2019;570(7762):523–527. https://doi.org/10.1038/s41586-019-1289-x.
92. Quadrato G, Nguyen T, Macosko EZ, et al. Cell diversity and network dynamics in photosensitive human brain organoids. *Nature*. 2017;545(7652):48–53. https://doi.org/10.1038/nature22047.

93. Sloan SA, Andersen J, Paşca AM, Birey F, Paşca SP. Generation and assembly of human brain region–specific three-dimensional cultures. *Nat Protoc*. 2018;13(9):2062–2085. https://doi.org/10.1038/s41596-018-0032-7.
94. Lancaster MA, Knoblich JA. Generation of cerebral organoids from human pluripotent stem cells. *Nat Protoc*. 2014;9(10):2329–2340. https://doi.org/10.1038/nprot.2014.158.
95. Luo C, Lancaster MA, Castanon R, Nery JR, Knoblich JA, Ecker JR. Cerebral organoids recapitulate epigenomic signatures of the human fetal brain. *Cell Rep*. 2016;17(12):3369–3384. https://doi.org/10.1016/j.celrep.2016.12.001.
96. Nageshappa S, Carromeu C, Trujillo CA, et al. Altered neuronal network and rescue in a human MECP2 duplication model. *Mol Psychiatr*. 2016;21(2):178–188. https://doi.org/10.1038/mp.2015.128.
97. Amiri A, Coppola G, Scuderi S, et al. Transcriptome and epigenome landscape of human cortical development modeled in organoids. *Science*. 2018;362(6420). https://doi.org/10.1126/science.aat6720.
98. Mariani J, Coppola G, Zhang P, et al. FOXG1-dependent dysregulation of GABA/glutamate neuron differentiation in autism spectrum disorders. *Cell*. 2015;162(2):375–390. https://doi.org/10.1016/j.cell.2015.06.034.
99. Culotta L, Penzes P. Exploring the mechanisms underlying excitation/inhibition imbalance in human iPSC-derived models of ASD. *Mol Autism*. 2020;11(1). https://doi.org/10.1186/s13229-020-00339-0.
100. Birey F, Andersen J, Makinson CD, et al. Assembly of functionally integrated human forebrain spheroids. *Nature*. 2017;545(7652):54–59. https://doi.org/10.1038/nature22330.
101. Bortone D, Polleux F. KCC2 expression promotes the termination of cortical interneuron migration in a voltage-sensitive calcium-dependent manner. *Neuron*. 2009;62(1):53–71. https://doi.org/10.1016/j.neuron.2009.01.034.
102. Marín O. Interneuron dysfunction in psychiatric disorders. *Nat Rev Neurosci*. 2012;13(2):107–120. https://doi.org/10.1038/nrn3155.
103. Bagley JA, Reumann D, Bian S, Lévi-Strauss J, Knoblich JA. Fused cerebral organoids model interactions between brain regions. *Nat Methods*. 2017;14(7):743–751. https://doi.org/10.1038/nmeth.4304.
104. Xiang Y, Tanaka Y, Patterson B, et al. Fusion of regionally specified hPSC-derived organoids models human brain development and interneuron migration. *Cell Stem Cell*. 2017;21(3):383.e7–398.e7. https://doi.org/10.1016/j.stem.2017.07.007.
105. Xiang Y, Tanaka Y, Cakir B, et al. hESC-derived thalamic organoids form reciprocal projections when fused with cortical organoids. *Cell Stem Cell*. 2019;24(3):487.e7–497.e7. https://doi.org/10.1016/j.stem.2018.12.015.
106. TCftSoD D. CNS drugs take 20% longer to develop and to approve vs. non-CNS drugs. *Tufts CSDD Impact Rep*. 2018;20(5).
107. Barber SC, Higginbottom A, Mead RJ, Barber S, Shaw PJ. An in vitro screening cascade to identify neuroprotective antioxidants in ALS. *Free Radic Biol Med*. 2009;46(8):1127–1138. https://doi.org/10.1016/j.freeradbiomed.2009.01.019.
108. Pollitt SK, Pallos J, Shao J, et al. A rapid cellular FRET assay of polyglutamine aggregation identifies a novel inhibitor. *Neuron*. 2003;40(4):685–694. https://doi.org/10.1016/S0896-6273(03)00697-4.
109. Swindell WR, Bojanowski K, Kindy MS, Chau RMW, Ko D. Correction: GM604 regulates developmental neurogenesis pathways and the expression of genes associated with amyotrophic lateral sclerosis. *Transl Neurodegener*. 2020. https://doi.org/10.1186/s40035-020-00207-0.
110. Christensen DH, Horváth-Puhó E, Thomsen RW, et al. Venous thromboembolism and risk of cancer in patients with diabetes mellitus. *J Diabetes Complicat*. 2016;30(4):603–607. https://doi.org/10.1016/j.jdiacomp.2016.01.011.
111. Pouton CW, Haynes JM. Embryonic stem cells as a source of models for drug discovery. *Nat Rev Drug Discov*. 2007;6(8):605–616. https://doi.org/10.1038/nrd2194.
112. Allsopp TE, Ebneth A, Cabrera-Socorro A. Deploying human pluripotent stem cells to treat central nervous system disorders: facts, challenges and realising the potential. *Stem Cell Res*. 2019;41. https://doi.org/10.1016/j.scr.2019.101581.
113. Sridharan B, Hubbs C, Llamosas N. A simple procedure for creating scalable phenotypic screening assays in human neurons. *Sci Rep*. 2019;9(1).
114. Horvath P, Aulner N, Bickle M, et al. Screening out irrelevant cell-based models of disease. *Nat Rev Drug Discov*. 2016;15(11):751–769. https://doi.org/10.1038/nrd.2016.175.
115. Christensen K, Roudnicky F, Patsch C, Burcin M. Requirements for using IPSC-based cell models for assay development in drug discovery. In: *Advances in Biochemical Engineering/Biotechnology*. 163. Switzerland: Springer Science and Business Media Deutschland GmbH; 2018:207–220. https://doi.org/10.1007/10_2017_23.

116. Musunuru K. Genome editing of human pluripotent stem cells to generate human cellular disease models. *Dis Models Mech*. 2013:896–904. https://doi.org/10.1242/dmm.012054.
117. Gurwitz D. Human iPSC-derived neurons and lymphoblastoid cells for personalized medicine research in neuropsychiatric disorders. *Dialogues Clin Neurosci*. 2016;18(3):267–276. https://www.ncbi.nlm.nih.gov/pmc/articles/PMC5067144/pdf/DialoguesClinNeurosci-18-267.pdf.
118. Quadrato G, Brown J, Arlotta P. The promises and challenges of human brain organoids as models of neuropsychiatric disease. *Nat Med*. 2016;22(11):1220–1228. https://doi.org/10.1038/nm.4214.
119. Soliman MA, Aboharb F, Zeltner N, Studer L. Pluripotent stem cells in neuropsychiatric disorders. *Mol Psychiatr*. 2017;22(9):1241–1249. https://doi.org/10.1038/mp.2017.40.
120. Bar-Nur O, Russ HA, Efrat S, Benvenisty N. Epigenetic memory and preferential lineage-specific differentiation in induced pluripotent stem cells derived from human pancreatic islet beta cells. *Cell Stem Cell*. 2011;9(1):17–23. https://doi.org/10.1016/j.stem.2011.06.007.
121. Kim K, Zhao R, Doi A, et al. Donor cell type can influence the epigenome and differentiation potential of human induced pluripotent stem cells. *Nat Biotechnol*. 2011;29(12):1117–1119. https://doi.org/10.1038/nbt.2052.
122. Perrera V, Martello G. How does reprogramming to pluripotency affect genomic imprinting? *Front Cell Dev Biol*. 2019;7:76. https://doi.org/10.3389/fcell.2019.00076.
123. Kilpinen H, Goncalves A, Leha A, et al. Common genetic variation drives molecular heterogeneity in human iPSCs. *Nature*. 2017;546(7658):370–375. https://doi.org/10.1038/nature22403.
124. Annaluru N, Ramalingam S, Chandrasegaran S. Rewriting the blueprint of life by synthetic genomics and genome engineering. *Genome Biol*. 2015;16(1). https://doi.org/10.1186/s13059-015-0689-y.
125. Martella A, Pollard SM, Dai J, Cai Y. Mammalian synthetic biology: time for big MACs. *ACS Synth Biol*. 2016;5(10):1040–1049. https://doi.org/10.1021/acssynbio.6b00074.
126. de Souza JS, Carromeu C, Torres LB, et al. IGF1 neuronal response in the absence of MECP2 is dependent on TRalpha 3. *Hum Mol Genet*. 2017;26(2):270–281. https://doi.org/10.1093/hmg/ddw384.
127. Williams EC, Zhong X, Mohamed A, et al. Mutant astrocytes differentiated from Rett syndrome patients-specific iPSCs have adverse effects on wildtype neurons. *Hum Mol Genet*. 2014;23(11):2968–2980. https://doi.org/10.1093/hmg/ddu008.
128. Ip JPK, Mellios N, Sur M. Rett syndrome: insights into genetic, molecular and circuit mechanisms. *Nat Rev Neurosci*. 2018;19(6):368–382. https://doi.org/10.1038/s41583-018-0006-3.
129. Sahin M, Sur M. Genes, circuits, and precision therapies for autism and related neurodevelopmental disorders. *Science*. 2015;350(6263). https://doi.org/10.1126/science.aab3897.
130. Schöndorf DC, Aureli M, McAllister FE, et al. iPSC-derived neurons from GBA1-associated Parkinson's disease patients show autophagic defects and impaired calcium homeostasis. *Nat Commun*. 2014. https://doi.org/10.1038/ncomms5028.

CHAPTER

10

Probing the electrophysiological properties of patient-derived neurons across neurodevelopmental disorders

Brooke Latour[1], *Teun Klein Gunnewiek*[1] *and Nael Nadif Kasri*[1,2]

[1]Department of Human Genetics, Radboudumc, Donders Institute for Brain, Cognition and Behaviour, Nijmegen, Netherlands; [2]Department of Cognitive Neuroscience, Radboudumc, Donders Institute for Brain, Cognition and Behaviour, Nijmegen, Netherlands

Induced pluripotent stem cells and modeling brain disorders

Modeling neurodevelopmental and neuropsychiatric disorders has long been accomplished through the use of animal models and heterologous expression systems. Despite groundbreaking discoveries, these systems present many challenges to accurately recapitulate human disease mechanisms and test patient-specific treatments. In 2006, Yamanaka et al. revolutionized the disease-modeling field by successfully reprogramming first murine and later human fibroblasts to a pluripotent cell-type, so called induced pluripotent stem cells (iPSC).[1,2] Since then, disorders of the central nervous system (CNS) have comprised a large portion of the published iPSC disease models, mostly because of the limited access, availability, and growth potential of primary patient-derived brain samples. In the past decade, technical advances have enabled the differentiation from iPSCs toward relatively pure populations of specific neuronal and non-neuronal cell subtypes, including excitatory neurons, specific subtypes of inhibitory neurons, astrocytes, oligodendrocytes, and microglia.[3] Depending on the differentiation protocols used, in vitro differentiation processes appear to follow physiological developmental stages of progression, lineage restriction, and neuronal/synaptic maturation that can be directly observed using appropriate reporter lines, live cell imaging modalities, as well as single-cell and neural network electrophysiological

Phenotyping of Human iPSC-derived Neurons
https://doi.org/10.1016/B978-0-12-822277-5.00010-9

recordings. The combination of these technologies allows for the recapitulation of step-by-step disease progression, with the added advantage of studying the cell-type-specific contributions to a disorder, both in 2D and 3D. Brain organoids, for example, have emerged as an exciting new system to model human brain development, recapitulating the early steps of brain development in health and disease in 3D.[4–6] Brain organoids can be generated through self-patterning or by directed differentiation.[7,8] Notably, at the genetic and epigenetic level, these brain organoids, when grown over an extended amount of time, show striking resemblance to human fetal brain, with many common epigenomic and transcriptional features.[9–12] Because brain organoids faithfully capture aspects of the structure and cytoarchitecture of the developing human brain, they are well suited to study early aspects of normal human brain development, and are also very useful to decipher mechanisms underlying genetic and nongenetic forms of early neurodevelopmental disorders, such as microcephaly and lissencephaly.[6,13,14] The use of patient- and control-derived iPSCs further provides researchers with an unlimited supply of material for molecular, proteomic, and electrophysiological studies, as well enabling the observation of the normally inaccessible neurodevelopmental processes. An important additional technological development closely linked to the success of iPSC methodology is the ability to edit the genome using genome-editing tools like clustered regularly interspaced palindromic repeats/CRISPR associated protein (CRISPR/Cas9).[15] These are powerful tools, as they enable us to generate iPSCs that are genetically identical, except for the target loci of interest, i.e., isogenic. In an ideal situation, isogenic lines should be used whenever possible, to show the causal link between genotype and phenotype.[16,17] This is because even when focusing on single gene variants with large effect sizes, uncovering disease-specific phenotypes can be complicated by commonly occurring genetic variation between donors [18] (Fig. 10.1).

The theoretical potential of using iPSCs for disease modeling is thus apparent[19] and mainly relies on coupling genetic variations to clinical data through molecular phenotypes in vitro. However, to fully harness the potential of iPSC-based models, there are remaining challenges: (i) To define the extent to which cell lineages are informative about disease-relevant pathways and phenotypes affected by genetic mutations. One of the main advantages of iPSC technology is the ability to generate several neural and non-neural cell-types starting from each individual iPSC line (i.e., control vs. patients). For most neurological disorders, clinical phenotypes are not due to deficits in single-cell-types or brain regions, but rather, they are a result of an intricate interaction between several cell-types within or outside interconnected brain regions. For example, microglia and astrocytes contribute to the pathophysiology of several disorders, including schizophrenia and ASD.[20–22] Therefore, generating iPSC-derived models that include all affected cell-types would be essential to understand the full spectrum of these disorders. (ii) To assess the feasibility of reliably identifying pathways and phenotypes beyond the variability inherent in the iPSC-based approach. (iii) To identify the molecular and cellular phenotypes that are most relevant and predictive for each individual disorder. Once these challenges are addressed, the full potential of iPSC-derived neuronal modeling can be realized, including improved success rates from drug screening efforts transitioning to the clinic to identify novel therapeutics.

Here, we will discuss some of the approaches that can be utilized to tackle the aforementioned challenges. In particular, we will focus on the methodologies that can be leveraged to functionally characterize iPSC-based neuronal model systems.

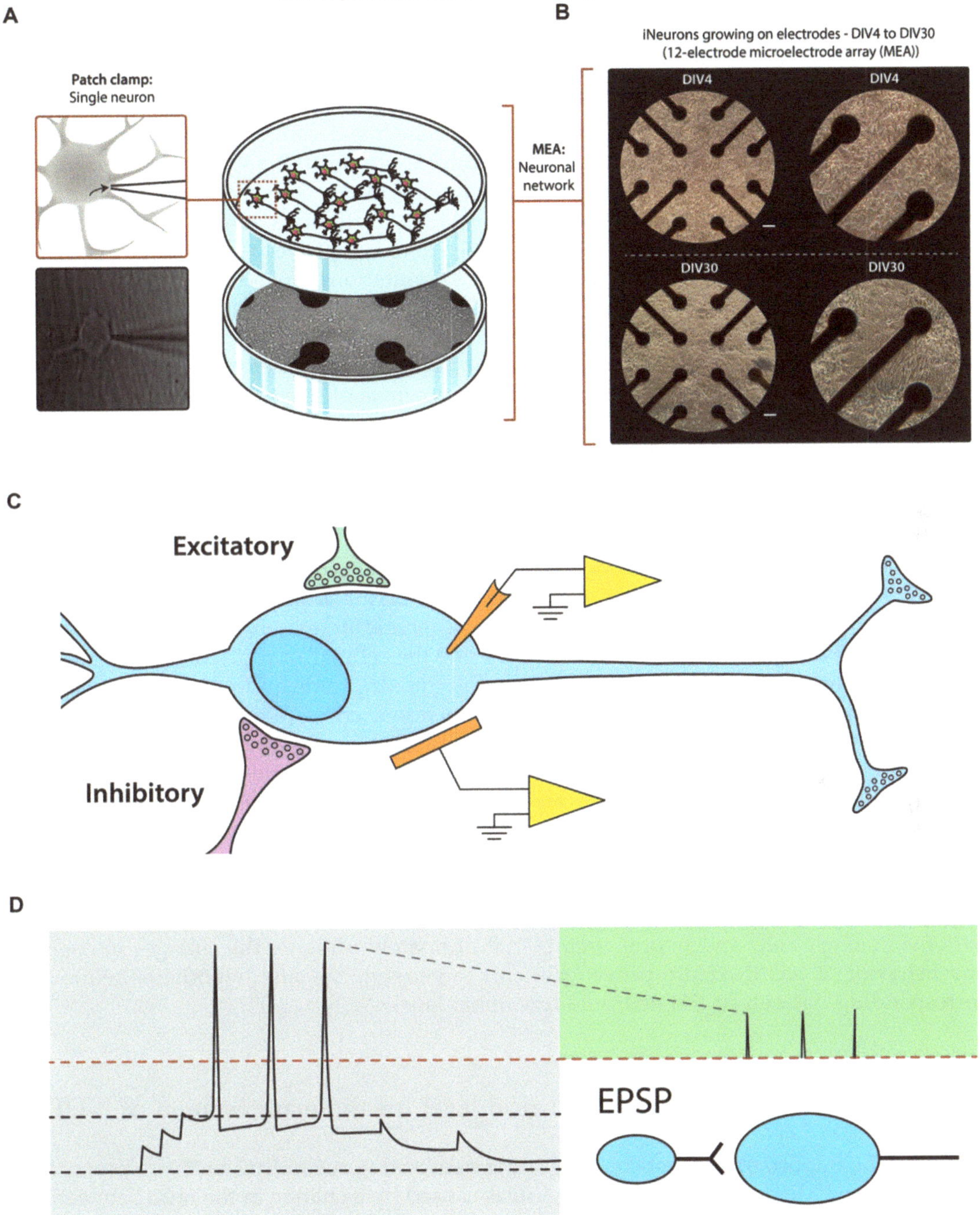

FIGURE 10.1 **Schematic overview of MEA recordings.** (A) Schematic image showing neurons grown on MEAs (B) Image showing iPSC-derived neurons grown on MEAs at two developmental stages, days in vitro (DIV) 4 and 30 (C) Schematic drawing of an intracellular (upper orange electrode) and extracellular (lower orange electrode) recording of a neuron. (D) A neuron endogenously generates a train of action potentials (APs) as shown in the intracellular recording (left). The extracellular recording (right) picks up the field potential generated by the APs, but cannot record subthreshold events due to attenuation of the signal.

Progressing from gene discovery to functional gene groupings to pathophysiology

Undoubtedly, the most significant progress over the last decade in the field of neurodevelopmental disorders (NDDs), including intellectual disability (ID) and autism spectrum disorders (ASD), has been the identification of individual genes for which Mendelian mutations are sufficient to cause NDDs. Currently, mutations have been identified in >1000 genes that give rise to distinct phenotypes and symptoms associated with ID and/or syndromic ASD.[23] Although the frequency of mutations in individual NDD genes is low, they collectively amount to more than 2%–3% of the general population. These NDD genes provide us with a molecular entry point into these disorders. They are involved in numerous functions, yet they seem to converge on a limited number of molecular complexes and cellular pathways. One of the most striking outcomes of these sequencing studies is the emergence of chromatin regulation as one of the two domains most affected, the other being synaptic function, whose involvement in NDDs had been largely predicted.[24–26] Interestingly, these enrichments are seen both in ID and ASD, which might reflect the high co-occurrence of the two cognitive impairments. Currently, over 100 chromatinopathy genes have been identified,[27] including genes that code for proteins with a direct enzymatic effect on histones or DNA, such as writers and erasers, but also more indirect regulators that are part of larger chromatin-remodeling complexes (e.g., BAF complex).[28]

Despite considerable progress in elucidating the genetic architecture of NDDs, a major gap exists between the genetic findings and deciphering the cellular or molecular pathobiology.[29] In particular, understanding which changes in the transcriptome and proteome downstream of the genetic insult have the most relevant functional consequences and whether those changes overlap in different patients is of great functional importance. Over the past years, many studies have been devoted to the categorization of NDD genes according to their biological and cellular function, protein-protein network, and human cell type expression, with the aim to identify points of convergence.[29,30] These studies provide valuable groupings of genes/proteins, but not an explicit hypothesis about their mechanistic relationship to the disorder as a whole. For example, grouping of NDD genes according to their biological function (e.g., chromatin remodeling), does not tell us how these genes lead to alterations in activity patterns in single cells and neural circuits that ultimately underlie the changes in cognition and behavior. It has therefore proven difficult to generate specific hypotheses without the corresponding datasets at the neuronal functional level.

Neuronal networks represent a logical level for the manifestation of NDDs

One of the important concepts of integrating neuronal function that first was proposed by Rubenstein and Merzenich,[31] posits that ASD is caused by a change in the ratio between excitation and inhibition, called the E/I balance. Since then, this hypothesis has been substantiated by a vast number of studies, and has also been implicated in other NDDs, such as ID and epilepsy.[32] Although relatively ambiguous in its meaning, the general concept of E/I balance is an attractive concept because experimental evidence supports the idea that cortical networks require a finely tuned coordination of excitatory and inhibitory inputs in order to function normally.[33] A failure to maintain E/I balance, independent of its molecular and/or cellular origin, would then underlie neuronal circuit dysfunction, as seen in many NDDs.[34] This concept was initially supported by the observations that (i) many NDD genes code for pre- or postsynaptic proteins,[26] and (ii) regardless of the primary function of the affected gene, structural and functional excitatory synapse phenotypes have been observed in mouse models for NDD genes.[35] These changes go in both direction (increasing and decreasing synaptic numbers and/or strength), suggesting that in NDDs there is not a specific preference in the directionality of the E/I imbalance. More recently, the focus has shifted to the inhibitory side of the E/I balance, in particular to fast-spiking parvalbumin-positive (PV^+) GABAergic neurons.[32] Alterations in E/I balance have mostly been studied at the single cell level and it has proven challenging to probe and understand E/I imbalance at the circuit level. For example, in a very elegant study, they demonstrated that in four well-validated mouse models for ASD (Pitt Hopkins-like syndrome caused by mutations in Contactin associated protein 2 (*Cntnap2*)), Fragile X syndrome caused by mutations in FMRP Translational Regulator 1 (*Fmr1*), Tuberous Sclerosis Complex caused by mutations in Tuberin (*Tsc2*), and *16p11.2* deletion syndrome mice, there was a marked reduction of inhibition and a smaller reduction of excitation in feed-forward circuits of the visual cortex, which led to an increase in the E/I ratio.[36] At the circuit level, however, they did not find that increased E/I ratio translated into a greater firing rate. They therefore concluded that the E/I alterations could be caused by a compensatory homeostatic response. This study illustrates the importance of understanding the impact of single cell E/I changes at the network level. The network level might also represent a logical level for the manifestation of disease-linked phenotypes, in which changes in diverse genes, protein networks, cell types, or developmental stages may elicit specific or similar changes in neuronal network function, in other words, dysfunctional neuronal networks may show common modes of dysfunction that arise from very different molecular and cellular level pathways.

Micro-electrode arrays as a scalable high-throughput functional assay

The advent of iPSCs now offers unparalleled opportunities for research into the key molecular and cellular pathophysiological mechanisms of NDDs at the level of synapse and network dysfunction.[37] The advantage is that disease-relevant cells can be generated to track disease progression in the genetic background of each patient. The main challenges in modeling NDDs with iPSCs have been the generation of faithful and robust differentiation protocols, together with the establishment of reliable and meaningful readouts.

A scalable functional assay that can be used to detect synaptic and network activities is critical for categorizing synaptic and network phenotypes and for identifying potential converging functional pathways among the hundreds of NDD genes. Widely used state-of-the-art technologies present various limitations due to a trade-off between throughput and resolution. For example, whole-cell patch-clamp technique is characterized by a very low experimental throughput, being able to record precisely from a single neuron, but only from few neurons at a time. In addition, patch-clamp recordings are invasive and only allow for the recording of a neuron for a brief period of time (maximum a few hours), making it difficult to investigate long-term neurodevelopmental processes. On the other end of the spectrum, calcium imaging of neuronal populations allows us to measure network activity at a high spatial resolution. Multi-well systems enable high-throughput screening of iPSC-derived neuronal cultures[38]; however, the temporal limitations of calcium imaging often limit the acquisition of information about firing properties of single neurons. Moreover, fluorescent imaging can only record from neurons for a brief time span, due to the intrinsic phototoxicity and photobleaching of fluorescent indicators. In addition, the platforms available for calcium imaging analysis aren't always easy to adapt to handling long, repeated recordings, or recordings across multi-well plates. An intermediate, label-free, and long-term recording approach consists of using micro-electrode arrays (MEAs) to record extracellular potentials generated by neuronal spiking activity (Fig. 10.1).

MEAs are devices containing a flat array of electrodes on which cells can be grown. The electrodes can detect large deflections in voltage potential of nearby cells, as occurring during action potential firing of neurons. MEAs are capable of measuring the activity of hundreds of neurons simultaneously (Fig. 10.1). The first MEAs consisted of 30–36 electrodes that were created by the etching of thin gold films on glass and were capable of picking up spikes of up to 360 μV with a signal-to-noise ratio as high as 15 to 1.[39–41] Since then, many improvements have been made and MEAs can now have thousands of electrodes (high density (HD)-MEAs)[42–46] and be built on multi-well plates in which each well is embedded with electrodes capable of recording and/or stimulating neuronal networks.

There are several advantages that make MEA recording an attractive readout system for iPSC-derived neuronal cultures: (i) neuronal network measurements with MEAs allow for non-invasive and repetitive neuronal phenotyping while the neurons traverse distinct neurodevelopmental stages. These recordings can be performed at high temporal resolution (10 kHz), a level comparable to patch clamp recordings. (ii) MEA recordings provide an integrative measurement as changes in gene expression, neuronal morphology, intrinsic properties, synaptic connectivity or E/I balance are reflected at the network level, providing an unbiased assessment of the neuronal properties, in which no prior assumption or hypothesis is required. Early reports have shown that rat cortical neurons cultured on MEAs develop various patterns of spontaneous behaviors, including network oscillations at frequencies similar to those found in vivo.[47,48] In particular, neurons cultured in vitro develop, over time, neuronal networks that exhibit synchronized bursting events, wherein most of the neurons in the network spike together. The developmental trajectories seen in rodent cultures have also been observed in iPSC-derived neuronal cultures. In particular, neuronal networks from healthy controls exhibit random events in the form of action potential spikes and bursts during early development.[49–51] As the cultures mature, bursts appear simultaneously in most of the channels, forming synchronous events, i.e., network bursts. Furthermore, the neuronal network synchronous activity develops from stochastic to a typical pattern with a stable state of activity, indicating a functionally "mature" neuronal network. These stereotyped developmental trajectories depend on the cell type and differentiation protocol used, where neuronal cultures can reach a maturation stage after more than 100 days in culture, or as fast as 3 weeks.[50,52] It is therefore important to establish the developmental trajectory for each of the chosen protocols. Mossink et al. recently showed that by using a widely used direct differentiation protocol involving the overexpression of *Ngn2*, a consistent developmental profile was observable when measured on MEAs.[49] Additionally, they observed little difference between 10 different control iPSC lines (five males, five females) when comparing 17 independent MEA parameters, making this system ideally suited to phenotype iPSCs from patients with CNS developmental disorders and as a drug screening platform.

Although MEAs can reliably record action potentials from a neuronal culture, they do not detect subthreshold activity in neurons (Fig. 10.1). Therefore, it can be useful to combine MEA recordings with other techniques, like calcium imaging[53] and single-cell patch-clamp electrophysiology. Vardi et al. performed single-cell patch clamp and MEA recordings simultaneously to discover neurons that increase their firing rate within the downtime of a bursts in which most of the neurons participating in the burst show reduced firing.[54] Hondebrink et al. used MEAs to investigate the effect of drugs on neuronal activity and then used calcium imaging to investigate the mechanisms of the drugs, by looking at the effect of the drug upon calcium transients after the addition of different receptor agonists and antagonists.[55] Kumar et al. combined simultaneous recording of MEA with whole-field synaptic glutamate imaging (iGluSnFr) to reveal astrocyte regulation of neuronal network bursts.[56] Together, these studies show how MEA recordings can complement other functional endpoints to further examine mechanisms behind network dysfunction in neuronal networks from patients with neurological disorders.

Phenotyping NDD patient-derived neurons using MEA recordings

Because of the abovementioned advantages of neuronal network recordings using MEAs, they have become an attractive tool in the field to study the functional consequences of NDD mutations on neuronal activity and as a semi-high throughput drug testing platform. Several studies have been performed on iPSC-derived neuronal network from NDDs. Below we will highlight some of these studies.

Fragile X and Rett syndrome

Fragile X syndrome (FXS) and Rett syndrome are among the two most common monogenic NDDs caused by mutations in respectively Fragile X mental retardation protein (FMRP) Translational Regulator 1 (*FMR1*) and Methyl-CpG Binding Protein 2 (*MECP2*). Loss of *FMR1* (FXS) in iPSC-derived neuronal networks resulted in an increased firing rate, and reactivation of *FMR1* reversed this hyperactive phenotype in FXS neurons on MEA.[57–59] A similar increase in the neuronal firing rate was found on MEA using neurons from $Fmr1^{-/-}$ knockout mice, as well as knock-in mice carrying 170 CGG repeats in the *Fmr1* promoter.[60,61] Neuronal networks derived from patients carrying a *MECP2* duplication[62] showed an increased firing rate and network burst rate as compared to controls. This increase in network activity correlated well with an increase in glutamatergic synapses. Conversely, the same research group recently found reduced network activity (reduced spike frequency) in patient-derived networks in 2D cultures, but also in 3D brain organoids, of patients with classical Rett syndrome due to MECP2 loss of function. Using MEAs as screening platform, they confirmed the effectiveness of two compounds (Nefiracetam and PHA 543613) in increasing spike frequency in MECP2KO neuronal networks.[63] In idiopathic cases of ASD with associated macrocephaly, Marchetto et al. showed that ASD neuronal networks were hypoactive and that this phenotype could be rescued with IGF-1 in a patient-specific manner.[64] MEA recording of these patient lines also revealed the variability that can be observed between patients in terms of phenotypes and drug responses, highlighting the sensitivity of MEA recordings.

Kleefstra syndrome

Similar as described above, Frega et al. used MEAs to study how heterozygous loss of Euchromatic histone-lysine N-methyltransferase 1 (*EHMT1*), a gene causative for Kleefstra syndrome, affects circuitry function in both human and rodent neuronal networks.[65] Kleefstra syndrome is characterized by ID, ASD, hypotonia, and dysmorphic features.[66] Knockdown of *Ehmt1* in rat neuronal networks resulted in hyperactive neuronal networks that showed an irregular network organization and long network bursts.[67] A similar neuronal network phenotype was described in *Ehmt1*$^{+/-}$ mice neuronal networks.[65] Interestingly, human neuronal networks composed of only glutamatergic neurons shared the same irregular network organization and long burst phenotype, even though these networks lacked several neuronal subtypes that are present in rodent networks.[65] This suggests that the pathogenic mechanism causing these long and irregular network bursts is evolutionarily conserved between mice, rats, and humans, and this phenotype is potentially driven by glutamatergic neuron dysfunction. Further work is needed to determine the contribution of GABAergic neurons or other cell types in the neuronal network disruptions likely occurring in patients. In line with pharmacological studies performed in rodents, where NMDA receptors were shown to regulate the duration and extinction of the network burst, the study utilizing human iPSC-derived neurons also related the network phenotypes to an increase in NMDAR function.[65] These finds are critical to begin thinking about pathways and mechanisms available for reversing network-level dysfunction in patients.

Neuronal networks as converging pathways?

The abovementioned studies illustrate that MEAs have the sensitivity to detect distinct network phenotypes between healthy and disease networks, and these phenotypes may be conserved between human and rodent neurons. In many cases, synapse number and function (e.g., AMPAR and NMDAR) correlated well with functional synchronized activity in human iPSC-derived neurons, revealing the potential of MEAs to study NDDs that involve synaptic deficits (e.g., E/I imbalances). The ability to generate composite networks composed of several types of neurons, including glutamatergic and GABAergic neurons,[68] further allows us to interrogate the contribution of each cell type to a network phenotype and, for example, measure E/I balance at the network level. Using networks consisting of different ratios of glutamatergic and GABAergic neurons Mossink et al. recently demonstrated that basal network properties were scalable with the number of GABAergic neurons present.[69] They identified specific network parameters, including mean firing rate and burst duration, that were dependent on this ratio, which they could leverage to study E/I imbalance at the network level in networks lacking the ADHD and ASD-associated gene Cadherin13 (*CDH13*).

MEA recordings, which measure the output (spike behavior) of thousands of cells, can capture changes at different functional levels of a network. Changes in cellular parameters, such as neuronal morphology, intrinsic excitability, and synapse function will ultimately affect how these neurons will fire action potentials. Each of these cellular parameters can be affected by shared or distinct molecular mechanisms. For example, synaptic E/I imbalance can be caused by hundreds of different molecular mechanisms affecting the pre- and/or post-synaptic function of excitatory, inhibitory neuron, or even glial cells. Neuronal network measurements are thus ideally suited for the investigation of whether there are points of convergence between the different NDDs and whether there is an evolutionary conserved molecular mechanism explaining specific network phenotypes. Recent studies have started to investigate whether different genes associated with the same or different NDD showed a similar or distinct neuronal network phenotype. For example, Frega et al. found that in rat cortical networks deficient for genes that are associated with Kleefstra spectrum syndrome,[70] *Ehmt1*, *Mll3*, *Mbd5*, and *Smarcb1*, all genotypes displayed hyperactive neuronal networks, albeit with some differences in individual network parameters between each of the genes.[71] Interestingly, they showed that the convergence at the network level was not fully recapitulated at the transcriptomic level, even seeing opposite changes in gene transcription between the individual Kleefstra genes (i.e., *Ehmt1* vs. *SmarcB1* and *MLL3*). This illustrates that distinct changes in gene expression and/or molecular mechanisms caused by different NDD genes might still converge at the neuronal network level.

Likewise, in a systematic screen of 10 ASD susceptibility genes, using MEA recordings, Denault et al. detected reduced network activity in a subset of gene models including *ATRX*, *AFF2*, *KCNQ2*, and *SCN2A*.[72] In a more recent study, they investigated *Ngn2*-induced neuronal network behavior of 53 different iPSC-derived glutamatergic neuronal lines from 25 participants from 12 unrelated families with ASD and identified a hyperactive network signature for patients with mutations in either *CNTN5* or *EHMT2*.[73] In neuronal networks with a *CNTN5* mutation, the network phenotype exhibited differences in network burst rate, whereas in networks with an *EHMT2* mutation differences in the mean firing rate were detected.[73] Together, this strengthens the evidence that early disease-associated network phenotypes can be revealed using MEAs. These platforms also allow for pharmacological manipulation of the cultures to better understand the molecular mechanisms driving the changes in neuronal activity.

The way forward

The combination of MEA recordings with patient-specific neuronal model systems and genome editing is an extremely powerful tool to understand the pathophysiology of neurological disorders and has great promise for translational applications.[74] However, it remains challenging to relate changes in neuronal network activity, measured in vitro, to the cognitive deficits observed in patients. Treatment strategies identified to be successful in restoring network activities will ultimately reveal the real predictive validity of these model systems.

In an effort to bridge this translational gap, Trujillo et al. recently used MEA recordings to measure field potentials in several month-old developing brain organoids. They showed that using the complex oscillatory waves that emerged from these organoids, they could model some aspects of early human brain network development.[51]

Epilepsy is one specific group of NDDs for which the gap between in vitro network activity and human phenotypes might be easier to bridge. Epilepsy is characterized by hyperactive neuronal networks and recent studies modeling genetic causes of epilepsies have been able to recapitulate these hyperactive network phenotypes.[74,75] For example, iPSC-derived neuronal networks from patients with mutations in *SCN8A* showed hyperactive network phenotypes and complex burst activities.[76] More importantly, using the MEA recordings they could test the efficiency of a novel anti-epileptic drug, riluzole, in a patient-specific manner, and predict its efficacy for the corresponding patients. This illustrates the potential of using this type of strategy for predicting the efficacy of anti-epileptic drugs for epilepsy patients. Trial-and-error testing of existing anti-epileptic drugs could then be done in vitro instead of on patients.

One can anticipate that the number of studies using the MEA platform to phenotype iPSC-derived models and to screen drugs will continue increasing. Advances in MEA technology will make high-throughput experiments easier, especially in multi-well systems and with newer high density electrode systems. As analysis of complex network activity will become more accessible, we will also see an increase in the quality and richness of MEA datasets, which are currently not yet analyzed to their fullest potential. In most studies, MEA data analysis is often limited to a description of the parameters describing the general activity (mean firing rate), leaving aside the more complex, but relevant network organization and network activity. Because tens of independent MEA parameters, that together describe a neuronal network phenotype, can be extracted from MEA recordings, one could use the ensemble of MEA parameters to create a detailed phenotypic score map in which all independent MEA parameters are included. These could also be complemented with plasticity experiments, such as Hebbian and homeostatic forms of plasticity.[77] This approach might be useful to systematically phenotype a broad collection of NDD genes in an effort to define subsets of interrelated genes and assess how neuronal networks are influenced by these gene sets.

Acknowledgments

This work was supported by grants from: the Netherlands Organization for Health Research and Development ZonMw grant 91217055 (to N.N.K); SFARI grant 610264 (to N.N.K); ERA-NET NEURON DECODE! grant (NWO) 013.18.001 (to N.N.K).

References

1. Takahashi K, Yamanaka S. Induction of pluripotent stem cells from mouse embryonic and adult fibroblast cultures by defined factors. *Cell*. 2006;126(4):663–676. https://doi.org/10.1016/j.cell.2006.07.024.
2. Takahashi K, Tanabe K, Ohnuki M, et al. Induction of pluripotent stem cells from adult human fibroblasts by defined factors. *Cell*. 2007;131(5):861–872. https://doi.org/10.1016/j.cell.2007.11.019.
3. Mertens J, Marchetto MC, Bardy C, Gage FH. Evaluating cell reprogramming, differentiation and conversion technologies in neuroscience. *Nat Rev Neurosci*. 2016;17(7):424–437. https://doi.org/10.1038/nrn.2016.46.
4. Camp JG, Treutlein B. Human development: advances in mini-brain technology. *Nature*. 2017;545(7652):39–40. https://doi.org/10.1038/545039a.

5. Chiaradia I, Lancaster MA. Brain organoids for the study of human neurobiology at the interface of in vitro and in vivo. *Nat Neurosci*. 2020;23(12):1496–1508. https://doi.org/10.1038/s41593-020-00730-3.
6. Lancaster MA, Renner M, Martin CA, et al. Cerebral organoids model human brain development and microcephaly. *Nature*. 2013;501(7467):373–379. https://doi.org/10.1038/nature12517.
7. Amin ND, Pasca SP. Building models of brain disorders with three-dimensional organoids. *Neuron*. 2018;100(2):389–405. https://doi.org/10.1016/j.neuron.2018.10.007.
8. Sloan SA, Andersen J, Pasca AM, Birey F, Pasca SP. Generation and assembly of human brain region–specific three-dimensional cultures. *Nat Protoc*. 2018;13(9):2062–2085. https://doi.org/10.1038/s41596-018-0032-7.
9. Luo C, Lancaster MA, Castanon R, Nery JR, Knoblich JA, Ecker JR. Cerebral organoids recapitulate epigenomic signatures of the human fetal brain. *Cell Rep*. 2016;17(12):3369–3384. https://doi.org/10.1016/j.celrep.2016.12.001.
10. Pollen AA, Bhaduri A, Andrews MG, et al. Establishing cerebral organoids as models of human-specific brain evolution. *Cell*. 2019;176(4). https://doi.org/10.1016/j.cell.2019.01.017, 743.e17–756.e17.
11. Quadrato G, Nguyen T, Macosko EZ, et al. Cell diversity and network dynamics in photosensitive human brain organoids. *Nature*. 2017;545(7652):48–53. https://doi.org/10.1038/nature22047.
12. Trevino AE, Sinnott-Armstrong N, Andersen J, et al. Chromatin accessibility dynamics in a model of human forebrain development. *Science*. 2020;367(6476). https://doi.org/10.1126/science.aay1645.
13. Iefremova V, Manikakis G, Krefft O, et al. An organoid-based model of cortical development identifies non-cell-autonomous defects in Wnt signaling contributing to Miller-Dieker syndrome. *Cell Rep*. 2017;19(1):50–59. https://doi.org/10.1016/j.celrep.2017.03.047.
14. Qian X, Nguyen HN, Song MM, et al. Brain-Region-specific organoids using mini-bioreactors for modeling ZIKV exposure. *Cell*. 2016;165(5):1238–1254. https://doi.org/10.1016/j.cell.2016.04.032.
15. Jinek M, Chylinski K, Fonfara I, Hauer M, Doudna JA, Charpentier E. A programmable dual-RNA-guided DNA endonuclease in adaptive bacterial immunity. *Science*. 2012;337(6096):816–821. https://doi.org/10.1126/science.1225829.
16. Fernando MB, Ahfeldt T, Brennand KJ. Modeling the complex genetic architectures of brain disease. *Nat Genet*. 2020;52(4):363–369. https://doi.org/10.1038/s41588-020-0596-3.
17. Michael Deans PJ, Brennand KJ. Applying stem cells and CRISPR engineering to uncover the etiology of schizophrenia. *Curr Opin Neurobiol*. 2021;69:193–201. https://doi.org/10.1016/j.conb.2021.04.003.
18. Hoffman GE, Schrode N, Flaherty E, Brennand KJ. New considerations for hiPSC-based models of neuropsychiatric disorders. *Mol Psychiatr*. 2019;24(1):49–66. https://doi.org/10.1038/s41380-018-0029-1.
19. Yamanaka S. A fresh look at iPS cells. *Cell*. 2009;137(1):13–17. https://doi.org/10.1016/j.cell.2009.03.034.
20. Liddelow S, Barres B. SnapShot: astrocytes in health and disease. *Cell*. 2015;162(5). https://doi.org/10.1016/j.cell.2015.08.029, 1170.e1–1170.e1.
21. Salter MW, Stevens B. Microglia emerge as central players in brain disease. *Nat Med*. 2017;23(9):1018–1027. https://doi.org/10.1038/nm.4397.
22. Scuderi C, Verkhratsky A. The role of neuroglia in autism spectrum disorders. In: *Progress in Molecular Biology and Translational Science*. 173. Italy: Elsevier B.V.; 2020:301–330. https://doi.org/10.1016/bs.pmbts.2020.04.011.
23. Kochinke K, Zweier C, Nijhof B, et al. Systematic phenomics analysis deconvolutes genes mutated in intellectual disability into biologically coherent modules. *Am J Hum Genet*. 2016;98(1):149–164. https://doi.org/10.1016/j.ajhg.2015.11.024.
24. De Rubeis S, He X, Goldberg AP, et al. Synaptic, transcriptional and chromatin genes disrupted in autism. *Nature*. 2014;515(7526):209–215. https://doi.org/10.1038/nature13772.
25. Satterstrom FK, Kosmicki JA, Wang J, et al. Large-scale exome sequencing study implicates both developmental and functional changes in the neurobiology of autism. *Cell*. 2020;180(3). https://doi.org/10.1016/j.cell.2019.12.036, 568.e23–584.e23.
26. Van Bokhoven H. Genetic and epigenetic networks in intellectual disabilities. *Annu Rev Genet*. 2011;45:81–104. https://doi.org/10.1146/annurev-genet-110410-132512.
27. Ciptasari U, van Bokhoven H. The phenomenal epigenome in neurodevelopmental disorders. *Hum Mol Genet*. 2020;29(1):R42–R50. https://doi.org/10.1093/hmg/ddaa175.
28. Iwase S, Bérubé NG, Zhou Z, et al. Epigenetic etiology of intellectual disability. *J Neurosci*. 2017;37(45):10773–10782. https://doi.org/10.1523/JNEUROSCI.1840-17.2017.

29. Willsey AJ, Morris MT, Wang S, et al. The psychiatric cell map initiative: a convergent systems biological approach to illuminating key molecular pathways in neuropsychiatric disorders. *Cell*. 2018;174(3):505–520. https://doi.org/10.1016/j.cell.2018.06.016.
30. Iakoucheva LM, Muotri AR, Sebat J. Getting to the cores of autism. *Cell*. 2019;178(6):1287–1298. https://doi.org/10.1016/j.cell.2019.07.037.
31. Rubenstein JLR, Merzenich MM. Model of autism: increased ratio of excitation/inhibition in key neural systems. *Gene Brain Behav*. 2003;2(5):255–267. https://doi.org/10.1034/j.1601-183X.2003.00037.x.
32. Selten M, van Bokhoven H, Nadif Kasri N. Inhibitory control of the excitatory/inhibitory balance in psychiatric disorders. *F1000Research*. 2018;7:23. https://doi.org/10.12688/f1000research.12155.1.
33. He H, Cline HT. What is excitation/inhibition and how is it regulated? A case of the elephant and the Wisemen. *J Exp Neurosci*. 2019;13. https://doi.org/10.1177/1179069519859371, 117906951985937.
34. Mullins C, Fishell G, Tsien RW. Unifying views of autism spectrum disorders: a consideration of autoregulatory feedback loops. *Neuron*. 2016;89(6):1131–1156. https://doi.org/10.1016/j.neuron.2016.02.017.
35. Pavlowsky A, Chelly J, Billuart P. Major synaptic signaling pathways involved in intellectual disability. *Mol Psychiatr*. 2012;17(7):663. https://doi.org/10.1038/mp.2012.79.
36. Antoine MW, Langberg T, Schnepel P, Feldman DE. Increased excitation-inhibition ratio stabilizes synapse and circuit excitability in four autism mouse models. *Neuron*. 2019;101(4). https://doi.org/10.1016/j.neuron.2018.12.026, 648.e4–661.e4.
37. Linda K, Fiuza C, Nadif Kasri N. The promise of induced pluripotent stem cells for neurodevelopmental disorders. *Prog Neuro Psychopharmacol Biol Psychiatr*. 2018;84:382–391. https://doi.org/10.1016/j.pnpbp.2017.11.009.
38. Negraes PD, Trujillo CA, Yu NK, et al. Altered network and rescue of human neurons derived from individuals with early-onset genetic epilepsy. *Mol Psychiatr*. 2021;26(11):7047–7068. https://doi.org/10.1038/s41380-021-01104-2.
39. Gross GW, Williams AN, Lucas JH. Recording of spontaneous activity with photoetched microelectrode surfaces from mouse spinal neurons in culture. *J Neurosci Methods*. 1982;5(1–2):13–22. https://doi.org/10.1016/0165-0270(82)90046-2.
40. Pine J. Recording action potentials from cultured neurons with extracellular microcircuit electrodes. *J Neurosci Methods*. 1980;2(1):19–31. https://doi.org/10.1016/0165-0270(80)90042-4.
41. Thomas CA, Springer PA, Loeb GE, Berwald-Netter Y, Okun LM. A miniature microelectrode array to monitor the bioelectric activity of cultured cells. *Exp Cell Res*. 1972;74(1):61–66. https://doi.org/10.1016/0014-4827(72)90481-8.
42. Gandolfo M, Maccione A, Tedesco M, Martinoia S, Berdondini L. Tracking burst patterns in hippocampal cultures with high-density CMOS-MEAs. *J Neural Eng*. 2010;7(5):056001. https://doi.org/10.1088/1741-2560/7/5/056001.
43. Obien MEJ, Deligkaris K, Bullmann T, Bakkum DJ, Frey U. Revealing neuronal function through microelectrode array recordings. *Front Neurosci*. 2015;9:423. https://doi.org/10.3389/fnins.2014.00423.
44. Ronchi S, Buccino AP, Prack G, et al. Electrophysiological phenotype characterization of human iPSC-derived neuronal cell lines by means of high-density microelectrode arrays. *Adv Biol*. 2021;5(3). https://doi.org/10.1002/adbi.202000223.
45. Spira ME, Hai A. Multi-electrode array technologies for neuroscience and cardiology. *Nat Nanotechnol*. 2013;8(2):83–94. https://doi.org/10.1038/nnano.2012.265.
46. Yuan X, Schröter M, Obien MEJ, et al. Versatile live-cell activity analysis platform for characterization of neuronal dynamics at single-cell and network level. *Nat Commun*. 2020;11(1). https://doi.org/10.1038/s41467-020-18620-4.
47. Hales CM, Zeller-Townson R, Newman JP, Shoemaker JT, Killian NJ, Potter SM. Stimulus-evoked high frequency oscillations are present in neuronal networks on microelectrode arrays. *Front Neural Circ*. 2012;2012:1–10. https://doi.org/10.3389/fncir.2012.00029.
48. Wagenaar DA, Pine J, Potter SM. An extremely rich repertoire of bursting patterns during the development of cortical cultures. *BMC Neurosci*. 2006;7. https://doi.org/10.1186/1471-2202-7-11.
49. Mossink B, Verboven AHA, van Hugte EJH, et al. Human neuronal networks on micro-electrode arrays are a highly robust tool to study disease-specific genotype-phenotype correlations in vitro. *bioRxiv*. 2021. https://doi.org/10.1101/2021.01.20.427439.

50. Odawara A, Katoh H, Matsuda N, Suzuki I. Physiological maturation and drug responses of human induced pluripotent stem cell-derived cortical neuronal networks in long-term culture. *Sci Rep*. 2016;6. https://doi.org/10.1038/srep26181.

51. Trujillo CA, Gao R, Negraes PD, et al. Complex oscillatory waves emerging from cortical organoids model early human brain network development. *Cell Stem Cell*. 2019;25(4). https://doi.org/10.1016/j.stem.2019.08.002, 558.e7–569.e7.

52. Frega M, Van Gestel SHC, Linda K, et al. Rapid neuronal differentiation of induced pluripotent stem cells for measuring network activity on micro-electrode arrays. *JoVE*. 2017;2017(119). https://doi.org/10.3791/54900.

53. Shew WL, Bellay T, Plenz D. Simultaneous multi-electrode array recording and two-photon calcium imaging of neural activity. *J Neurosci Methods*. 2010;192(1):75–82. https://doi.org/10.1016/j.jneumeth.2010.07.023.

54. Vardi R, Goldental A, Sardi S, Sheinin A, Kanter I. Simultaneous multi-patch-clamp and extracellular-array recordings: single neuron reflects network activity. *Sci Rep*. 2016;6:36228. https://doi.org/10.1038/srep36228.

55. Hondebrink L, Kasteel EEJ, Tukker AM, Wijnolts FMJ, Verboven AHA, Westerink RHS. Neuropharmacological characterization of the new psychoactive substance methoxetamine. *Neuropharmacology*. 2017;123:1–9. https://doi.org/10.1016/j.neuropharm.2017.04.035.

56. Kumar R, Huang Y-T, Chen C-C, Tzeng S-F, Chan C-K. Astrocytic regulation of synchronous bursting in cortical cultures: from local to global. *Cerebral Cortex Commun*. 2020;1(1). https://doi.org/10.1093/texcom/tgaa053.

57. Graef JD, Wu H, Ng C, et al. Partial FMRP expression is sufficient to normalize neuronal hyperactivity in Fragile X neurons. *Eur J Neurosci*. 2020;51(10):2143–2157. https://doi.org/10.1111/ejn.14660.

58. Liu XS, Wu H, Krzisch M, et al. Rescue of fragile X syndrome neurons by DNA methylation editing of the FMR1 gene. *Cell*. 2018;172(5). https://doi.org/10.1016/j.cell.2018.01.012, 979.e6–992.e6.

59. Utami KH, Skotte NH, Colaço AR, et al. Integrative analysis identifies key molecular signatures underlying neurodevelopmental deficits in fragile X syndrome. *Biol Psychiatr*. 2020;88(6):500–511. https://doi.org/10.1016/j.biopsych.2020.05.005.

60. Cao Z, Hulsizer S, Tassone F, et al. Clustered burst firing in FMR1 premutation hippocampal neurons: amelioration with allopregnanolone. *Hum Mol Genet*. 2012;21(13):2923–2935. https://doi.org/10.1093/hmg/dds118.

61. Moskalyuk A, Van De Vijver S, Verstraelen P, De Vos WH, Kooy RF, Giugliano M. Single-cell and neuronal network alterations in an in vitro model of fragile X syndrome. *Cerebr Cortex*. 2020;30(1):31–46. https://doi.org/10.1093/cercor/bhz068.

62. Nageshappa S, Carromeu C, Trujillo CA, et al. Altered neuronal network and rescue in a human MECP2 duplication model. *Mol Psychiatr*. 2016;21(2):178–188. https://doi.org/10.1038/mp.2015.128.

63. Trujillo CA, Adams JW, Negraes PD, et al. Pharmacological reversal of synaptic and network pathology in human MECP2-KO neurons and cortical organoids. *EMBO Mol Med*. 2021;13(1). https://doi.org/10.15252/emmm.202012523.

64. Marchetto MC, Belinson H, Tian Y, et al. Altered proliferation and networks in neural cells derived from idiopathic autistic individuals. *Mol Psychiatr*. 2017;22(6):820–835. https://doi.org/10.1038/mp.2016.95.

65. Frega M, Linda K, Keller JM, et al. Neuronal network dysfunction in a model for Kleefstra syndrome mediated by enhanced NMDAR signaling. *Nat Commun*. 2019;10(1). https://doi.org/10.1038/s41467-019-12947-3.

66. Kleefstra T, Brunner HG, Amiel J, et al. Loss-of-function mutations in Euchromatin histone methyl transferase 1 (EHMT1) cause the 9q34 subtelomeric deletion syndrome. *Am J Hum Genet*. 2006;79(2):370–377. https://doi.org/10.1086/505693.

67. Martens MB, Frega M, Classen J, et al. Euchromatin histone methyltransferase 1 regulates cortical neuronal network development. *Sci Rep*. 2016;6. https://doi.org/10.1038/srep35756.

68. Yang N, Chanda S, Marro S, et al. Generation of pure GABAergic neurons by transcription factor programming. *Nat Methods*. 2017;14(6):621–628. https://doi.org/10.1038/nmeth.4291.

69. Mossink B, van Rhijn JR, Wang S, et al. Cadherin-13 is a critical regulator of GABAergic modulation in human stem-cell-derived neuronal networks. *Mol Psychiatr*. January 2021. https://doi.org/10.1038/s41380-021-01117-x.

70. Kleefstra T, Kramer JM, Neveling K, et al. Disruption of an EHMT1-associated chromatin-modification module causes intellectual disability. *Am J Hum Genet*. 2012;91(1):73–82. https://doi.org/10.1016/j.ajhg.2012.05.003.

71. Frega M, Selten M, Mossink B, et al. Distinct pathogenic genes causing intellectual disability and autism exhibit a common neuronal network hyperactivity phenotype. *Cell Rep*. 2020;30(1). https://doi.org/10.1016/j.celrep.2019.12.002, 173.e6–186.e6.

72. Deneault E, White SH, Rodrigues DC, et al. Complete disruption of autism-susceptibility genes by gene editing predominantly reduces functional connectivity of isogenic human neurons. *Stem Cell Rep*. 2018;11(5):1211–1225. https://doi.org/10.1016/j.stemcr.2018.10.003.
73. Deneault E, Faheem M, White SH, et al. CNTN5$^{-/+}$ or EHMT2$^{-/+}$ human iPSC-derived neurons from individuals with autism develop hyperactive neuronal networks. *Elife*. 2019;8. https://doi.org/10.7554/eLife.40092.
74. Tidball AM, Parent JM. Concise review: exciting cells: modeling genetic epilepsies with patient-derived induced pluripotent stem cells. *Stem Cell*. 2016;34(1):27–33. https://doi.org/10.1002/stem.2203.
75. Winden KD, Sundberg M, Yang C, et al. Biallelic mutations in TSC2 lead to abnormalities associated with cortical tubers in human ipsc-derived neurons. *J Neurosci*. 2019;39(47):9294–9305. https://doi.org/10.1523/JNEUROSCI.0642-19.2019.
76. Tidball AM, Lopez-Santiago LF, Yuan Y, et al. Variant-specific changes in persistent or resurgent sodium current in SCN8A-related epilepsy patient-derived neurons. *Brain*. 2020;143(10):3025–3040. https://doi.org/10.1093/brain/awaa247.
77. Odawara A, Katoh H, Matsuda N, Suzuki I. Induction of long-term potentiation and depression phenomena in human induced pluripotent stem cell-derived cortical neurons. *Biochem Biophys Res Commun*. 2016;469(4):856–862. https://doi.org/10.1016/j.bbrc.2015.12.087.

CHAPTER

11

Advantages and limitations of hiPSC-derived neurons for the study of neurodegeneration

Swati Mishra[1,2], Coen Van Deursen[3], Aiko Robert[3], Rik Van Der Kant[3] and Jessica E. Young[1,2]

[1]Laboratory Medicine and Pathology, University of Washington, Seattle, WA, United States; [2]Institute for Stem Cell and Regenerative Medicine, University of Washington, Seattle, WA, United States; [3]Center for Neurogenomics and Cognitive Research, Amsterdam Neuroscience, Alzheimer Center Amsterdam, Department of Neurology, Vrije Universiteit Amsterdam, Amsterdam UMC, Amsterdam, the Netherlands

Introduction

Human-induced pluripotent stem cell (hiPSC) technology has revolutionized preclinical research into neurodegenerative diseases, as it has provided the ability to dynamically study what was previously static: cellular interactions in living human central nervous system (CNS) cell types. Prior to hiPSC technology, most research on neurodegenerative conditions relied on postmortem tissue, animal models, or transformed non-neural cell lines. While important, these models have several limitations, including being conducted at the end-stage of decade's long disease, as well as insufficient recapitulation of both the complexity of the human brain and the human genetic architecture that contributes to risk for developing neurodegeneration. Human stem cell models now provide an experimental platform to interrogate disease mechanisms in living cells that harbor patient-specific genetic backgrounds. Furthermore, advances in gene-editing technology have provided evidence for the function of disease-causing mutations and disease-associated variants in isogenic systems. Protocols are available for the generation of all major CNS cell types and, in the case of neurons, specific neuronal subtypes. Three-dimensional multi-cellular brain organoids are providing clues to disease-specific interactions between distinct cellular populations. Differentiation protocols

Phenotyping of Human iPSC-derived Neurons
https://doi.org/10.1016/B978-0-12-822277-5.00016-X

are becoming more robust, allowing for high-throughput screening assays for novel therapeutics to be feasible. In this chapter, we will discuss neurodegenerative disease modeling using hiPSCs with the main focus on Alzheimer's disease (AD). We will discuss biological pathways in AD and how the advantages of generating the diverse living cell types of the CNS are crucial for dissecting cell-type specific functions of these pathways that contribute to disease pathogenesis in both familial AD (FAD) and sporadic AD (SAD) models. We will discuss 3D modeling and drug screening and we will address key challenges still present in these models.

Biology of Alzheimer's disease

Alzheimer's disease (AD) is the most common neurodegenerative disease affecting the elderly population. AD leads to progressive cognitive decline and, to date, there is no disease modifying treatment. AD neuropathology is characterized by the presence of extracellular senile plaques composed of the small peptide amyloid beta (Aβ) and intracellular accumulations of hyperphosphorylated tau protein, called neurofibrillary tangles. Other pathologies seen in the AD brain include activated microglia, indicative of neuroinflammation, lipid inclusions indicative of metabolic changes, and dystrophic neurites, indicative of synaptic loss. AD is genetically and phenotypically heterogenous. Rare, early-onset, forms of AD, called Familial AD (FAD), are autosomal dominantly inherited and are driven by mutations in genes that impact the generation of Aβ peptide. In the last decade or so, large population genome wide association studies (GWAS) have implicated several dozen loci that are implicated in increased risk for late-onset, SAD. These risk variants are common, have low effect size, and encompass pathways such as amyloid precursor protein and tau metabolism, lipid and cholesterol metabolism, endocytosis and vesicular trafficking, and immune function.[1,2] In addition to GWAS, whole exome sequencing studies can identify rare, but penetrant, variants for SAD. The most common risk factor for SAD is the epsilon four allele of the Apolipoprotein E gene, *APOE*, discussed in a subsequent section of this chapter. *APOE e4* is an example of a relatively common variant with a large effect size. Functionally, *APOE* is implicated in many of the biological pathways identified by GWAS. Importantly, these pathways function differently in the diverse cell types of the CNS. This underscores the need for hiPSC-derived models where patient genetic background, direct manipulation of mutations or variants, as well as the ability to generate multiple CNS cell types is imperative to understanding cellular dysfunction in AD and for uncovering pathways for novel therapeutics.

Alzheimer's disease hiPSC models

Familial AD (FAD)

Approximately 1%–5% of all AD cases are familiar (familial AD, FAD), with an early onset (<65-year-old) of disease. A number of mutations in amyloid precursor protein, APP, or its processing enzymes presenilins 1 and 2, PSEN1 and PSEN2, are known to cause familial AD. Much of the early research with AD stem cell models has been performed in iPSC-derived neurons from patients harboring mutations in APP or PSEN1, in order to generate proof of concept that human iPSC-models can recapitulate AD phenotypes "in a dish." The interest in these models has led to the generation of an important resource for researchers working on iPSC-modeling of FAD from the Dominantly Inherited Alzheimer Network (DIAN), which encompasses fibroblasts and iPSCs from more than 42 FAD families with different APP, PSEN1 and PSEN2 mutations.[3] A depiction of the FAD genes *APP, PSEN1,* and *PSEN2* are shown in Fig. 11.1.

Yagi and colleagues were among the first to show that iPSC-derived neurons from PS1 and PS2 (N141)-mutation carriers produce increased levels of Aβ peptides. Aβ is the main component of senile plaques and is generated by enzymatic cleavage of APP, first by β-secretase and then by γ-secretase, which is comprised of PS1/2 (Fig. 11.1). Yagi et al. reported higher levels of Aβ42, and an increased Aβ42/40 ratio. These results recapitulated important aspects of AD caused by PSEN mutations.[4] Others later reported similar changes in Aβ42/40 ratio in either patient-derived or gene-edited PSEN1-mutant iPSC-derived neurons (PSEN delta E9[5], A246E,

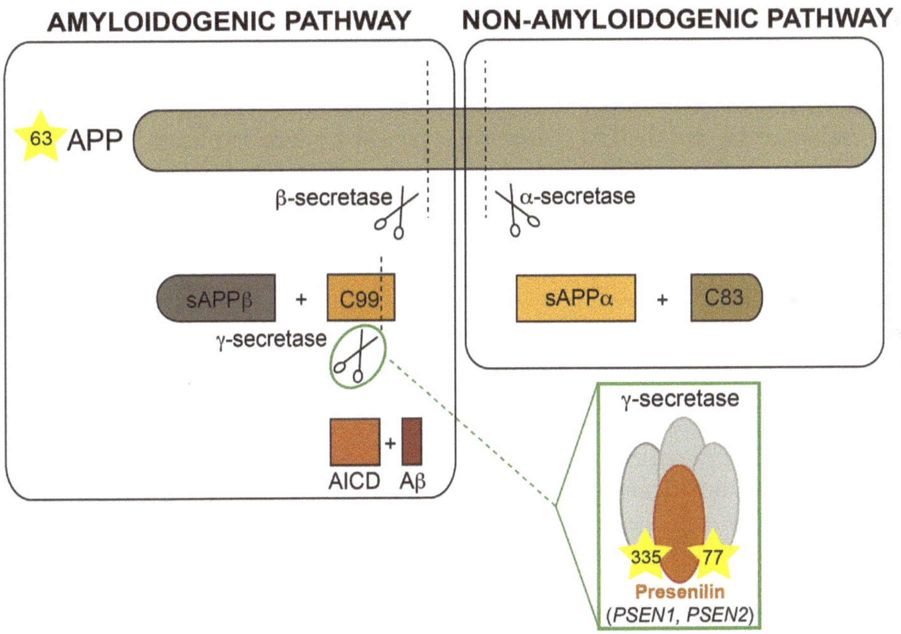

FIGURE 11.1 Familial AD risk genes. FAD results from mutations inherited in an autosomal dominant manner located on APP, PSEN1, and PSEN2. These proteins are involved in the cascade of events leading up to a number of splice products including Ab, known as the amyloidogenic pathway. This is initiated by cleavage of APP by the enzyme b-secretase resulting in the generation of the soluble APP fragment sAPPb and the 99 amino acid APP C-terminal fragment C99. The enzyme g-secretase further cleaves the latter fragment to generate AICD and Ab (left panel). g-secretase is a multiprotein complex including the catalytic active presenilin proteins, PSEN1 or PSEN2. An alternative pathway, the non-amyloidogenic pathway, consists of APP cleavage by a-secretase, resulting in generation of sAPPa and C83 fragments, preventing the generation of Ab. Stars represent the number of known AD-linked mutations in each protein, including benign and pathogenic mutations, and mutations of unknown significance. *Created with Adobe Illustrator.*

H163R, and M146L,[5] PSEN M146V, PSEN G384A).[6] FAD can also be caused by increased genetic dose of APP, as occurs in individuals in which the APP gene is duplicated or in subjects with Down syndrome due to a trisomy of chromosome 21, on which APP is located. Due to an extra copy of APP, these patients tend to have a very early onset of disease with strong amyloid plaque deposition and neurofibrillary tangle (NFT) formation. This is reflected in iPSC-derived neurons derived from patients with an APP-duplication,[7] or trisomy of chromosome 21,[8] which strongly overproduce both Aβ1-40 and 1-42. In addition to increased levels of Aβ, both of these studies also showed accumulation of phosphorylated Tau, indicating that AD-derived iPSC-neurons can recapitulate both classical AD pathologies. Furthermore, in both PSEN- and APP-mutant neurons it was shown that β- and/or γ-secretase inhibitors can reduce the levels of Aβ that are produced by these neurons, providing the first proof-of-principle that these AD iPSC-derived neurons may also be used for drug discovery. By manipulating Aβ production through inhibitors, these iPSC-derived neurons can now also be used to study the connection between Aβ production in APP and PSEN1-mutant neurons, and Tau accumulation. Surprisingly, two studies that evaluated the effect of β- and γ-secretase inhibitors on Aβ and pTau levels found that while both β-secretase and γ-secretase inhibition reduced Aβ levels, only β-secretase inhibition also reduced Tau hyperphoshorylation in APP mutant patient iPSC-derived neurons.[7,9] This data seems to indicate that in addition to Aβ, other APP fragments such as APP β-CTF may contribute to Tau pathology in FAD neurons. Increased levels of pTau and Aβ42 were also observed in iPSC-derived neurons from subjects with the familial APP mutation V717I by Muratore and colleagues.[10] While the effect of β/γ-secretase inhibitors on pTau levels was not evaluated in this study, the authors did show that treatment with Aβ-specific antibodies decreased Tau levels. Together these data indicate that both Aβ-dependent, and Aβ-independent[11] pathways may contribute to Tau pathology in FAD mutation carriers.

Interestingly, while increased levels of pTau and total Tau seem to be consistently observed in neurons from FAD patients with an APP duplication or APP mutations, this phenotype does not seem to be as pronounced in iPSC-derived neurons from PSEN mutation carriers in similar experimental paradigms. For example, Moore et al. examined PSEN1 Y115C and intron four mutations and Woodruff et al. studied PSEN1 delta E9 variants and did not observe increased pTau or Tau in PSEN mutant iPSC-derived neurons,[9] although Ochelek et al.[12] did find increased levels of pTau in a number of different PSEN1-mutant lines. Why differences in pTau are observed in some studies but not others is currently not well understood, but could depend on the specific mutation studied or the specific Tau epitopes examined.

In addition to accumulation of Aβ and Tau, endosomal dysfunction is also a major AD phenotype often reported in FAD-patient derived neurons. Enlarged RAB5-positive endosomes have been observed in APP duplication iPSC-derived neurons[7] and increases in RAB5 endosomal size were also observed in a large panel of isogenic APP and PSEN iPSC-derived neurons.[13] In the later study, CRISPR/Cas9 gene-editing was used to introduce heterozygous or homozygous APP or PSEN1 mutations in iPSCs derived from a healthy control subject. Neurons with heterozygous and homozygous mutations in APP all displayed increased endosomal size, whereas only homozygous PSEN1 mutations induced significantly increased endosome size. This fits with previous observations that phenotypes in PSEN-mutant neurons are in general less pronounced than phenotypes in APP-mutant neurons.

RAB5 endosomal enlargement downstream of APP mutations and homozygous PSEN1 mutations were shown to depend on accumulation of the APP β-CTF fragment of APP.[13] In a possibly connected pathway, the PSEN1 delta E9 mutation[14] was also previously shown to alter RAB11-dependent endosomal transport and transcytosis of LDL.[15] Overall, these studies show that endosomal enlargement phenotypes are also recapitulated in iPSC-derived neuronal models of AD, however with phenotypes overall more pronounced in APP-mutation carrier neurons compared to PSEN1-carrier neurons. Why these phenotypes are more pronounced in the APP mutant carriers is not known, as both patient populations develop similar symptoms and neuropathology. This is a potential limitation for the use of iPSC-derived neuronal cultures to study pathology for PSEN1 mutation carriers.

Astrocytic and microglial dysfunction are receiving considerable attention as major pathological events in AD, but to date relatively little work has been done to characterize glial (astrocytes and microglia) phenotypes in FAD patient-derived cultures. iPSC-derived astrocytes from APP E693delta mutation carriers were shown to accumulate intracellular Aβ and have increased levels of reactive oxygen species.[16] Oksanen and colleagues also reported a number of AD-related phenotypes in PSEN (delta E9)-mutant iPSC-derived astrocytes, including increased oxidative stress, increased Aβ production, enhanced cytokine release and dysregulated Ca^{2+} homeostasis and the mutant astrocytes were compromised in supporting neuronal function.[17] PSEN1 (deltaE9)- and APP (APPswe) mutant iPSC-derived microglia display deaccelerated chemokines and decreased cytokine release.[18] However, these observed effects were rather minor when compared to much stronger phenotypes observed in sporadic APOE ϵ4/ϵ4 iPSC-derived microglia.[18] While most of the studies described above have studied 2D mono-cultures of the different cell types, in recent years more advanced 3D-cultures and organoid models have been developed. These 3D cultures might recapitulate late-stage pathologies such as formation of mature Aβ-plaques and neurofibrillary tangles,[19] not typically seen in 2D cultures. Importantly, organoid cultures generated from FAD patient cells have been compared with pathology in human brain[20] as validation. The next step for the field will be focusing on disease-relevant biological pathways and screening new therapeutics in these models.

Sporadic Alzheimer's disease (SAD)

Late-onset, SAD comprises the majority of the disorder and includes genetic and environmental risk factors. Due to this complexity, there are fewer studies modeling SAD using hiPSCs and those that do acknowledge the variability in disease-related phenotypes and the necessity of using multiple lines and clones from patients. Despite these limitations however, several studies have uncovered novel disease mechanisms using hiPSCs from SAD patients. In an early study, Israel et al. cultured hiPSC-derived neurons from two SAD patients. While one SAD line tracked with the non-demented control cells in terms of cellular phenotypes, the other was similar to FAD neurons in terms of Aβ generation and Tau phosphorylation.[7] Interestingly, in SAD lines, as well as the FAD lines, it was shown that Tau phosphorylation was due to accumulations of the β-CTF of APP rather than the Aβ peptide itself.[7] Another early study, Kondo et al., 2013, described endoplasmic reticulum and oxidative stress in neurons with accumulated Aβ oligomers.[16] However, similarly to Israel et al., this study reported phenotypes in only one of two SAD patient lines, underscoring the inherent issues with phenotypic variability. Flamier et al., observed reduced expression of BMI1, a component of the Polycomb Repressive Complex (PRC), in cortical neurons from SAD patients and in AD brains and suggested that loss of BMI1 can induce tauopathy independently of the amyloid pathway.[21] More recent studies have begun to examine phenotypes beyond Aβ and pTau using neurons from SAD patients. For example, neural cells derived from SAD patients were shown to have altered gene expression network due to reduced levels of the transcriptional repressor REST (repressor element 1-silencing transcription factor), which also interacts with the PRC.[22] Neurons derived from these iPSC lines showed premature differentiation, reduced progenitor renewal, and increased electrical activity. These phenotypes were not reversed by reducing Aβ peptides in the cultures. Together these studies suggest that SAD genomes may promote epigenetic dysregulation that may even impact developmental programs of neural cells, thus providing early targets for therapeutic intervention.

Apolipoprotein E (APOE)

Apolipoprotein E (APOE) is the strongest risk factor for the development of SAD. Identified as a pivotal factor in SAD in 1993, the APOE protein makes up the protein core of lipoprotein particles after being loaded with cholesterol and phospholipids, making it a key player in the redistribution of lipids. The human *APOE* gene has three polymorphisms: ϵ2 (~8%), ϵ3 (~75%), and ϵ4 (~15%).[23,24] The APOE ϵ4 allele is associated with a higher risk of SAD, compared to the predominant ϵ3 allele: one ϵ4 allele increases the risk of SAD approximately fourfold, whereas two ϵ4 alleles increase the risk more than ten-fold. Carriers of an ϵ2 allele however, have a decreased risk of developing the disease.[25,26] Despite these significant differences in risk, only two amino acid residues vary in the resulting APOE proteins: 112 and 158 (Fig. 11.2). These two amino acids have been shown to impact the structure of the APOE protein and its potential to interact with lipids and receptors, such as the low-density lipoprotein (LDL) receptor family members. Mounting evidence suggests that APOE ϵ4 impairs clearance of Aβ, while promoting the aggregation of Aβ, but the exact

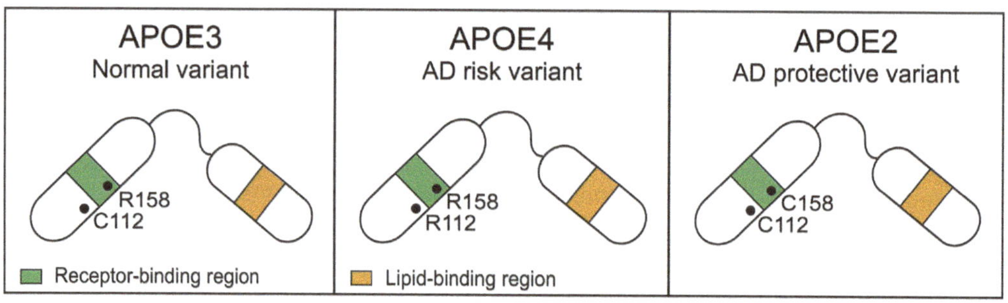

FIGURE 11.2 APOE allelic variants. APOE proteins comprise a receptor-binding region and a lipid-binding region. AD-linked APOE mutations are located in the receptor-binding arm. The three APOE variants only differ by two amino acids, which confer contrasting risk for LOAD, with the APOE e4 variant strongly conferring increasing risk for AD, and the APOE e2 variant reducing risk. *Created with Adobe Illustrator.*

effects of the allelic variants and the mechanisms behind these effects have not yet been fully elucidated.[24] Duan and colleagues were the first to investigate the effects of different APOE genotypes in hiPSC-derived neurons, derived from patients, comparing APOE ϵ3/ϵ4 and APOE ϵ3/ϵ3 basal forebrain cholinergic neurons and neurons carrying the ϵ4 allele showed elevated Aβ42/40 ratios.[27] In addition, while FAD and control lines showed a decrease in Aβ40 production after treatment with γ-secretase inhibitors, most ϵ4 allele carrying neurons showed an increased Aβ40 production.[27] These data indicate specific differences between APOE ϵ4-associated phenotypes and phenotypes observed in FAD neurons. In 2018, two landmark studies using isogenic gene-edited stem cell lines to investigate the differences between the APOE ϵ3/ϵ3 and APOE ϵ4/ϵ4 genotypes were published.[28,29] Wang et al. showed that APOE ϵ4/ϵ4, compared to APOE ϵ3/ϵ3, neurons produced less full length APOE and more APOE fragments. Furthermore, APOE ϵ4/ϵ4 neurons had higher phosphorylated Tau, Aβ40, and Aβ42 levels. CRISPR/Cas9 gene editing-mediated conversion of the APOE ϵ4/ϵ4 to an APOE ϵ3/ϵ3 genotype completely rescued the observed phenotypes, including a rescue of full length APOE, a reduction of fragmented APOE and lower phosphorylated Tau, Aβ40, and Aβ42 levels. Moreover, the authors also showed that overexpression of APOE ϵ4 in APOE knock-out neurons lead to higher phosphorylated Tau, Aβ40, and Aβ42 levels compared to APOE ϵ3/ϵ3 or control knock-out lines. Finally, PH002, a small molecule that corrects the protein structure of APOE ϵ4 to that of APOE ϵ3, also reversed the effects of APOE ϵ4: it reduced fragmented APOE and phosphorylated Tau levels, as well as decreasing Aβ40 and Aβ42 production and secretion, in a dose-dependent manner.[28] Lin et al. also used gene-editing to study the effects of APOE ϵ3/ϵ3 or APOE ϵ4/ϵ4 genotypes describing a host of APOE-ϵ4 transcriptomic changes in iPSC-derived neurons, astrocytes, and microglia.[29] In this study, they documented detrimental effects of APOE ϵ4 on neuronal synapses, astrocytic clearance of Aβ and cholesterol transport as well as increased inflammatory gene expression in microglia. In another study, APOE ϵ4/ϵ4 neurons were shown to be more susceptible to ionomycin-induced cytotoxicity, had higher phosphorylated tau and phosphorylated ERK1/2 levels, and secreted more phosphorylated Tau.[30]

Although the molecular differences between APOE ϵ3/ϵ3 and APOE ϵ4/ϵ4 neurons have not been fully elucidated yet, these iPSC-models have provided important new insights. APOE ϵ4 appears to be less stable, reducing the amount of full length APOE protein, increased Tau phosphorylation and Aβ production, as well as changes in synaptogenesis and susceptibility to cytotoxicity. This is likely the result of conformation changes in the APOE protein leading to changed affinity for receptors, ligands, and lipids. Advancements in hiPSC differentiation techniques and co-culturing neurons with astrocytes and microglia, in which ApoE is expressed at much higher levels, will help unravel more complex effects of APOE on neuronal pathology and overall cellular functioning. The different isoforms of APOE are shown in Fig. 11.2.

Other AD risk factors

Genome wide association studies (GWAS) have identified nearly 40 loci associated with increased risk for SAD, although a limited number of putative causal genes in these loci have been suggested.[31] Biological pathways implicated include APP processing, Tau, endolysosomal function, immunity, and lipid processing. For functional assessment of how risk loci may alter the function of these pathways in distinct cell types of the CNS, hiPSC models are beneficial in that they can be derived from patient cells that harbor these variants or be engineered to express the variants using genome editing. hiPSC-derived neurons from patients with AD risk variants in the Sortilin-related receptor gene, SORL1, demonstrate differential responses in SORL1 expression induction after treatment with neurotrophic factors and that this impacted Aβ secretion.[32] Furthermore, loss of SORL1 expression in hiPSC neurons leads to enlarged early endosomes and sequestration of APP in early endosomal compartments, suggesting that SORL1 is an integral component of neuronal endolysosomal sorting.[33] A similar gene-editing approach generated hiPSC-neurons deficient in the AD risk gene Clusterin (CLU).[34] Clusterin protein levels have been shown to be increased in AD brains and loss of CLU in hiPSC neurons was shown to be protective against Aβ insult.[34] Increased levels of clusterin, APOE and Aβ, as well as clusterin complexed with Ab, has been observed in synapses in human AD brains and is believed to contribute to synapse degeneration in AD.[35] Neuroimmune pathways are also highly implicated in increased AD risk and variants in the triggering receptor expressed on myeloid cells 2 (TREM2) gene are associated with a two- to threefold increase in AD risk in European and North American populations.[36] TREM2 can regulate uptake of Aβ by either directly binding to Aβ[37] or by binding to APOE-Aβ and/or Clusterin-Aβ complex.[38] Genetic mutations in TREM2 such as R47H observed in AD, significantly reduce interaction of TREM2 with Aβ by both direct and indirect pathways thus reducing Aβ phagocytosis and contributing to Aβ pathology.[37,38] Brownjohn et al., derived hiPSC microglia, the innate immune cells of the CNS, with TREM2 missense mutations.[39] Interestingly, these mutations appeared to reduce the maturation of TREM2 and lead to mislocalization away from the plasma membrane. Despite this, however, these TREM2 mutant microglia responded to pathogenic stimuli and were still able to sufficiently phagocytose substrates, indicating that this missense mutation does not cause a complete loss of function of TREM2.[31] Rather, this data represents a more subtle effect of TREM2 variants in hiPSC-derived microglia under the current culturing conditions, which might raise the issue of whether cellular aging or longer maturation would be required to detect more overt AD-associated phenotypes in iPSC-derived microglia. Recently, Martins et al., generated iPSCs from patients with the TREM2 R47H variant and analyzed gene expression profiles in differentiated neurons.[30] Transcriptomic changes between variant neurons and controls included metabolism and immune-related pathways, in particular ER stress and unfolded protein response pathways.[30]

Risk factors that contribute to SAD comprise multiple pathways that may have different functions in the unique cell types of the brain, such as neuroinflammatory pathways. Together, these studies demonstrate that by either using patient-derived cell lines or genetically manipulating hiPSCs to derive isogenic cell lines, the role of AD risk genes, targeted variants can be assessed in cell-type specific assays. Future experiments will help elucidate the complex and differing roles of these molecular players in the diverse cell types of the brain. How different AD risk genes may impact the function of CNS cells is shown in Fig. 11.3.

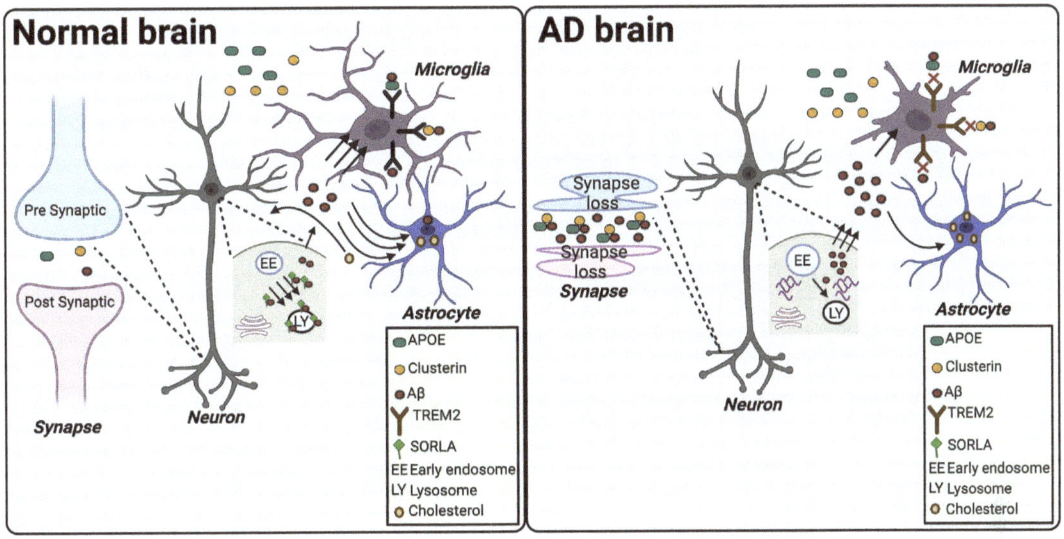

FIGURE 11.3 AD risk genes. Cellular pathways impacted by AD risk genes include cholesterol and lipid metabolism, neuroinflammation, endocytosis, and vesicular trafficking, and synaptic function. Key AD risk genes have been shown to impact these pathways. For example, SORL1 is an endocytic receptor that functions in trafficking cargo, such as Ab, to lysosomes and cell surface recycling pathways in the healthy brain. These functions are impaired in the AD brain. APOE is a main cholesterol transporter from astrocytes to neurons and this function is reduced in cells with APOE e4 alleles, an increased risk for AD. Clusterin is also involved in the clearance of Ab. APOE and Clusterin can accumulate along with Ab at the neuronal synapse and induce synapse loss in the AD brain. TREM2 regulates Ab uptake in microglia by direct binding or in complexes with APOE and/or Clusterin. AD-associated variants in TREM2 can reduce these interactions and impact Ab phagocytosis. *Created with BioRender.com.*

Tauopathies: Alzheimer's disease related dementias

Tau proteins, the products of the *MAPT* gene, are well known for their hyperphosphorylation and aggregates in structures called neurofibrillary tangles found in AD, as well as in other tauopathies.[40] As discussed earlier, accumulation of phosphorylated Tau can be recapitulated in iPSC-derived models of AD. Yet, iPSC-modeling of tauopathies is, to some extent, limited by the fact that iPSC-derived neurons mostly express a more immature variant of Tau (Fig. 11.4).

There are six isoforms of the microtubule stabilizing Tau protein in the human brain as the result of alternative splicing of *MAPT* exons 2, 3, and 10. Exons 2 and 3 give rise to 29 amino acid long inserts that are located in the N-terminal region of Tau, but exon 3 is exclusively transcribed in conjunction with exon 2, leading to either 0N, 1N, or 2N Tau. Exons 9 through 12 of Tau protein form microtubule-binding repeat domains. The presence or absence of exon 10 in the splicing product determines if the Tau protein has 3 or 4 of these repeat domains (3R or 4R tau). Because of the higher amount of microtubule-binding domains, the 4R form of Tau has higher microtubule assembly affinity compared to 3R.[40–42] The different isoforms of Tau are shown in Fig. 11.4. Different tauopathies are the result of aggregation of different

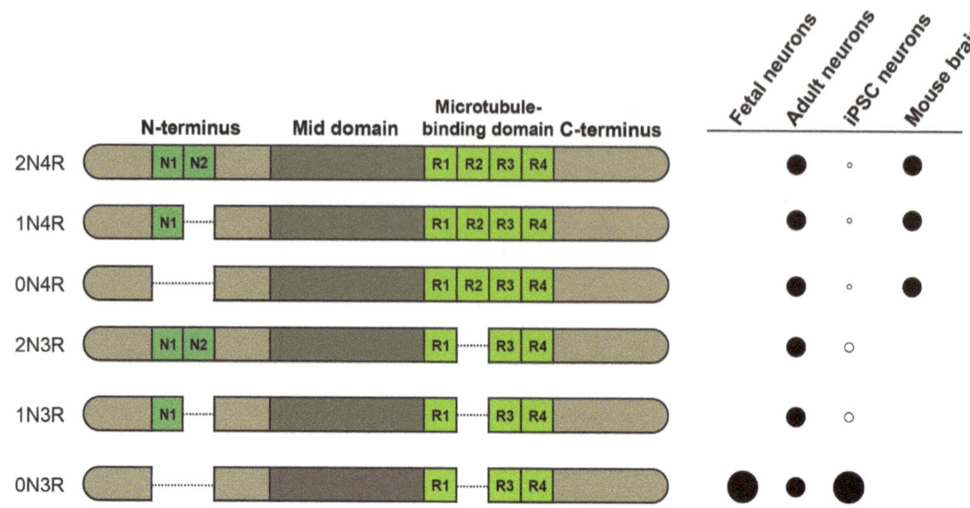

FIGURE 11.4 Tau isoforms. There exist six tau protein isoforms in the adult human brain (left panel). These differ in the number of N-terminal inserts (0-2 inserts), which stabilize the microtubule, and of microtubule-binding repeats (3 or 4 repeats). Adult neurons in the human brain express all six isoforms with an equal ratio of 3R and 4R tau. However, most human iPSC-derived neuronal and mouse models do not recapitulate this full range. While only 4R tau is expressed in the wild-type mouse brain, many iPSC models merely comprise the 0N3R isoform, mirroring the fetal neuronal composition. However, upon maturation, some of these iPSC-derived neuronal models bearing frontotemporal dementia (FTD)-linked mutations better recapitulate the range of tau isoforms expressed in the adult human brain. Importantly, while equal levels of 3R and 4R tau are expressed in the adult brain, mature iPSC-derived neurons often bear a higher 3R tau load compared to 4R tau. Dot size represents the proportion of the isoform present. Filled dots indicate the presence of the isoform. White dots represent their presence upon longer maturation or in some mutation backgrounds. *Created with Adobe Illustrator.*

tau isoforms and tauopathies are subdivided into three classes: 3R tauopathies (e.g., Pick's Disease), 4R tauopathies (e.g., progressive supranuclear palsy and Huntington's disease) and 3R+4R tauopathies (e.g., AD and chronic traumatic encephalopathy).[42,43] 3R and 4R isoform expression in the human brain is developmentally regulated. Human fetal neurons express only the shortest 3R Tau proteins, whereas all six isoforms are present in adult neurons, resulting in a 1:1 ratio of the 3R and 4R Tau isoforms in the adult human brain.[40,41] This shift from shorter to longer Tau isoforms is seen in all vertebrates studied to date, but not every species expresses equal amounts of 3R and 4R Tau in adult neurons. Adult mice for instance, express 4R Tau exclusively, whereas adult chicken neurons even express a five repeat domain (5R) Tau variant.[44,45] In a study by Hefti et al., it was shown that the expression of *MAPT* exon 2 and exon 10 increases sharply in the perinatal period, as well as a smaller but significant increase of exon 3 expression. Interestingly fetal neurons also appeared to have higher levels of Tau phosphorylation compared to adult neurons.[46]

Why and how Tau splicing is different during development is not known, but Tau isoforms need to be taken into account when studying Tau in iPSC-models. Similar to neurons in early development, iPSC-derived neurons mostly express 0N3R Tau isoforms.[47–49] However, iPSC-derived neurons with mutations in the *MAPT* gene associated with frontotemporal dementia have been shown to drive higher expression of 4R Tau.[47,50,51] Maturing iPSC-derived neurons for longer periods (up to a year) increases 4R isoform expression, but does not recapitulate the full adult neuron Tau isoform expression.[51] A method that did result in a 3R:4R Tau ratio resembling that of adult neurons was grafting iPSC-derived neurons into an AD mouse model. In this study, the expression of 4R Tau reached a similar level as the 3R Tau 8 months after transplantation, indicating that certain factors in the mouse brain environment help the iPSC-derived neurons to mature and start expressing longer Tau isoforms.[52] Finally, a different in vitro culturing method that shows some promise was published by Miguel and colleagues.[53] They argued that iPSC-derived neurons mature better in a 3D microenvironment, which would lead to an increase of 4R Tau isoforms. In order to test this, a new 3D neuronal culture system using BrainPhys media and Matrigel-coated capsules was used together with a new Tau isoform analysis assay. After 25 weeks in culture, the iPSC-derived cortical neurons cultured in BrainPhys showed expression of all Tau isoforms, including the 2N isoforms. However, the 4R isoforms only constituted 12% of the total Tau, rather than the approximately 50% observed in the adult human brain.[53]

In conclusion, current culturing or maturation methods for iPSC-derived neurons that lead to an expression of Tau isoforms identical to adult human neurons still need development. Several studies have shown that longer in vitro maturation, 3D culturing, or transplantation into mouse brains can lead to more 4R Tau isoform being expressed, but these do not yet completely mimic the human in vivo physiological and pathophysiological situation. Advances in these current techniques or new approaches are necessary in the field in order to truly model the range of tauopathies in vitro.

The challenge of aging in hiPSC models of age-related disease

Aging is a multifactorial process that affects diverse pathways and is driven by both genetics and environment.[54] For human neurodegenerative disease, including AD, brain aging is the largest risk factor.[55] Because hiPSCs are fetal in nature, the immaturity of cell types and the lack of age-accumulated epigenetic changes has presented some challenges in modeling late-onset neurodegenerative disease and prompted several groups to attempt to introduce the effects of aging into their in vitro models. One significant milestone for these efforts was the introduction of the progerin protein into hiPSC-derived neurons.[56] Progerin is a pathogenic protein encoded by LMNA gene that results in the premature aging syndrome Hutchinson–Gilford progeria syndrome (HGPS). In 2013, Miller and colleagues documented more significant cellular phenotypes in their hiPSC-derived model of Parkinson's disease, including loss of tyrosine hydroxylase expression, enlarged mitochondria, and dendritic degeneration, when progerin was overexpressed in the cells.[56] More recently, a study revealed significant differences in IGF signaling and neuronal development genes in an hiPSC model of Huntington's disease upon progerin overexpression in these cells.[57] To date, however, there has not been a study using progerin expression in AD hiPSC models. HGPS is a specific type of premature aging disorder and progerin overexpression may not be the most physiologic representation of typical aging, therefore a significant need exists to identify drivers of non-syndromic aging to use in hiPSC models. Tagliafierro et al. documented aging phenotypes in high passage neural progenitor cells compared to low passage cells and also showed that overexpression of the

SNCA gene, which encodes for alpha-synuclein, induced nuclear aging phenotypes in hiPSC neurons.[58] Finally, direct conversion of neurons from fibroblasts in a process that by-passes reprogramming to a pluripotent state has been shown to maintain an age-related transcriptomic signature.[59] Directly converted neurons, or induced neurons (iNs) represent another in vitro model for the study of neurodegenerative disorders and comparing iNs to hiPSC-derived neurons has been recently reviewed.[60]

Modeling Alzheimer's disease with cerebral organoids

hiPSC-derived cerebral organoids are a widely used research tool to model human brain disorders in vitro since hiPSCs contain the genetic heterogeneity of patients and hiPSCs are easily accessible. Extensive optimization of organoid differentiation protocols over the years led to protocols with increased survival and differentiation into diverse and mature neural and non-neural cell types, which allowed cerebral organoids to be used to study adult neurodegeneration.[61] These human 3D brain-like structures are especially important for modeling disorders like AD because animal models of AD have increasingly been shown to fail in reproducing human-like pathophysiology.[62,63] In addition, 2D monolayer neuronal cultures from AD patient-derived hiPSCs yield increased levels of Amyloid β (Aβ) and phosphorylated tau, but aggregates of these molecules, including extracellular Aβ plaques and tau fibrils, are not observed within neurons. The first to study AD using 3D modeling was Choi et al. in 2014. This group used a matrigel-based matrix system to develop 3D structures comprising neurons and astrocytes harboring FAD mutations in APP and PS1. Their data revealed that these 3D structures not only showed increased neuronal maturation, but were able to recapitulate AD-like Aβ and tau pathology.[19] A similar study modeled FAD in a 3D cell culture system that utilized microfluidics, but their 3D system also allowed for the co-culture of adult microglia in addition to neurons and astrocytes, enabling study of neuron-glia interactions in an AD environment.[64] Interestingly, this study reported an increase in pro-inflammatory cytokines secreted by microglial cells in FAD conditions. However, these methods were based on overexpression of AD-relevant factors and forced induction of AD pathology by exogenous application of Aβ on hiPSCs derived from healthy humans. While informative, these conditions do not necessarily mimic the environment of the brain of a SAD patient that does not harbor mutations in *APP* or *PSEN1/2*. Moreover, these 3D cell culture systems did not mimic the highly organized and complex development of neurons and glial cells seen in vivo. To better recapitulate AD pathology, Raja et al. generated cerebral organoids from hiPSCs derived from Familial AD (FAD) patients that exhibited pathological hallmarks of AD, including age-dependent accumulation of Aβ plaques and accumulation of hyperphosphorylated tau in the form of tau fibrils within neurons.[65] This AD-like pathology was consistent across distinct FAD patient lines harboring different FAD mutations, including Amyloid Precursor Protein (*APP*) and Presenilin 1 (*PSEN1*). More importantly, attenuation of AD pathology was observed within these organoids on treatment with β- and γ-secretase inhibitors, which suggests that these organoids are amenable to drug treatments in vitro.

An elegant study by Gonzales et al. in 2018 followed, wherein they also studied organoids derived from FAD hiPSCs, but additionally showed that development of AD-like phenotypes is very specific to cerebral organoids generated from FAD hiPSCs, and was not observed in organoids generated from mESCs, healthy hiPSCs or hiPSCs derived from patients with Creutzfeldt-Jakob disease, another neurodegenerative disorder involving abnormal protein accumulation. This group also tested cerebral organoids derived from Down syndrome (DS) patient hiPSCs because most Down syndrome patients go on to develop AD due to the duplication of the *APP* gene on chromosome 21. Comparison of cerebral organoids derived from FAD and DS patient hiPSCs revealed that in both of these conditions, accumulation of Aβ and Tau aggregates were present, which suggests that longer term culture of organoids can lead to recapitulation of AD pathology seen in later stages of AD.[20]

Another study combined CRISPR/Cas9 genetic engineering technology with 3D modeling to elucidate the role of genetic variants known to increase AD risk.[29] Specifically, hiPSCs from Sporadic AD (SAD) patients harboring the known AD risk factor, APOE ϵ4, were genetically manipulated by CRISPR/Cas9 to switch APOE ϵ4 allele to APOE ϵ3 allele. This manipulation was sufficient to ameliorate AD pathology, including accumulation of Aβ and phosphorylated Tau within organoids. Furthermore, they co-cultured APOE ϵ4 and APOE ϵ3 microglia with FAD mutant organoids and showed that APOE ϵ3 comprising organoids exhibited decreased accumulated Aβ as compared to organoids with APOE ϵ4 microglia, suggesting that APOE ϵ3 microglia have increased capability for Aβ uptake and clearance as compared to APOE ϵ4 microglia. This study highlights a possible crucial protective mechanism in SAD.[29] Overall, these studies cumulatively have shown that cerebral organoids can both recapitulate features of human brain development and with increased survival in culture can also mimic some aspects of adult brain function and dysfunction. Additionally, amenability of cerebral organoids to genetic and drug manipulation makes them a great tool for preclinical drug discovery.

Current challenges of 3D modeling and possible solutions

Although cerebral organoids have been shown to be an invaluable research tool to model adult neurodegeneration, these culture systems are not devoid of limitations. Batch to batch variability, occasional incomplete differentiation and neuronal maturation, decreased longevity in culture, necrosis in the center of organoids and absence of developmentally appropriate number of glial cells are some of the general drawbacks. Innervation with blood vessels or vascularization is important for neuronal and glial cell maturation and survival during brain development and in the adult brain. Since cerebral organoids lack vascularization, protocols which enable development of blood vessels within organoids will overcome at least some of the technical limitations associated with cerebral organoids. To this end, Pham et al. developed vascularized cerebral organoids by co-culturing endothelial cells and cerebral organoids differentiated from hiPSCs from the same patient.[66] The hiPSC-derived endothelial cells were able to penetrate the outer layers of the cerebral organoids in vitro and successfully traveled to the center of organoids when transplanted into mouse brains. Even though these endothelial cells expressed the specific endothelial marker CD31 within organoids, they did not show any functional characteristics of blood vessels like complex endothelial tubular formation, blood flow, or expression of markers specific to the blood–brain barrier (BBB). The BBB is a highly selective membrane-like structure formed primarily by endothelial cells, which facilitates transport of molecules between blood and the brain and is especially important in the context of AD. The BBB has been shown to be involved in clearance of Aβ[67] and dysfunction of BBB has been correlated with AD.[68,69] In 2019, Cakir et al. developed a model where overexpression of endothelial transcription factor *ETV2* induced endothelial cells to develop within cerebral organoids. These structures demonstrated some BBB characteristics, such as expression of tight junction markers and astrocytic and pericytic proteins. In addition, when they measured the trans-endothelial electrical resistance, they found it significantly increased compared to control organoids. While extremely promising, this method does not tightly control for vessel geometry, which may be important in allowing perfusion throughout the entire organoid.[70]

Another limitation of 3D modeling is the development of glial cells and interneurons within cerebral organoids. Initially, many of the protocols generated organoids comprising primarily of neurons. Importantly, the development and function of neurons in vivo is heavily dependent on the interaction of neurons with glial cells, including astrocytes, microglia, and oligodendrocytes. Lack of glial cells can account for restricted neuronal maturity and lack of complex synapses within organoids because glial cells are required for synapse formation and synaptic pruning. Recently, several research groups addressed this gap by generating cerebral organoids comprising of astrocytes,[71,72] myelinating oligodendrocytes[73] and microglia.[74] Transcriptome analysis and electrophysiology in these studies revealed that neurons within cerebral organoids had increased maturity and formed functional synapses, as compared to neurons within organoids generated in previous studies, corroborating the idea that presence of glial cells within cerebral organoids is essential for robust 3D modeling of brain function in vitro. Even though the presence of these glial cell types led to robust expression of synaptic markers and electrophysiologically mature neurons, recapitulation of fully formed complex synapses as seen in vivo was not observed. This could be because the development of each of these glial cell types within organoids was shown in isolated studies and for efficient synapse formation, the presence of all the glial cell types and blood vessels within cerebral organoids is essential. Interneurons are another cell type that are critical for neuron circuit-level function. However, none of the cerebral organoid protocols enabled growth of these cell types until research groups developed fusion protocols. These protocols allowed fusion of either dorsal and ventral cerebral organoids,[75,76] or cortical and thalamic organoids,[77] which allowed for formation of migratory interneurons that strengthened connections and hence functions of neurons in these regions.

In general, cerebral organoid modeling for AD can be promising in recapitulating some of the neuropathological features of the disorder. However, more work must still be done to uncover additional mechanisms of disease drivers. Future studies should explore molecular mechanisms regulating neuron-glia, neuron-blood vessels, and neuron-interneuron interactions leading to efficient synaptogenesis and incorporate this knowledge to develop fully functional synapses within cerebral organoids.

Using hiPSC models for drug discovery

Most drug-discovery efforts for AD have traditionally been performed in mouse models of AD. These models provide relatively low-throughput methods for drug screening. In addition, a large number of drugs that have been found to be effective in mouse models of AD, have later failed in the clinic, indicating that the translationability of mouse studies to humans may be low. Human iPSC-derived neural systems therefore provide an alternative high-throughput drug discovery platform. To discover novel modulators of Aβ generation, Brownjohn et al. used iPSC-derived neurons from a trisomy 21 patient to perform a high-throughput screen. Among 1200 compounds, they found that Avermectins, a group of commonly used antiparasitic drugs, were able to shift the production of Aβ from longer to shorter peptides.[78]Kondo et al. used iPSC-derived neurons with a PSEN1 G384A to screen a number of compounds to find drugs that reduced Aβ secretion and the Aβ42/40 ratio.[79] They identified 27 Aβ-lowering drugs, three of which they combined in an Aβ-lowering cocktail that they showed were also effective in lowering Aβ in four additional PSEN FAD neurons, as well as in SAD and healthy control neurons. However, lowering Aβ levels to improve cognition in patients may not be as trivial as previously assumed. Inhibition of the beta-secretase enzyme, BACE1 which decreases Aβ peptide generation, clinically worsened cognition in clinical trials.[80] Whether this is due to off-target effects of the drugs or whether it shows that lowering Aβ production by itself is not beneficial is still out for judgment, but might be addressed with these other compounds.

More recently, a screen to identify drugs that lower Tau hyperphosphorylation in APPdp iPSC-derived neurons was performed.[81] In this screen >1600 compounds were tested, discovering 42 phosphorylated Tau-lowering drugs, among which a group of drugs that reduce phosphorylated Tau in a cholesterol- and proteasome-dependent manner. Among the drugs identified was efavirenz, an HIV/AIDS drug that is currently being clinically evaluated for the treatment of AD.[82]

Due to the long and progressive nature of AD, most drug development efforts are aimed at preventing the accumulation of pathologic proteins, such as Aβ or Tau. By the time pathology is evident in the human brain, it is possible that too much cellular damage has already occurred and will not be reversible. However, as discussed above, advances in 3D organoid modeling, where the development of pathologic protein accumulation can be tracked in vitro, and by understanding mechanisms by which cells such as microglia can potentially clear these proteins, hiPSC-based models have the potential to uncover novel therapeutic pathways.

Conclusions

As with other complex disorders, many model systems are necessary to fully understand the mechanisms of neurodegenerative disease, in particular AD pathogenesis. The studies reviewed in the chapter show that neural cells derived from hiPSCs can generate cellular AD phenotypes in 2D cultures and in 3D organoid cultures. These iPSC-derived neuron model systems utilize either the unique genetic background from a patient affected by AD or have been generated using genome-editing to create isogenic series of cells with disease-associated mutations or variants. Despite the challenges discussed, such as the fetal nature or as-yet incomplete nature of cerebral organoids, these models produce disease-relevant, modifiable phenotypes that are used to develop assays that have the capability to uncover novel disease mechanisms and test new therapeutics. With the rapid advances of stem cell science, the current limitations are surmountable and hiPSC models will remain a viable experimental platform for the study of neurodegenerative disease.

References

1. Scheltens P, De Strooper B, Kivipelto M, et al. Alzheimer's disease. *Lancet*. 2021;397(10284):1577–1590. https://doi.org/10.1016/S0140-6736(20)32205-4.
2. Small SA, Simoes-Spassov S, Mayeux R, Petsko GA. Endosomal traffic Jams represent a pathogenic Hub and therapeutic target in Alzheimer's disease. *Trends Neurosci*. 2017;40(10):592–602. https://doi.org/10.1016/j.tins.2017.08.003.
3. Moulder KL, Snider BJ, Mills SL, et al. Dominantly Inherited Alzheimer Network: facilitating research and clinical trials. *Alzheimer's Res Ther*. 2013;5(5):48. https://doi.org/10.1186/alzrt213.
4. Yagi T, Ito D, Okada Y, et al. Modeling familial Alzheimer's disease with induced pluripotent stem cells. *Hum Mol Genet*. 2011;20(23):4530–4539. https://doi.org/10.1093/hmg/ddr394.
5. Liu Q, Waltz S, Woodruff G, et al. Effect of potent gamma-secretase modulator in human neurons derived from multiple presenilin 1-induced pluripotent stem cell mutant carriers. *JAMA Neurol*. 2014;71(12):1481–1489. https://doi.org/10.1001/jamaneurol.2014.2482.
6. Paquet D, Kwart D, Chen A, et al. Efficient introduction of specific homozygous and heterozygous mutations using CRISPR/Cas9. *Nature*. 2016;533(7601):125–129. https://doi.org/10.1038/nature17664.
7. Israel MA, Yuan SH, Bardy C, et al. Probing sporadic and familial Alzheimer's disease using induced pluripotent stem cells. *Nature*. 2012;482(7384):216–220. https://doi.org/10.1038/nature10821.
8. Shi Y, Kirwan P, Smith J, MacLean G, Orkin SH, Livesey FJ. A human stem cell model of early Alzheimer's disease pathology in Down syndrome. *Sci Transl Med*. 2012;4(124):124ra29. https://doi.org/10.1126/scitranslmed.3003771.
9. Moore S, Evans LDB, Andersson T, et al. APP metabolism regulates tau proteostasis in human cerebral cortex neurons. *Cell Rep*. 2015;11(5):689–696. https://doi.org/10.1016/j.celrep.2015.03.068.
10. Muratore CR, Rice HC, Srikanth P, et al. The familial Alzheimer's disease APPV717I mutation alters APP processing and Tau expression in iPSC-derived neurons. *Hum Mol Genet*. 2014;23(13):3523–3536. https://doi.org/10.1093/hmg/ddu064.
11. van der Kant R, Goldstein LSB, Ossenkoppele R. Amyloid-beta-independent regulators of tau pathology in Alzheimer disease. *Nat Rev Neurosci*. 2020;21(1):21–35. https://doi.org/10.1038/s41583-019-0240-3.
12. Ochalek A, Mihalik B, Avci HX, et al. Neurons derived from sporadic Alzheimer's disease iPSCs reveal elevated TAU hyperphosphorylation, increased amyloid levels, and GSK3B activation. *Alzheimer's Res Ther*. 2017;9(1):90. https://doi.org/10.1186/s13195-017-0317-z.
13. Kwart D, Gregg A, Scheckel C, et al. A large panel of isogenic APP and PSEN1 mutant human iPSC neurons reveals shared endosomal Abnormalities mediated by APP beta-CTFs, not Abeta. *Neuron*. 2019;104(2):256–270. https://doi.org/10.1016/j.neuron.2019.07.010. e5.
14. Woodruff G, Young JE, Martinez FJ, et al. The presenilin-1 DeltaE9 mutation results in reduced gamma-secretase activity, but not total loss of PS1 function, in isogenic human stem cells. *Cell Rep*. 2013;5(4):974–985. https://doi.org/10.1016/j.celrep.2013.10.018.
15. Woodruff G, Reyna SM, Dunlap M, et al. Defective transcytosis of APP and lipoproteins in human iPSC-derived neurons with familial Alzheimer's disease mutations. *Cell Rep*. 2016;17(3):759–773. https://doi.org/10.1016/j.celrep.2016.09.034.
16. Kondo T, Asai M, Tsukita K, et al. Modeling Alzheimer's disease with iPSCs reveals stress phenotypes associated with intracellular Abeta and differential drug responsiveness. *Cell Stem Cell*. 2013;12(4):487–496. https://doi.org/10.1016/j.stem.2013.01.009.
17. Oksanen M, Petersen AJ, Naumenko N, et al. PSEN1 mutant iPSC-derived model reveals Severe astrocyte pathology in Alzheimer's disease. *Stem Cell Rep*. 2017;9(6):1885–1897. https://doi.org/10.1016/j.stemcr.2017.10.016.

18. Konttinen H, Cabral-da-Silva MEC, Ohtonen S, et al. PSEN1DeltaE9, APPswe, and APOE4 confer disparate phenotypes in human iPSC-derived microglia. *Stem Cell Rep*. 2019;13(4):669–683. https://doi.org/10.1016/j.stemcr.2019.08.004.
19. Choi SH, Kim YH, Hebisch M, et al. A three-dimensional human neural cell culture model of Alzheimer's disease. *Nature*. 2014;515(7526):274–278. https://doi.org/10.1038/nature13800.
20. Gonzalez C, Armijo E, Bravo-Alegria J, Becerra-Calixto A, Mays CE, Soto C. Modeling amyloid beta and tau pathology in human cerebral organoids. *Mol Psychiatr*. 2018;23(12):2363–2374. https://doi.org/10.1038/s41380-018-0229-8.
21. Flamier A, El Hajjar J, Adjaye J, Fernandes KJ, Abdouh M, Bernier G. Modeling late-onset sporadic Alzheimer's disease through BMI1 deficiency. *Cell Rep*. 2018;23(9):2653–2666. https://doi.org/10.1016/j.celrep.2018.04.097.
22. Meyer K, Feldman HM, Lu T, et al. REST and neural gene network dysregulation in iPSC models of Alzheimer's disease. *Cell Rep*. 2019;26(5):1112–1127. https://doi.org/10.1016/j.celrep.2019.01.023. e9.
23. Strittmatter WJ, Saunders AM, Schmechel D, et al. Apolipoprotein E: high-avidity binding to beta-amyloid and increased frequency of type 4 allele in late-onset familial Alzheimer disease. *Proc Natl Acad Sci USA*. 1993;90(5):1977–1981. https://doi.org/10.1073/pnas.90.5.1977.
24. Yamazaki Y, Zhao N, Caulfield TR, Liu CC, Bu G. Apolipoprotein E and Alzheimer disease: pathobiology and targeting strategies. *Nat Rev Neurol*. 2019;15(9):501–518. https://doi.org/10.1038/s41582-019-0228-7.
25. Corder EH, Saunders AM, Risch NJ, et al. Protective effect of apolipoprotein E type 2 allele for late onset Alzheimer disease. *Nat Genet*. 1994;7(2):180–184. https://doi.org/10.1038/ng0694-180.
26. Farrer LA, Cupples LA, Haines JL, et al. Effects of age, sex, and ethnicity on the association between apolipoprotein E genotype and Alzheimer disease. A meta-analysis. APOE and Alzheimer disease meta analysis consortium. *JAMA*. 1997;278(16):1349–1356. https://www.ncbi.nlm.nih.gov/pubmed/9343467.
27. Duan L, Bhattacharyya BJ, Belmadani A, Pan L, Miller RJ, Kessler JA. Stem cell derived basal forebrain cholinergic neurons from Alzheimer's disease patients are more susceptible to cell death. *Mol Neurodegener*. 2014;9:3. https://doi.org/10.1186/1750-1326-9-3.
28. Wang C, Najm R, Xu Q, et al. Gain of toxic apolipoprotein E4 effects in human iPSC-derived neurons is ameliorated by a small-molecule structure corrector. *Nat Med*. 2018;24(5):647–657. https://doi.org/10.1038/s41591-018-0004-z.
29. Lin YT, Seo J, Gao F, et al. APOE4 causes Widespread molecular and cellular Alterations associated with Alzheimer's disease phenotypes in human iPSC-derived brain cell types. *Neuron*. 2018;98(6):1141–1154. https://doi.org/10.1016/j.neuron.2018.05.008. e7.
30. Wadhwani AR, Affaneh A, Van Gulden S, Kessler JA. Neuronal apolipoprotein E4 increases cell death and phosphorylated tau release in alzheimer disease. *Ann Neurol*. 2019;85(5):726–739. https://doi.org/10.1002/ana.25455.
31. Andrews SJ, Fulton-Howard B, Goate A. Interpretation of risk loci from genome-wide association studies of Alzheimer's disease. *Lancet Neurol*. 2020;19(4):326–335. https://doi.org/10.1016/S1474-4422(19)30435-1.
32. Young JE, Boulanger-Weill J, Williams DA, et al. Elucidating molecular phenotypes caused by the SORL1 Alzheimer's disease genetic risk factor using human induced pluripotent stem cells. *Cell Stem Cell*. 2015;16(4):373–385. https://doi.org/10.1016/j.stem.2015.02.004.
33. Knupp A, Mishra S, Martinez R, et al. Depletion of the AD risk gene SORL1 selectively impairs neuronal endosomal traffic independent of amyloidogenic APP processing. *Cell Rep*. 2020;31(9):107719. https://doi.org/10.1016/j.celrep.2020.107719.
34. Robbins JP, Perfect L, Ribe EM, et al. Clusterin is required for beta-amyloid toxicity in human iPSC-derived neurons. *Front Neurosci*. 2018;12:504. https://doi.org/10.3389/fnins.2018.00504.
35. Jackson RJ, Rose J, Tulloch J, Henstridge C, Smith C, Spires-Jones TL. Clusterin accumulates in synapses in Alzheimer's disease and is increased in apolipoprotein E4 carriers. *Brain Commun*. 2019;1(1):fcz003. https://doi.org/10.1093/braincomms/fcz003.
36. Martins S, Müller-Schiffmann A, Erichsen L, et al. R47H variant enables the construction of an Aβ-induced gene regulatory network. *Int J Mol Sci*. 2020;21(12). https://doi.org/10.3390/ijms21124516.
37. Zhao Y, Wu X, Li X, et al. TREM2 is a receptor for β-amyloid that Mediates microglial function. *Neuron*. 2018;97(5):1023–1031. https://doi.org/10.1016/j.neuron.2018.01.031. e7.

38. Yeh FL, Wang Y, Tom I, Gonzalez LC, Sheng M. TREM2 Binds to apolipoproteins, including APOE and CLU/APOJ, and Thereby facilitates uptake of amyloid-beta by microglia. *Neuron*. 2016;91(2):328–340. https://doi.org/10.1016/j.neuron.2016.06.015.
39. Brownjohn PW, Smith J, Solanki R, et al. Functional studies of missense TREM2 mutations in human stem cell-derived microglia. *Stem Cell Rep*. 2018;10(4):1294–1307. https://doi.org/10.1016/j.stemcr.2018.03.003.
40. Wang Y, Mandelkow E. Tau in physiology and pathology. *Nat Rev Neurosci*. 2016;17(1):5–21. https://doi.org/10.1038/nrn.2015.1.
41. Goedert M, Jakes R. Expression of separate isoforms of human tau protein: correlation with the tau pattern in brain and effects on tubulin polymerization. *EMBO J*. 1990;9(13):4225–4230. https://www.ncbi.nlm.nih.gov/pubmed/2124967.
42. Goedert M, Spillantini MG. Ordered assembly of tau protein and neurodegeneration. *Adv Exp Med Biol*. 2019;1184:3–21. https://doi.org/10.1007/978-981-32-9358-8_1.
43. Liu F, Gong CX. Tau exon 10 alternative splicing and tauopathies. *Mol Neurodegener*. 2008;3:8. https://doi.org/10.1186/1750-1326-3-8.
44. Takuma H, Arawaka S, Mori H. Isoforms changes of tau protein during development in various species. *Brain Res Dev Brain Res*. 2003;142(2):121–127. https://doi.org/10.1016/s0165-3806(03)00056-7.
45. Yoshida H, Goedert M. Molecular cloning and functional characterization of chicken brain tau: isoforms with up to five tandem repeats. *Biochemistry*. 2002;41(51):15203–15211. https://doi.org/10.1021/bi026464m.
46. Hefti MM, Farrell K, Kim S, et al. High-resolution temporal and regional mapping of MAPT expression and splicing in human brain development. *PLoS One*. 2018;13(4):e0195771. https://doi.org/10.1371/journal.pone.0195771.
47. Iovino M, Agathou S, Gonzalez-Rueda A, et al. Early maturation and distinct tau pathology in induced pluripotent stem cell-derived neurons from patients with MAPT mutations. *Brain*. 2015;138(Pt 11):3345–3359. https://doi.org/10.1093/brain/awv222.
48. Sato C, Barthelemy NR, Mawuenyega KG, et al. Tau kinetics in neurons and the human central nervous system. *Neuron*. 2018;97(6):1284–1298. https://doi.org/10.1016/j.neuron.2018.02.015. e7.
49. Silva MC, Cheng C, Mair W, et al. Human iPSC-derived neuronal model of tau-A152T frontotemporal dementia reveals tau-mediated mechanisms of neuronal vulnerability. *Stem Cell Rep*. 2016;7(3):325–340. https://doi.org/10.1016/j.stemcr.2016.08.001.
50. Wren MC, Zhao J, Liu CC, et al. Frontotemporal dementia-associated N279K tau mutant disrupts subcellular vesicle trafficking and induces cellular stress in iPSC-derived neural stem cells. *Mol Neurodegener*. 2015;10:46. https://doi.org/10.1186/s13024-015-0042-7.
51. Sposito T, Preza E, Mahoney CJ, et al. Developmental regulation of tau splicing is disrupted in stem cell-derived neurons from frontotemporal dementia patients with the 10 + 16 splice-site mutation in MAPT. *Hum Mol Genet*. 2015;24(18):5260–5269. https://doi.org/10.1093/hmg/ddv246.
52. Espuny-Camacho I, Arranz AM, Fiers M, et al. Hallmarks of Alzheimer's disease in stem-cell-derived human neurons transplanted into mouse brain. *Neuron*. 2017;93(5):1066–1081. https://doi.org/10.1016/j.neuron.2017.02.001. e8.
53. Miguel L, Rovelet-Lecrux A, Feyeux M, et al. Detection of all adult Tau isoforms in a 3D culture model of iPSC-derived neurons. *Stem Cell Res*. 2019;40:101541. https://doi.org/10.1016/j.scr.2019.101541.
54. Lopez-Otin C, Blasco MA, Partridge L, Serrano M, Kroemer G. The hallmarks of aging. *Cell*. 2013;153(6):1194–1217. https://doi.org/10.1016/j.cell.2013.05.039.
55. Hou Y, Dan X, Babbar M, et al. Ageing as a risk factor for neurodegenerative disease. *Nat Rev Neurol*. 2019;15(10):565–581. https://doi.org/10.1038/s41582-019-0244-7.
56. Miller JD, Ganat YM, Kishinevsky S, et al. Human iPSC-based modeling of late-onset disease via progerin-induced aging. *Cell Stem Cell*. 2013;13(6):691–705. https://doi.org/10.1016/j.stem.2013.11.006.
57. Cohen-Carmon D, Sorek M, Lerner V, et al. Progerin-induced transcriptional changes in Huntington's disease human pluripotent stem cell-derived neurons. *Mol Neurobiol*. 2020;57(3):1768–1777. https://doi.org/10.1007/s12035-019-01839-8.
58. Tagliafierro L, Zamora ME, Chiba-Falek O. Multiplication of the SNCA locus exacerbates neuronal nuclear aging. *Hum Mol Genet*. 2019;28(3):407–421. https://doi.org/10.1093/hmg/ddy355.

59. Mertens J, Paquola ACM, Ku M, et al. Directly reprogrammed human neurons Retain aging-associated transcriptomic signatures and reveal age-related nucleocytoplasmic Defects. *Cell Stem Cell.* 2015;17(6):705–718. https://doi.org/10.1016/j.stem.2015.09.001.
60. D'Souza GX, Rose SE, Knupp A, Nicholson DA, Keene CD, Young JE. The application of in vitro-derived human neurons in neurodegenerative disease modeling. *J Neurosci Res.* 2020. https://doi.org/10.1002/jnr.24615.
61. Amin ND, Pasca SP. Building models of brain disorders with three-dimensional organoids. *Neuron.* 2018;100(2):389–405. https://doi.org/10.1016/j.neuron.2018.10.007.
62. Dawson TM, Golde TE, Lagier-Tourenne C. Animal models of neurodegenerative diseases. *Nat Neurosci.* 2018;21(10):1370–1379. https://doi.org/10.1038/s41593-018-0236-8.
63. Drummond E, Wisniewski T. Alzheimer's disease: experimental models and reality. *Acta Neuropathol.* 2017;133(2):155–175. https://doi.org/10.1007/s00401-016-1662-x.
64. Park J, Wetzel I, Marriott I, et al. A 3D human triculture system modeling neurodegeneration and neuroinflammation in Alzheimer's disease. *Nat Neurosci.* 2018;21(7):941–951. https://doi.org/10.1038/s41593-018-0175-4.
65. Raja WK, Mungenast AE, Lin YT, et al. Self-organizing 3D human neural tissue derived from induced pluripotent stem cells recapitulate Alzheimer's disease phenotypes. *PLoS One.* 2016;11(9):e0161969. https://doi.org/10.1371/journal.pone.0161969.
66. Pham MT, Pollock KM, Rose MD, et al. Generation of human vascularized brain organoids. *Neuroreport.* 2018;29(7):588–593. https://doi.org/10.1097/WNR.0000000000001014.
67. Ballabh P, Braun A, Nedergaard M. The blood-brain barrier: an overview: structure, regulation, and clinical implications. *Neurobiol Dis.* 2004;16(1):1–13. https://doi.org/10.1016/j.nbd.2003.12.016.
68. Nation DA, Sweeney MD, Montagne A, et al. Blood-brain barrier breakdown is an early biomarker of human cognitive dysfunction. *Nat Med.* 2019;25(2):270–276. https://doi.org/10.1038/s41591-018-0297-y.
69. Sweeney MD, Montagne A, Sagare AP, et al. Vascular dysfunction-The disregarded partner of Alzheimer's disease. *Alzheimers Dement.* 2019;15(1):158–167. https://doi.org/10.1016/j.jalz.2018.07.222.
70. Cakir B, Xiang Y, Tanaka Y, et al. Engineering of human brain organoids with a functional vascular-like system. *Nat Methods.* 2019;16(11):1169–1175. https://doi.org/10.1038/s41592-019-0586-5.
71. Pasca AM, Sloan SA, Clarke LE, et al. Functional cortical neurons and astrocytes from human pluripotent stem cells in 3D culture. *Nat Methods.* 2015;12(7):671–678. https://doi.org/10.1038/nmeth.3415.
72. Sloan SA, Darmanis S, Huber N, et al. Human astrocyte maturation captured in 3D cerebral cortical spheroids derived from pluripotent stem cells. *Neuron.* 2017;95(4):779–790 e6. https://doi.org/10.1016/j.neuron.2017.07.035.
73. Madhavan M, Nevin ZS, Shick HE, et al. Induction of myelinating oligodendrocytes in human cortical spheroids. *Nat Methods.* 2018;15(9):700–706. https://doi.org/10.1038/s41592-018-0081-4.
74. Ormel PR, Vieira de Sa R, van Bodegraven EJ, et al. Microglia innately develop within cerebral organoids. *Nat Commun.* 2018;9(1):4167. https://doi.org/10.1038/s41467-018-06684-2.
75. Birey F, Andersen J, Makinson CD, et al. Assembly of functionally integrated human forebrain spheroids. *Nature.* 2017;545(7652):54–59. https://doi.org/10.1038/nature22330.
76. Dutta D, Heo I, Clevers H. Disease modeling in stem cell-derived 3D organoid systems. *Trends Mol Med.* 2017;23(5):393–410. https://doi.org/10.1016/j.molmed.2017.02.007.
77. Xiang Y, Tanaka Y, Cakir B, et al. hESC-derived thalamic organoids form reciprocal Projections when fused with cortical organoids. *Cell Stem Cell.* 2019;24(3):487–497. https://doi.org/10.1016/j.stem.2018.12.015. e7.
78. Brownjohn PW, Smith J, Portelius E, et al. Phenotypic screening Identifies modulators of amyloid precursor protein processing in human stem cell models of Alzheimer's disease. *Stem Cell Rep.* 2017;8(4):870–882. https://doi.org/10.1016/j.stemcr.2017.02.006.
79. Kondo T, Imamura K, Funayama M, et al. iPSC-based compound screening and in vitro trials identify a synergistic Anti-amyloid beta combination for Alzheimer's disease. *Cell Rep.* 2017;21(8):2304–2312. https://doi.org/10.1016/j.celrep.2017.10.109.
80. Imbimbo BP, Watling M. Investigational BACE inhibitors for the treatment of Alzheimer's disease. *Expet Opin Invest Drugs.* 2019;28(11):967–975. https://doi.org/10.1080/13543784.2019.1683160.
81. van der Kant R, Langness VF, Herrera CM, et al. Cholesterol metabolism is a druggable Axis that independently regulates tau and amyloid-beta in iPSC-derived Alzheimer's disease neurons. *Cell Stem Cell.* 2019;24(3):363–375. https://doi.org/10.1016/j.stem.2018.12.013. e9.
82. Efavirenz for patients with Alzheimer's Disease (EPAD). https://clinicaltrials.gov/ct2/show/NCT03706885.

SECTION III

New technology, industry perspective, and transitioning to the clinic

CHAPTER

12

Developing clinically translatable screens using iPSC-derived neural cells

Sandra J. Engle[1], *Patrick Faloon*[2,3], *Abigail Mariga*[3,4] and *Robin J. Kleiman*[1,5]

[1]Biogen, Translational Biology, Cambridge, MA, United States; [2]Vesalius Therapeutics, Cambridge, MA, United States; [3]Biogen, Biotherapeutics and Medicinal Sciences, Cambridge, MA, United States; [4]Bristol Myers Squibb, Neuroscience Research and Early Development, Cambridge, MA, United States; [5]Alkermes Inc., Research, Waltham MA, United States

Introduction

Discussions of drug discovery often start with dismal statistics. The probability of success is low, particularly in neuroscience, and the cost is high.[1–3] Productivity declines in the late 20th and early 21st century undermined optimism that advances in genetics and genomics, coupled with advances in automation and computation, would lead to unprecedented medical advances.[4] Drug discovery scientists, keenly aware of the shortfall in industry productivity, have repeatedly engaged in systematic evaluation of the strengths and weakness of the drug discovery process.[5–8]

Four principles emerged from those evaluations. First, human drug discovery should be driven by human data. The over reliance on rodent models often does not consider the physiological differences between species, which can lead to inaccurate conclusions.[9,10] This is particularly true in neuroscience drug discovery, where rodents do not naturally develop the neurodevelopmental, neurodegenerative, and neuropsychiatric disorders that afflict humans. Second, physiologically relevant cells should be used to generate preclinical data. While cancer-derived, immortalized cell lines are easily accessible and have been used heavily in drug discovery, they do not represent normal cell biology and cannot replicate the functional specialization inherent to cells of the nervous system such as neurons, astrocytes,

Phenotyping of Human iPSC-derived Neurons
https://doi.org/10.1016/B978-0-12-822277-5.00011-0

oligodendrocytes, and microglia. Third, the parameter that is measured should be directly related to the phenotype to be modulated. Surrogate markers and downstream signals are often easier to assess but can be impacted by artifacts stemming from in vitro or species-dependent systems. Readouts from in vitro signaling cascades do not always result in the desired, more complex, physiological process. To ensure a physiologically relevant result, the cell type used for the assay must be capable of producing the phenotype of interest.[11] This is especially true when the underlying biology is not well understood, since contrived cell systems can lead to artificial results. Finally, drug discovery programs that are underpinned by human genetic data have a higher probability of success.[12,13] With the advent of lower cost genome sequencing has come large sequencing projects which marry clinical information with genetic information at both the population and patient level.[14] This information can be used to stratify patients to ensure that therapies are targeted to patients with the highest likelihood of receiving benefit and the lowest probability of suffering no benefit or an adverse effect.

These principles have become more urgent as new therapeutic modalities have become increasingly human-specific. Within the relatively narrow realm of small molecule drug discovery, there is a reasonable expectation that a compound can be found that interacts with a conserved binding site on the same protein in different species. However, this becomes more challenging with antibody-based therapies as epitopes can differ substantially between species, or with antisense oligonucleotide therapies that require that the agent match at the highly variable nucleotide sequence level. Gene therapy also requires human sequences to substitute for human mutations, and noncoding RNAs such as micro RNAs are notoriously differentially regulated among species. In protein replacement therapies, an immune response to nonhuman or human modified proteins becomes a concern, and this concern is amplified in gene therapies where protein expression is permanently modified. Cell-based therapies are the ultimate human therapy and can only be appropriately investigated and modeled using human cells.

The convergence of humanized drug discovery and the human specificity of new therapeutic modalities have placed induced pluripotent stem cells (iPSCs) at the center of patient-driven research. From their initial discovery, iPSCs were widely seen as a solution to poorly predictive nonhuman models.[15–17] Derived directly from patients, often with known genetic mutations and accompanying clinical phenotypes, these cells are capable of differentiation into potentially unlimited quantities of physiologically relevant cell types. Thus, iPSCs offer a foundation upon which to build a clinically translatable drug discovery platform. This is particularly important in neuroscience, where access to primary human cells is limited and normal cellular function is complex. If done correctly, these cells can offer the first step in building a "chain of translatability" that links the underlying mechanism of disease from the in vitro assay directly to the same molecular drivers measured in the preclinical animal disease model and ultimately to the clinical features to be measured in patients.[18] A strong "chain of translatability" increases the probability that a novel therapeutic will yield a beneficial effect in the clinic. A weak "chain of translatability" creates gaps in understanding the relationship between the parameters measured in vitro and parameters measured in vivo. In the past, weak "chains of translatability" have created the uncertainties that, in hindsight, have led to clinical failure.

In this chapter, we will discuss some of the key considerations for establishing meaningful and reproducible screens that support drug discovery programs using iPSC-derived cell types unique to the central nervous system. This includes careful consideration of whether an iPSC-derived model is best suited to the problem being asked, followed by careful selection of the appropriate endpoint and detection platform, minimizing the variability of the assay and finally scale-up and automation of the assay. For clarity, we will use the following terminology: A platform serves as an organizing experimental paradigm that can contain screens and/or assays. An assay is defined as a reproducible protocol to measure a feature of interest. A screen is a specific effort to use an assay or a series of assays to search for a molecule with prespecified properties. A screening funnel may contain a high throughput screening (HTS) assay and multiple counter assays and/or secondary assays.

Is an iPSC-derived platform right for the application?

The Law of Parsimony states that the best explanation is the one that requires the fewest possible assumptions. Similarly, the best platform for identifying new therapies for a disease is often the simplest. Just because one can use iPSC-derived cells and models does not mean that these will always be the best choice. A platform and the screens and assays contained within it should be fit-for-purpose and drive decision-making for selection of the therapeutic entity.

Building an effective iPSC-based platform is not a trivial undertaking. While the obstacles to working with iPSCs have lessened in recent years, even the most enthusiastic supporters of iPSC technology recognize the current challenges and limitations. iPSCs are generally more expensive, time consuming, and labor intensive to work with than either immortalized cell lines or rodent primary cells. As discussed in detail below, an iPSC-based platform requires consideration of factors that have only marginal impact when working with cell lines. Furthermore, iPSCs remain an imperfect replica of in vivo biology, as iPSC-derived neurons can fail to achieve the full gene and protein expression profiles of cells found in endogenous human healthy or diseased primary tissue. Late-onset, low penetrant, or environmentally induced phenotypes may be difficult to reproduce. Central nervous system disorders often have limited understanding of underlying biology and require multiple interacting cell types to manifest a disease phenotype, making them particularly challenging to recreate in vitro.

Despite these caveats, iPSC-derived cells do capture fundamental features of human neurobiology that cannot be replicated by any other method in a robust, reproducible, and scalable manner. This is particularly important in neuroscience where human primary tissue is particularly limiting. iPSC-derived cells represent a native cell type with high disease relevance that can respond to disease appropriate stimuli, such as a patient mutation and/or genetic background, to produce a functional phenotype that correlates with a clinical endpoint. iPSC-derived neurons need not replicate every feature of the biology being investigated to be useful; only those features that are required to test the hypothesis or define the features being sought in the therapeutic modality. When applied appropriately, an iPSC-based drug discovery platform captures the critical complexity of the human in vivo phenotype to allow for the development of quantifiable and biologically useful assays.

When considering if an iPSC-based platform is suitable for a potential disease application, several criteria should be reviewed. Consider whether iPSC-derived screens and assays will be used to support the entire platform or used selectively to confirm findings obtained from less physiologically relevant but potentially higher throughput or cost-effective screens and assays. Determine at a high level whether an appropriate cell type can be made at the necessary scale which captures the relevant biology and the endpoint of interest. Review the skill set of those expected to carry out the work, as iPSC culture and *in vitro* differentiation require advanced cell culture skills, significant attention to detail and adherence to standardized procedures. Verify whether project leadership is aligned with the likely increased investment in resources and timelines. If, after thoughtful consideration, an iPSC-based platform, either in total or part, is still deemed appropriate, attention should be paid to the specifics of assay design.

What factors should be considered in developing iPSC-based assays?

At a minimum, all assays are composed of a starting material, the molecule being tested, and an endpoint measurement. Each is important and interdependent from the other. For iPSC-based assays, there are many choices that can be made at each point in the process and these choices can significantly impact the likelihood of obtaining a meaningful outcome. To design a biologically meaningful assay, it is important to select iPSC lines and terminal cell types that are most suitable for the disease to be studied and to choose endpoints that measure disease-relevant phenotypes and support the chain of translatability.

iPSC line selection

One of the strongest justifications for using iPSC-derived cells is the disease relevance inherent in a reagent prepared from patients. Patient-derived iPSCs offer the benefit of capturing the genetic complexity that led to disease even when the underlying genetic cause of the disease is not known, such as in sporadic disease. iPSC lines from clinically well-characterized patients, for whom disease progression and response to therapeutic interventions is well-documented, can allow grouping of iPSC lines by clinical symptoms, which may have relevance to the underlying mechanism of disease. In disorders for which causal or contributing genetic variants are known, patients can be genotyped prior to iPSC generation to ensure appropriate classification and comparison. When patient populations are rare or difficult to access, genomic modification techniques can be used to introduce known, disease-causing mutations into an iPSC line from a healthy control individual.[19] The advantage of this approach is that it increases the diversity of mutations that can be studied. Additionally, it creates isogenic pairs in which the genetic background is constant, and the predominant source of variation is the mutation. This can be helpful in assessing the impact of different mutations on cellular phenotype and therapeutic intervention. Genomic modification techniques can also be used to restore patient mutations to the canonical sequence with the expectation that cellular phenotypes associated with the disease are rescued.[20] Age-, sex-, and, preferably, ethnic background-matched healthy iPSC lines are necessary to establish a well-controlled assay.

In a well-designed iPSC-based platform, it is likely that all four types of iPSC lines—patient lines, mutation-corrected patient lines, engineered mutant lines, and healthy controls—will be necessary at some point in the drug discovery process as each provides unique information. From a practical perspective, the primary assay in a screen is likely to focus on only a single iPSC line or pair of iPSC lines. Additional iPSC lines with varying mutational burden, sex, or ethnic background will be incorporated at later stages of the process to confirm the initial results. Numerous studies have shown that the donor's genetic make-up is the predominant driver of phenotype.[15,21–23] This may require the testing of several healthy control and patient iPSC lines to understand the range of phenotypes and donor to donor variability inherent in measured endpoints. When an assay is reliant on comparing unrelated healthy and disease iPSCs, four to six individual lines per group are likely necessary to achieve reasonable sensitivity and specificity.[24] The use of isogenic iPSC lines may require only three to four pairs to achieve the same sensitivity and specificity.[24] Sporadic diseases, diseases with significant contribution from modifying genes, or diseases in which genotype-phenotype correlation is weak may require additional iPSC lines to achieve meaningful results. Given the time and expense associated with acquiring iPSC lines, careful consideration should be given to the information each type of iPSC line will provide and how many individual lines will be necessary within a category.

Notwithstanding the number of lines needed, iPSC lines should be of the highest quality. Several large, reputable iPSC repositories provide an increasing diversity of healthy control and patient-derived iPSC lines.[11,25] These lines are readily available to academia and industry for a nominal fee under a standard material transfer agreement (MTA). Additionally, these lines have been characterized for the key iPSC attributes of pluripotency, genetic stability via karyotyping, negative mycoplasma status and microbial sterility, thus making them a good starting point for assay and platform development. If the desired patient population is not represented in iPSC repositories, it may be necessary to consider alternative sourcing options. One option is to acquire relevant iPSC lines from individual laboratories or small institutional repositories. This can be time consuming and challenging as individual MTAs will need to be negotiated and the quality of iPSC can vary significantly.[26,27] A second option is to generate new iPSC lines. Several non-integrating options such as episomal vectors, Sendai virus, and RNA transfection which produce fully pluripotent cells are available for reprogramming and are preferred over older technologies such as lentivirus that result in residual integration of reprogramming factors.[28] While a variety of somatic tissues can serve as the starting material for iPSC generation, peripheral blood mononuclear cells, properly collected and preserved, have rapidly become the preferred choice due to the relative ease of acquisition from the patient.[26] Generation and characterization of new iPSC lines can take well over 4 months, not including the time to identify and properly consent the appropriate patient population, and this should be factored into timelines when considering this option. A third option, as mentioned previously, is to engineer a healthy iPSC line to carry the mutation of interest. CRISPR technology has revolutionized the introduction of genetic changes into iPSCs; however, it is still a 4-6-month process to generate and characterize the resulting clones. The time needed to reprogram or edit one's own iPSC lines may be worth the investment to ensure that the foundation of an iPSC platform is built on quality, which will be discussed in more detail below.

Cell type selection

Equally important as the choice of iPSC lines to be evaluated is the selection of cell type to be assayed. This is intimately intertwined with the identification of an appropriate assay endpoint. As noted in the introduction, to build a strong chain of translatability requires that the same underlying mechanism of disease be evaluated in the patient and in vitro. To do this effectively, the cell type used in the assay must be able to demonstrate the relevant phenotype. However, biological relevance is a continuum and feasibility must be taken into consideration. If the number of molecules to be assayed is large and the yield of the desired cell type during production is low or heterogeneous, it may not be possible to capture both the most relevant cell type and the most relevant phenotype in the same assay. For example, spinal muscular atrophy (SMA) is caused by a mutation in the SMN1 gene which reduces total SMN protein in motor neurons and can be mitigated by increasing productive splicing of a paralogue gene, SMN2. While an assay to quantify SMN protein in SMA patient-derived motor neurons would seem to be the most physiologically relevant choice, the ability to generate large numbers of patient-specific iPSC-derived motor neurons and measure meaningful changes in low levels of SMN protein could present a technical challenge. It may be more feasible to develop a protein quantitation assay in SMA patient-derived undifferentiated iPSCs or neural progenitors, which would preserve many aspects of iPSC value such as the human and patient context, while accommodating the need for robust and reproducible scaling of cell production. Addition of an orthogonal assay to the screening funnel that includes iPSC-derived motor neurons measuring SMN2 splicing, while potentially lower in throughput, would address both the need for physiological relevance and feasibility. Conversely, if the primary goal is to identify molecules that promote neuronal axon myelination via phenotypic screening, then the most appropriate cell types for use in the assay may be axon-generating neurons in co-culture with oligodendrocytes capable of myelination. The number of molecules to be screened would need to be adjusted to accommodate the more prolonged and potentially limited cell production.

In some instances, it may be worthwhile to consider an alternative physiologically relevant cell choice. For example, Friedrich's Ataxia patients have autosomal recessive loss of function mutations that reduce expression of frataxin protein and, as a result, they suffer both neurological and cardiovascular symptoms. Clinically, it may be more tractable to measure acute changes in cardiovascular phenotypes using well-established, quantitative measures as compared to neurological phenotypes. In this instance, developing an *in vitro* assay in patient iPSC-derived cardiomyocytes, instead of neurons, may provide a more robust primary assay with a physiologically relevant measure of therapeutic modulation. Secondary in vitro assays using iPSC-derived neurons could potentially provide physiologically relevant information on the neurologic impacts.

A key consideration regarding the choice of assay cell type is the ability to generate the cell type in a robust and reproducible manner in quantities that support the intended assay. iPSC-derived neuron and glia differentiation protocols generally fall into two broad categories: culture-driven protocols or transcription factor driven protocols. Culture-driven protocols attempt to recapitulate embryonic development through use of a series of cell culture media supplemented with growth factors or small molecule inhibitors and/or agonists of developmental pathways. These protocols step through the various cell stages of differentiation and often provide points for cryopreservation of expanded cell populations, such as neural progenitor cells, which can facilitate batch production and handling.[29–33] Drawbacks to these approaches are that culture-driven protocols can be prolonged (>60 days) and subtle differences in the quality of the starting undifferentiated iPSCs, the addition of patterning factors, or cell densities can lead to alterations in cell fate.[34,35] Low conversion to the desired intermediary at the initial stages of differentiation can result in heterogeneity at terminal stages. This heterogeneity can make it difficult to obtain meaningful results from the cell type of interest. The longer the differentiation protocol, the more opportunity for batch to batch variation and heterogeneity. Recently, design of experiment (DoE), a systematic method for evaluating the relationship between factors to determine cause-and-effect relationships, has been applied to developing iPSC-differentiation protocols and shown success in increasing conversion rates at each stage to yield more specific and homogenous cells.[36] Culture-driven protocols are useful when multiple iPSC lines need to be evaluated and the protocol has been well-optimized and confirmed with equal robustness across multiple donor cell lines.

An alternative approach is the use of cell-fate driving transcription factor expression to ensure iPSC differentiation to a specified terminal cell type.[37–40] Transcription factor-driven differentiation approaches tend to be faster and produce a more homogenous cell population. A drawback of this approach is that normal development is by-passed, which may mask subtle phenotypes associated with the development process or a particular stage of differentiation.[41,42] Transcription factors can be introduced through genetic engineering or viral transduction. Engineered iPSC lines are generally inducible and require more up-front effort to produce and characterize than viral transduction, thus making them less suitable when many lines need to be evaluated. However, once the induction and differentiation protocol is optimized, the relative ease of differentiation from engineered iPSC lines often makes this approach most suitable for supporting assays requiring large quantities of cells.[43] Viral transduction can be applied across many cell lines with less upfront work and is technically easier.[40,44] Drug selection can be used to eliminate cells that are not virally transduced, though cells that do not express enough transcription factor to fully convert to the terminal cell type may survive and contribute to cellular heterogeneity. Furthermore, given that transcription factor driven protocols can yield neurons within 1–2 weeks, the cost of labor and reagents required to maintain these cell types makes them a more cost-efficient reagent for large screening campaigns.

Regardless of the cell type or the differentiation method chosen, the careful characterization and qualification of the cells is essential prior to final selection. Characterization by gene and protein expression should be carried out at key transition points in the differentiation protocol and metrics should be established to identify minimal acceptable criteria for progression to the next stage. An unbiased assessment of the intended cell type should be used to determine empirically how closely the in vitro differentiated cells resemble the endogenous human target cell type. As mentioned previously, the cells do not need to recapitulate every feature of the endogenous tissue; however, at a minimum, the in vitro differentiated cells should express the target protein and key pathways involved in the disease process. Additionally, the terminal cells should be characterized to determine if they can respond to relevant stimuli with an appropriate response.

Assay endpoint selection

Having weighed the choices of iPSC lines and the cell type to assay, the assay endpoints are the third critical component to consider. To maintain a strong chain of translatability, it is important to ensure that the underlying mechanism of disease biology that will be measured in the clinic is captured by the set of assays that define the screening funnel. The specific parameter to be measured should be closely linked to the phenotype to be modulated to prevent the introduction of extraneous influences on the outcome and any stimulus used to induce a disease state should be relevant to the disease.[11]

Primary (first tier) assays are used to efficiently identify low potency molecules (Hits) for further optimization. Standard primary assay endpoints include quantitative measures of protein, post-translational modification (e.g., phosphorylation), RNA expression/splicing or analysis of image-based endpoints. Because primary assays are intended to test many molecules, cost, ease of handling, and a robust dynamic measurement range are often key considerations. In some situations, this may mean quantifying the level, location, or phosphorylation of a protein without the need for a functionally mature cellular activity such as action potential firing. Primary assays are generally set up to be more tolerant of false positive molecules with the assumption that these molecules will be identified and eliminated by more stringent, secondary assays.

Primary screens can be broadly classified into target-based screens and phenotypic screens.[45] Target-based screens, as the name implies, seek to inhibit or activate a specific gene, RNA, or protein that has previously been identified through some other mechanism. A phenotypic screen seeks to modulate an observable characteristic that is linked to the disease of interest and provides the opportunity to identify novel targets or pathways. High content screening is a subtype of phenotypic screen that uses automated microscopy and image analysis to quantify observable changes in cell morphology or protein localization. The unique morphology and discrete anatomical location of RNAs and proteins within neuronal cells makes this a potentially powerful technique for use with iPSC-derived cell types. For any type of screen, especially neuronal subtype-specific screens, iPSC-derived neuronal cell types provide cellular context in which cell-specific endogenous isoforms of RNAs or proteins are expressed at native levels with relevant partners and pathways which should be more representative of the *in vivo* situation.

Secondary (second tier) assays, which are used to discriminate among the properties of molecules, have different considerations than primary assays. In addition to measuring basic cellular processes such as viability, secondary assays can be set-up to confirm the activity of the molecule with a different technology than used in the primary screen (orthogonal assays) or to measure off-target effects. These assays can be lower throughput and more labor intensive. They often focus on measuring functional attributes of the cells such as electrophysiological properties, myelination, or protein secretion. It is at this stage of screening that iPSC lines representing different patient mutations or genetic backgrounds may be evaluated to ensure the broad applicability of modulation.

In an ideal situation, the changes measured in a primary assay would translate into a functional phenotype that could be assayed in secondary assays. For example, increased FMRP protein expression in iPSC-derived glutamatergic neurons rescues hyperexcitable neuronal firing.[46,47] Thus, a target-based phenotypic screening campaign focused on developing therapeutic approaches for Fragile X syndrome (FXS) may develop a primary screen with patient iPSC-derived neural progenitor cells to detect increased levels of FMRP expression by time-resolved fluorescence resonance energy transfer (TR-FRET)-based assay,[48] followed by a secondary functional assay using glutamatergic neurons in an automated multiple electrode array (MEA). An unbiased phenotypic screen for FXS may start with an MEA-based screen in patient-derived iPSC-neurons and use chemogenomic tools to deconvolute the mechanism of phenotypic rescue. The proper sequencing of assays ensures that molecules that emerge from the first-tier assays are identified with the highest throughput and lowest cost, while second tier assays derived in more complex (and expensive) cell types or cell systems, ensure that emerging molecules will support function in a relevant cellular context using a lower throughput assay. Each assay in the screen should allow clear decision-making regarding which molecules should progress.

What factors should be considered when running an iPSC-based screen?

Overall, the design and considerations required for working with iPSC-derived cells are not fundamentally different from best practices identified for developing robust assays with cell lines. Good references exist for weighing the pros and cons of different assay formats and developing appropriate screening statistics as well as implementation of automation to support high throughput screening.[49,50] What differentiates iPSC-based platforms from other cell-based assays is the extra time and effort required to ensure high quality iPSCs and differentiated cells. In the following section, guidance will be given to the practical considerations of using iPSC-derived cell types in a small molecule HTS. However, most of these considerations are broadly applicable to a broad range of assays and therapeutic modalities.

Cell culture

Undifferentiated iPSC culture and scale-up

Careful adherence to stem cell culture best practices[51,52] is critical to success as undifferentiated iPSCs are particularly sensitive to poor handling that can affect their genetic stability, pluripotency, and ability to differentiate consistently.[53] Cultures should be monitored routinely for normal morphology and karyotype, the expression of markers of pluripotency, and the presence of mycoplasma.[26] Fully characterized master banks of 10–20 cryovials should be prepared from newly received iPSC lines and working stocks of 50–100 cryovials should be derived from the master stock. Prolonged culture (>5 passages) of undifferentiated iPSC should be avoided to prevent the accumulation and propagation of genetically unstable clones. Standardized nomenclature of iPSC lines should be used to provide clarity in tracking cell lines.[54,55] Short tandem repeat (STR) PCR analysis or an equivalent should be implemented early in platform development to enable routine cell line authentication.[56]

Undifferentiated iPSCs can be maintained and expanded in a variety of formats. The most common is a two-dimensional, matrix-coated multilayer flask which yields sufficient cells to support weekly screening efforts. For example, a single triple-layered flask where each layer has the same surface area as a T175 flask can yield 1.0×10^9 cells per flask at 80% confluency.

This scale can support a modest number of microtiter plates (10–20) per week depending on assay requirements. Other multilayer formats that can be used have been reported previously.[57] Incubation and automated maintenance of iPSCs in flasks can be accomplished using the CompacT select (Sartorius) or similar robotic platforms, which supports media changing, viability monitoring, and cell harvesting.[58] Suspension cultures using microcarrier beads and hydrogels can also be used for expansion of undifferentiated iPSCs and iPSC-derived neural cell types.[57,59,60] Drawbacks to this approach include challenges with heterogeneity of iPSC aggregate size, reduced expansion rates, and spontaneous differentiation, therefore, careful assessment of cell integrity is critical.[61] Monitoring cell quality for iPSCs in suspension cultures can be achieved by sampling a portion of the batch to evaluate karyotype and expression of pluripotent markers using qPCR. Other approaches include dye exclusion assays and flow cytometry to monitor cell viability, flow cytometry coupled with fluorescence cell barcoding, and mass spectrometry for metabolomic/proteomic profiling for metabolic markers that have an influence on pluripotency.[62,63] Processing time between sample collection and obtaining results may limit the benefit of some of these approaches. Ideally, iPSC scale-up approaches should closely match routine iPSC culture standard operating protocols to avoid significant drift in assay performance. Although labor-intensive, two-dimensional (2D) iPSC scale-up approaches are advantageous for monitoring cell quality during HTS as one can visually inspect colony morphology on each passage cycle in addition to characterizing the expression of pluripotent markers with immunocytochemistry and qPCR.

Differentiated iPSC culture and scale-up

Differentiation of iPSCs into specific cell types in sufficient numbers to support an HTS can occur as adherent cultures in 2D, or as 3D organoid-like preparations. The advantage of 3D preparations is that they can be patterned to produce regional CNS cell types and their supporting cells in large numbers and these preparations will attain maturation states not typically achievable in 2D culture. However, many desired endpoints for assay development are not currently accessible in 3D preparations. An obvious solution is to dissociate 3D preparations and re-plate them in conventional multi-well plates as a cell source for conducting screens.[64] Protocols for large scale differentiation and maturation of iPSC-derived cortical and motor neurons in suspension cultures have been described for producing large yields of cortical, dopaminergic, and motor neurons via differentiation of 3D stem cell spheres in spinner flasks.[65,66] The disadvantage of 3D organoid-like preparations is that while they yield a heterogeneous cell mixture that is more reflective of the brain tissue, this may not be desired if the screen requires a relatively pure target cell type.

To support screens of relatively pure cell types, a transcription factor-driven protocol is typically chosen for all the reasons described previously. If the iPSC line has been genome engineered to express a tetracycline-inducible transcription factor or set of factors required to produce the cell type of interest, then scale up of the engineered or transduced iPSC can often be accomplished by expanding the undifferentiated cells in a progenitor state, which is easier to work with than the iPSC state. Once a sufficient number of progenitor cells have been scaled up, the cells can be plated, and differentiation is triggered with appropriate induction stimuli on the substrate of choice for the final assay.

Cell plating, incubation, and long-term maintenance

Cells can be uniformly plated in many formats using the Multidrop Combi reagent dispenser (ThermoFisher) or the Bioraptor multi-reagent dispenser (Beckman Coulter). When plating large batches of plates, the neuronal suspension should be gently mixed every fifth plate to prevent cells from settling on the bottom of the flask, which can result in uneven dispensing across plates. Plating iPSC-derived neurons in microtiter plates often requires more than one coating, with incubation and wash steps that add time and reduce screening throughput. It is worth testing assay performance in microtiter plates that are pre-coated to reduce plate preparation time. Pre-coated plates are standard offerings from several vendors or can be generated via custom order. Additional substrates for supporting neural attachment and growth, such as laminin or vitronectin, can also be added to the plating media to minimize handling steps during the miniaturization process.[67]

Choice of coating substrate can greatly influence assay outcome and should be carefully assessed. For example, adding laminin allows neuronal processes to grow evenly with longer neurite extensions,[68] and this can increase the dynamic range of assays designed to identify compounds that inhibit axonal growth.[69] Conversely, if one is developing a neurite outgrowth assay, addition of chondroitin sulfate proteoglycans and myelin-derived proteins can be used to inhibit neurite outgrowth, thus increasing the assay window for detecting neurite growth.[70]

Cell clumping can become a chronic challenge when working with iPSC-derived neurons. Adequate plate coating and allowing plates to sit at room temperature for approximately half an hour at room temperature post-plating, but prior to transferring to an incubator can improve even extension of projections over long periods in culture. Another approach is to plate neurons on astrocyte monolayers,[71] but one needs to tailor their readout to minimize any influence by non-neuronal cells. Detailed protocols on seeding iPSC-derived neurons in microtiter plates have recently been published.[72] Minimized cell clumping is most critical for imaging assays where measurements are often on a per cell basis but may be less problematic for assays with a lysing endpoint. Image-based assays may also benefit from specialized plates that offer improved optical clarity, especially those with cyclic olefin-based materials.[73] Additional improvements in flatness and plate uniformity can boost the performance of image acquisition through minimization in z-plate variability (e.g., Aurora Microplate technologies IQ plates and PerkinElmer Cell Carrier Ultra plates).

Long-term maintenance of cultures requires cell culture incubators engineered to maintain a homogenous, high humidity chamber, and minimize assay variability. This variability includes edge effects when disproportionate evaporation of media in the outer wells leads to changes in cell performance and assay outcome. Several incubators are on the market using plate racks in a carousel configuration that can interface with automation, such as ThermoFisher Cytomat, Liconic StoreX, and HighRes Biosolutions SteriStore incubators. Additional ways to minimize edge effect include the application of plate seals, specialized plate lids (Labcyte MicroClime) and incorporating media changes or partial additions into protocols.

Media changes

iPSC-derived neurons often require media changes a few days after plating and throughout the duration of the culture period. However, many neurons tend to lift easily when cultured for long periods, so care should be taken when introducing any media change or washing steps. Determining the minimum media required for optimal growth and adding fresh media on top of existing media, instead of drawing out media, can preserve the neuronal integrity. Additionally, media changes come with the risk of modulating compound concentration, diluting paracrine components secreted within the culture or disrupting cell monolayers. While consideration should be made to minimize regular media changes and maintain sterility, particularly for long-term cultures, there are ways to automate media changes. Traditional plate washers use arrayed pins to remove and add reagents to wells, such as the BioTek line of plate washers and the GNF Systems WDII. The BioTek EL406 plate washer is a versatile liquid handler that has both media removal and addition capabilities and can be used in walk up mode or can easily be integrated to any automation system. If properly configured for media removal, BioTek plate washers can leave little to no residual media. BlueCat Bio BlueWasher plate washers use centrifugation rather than evacuation, which can reduce the number of washing steps and accommodate all microtiter plate types and configurations. They have been successfully implemented to support 1536 well plate-based screens and iPSC-derived neurons in 384-well format, but one must confirm that its use does not impact the biology of the assay, including disturbance of the monolayer.[74–76]

Assay optimization

Assay formats

Protein readouts can be measured with a variety of technologies, such as AlphaLISA, HTRF (Homogeneous Time Resolved Fluorescence), or older methods relying upon artificial reporter systems or GFP-tagged proteins. The use of engineered assay components, such as reporter genes or fusion proteins, has the potential to introduce artifacts. Therefore, inclusion of relevant counterscreens into a triage plan is needed to rule out technical artifacts.

Mass spectrometry-based assays allow for quantification of post-translational modifications and other cellular components such as lipids and metabolites. Advances have been made with mass spectrometry autosamplers to facilitate large scale HTS campaigns with platforms such as the Bruker RapifleX, Apricot Designs ADDA, and even customized Labcyte Echos (Sinclair et al., 2019). The BMG Pherastar and PerkinElmer EnVision plate readers are versatile and sensitive to quantify a wide array of measurements. Transcript levels of mRNA can be measured by quantitative RT-PCR with the ABI QuantStudio or Roche 480 thermocyclers.

Image-based microscopy assays, also called high content screening (HCS), is a key platform for iPSC-derived cells because of the multitude of phenotypes that can be assessed. For neurons, one can measure changes in protein levels, protein-protein relationships, subcellular localization, neuronal morphology or other parameters. These assays are imaged using plate-based microscopes such as the PerkinElmer Opera/Phenix, Yokogawa CT8000 or similar scopes that can handle microtiter plates at scale, have confocal capabilities, flexible fluorescence range and fast acquisition time. Specialty microscopes with ultra-fast acquisition speed and integrated software such as the Vala Sciences KIC200 can capture transient phenomena, including calcium waves following electrical stimulation. Use of DNA dyes or inclusion of multiple parameters in the screen readout, including subcellular localization, can potentially improve assay metrics and is a key part of assay development for HCS. There are several special considerations for imaging neurons, including maintaining cultures for extended periods of time, cell adherence with protocols that require washing of cells and difficulties of parsing out individual cells with imaging software. Neurons extend processes throughout a well. These processes overlap with neighboring cells and are almost impossible to trace with current software. There are some potential modifications to alleviate this issue. One can plate cells at a low cell density to prevent overlapping processes, label a subset of cells with fluorescent reporters, or apply the equivalent of "brainbow" mice where neighboring cells express distinct fluorescent proteins and can be parsed based upon their fluorescent signature.[77] Another approach is testing numerous isogenic cell lines in parallel, which can be done to compare diseased against a corrected line or numerous lines.[78] Pooling of cell lines can be done at larger scale depending upon the assay readout. A standard readout such as cell survival can be paired with barcoding of lines to allow for hundreds of lines to be tested.[79] Of course, one must balance the complexity of executing screens against the size of a compound library, throughput of an assay, and what makes sense to be interrogated as secondary and counterscreen assays following the HTS.

Miniaturization

A robust cell-based assay that is amenable to miniaturization for HTS and automation is key to a successful screen. Assays established in low to medium throughput in 6- to 24-well formats should be miniaturized to formats that increase throughput during screening. Use of low volume microtiter plates will reduce reagents needed to run the assay. For iPSC-derived neurons, cell production can be costly and time-consuming, therefore an assay that has a good signal window with fewer cells per well can be beneficial. Assay workflows using 96- or 384-well formats with iPSCs and iPSC-derived neurons are increasingly being used and have been shown in recent studies.[80]

The 384-well plate-based assays is most widely used for HTS and is the focus of this discussion. However, much of the same instruments and guidance applies to 96-well formats. 1536-well formats can use most of the same equipment with additional considerations.[76] Depending upon the scale of a screening campaign, available funds, and infrastructure one may be working with a collection of instruments in a lab or using those same instruments on fully integrated automated platforms to complete assays (e.g., HighRes Biosolutions, GNF, ThermoFisher, PAA). For smaller scale efforts, use of instruments with plate stackers or robotic arms to feed plates allows the researcher to process an appreciable number of plates and multitask. A key part of manual work is to maintain plate order so that there is minimal difference in the timing of each step and/or incubation times for individual steps across all plates. Plate barcoding and maintaining plates in sequential order eases this process. Plate orientation and an integral system of plate maps (i.e., sample management/informatics) must be maintained to ensure accurate sample tracking throughout lengthy workflows.

There are some specialized considerations for integrating iPSC-derived cells into standard assay formats. For example, in some cases, the specialized plate coating required by iPSC-derived cells can interfere with the assay technologies, resulting in high background. An all-in-one assay format is often desired to maximize screening throughput for primary assays. This is ideal when the entire assay can be performed in the same assay plate. This assay format has worked well for high content imaging assays, but can be challenging for some plate-reader based assay technologies. Our lab found that an HTRF assay with iPSCs was incompatible with Matrigel plate coating. In such cases, a lysate transfer step becomes necessary and this can be easily done using tip-based, arrayed pipettors as described below. Conversely, plate coating has no effect on AlphaLISA immunoassay detection as observed in an assay using iPSC-neurons plated on PDL and laminin.

Executing a screening campaign can be very complex and needs careful planning and management. Using automation flowcharts can streamline managing the entire process. After the assay has been well vetted and adapted for HTS and all necessary instrumentation identified, it may be helpful to create a flowchart with details of the assay in high throughout format, operational steps (reagent addition, washing, incubation, imaging) and instrumentation settings. An example flowchart is illustrated in Fig. 12.1.

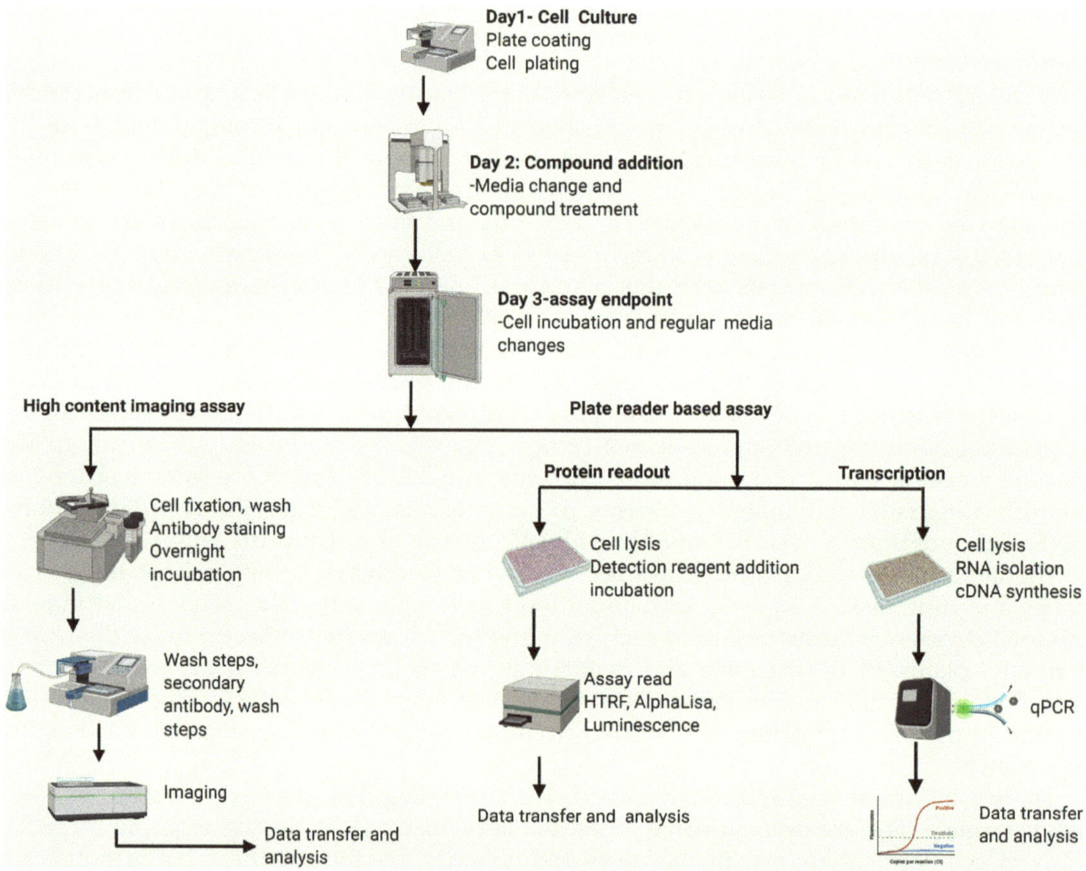

FIGURE 12.1 **HTS workflow using iPSC-derived cells.** Screening workflow can be complex with multiple interconnected steps. A clearly diagrammed process and checklist of steps can ensure smooth execution.

Balancing throughput with assay stability

Assay complexity plays a key role in determining screening throughput. It is important to identify key steps in an assay that may limit throughput. Compound transfers can be a limiting step for screens especially when an intermediate dilution step is required before the compounds are added to assay plates. To maximize throughput, runs can be staggered across multiple days to allow three or more runs per week depending on assay incubation time, the number of steps and the complexity of the takedown process, as exemplified in Fig. 12.1. For assays with a lysis step after assay incubation, lysed plates can be frozen. The assay will be resumed when plates are thawed and analyzed on a different day.

A staged HTS campaign requires understanding the stability of reagents and their impact on an assay. Some reagents may be light or temperature sensitive and this can result in plate to plate variability over time. Care should be taken to reduce exposure to light, by wrapping liquid dispensing tubing with light blocking foil throughout the duration of the screen. For temperature-sensitive reagents, reservoirs and dispenser tubing need to be kept on ice. In cases where reagents are needed for extended times, one needs to accommodate ice melting through careful setup of ice and reagent containers or through regular replenishment of ice. Other factors that affect reagent performance include reagent precipitation in tubing and clogging of dispenser tips. Fluid dispensers should be monitored routinely and fixed. Assay variability and subsequent changes in signal or background can occur over long periods of time and should be assessed in pilot runs. HTS protocols should ensure that all plates of a run are processed with minimal changes in timing for minimal impact on assay performance. HTS reagents should be aliquoted into single use tubes to avoid multiple freeze-thaw cycles, especially for proteins and antibodies. When possible, purchase reagents in bulk so that the same batch or lot number is used for the entire campaign and hit confirmation phase. Use of a single lot for the entire screen is ideal to minimize batch to batch variability, minimize timeline for execution and avoid work stoppage if a replacement lot fails to meet requirements mid-campaign. For iPSCs, variability can also occur between cycles of different cell thaws, and passage numbers or separate rounds of differentiation. Therefore, assay outcome should be evaluated with multiple thaws, passage numbers and differentiation cycles.

Small molecule considerations

Small molecule libraries are usually stored as millimolar stocks in dimethylsulfoxide (DMSO). The high concentration and subsequent dilutions minimize the amount of DMSO in the assay buffer or cell culture media well below 0.5% and ideally at 0.1%. As an example, neurons in microelectrode array studies tolerate 0.2% DMSO.[81] It is important to measure the tolerance of each assay to a titration of DMSO. This requires monitoring the assay outcome and cell viability, as solvents can have a variety of influence on assays[82] as well as impact on differentiation of iPSCs.[83,84]

Compound transfers require accurate volumes and methods to minimize unintentional carryover of compounds into subsequent plates. Volumes used are typically in the low μL to hundreds of nL range. Arrayed pin (e.g., Analytika Jena's CyBio Vario)[85] or disposable tip-based instruments are typically used for compound transfers. If limited to standard pipettors or workstations (e.g., Beckman Biomek i7, Hamilton MicroStar, HighRes Biosolutions Prime), one can perform intermediate dilutions to obtain appropriate concentrations and volumes. Compound addition has been revolutionized by efficient and accurate acoustic pipettors such as the Labcyte Echo (Beckman) that eliminate the need for disposable tips or solvent-based pin tool setups.[86] The key thing to accommodate acoustic transfers is the need to flip the plate upside down for compound additions. If media and cells are present, it should never exceed ~30 μL for a typical 384 well plate so that surface tension prevents the media from falling out of the plate. Targeting the Echo dispense to the corner of the well where the meniscus is deeper will minimize disruption of the cell monolayer.[87] As an alternative, one can dispense into a dry plate, known as an assay ready plate (ARP) [86], followed by the addition of cells and media, but this may not be feasible for long term cultures of already differentiated neurons. With most iPSC-based assays, a media change step may be necessary. In such cases, compounds can be added to ARPs followed by generation of intermediate source plates using an Agilent Bravo or Beckman BioMek or HighRes Biosolutions Prime liquid handler device to transfer assay media and compound to the assay plate. There are additional considerations when using ARPs, such as compound stability and solubility.[88]

A key aspect of handling small molecule compounds is sample integrity. DMSO is hygroscopic and solutions will absorb water over time that can lead to changes in small molecule properties.[89] It is best to store plates with DMSO-compatible seals in a dry environment, preferably under inert gas. Use of MicroClime lids saturated with DMSO was shown to maintain low water content[87] but these lids are expensive and require additional considerations when used in automation. Another factor to consider is storage temperature. Some compounds will become insoluble with multiple freeze-thaw cycles. Our lab stores actively used plates at room temperature and monitors water content with the Echo survey capacity, while longer term storage is at −80°C. Critically, the final screening concentration of compound must be established. Cell based assays are often testing compounds in the low μM range to strike a balance between maximizing likelihood of compound effect and minimizing the potential for cellular toxicity.

Assay analytics

Data analysis and hit determination

The use of screening software (e.g., Genedata Screener, Collaborative Drug Discovery Vault, Dotmatics Vortex, or TIBCO Spotfire) will be necessary to monitor assay outcome, plate effects, patterns or trends within a run and determination of active compounds when dealing with large numbers of compounds.[90,91] Data can easily be imported from instruments or web-based data storage for quick assessment of assay performance and overall analysis of data for hit determination. These tools can be integrated with high content imaging data analysis platforms such as PerkinElmer's Columbus and Harmony to define assay parameters and subsequent statistics for image-based screens. Cell Profiler is another open-source image analysis platform that can also handle high content imaging data.[92,93] Cell Profiler has the added benefit of community sharing methods and scripts.

A key step before hit determination is compound mapping where each compound is mapped to the corresponding assay plate well where it was delivered to establish compound identity. This can be achieved by creating compound mapping files in Python, Pipeline Pilot or R. Mapping of data is best managed with an integrated Laboratory Information Management System (LIMS), but there are few fully developed commercial options and many organizations resort to customized platforms[94] or coupling several commercially available tools to create an integrated environment.

Normalization methods and hit selection

Assay metrics have been covered in detail elsewhere[95–97] but an overview pertaining to cell-based assays follows. HTS data is prone to variability and data normalization is applied to minimize the impact of assay variability across plates and over multiple experiments to reduce the rate of false positives. Data normalization can incorporate the use of both positive and negative controls to calculate the signal as a percent of control populations.[98] The position of controls on a plate can impact the performance of the controls and the subsequent assay normalization.[99] Normalization methods that do not use controls exist, such as B-scores or Z-scores, where relative potency score is determined using raw values of every compound on the plate and has been applied for high throughput screening with iPSC-derived neurons.[43,95] Data analysis platforms such as Genedata Screener have built-in normalization approaches that let the user explore different methods. One of the most prevalent metrics used in HTS assays is Z' factor[100] which incorporates the mean values of signal and background for control populations along with the standard deviation of these same populations. Achieving a Z' of 0.five to one is ideal, but some readouts without a large window need minimal variability to achieve a Z' of ≥ 0.5. For assays below 0.5, hit calling becomes less reliable and the number of false positives will increase.

Use of a pilot screen prior to execution of the full-fledged HTS is a wise investment of time to inform on performance and determine how to advance an assay or program in a cost-effective manner. Pilot screens can validate all workflows, identify weak spots of the protocol, assess the sources of variability, and determine an initial hit rate. Ideally, pilot studies should have a considerable number of plates (approximately 50 plates with ~3000 compounds) where both technical and biological replicates can be incorporated. Chemogenomic libraries, sometimes referred to as bioactive sets,[101,102] or a chemical diversity set representative of the larger library, are reliable compound sets to use in a pilot run. In addition, use of a chemogenomics set may provide insight into a mechanism responsible for an assay phenotype or identify potential positive control compounds.[103] The pilot should be run twice on separate days to verify that hit rates are consistent between runs and there is good concordance between biological replicates. For clarity, a technical replicate is when a perturbation is tested at the same time in several wells with the same batch of cells and reagents either on the same plate or across multiple plates. A biological replicate is when a perturbation is tested with independent batches of cells on different days.

Determination of active compounds can be done in multiple ways[95] such as selecting a set number of compounds with the greatest activity, setting a percent activity threshold based upon controls, appointing a cutoff of a number of 3 times or greater standard deviation above the median DMSO value, or the Z score method.[104] It is quite common to have up to 50% of false positives that do not confirm in post-HTS studies. False positives can be partially avoided through screening of replicates[105] but that dramatically increases the overall workload. Factors that can determine when and how many replicates should be used include assay complexity, ease of cell maintenance and reagent costs. Finally, once active compounds that qualify as hits are identified, the number of compounds will be significantly reduced in number compared to the full screening collection, which should allow for testing with biological and technical replicates across 8–10 different concentrations to permit potency determination.

Summary

The use of human iPSC-derived cell types can provide a patient-relevant cellular context for screening activities aimed at the discovery of novel therapeutics to treat brain disorders. Key considerations for the development of assays include whether an iPSC-derived model is "fit for the purpose" and the careful selection of appropriate endpoints and detection platform. Great care needs to be taken to maximize the quality of the iPSC cells and to develop workflows for plating and long-term maintenance of cultures with protocols that minimize media changes and evaporation. With careful planning and use of automation equipment to minimize the number of hands-on steps for the scale up and assay optimization, the execution of a high throughput screen can be accomplished using these very specialized cells. In the long run, use of these platforms should increase the translational potential of emerging therapeutic candidate molecules.

References

1. DiMasi JA, Grabowski HG, Hansen RW. Innovation in the pharmaceutical industry: new estimates of R&D costs. *J Health Econ*. 2016;47:20–33. https://doi.org/10.1016/j.jhealeco.2016.01.012.
2. Khanna I. Drug discovery in pharmaceutical industry: productivity challenges and trends. *Drug Discov Today*. 2012;17(19–20):1088–1102. https://doi.org/10.1016/j.drudis.2012.05.007.
3. Wong CH, Siah KW, Lo AW. Estimation of clinical trial success rates and related parameters. *Biostatistics*. 2019;20(2):273–286. https://doi.org/10.1093/biostatistics/kxx069.
4. MacArron R, Banks MN, Bojanic D, et al. Impact of high-throughput screening in biomedical research. *Nat Rev Drug Discov*. 2011;10(3):188–195. https://doi.org/10.1038/nrd3368.
5. Cook D, Brown D, Alexander R, et al. Lessons learned from the fate of AstraZeneca's drug pipeline: a five-dimensional framework. *Nat Rev Drug Discov*. 2014;13(6):419–431. https://doi.org/10.1038/nrd4309.
6. Paul SM, Mytelka DS, Dunwiddie CT, et al. How to improve RD productivity: the pharmaceutical industry's grand challenge. *Nat Rev Drug Discov*. 2010;9(3):203–214. https://doi.org/10.1038/nrd3078.
7. Tollman P, Panier V, Dosik D, Biondi P, Cuss F. Organizational effectiveness: a key to R&D productivity. *Nat Rev Drug Discov*. 2016;15(7):441–442. https://doi.org/10.1038/nrd.2016.91.
8. Wang L, Plump A, Ringel M. Racing to define pharmaceutical R&D external innovation models. *Drug Discov Today*. 2015;20(3):361–370. https://doi.org/10.1016/j.drudis.2014.10.008.
9. Hodge RD, Bakken TE, Miller JA, et al. Conserved cell types with divergent features in human versus mouse cortex. *Nature*. 2019;573(7772):61–68. https://doi.org/10.1038/s41586-019-1506-7.
10. Knight A. Animal experiments scrutinised: systematic reviews demonstrate poor human clinical and toxicological utility. *ALTEX*. 2007;24(4):320–325. https://doi.org/10.14573/altex.2007.4.320.
11. Vincent F, Loria P, Pregel M, et al. Developing predictive assays: the phenotypic screening rule of 3. *Sci Transl Med*. 2015;7(293). https://doi.org/10.1126/scitranslmed.aab1201.
12. King EA, Davis JW, Degner JF. Are drug targets with genetic support twice as likely to be approved? Revised estimates of the impact of genetic support for drug mechanisms on the probability of drug approval. *PLoS Genet*. 2019;15(12):e1008489. https://doi.org/10.1371/journal.pgen.1008489.
13. Nelson MR, Tipney H, Painter JL, et al. The support of human genetic evidence for approved drug indications. *Nat Genet*. 2015;47(8):856–860. https://doi.org/10.1038/ng.3314.
14. Dubow TM. *Population-Scale Sequencing and the Future of Genomic Medicine*. 2016.
15. Takahashi K, Okita K, Nakagawa M, Yamanaka S. Induction of pluripotent stem cells from fibroblast cultures. *Nat Protoc*. 2007;2(12):3081–3089.

16. Takahashi K, Tanabe K, Ohnuki M, et al. Induction of pluripotent stem cells from adult human fibroblasts by defined factors. *Cell*. 2007;131(5):861–872. https://doi.org/10.1016/j.cell.2007.11.019.
17. Yu J, Vodyanik MA, Smuga-Otto K, et al. Induced pluripotent stem cell lines derived from human somatic cells. *Science*. 2007;318(5858):1917–1920. https://doi.org/10.1126/science.1151526.
18. Moffat JG, Vincent F, Lee JA, Eder J, Prunotto M. Opportunities and challenges in phenotypic drug discovery: an industry perspective. *Nat Rev Drug Discov*. 2017;16(8):531–543. https://doi.org/10.1038/nrd.2017.111.
19. Wang Y, Liang P, Lan F, et al. Genome editing of isogenic human induced pluripotent stem cells recapitulates long QT phenotype for drug testing. *J Am Coll Cardiol*. 2014;64(5):451–459. https://doi.org/10.1016/j.jacc.2014.04.057.
20. Jang YY, Ye Z. Gene correction in patient-specific iPSCs for therapy development and disease modeling. *Hum Genet*. 2016;135(9):1041–1058. https://doi.org/10.1007/s00439-016-1691-5.
21. Aasen T, Raya A, Barrero MJ, et al. Efficient and rapid generation of induced pluripotent stem cells from human keratinocytes. *Nat Biotechnol*. 2008;26(11):1276–1284. https://doi.org/10.1038/nbt.1503.
22. Staerk J, Dawlaty MM, Gao Q, et al. Reprogramming of human peripheral blood cells to induced pluripotent stem cells. *Cell Stem Cell*. 2010;7(1):20–24. https://doi.org/10.1016/j.stem.2010.06.002.
23. Zhou T, Benda C, Dunzinger S, et al. Generation of human induced pluripotent stem cells from urine samples. *Nat Protoc*. 2012;7(12):2080–2089. https://doi.org/10.1038/nprot.2012.115.
24. Germain PL, Testa G. Taming human genetic variability: transcriptomic meta-analysis guides the experimental design and interpretation of iPSC-based disease modeling. *Stem Cell Rep*. 2017;8(6):1784–1796. https://doi.org/10.1016/j.stemcr.2017.05.012.
25. Clemons PA. Complex phenotypic assays in high-throughput screening. *Curr Opin Chem Biol*. 2004;8(3):334–338. https://doi.org/10.1016/j.cbpa.2004.04.002.
26. Engle SJ, Blaha L, Kleiman RJ. Best practices for translational disease modeling using human iPSC-derived neurons. *Neuron*. 2018;100(4):783–797. https://doi.org/10.1016/j.neuron.2018.10.033.
27. Volpato V, Webber C. Addressing variability in iPSC-derived models of human disease: guidelines to promote reproducibility. *DMM Dis Models Mech*. 2020;13(1). https://doi.org/10.1242/dmm.042317.
28. Churko JM, Lee J, Ameen M, et al. Transcriptomic and epigenomic differences in human induced pluripotent stem cells generated from six reprogramming methods. *NatBiomed Eng*. 2017;1(10):826–837. https://doi.org/10.1038/s41551-017-0141-6.
29. Abud EM, Ramirez RN, Martinez ES, et al. iPSC-derived human microglia-like cells to study neurological diseases. *Neuron*. 2017;94(2):278–293. https://doi.org/10.1016/j.neuron.2017.03.042. e9.
30. Billon N, Jolicoeur C, Raff M. Generation and characterization of oligodendrocytes from lineage-selectable embryonic stem cells in vitro. *Methods Mol Biol*. 2006;330:15–32. https://doi.org/10.1385/1-59745-036-7:015.
31. Chambers SM, Fasano CA, Papapetrou EP, Tomishima M, Sadelain M, Studer L. Highly efficient neural conversion of human ES and iPS cells by dual inhibition of SMAD signaling. *Nat Biotechnol*. 2009;27(3):275–280. https://doi.org/10.1038/nbt.1529.
32. Chandrasekaran A, Avci HX, Leist M, Kobolák J, Dinnyés A. Astrocyte differentiation of human pluripotent stem cells: new tools for neurological disorder research. *Front Cell Neurosci*. 2016;10. https://doi.org/10.3389/fncel.2016.00215, 2016.
33. Mellough CB, Sernagor E, Moreno-Gimeno I, Steel DHW, Lako M. Efficient stage-specific differentiation of human pluripotent stem cells toward retinal photoreceptor cells. *Stem Cell*. 2012;30(4):673–686. https://doi.org/10.1002/stem.1037.
34. Tao Y, Zhang SC. Neural subtype specification from human pluripotent stem cells. *Cell Stem Cell*. 2016;19(5):573–586. https://doi.org/10.1016/j.stem.2016.10.015.
35. Volpato V, Smith J, Sandor C, et al. Reproducibility of molecular phenotypes after long-term differentiation to human iPSC-derived neurons: a multi-site omics study. *Stem Cell Rep*. 2018;11(4):897–911. https://doi.org/10.1016/j.stemcr.2018.08.013.
36. Walsh P, Truong V, Nayak S, et al. Accelerated differentiation of human pluripotent stem cells into neural lineages via an early intermediate ectoderm population. *Stem Cell*. 2020. https://doi.org/10.1002/stem.3260.
37. Busskamp V, Lewis NE, Guye P, et al. Rapid neurogenesis through transcriptional activation in human stem cells. *Mol Syst Biol*. 2014;10(11). https://doi.org/10.15252/msb.20145508.
38. Cheng C. Highly expandable human iPS cell-derived neural progenitor cells (NPC) and neurons for central nervous system disease modeling and high-throughput screening. *Curr Protoc Hum Genet*. 2017;92.

39. Ho SM, Hartley BJ, TCW J, et al. Rapid Ngn2-induction of excitatory neurons from hiPSC-derived neural progenitor cells. *Methods*. 2016;101:113–124. https://doi.org/10.1016/j.ymeth.2015.11.019.
40. Zhang Y, Pak CH, Han Y, et al. Rapid single-step induction of functional neurons from human pluripotent stem cells. *Neuron*. 2013;78(5):785–798. https://doi.org/10.1016/j.neuron.2013.05.029.
41. Schafer ST, Paquola ACM, Stern S, et al. Pathological priming causes developmental gene network heterochronicity in autistic subject-derived neurons. *Nat Neurosci*. 2019;22(2):243–255. https://doi.org/10.1038/s41593-018-0295-x.
42. Briggs JA, Li VC, Lee S, Woolf CJ, Klein A, Kirschner MW. Mouse embryonic stem cells can differentiate via multiple paths to the same state. *Elife*. 2017;6. https://doi.org/10.7554/eLife.26945.
43. Wang C, Ward ME, Chen R, et al. Scalable production of iPSC-derived human neurons to identify tau-lowering compounds by high-content screening. *Stem Cell Rep*. 2017;9(4):1221–1233. https://doi.org/10.1016/j.stemcr.2017.08.019.
44. Deneault E, White SH, Rodrigues DC, et al. Complete disruption of autism-susceptibility genes by gene editing predominantly reduces functional connectivity of isogenic human neurons. *Stem Cell Rep*. 2018;11(5):1211–1225. https://doi.org/10.1016/j.stemcr.2018.10.003.
45. Swinney DC. Phenotypic vs. Target-based drug discovery for first-in-class medicines. *Clin Pharmacol Therapeut*. 2013;93(4):299–301. https://doi.org/10.1038/clpt.2012.236.
46. Liu J, Kościelska KA, Cao Z, et al. Signaling defects in iPSC-derived fragile X premutation neurons. *Hum Mol Genet*. 2012;21(17):3795–3805. https://doi.org/10.1093/hmg/dds207.
47. Das Sharma S, Pal R, Reddy BK, et al. Cortical neurons derived from human pluripotent stem cells lacking FMRP display altered spontaneous firing patterns. *Mol Autism*. 2020;11(1):52. https://doi.org/10.1186/s13229-020-00351-4.
48. Kumari D, Swaroop M, Southall N, Huang W, Zheng W, Usdin K. High-throughput screening to identify compounds that increase fragile x mental retardation protein expression in neural stem cells differentiated from fragile x syndrome patient-derived induced pluripotent stem cells. *Stem Cells Transl Med*. 2015;4(7):800–808. https://doi.org/10.5966/sctm.2014-0278.
49. Grossman A, Brimacombe K, et al. Assay Guidance Manual [Internet]. *Bethesda (MD): Eli Lilly & Company and the National Center for Advancing Translational Sciences*; 2004. https://www.ncbi.nlm.nih.gov/books/NBK53196/.
50. Coussens NP, Sittampalam GS, Guha R, et al. Assay guidance manual: quantitative biology and pharmacology in preclinical drug discovery. *Clin Transl Sci*. 2018;11(5):461–470. https://doi.org/10.1111/cts.12570.
51. Pamies D, Hartung T. 21st century cell culture for 21st century toxicology. *Chem Res Toxicol*. 2017;30(1):43–52. https://doi.org/10.1021/acs.chemrestox.6b00269.
52. Pamies D, Bal-Price A, Simeonov A, et al. Good cell culture practice for stem cells & stem-cell-derived models. *ALTEX*. 2017;34(1):95–132. https://doi.org/10.14573/altex.1607121.
53. Keller A, Dziedzicka D, Zambelli F, et al. Genetic and epigenetic factors which modulate differentiation propensity in human pluripotent stem cells. *Hum Reprod Update*. 2018;24(2):162–175. https://doi.org/10.1093/humupd/dmx042.
54. Kurtz A, Seltmann S, Bairoch A, et al. A standard nomenclature for referencing and authentication of pluripotent stem cells. *Stem Cell Rep*. 2018;10(1):1–6. https://doi.org/10.1016/j.stemcr.2017.12.002.
55. Luong MX, Auerbach J, Crook JM, et al. A call for standardized naming and reporting of human ESC and iPSC lines. *Cell Stem Cell*. 2011;8(4):357–359. https://doi.org/10.1016/j.stem.2011.03.002.
56. Sarafian R, Morato-Marques M, Borsoi J, Pereira LV. Monitoring cell line identity in collections of human induced pluripotent stem cells. *Stem Cell Res*. 2018;28:66–70. https://doi.org/10.1016/j.scr.2018.01.030.
57. Le MNT, Hasegawa K. Expansion culture of human pluripotent stem cells and production of cardiomyocytes. *Bioengineering*. 2019;6(2). https://doi.org/10.3390/bioengineering6020048.
58. McLaren D, Gorba T, Marguerie De Rotrou A, et al. Automated large-scale culture and medium-throughput chemical screen for modulators of proliferation and viability of human induced pluripotent stem cell-derived neuroepithelial-like stem cells. *J Biomol Screen*. 2013;18(3):258–268. https://doi.org/10.1177/1087057112461446.
59. Badenes SM. Defined essential 8 medium and vitronectin efficiently support scalable xeno-free expansion of human induced pluripotent stem cells in stirred microcarrier culture systems. *PLoS One*. 2016;11(3).
60. Phillips BW, Horne R, Lay TS, Rust WL, Teck TT, Crook JM. Attachment and growth of human embryonic stem cells on microcarriers. *J Biotechnol*. 2008;138(1–2):24–32. https://doi.org/10.1016/j.jbiotec.2008.07.1997.

61. Polanco A, Kuang B, Yoon S. Bioprocess technologies that preserve the quality of iPSCs. *Trends Biotechnol*. 2020. https://doi.org/10.1016/j.tibtech.2020.03.006.
62. D'Antonio M, Woodruff G, Nathanson JL, et al. High-throughput and cost-effective characterization of induced pluripotent stem cells. *Stem Cell Rep*. 2017;8(4):1101–1111. https://doi.org/10.1016/j.stemcr.2017.03.011.
63. Ho MSH, Fryga A, Laslett AL. Flow cytometric analysis of human pluripotent stem cells. *Methods Mol Biol*. 2011;767:221–230. https://doi.org/10.1007/978-1-61779-201-4_16.
64. Xu M, Lee EM, Wen Z, et al. Identification of small-molecule inhibitors of Zika virus infection and induced neural cell death via a drug repurposing screen. *Nat Med*. 2016;22(10):1101–1107. https://doi.org/10.1038/nm.4184.
65. Ahfeldt T, Ordureau A, Bell C, et al. Pathogenic pathways in early-onset autosomal recessive Parkinson's disease discovered using isogenic human dopaminergic neurons. *Stem Cell Rep*. 2020;14(1):75–90. https://doi.org/10.1016/j.stemcr.2019.12.005.
66. Rigamonti A, Repetti GG, Sun C, et al. Large-scale production of mature neurons from human pluripotent stem cells in a three-dimensional suspension culture system. *Stem Cell Rep*. 2016;6(6):993–1008. https://doi.org/10.1016/j.stemcr.2016.05.010.
67. Dai S, Li R, Long Y, et al. One-step seeding of neural stem cells with vitronectin-supplemented medium for high-throughput screening assays. *J Biomol Screen*. 2016;21(10):1112–1124. https://doi.org/10.1177/1087057116670068.
68. Manthorpe M, Engvall E, Ruoslahti E, Longo FM, Davis GE, Varon S. Laminin promotes neuritic regeneration from cultured peripheral and central neurons. *JCB (J Cell Biol)*. 1983;97(6):1882–1890. https://doi.org/10.1083/jcb.97.6.1882.
69. High Content Screening with Primary Neurons. https://www.ncbi.nlm.nih.gov/books/NBK169433/. Published 2014. Accessed October 1, 2020.
70. Gopalakrishnan SM, Teusch N, Imhof C, et al. Role of Rho kinase pathway in chondroitin sulfate proteoglycan-mediated inhibition of neurite outgrowth in PC12 cells. *J Neurosci Res*. 2008;86(10):2214–2226. https://doi.org/10.1002/jnr.21671.
71. Noble M, Fok-Seang J, Cohen J. Glia are a unique substrate for the in vitro growth of central nervous system neurons. *J Neurosci*. 1984;4(7):1892–1903. https://doi.org/10.1523/jneurosci.04-07-01892.1984.
72. Little D, Luft C, Pezzini-Picart O, et al. Seeding induced pluripotent stem cell-derived neurons onto 384-well plates. *Methods Mol Biol*. 2019;1994:159–164. https://doi.org/10.1007/978-1-4939-9477-9_14. United Kingdom: Humana Press Inc.
73. Niles WD, Coassin PJ. Cyclic olefin polymers: innovative materials for high-density multiwell plates. *Assay Drug Dev Technol*. 2008;6(4):577–590. https://doi.org/10.1089/adt.2008.134.
74. Sridharan B. A simple procedure for creating scalable phenotypic screening assays in human neurons. *Sci Rep*. 2019;9.
75. Thomas-Fowlkes B, Cifelli S, Souza S, et al. Cell-based in vitro assay automation: balancing technology and data reproducibility/predictability. *SLAS Technology*. 2020;25(3):276–285. https://doi.org/10.1177/2472630320902095.
76. Knight S, Plant H, McWilliams L, et al. Enabling 1536-well high-throughput cell-based screening through the application of novel centrifugal plate washing. *SLAS Discov*. 2017;22(6):732–742. https://doi.org/10.1177/2472555216683650.
77. El-Nachef D, Shi K, Beussman KM, et al. A rainbow reporter tracks single cells and reveals heterogeneous cellular dynamics among pluripotent stem cells and their differentiated derivatives. *Stem Cell Rep*. 2020;15(1):226–241. https://doi.org/10.1016/j.stemcr.2020.06.005.
78. Cederquist GY. A multiplex human pluripotent stem cell platform defines molecular and functional subclasses of autism-related genes. *Cell Stem Cell*. 2020;27(1).
79. Corsello SM. Discovering the anti-cancer potential of non-oncology drugs by systematic viability profiling. *Nat Can (Que)*. 2020:235–248.
80. Burkhardt MF, Martinez FJ, Wright S, et al. A cellular model for sporadic ALS using patient-derived induced pluripotent stem cells. *Mol Cell Neurosci*. 2013;56:355–364. https://doi.org/10.1016/j.mcn.2013.07.007.
81. Bradley JA, Strock CJ. Screening for neurotoxicity with microelectrode array. *Curr Protocols Toxicol*. 2019;79(1). https://doi.org/10.1002/cptx.67.

82. Timm M, Saaby L, Moesby L, Hansen EW. Considerations regarding use of solvents in in vitro cell based assays. *Cytotechnology*. 2013;65(5):887–894. https://doi.org/10.1007/s10616-012-9530-6.
83. Czysz K, Minger S, Thomas N. Dmso efficiently down regulates pluripotency genes in human embryonic stem cells during definitive endoderm derivation and increases the proficiency of hepatic differentiation. *PLoS One*. 2015;10(2). https://doi.org/10.1371/journal.pone.0117689.
84. Li J. A transient DMSO treatment increases the differentiation potential of human pluripotent stem cells through the Rb family. *PLoS One*. 2018;13(12).
85. Cleveland PH, Koutz PJ. Nanoliter dispensing for uHTS using pin tools. *Assay Drug Dev Technol*. 2005;3(2):213–225. https://doi.org/10.1089/adt.2005.3.213.
86. Dawes TD, Turincio R, Jones SW, et al. Compound transfer by acoustic droplet ejection promotes quality and efficiency in ultra-high-throughput screening campaigns. *J Lab Autom*. 2016;21(1):64–75. https://doi.org/10.1177/2211068215590588.
87. Roberts K, Callis R, Ikeda T, et al. Implementation and challenges of direct acoustic dosing into cell-based assays. *J Lab Autom*. 2016;21(1):76–89. https://doi.org/10.1177/2211068215595212.
88. Smith T, Ho PI, Yue K, et al. Comparison of compound administration methods in biochemical assays: effects on apparent compound potency using either assay-ready compound plates or pin tool-delivered compounds. *J Biomol Screen*. 2013;18(1):14–25. https://doi.org/10.1177/1087057112455434.
89. Cheng X, Hochlowski J, Tang H, et al. Studies on repository compound stability in DMSO under various conditions. *J Biomol Screen*. 2003;8(3):292–304. https://doi.org/10.1177/1087057103008003007.
90. Clark R. A flexible high content imaging assay for profiling macrophage efferocytosis. *J Immunol Methods*. 2019;473.
91. Kossler N, Matheis KA, Ostenfeldt N, et al. Identification of specific mRNA signatures as fingerprints for carcinogenesis in mice induced by genotoxic and nongenotoxic hepatocarcinogens. *Toxicol Sci*. 2015;143(2):277–295. https://doi.org/10.1093/toxsci/kfu248.
92. Bray MA, Carpenter AE. Quality control for high-throughput imaging experiments using machine learning in cellprofiler. *Methods Mol Biol*. 2018;1683:89–112. https://doi.org/10.1007/978-1-4939-7357-6_7. United States: Humana Press Inc.
93. Shlevkov E. A high-content screen identifies TPP1 and Aurora B as regulators of axonal mitochondrial transport. *Cell Rep*. 2019;28(12).
94. Gubler H, Clare N, Galafassi L, Geissler U, Girod M, Herr G. Helios: history and anatomy of a successful in-house enterprise high-throughput screening and profiling data analysis system. *SLAS Discovery*. 2018;23(5):474–488. https://doi.org/10.1177/2472555217752140.
95. Brideau C, Gunter B, Pikounis B, Liaw A. Improved statistical methods for hit selection in high-throughput screening. *J Biomol Screen*. 2003;8(6):634–647. https://doi.org/10.1177/1087057103258285.
96. Caraus I, Alsuwailem AA, Nadon R, Makarenkov V. Detecting and overcoming systematic bias in highthroughput screening technologies: a comprehensive review of practical issues and methodological solutions. *Briefings Bioinf*. 2015;16(6):974–986. https://doi.org/10.1093/bib/bbv004.
97. HTS Assay Validation. https://www.ncbi.nlm.nih.gov/books/NBK83783/. Published 2012. Accessed October 1, 2020.
98. Makarenkov V, Zentilli P, Kevorkov D, Gagarin A, Malo N, Nadon R. An efficient method for the detection and elimination of systematic error in high-throughput screening. *Bioinformatics*. 2007;23(13):1648–1657. https://doi.org/10.1093/bioinformatics/btm145.
99. Mpindi JP, Swapnil P, Dmitrii B, et al. Impact of normalization methods on high-throughput screening data with high hit rates and drug testing with dose-response data. *Bioinformatics*. 2015;31(23):3815–3821. https://doi.org/10.1093/bioinformatics/btv455.
100. Zhang JH, Chung TDY, Oldenburg KR. A simple statistical parameter for use in evaluation and validation of high throughput screening assays. *J Biomol Screen*. 1999;4(2):67–73. https://doi.org/10.1177/108705719900400206.
101. Dandapani S, Rosse G, Southall N, Salvino JM, Thomas CJ. Selecting, acquiring, and using small molecule libraries for high-throughput screening. *Curr Protocols Chemical Biol*. 2012;4:177–191.
102. Bredel M, Jacoby E. Chemogenomics: an emerging strategy for rapid target and drug discovery. *Nat Rev Genet*. 2004;5(4):262–275. https://doi.org/10.1038/nrg1317.

103. Jones LH, Bunnage ME. Applications of chemogenomic library screening in drug discovery. *Nat Rev Drug Discov*. 2017;16(4):285–296. https://doi.org/10.1038/nrd.2016.244.
104. Maianti JP, Tan GA, Vetere A, et al. Substrate-selective inhibitors that reprogram the activity of insulin-degrading enzyme. *Nat Chem Biol*. 2019;15(6):565–574. https://doi.org/10.1038/s41589-019-0271-0.
105. Malo N, Hanley JA, Cerquozzi S, Pelletier J, Nadon R. Statistical practice in high-throughput screening data analysis. *Nat Biotechnol*. 2006;24(2):167–175. https://doi.org/10.1038/nbt1186.

CHAPTER

13

Gene editing hPSCs for modeling neurological disorders

Andrew Petersen

Madison, WI, United States

Introduction to limitations of iPSC-derived neuronal models that can be improved with gene editing

hPSC-based neuronal models are by necessity simplified relative to the human brain. From the difference in time scale of a differentiation relative to a person's life to the complexity of an organoid relative to the complexity of a mature brain, multiple factors must be considered while interpreting findings from hPSC-based models. Three primary limitations of this system can be identified, each of which is amenable to genetic modifications which help alleviate some of the limitation and increase the utility of the in vitro model system.

In vitro culture models are either artificially simplified or too complex for simple comparison

The generation of neural cell types is generally accomplished through two methods − directed differentiation to enriched populations of specific cell types using rigid paradigms or in situ generation of multiple cell types using a three-dimensional organoid system that self-differentiates and organizes. Both systems have their own caveats and benefits.

Directed differentiation can yield populations of cells with over 90% homogeneity in the cell type of interest. A pure population of cells is advantageous for protein and transcript analysis, and in drug discovery experiments where readouts could be confounded by the presence of alternative or contaminating cell types. As just one example, iPSCs from an Alzheimer's disease patient harboring a *Presenilin-1* mutation were differentiated to an enriched population of cortical neurons and subjected to a small compound library to identify factors which reduce amyloid β (Aβ) production.[1] This screen identified six compounds that reduced Aβ levels and did not display high toxicity, thus representing potential therapeutics for this disorder. The homogenous nature of the neurons used for this screen improved the overall robustness of the assay endpoint.

Phenotyping of Human iPSC-derived Neurons
https://doi.org/10.1016/B978-0-12-822277-5.00014-6

A drawback of directed, homogenous differentiation is the absence of external cues and exposure to other cell types. For neural cells, these cues are critical for regulating synaptic activity and pruning, promoting maturation, and helping long-term survival. In addition, most neurological disorders likely involve the interplay of neuronal subtypes and glial cells, so homogenous cultures will fall short of modeling all relevant aspects of the disorder. Neural differentiation via organoid systems gives rise to diverse neural subpopulations that interact with each other in a time course similar to neural development in vivo. When neural cells are allowed to differentiate in an undisrupted environment, they self-assemble into discernible layers recapitulating in vivo cortical layers.[2–5] Organoids are ideal for studying effects of neural development, specifically in the context of developmental disorders where the mutation likely affects neural populations at early stages of development. For example, a recent publication utilizing trisomy 21 iPSCs from a Down syndrome patient were found to generate higher levels of OLIG2+ progenitors when developing in an organoid platform, ultimately giving rise to altered levels of gamma aminobutyric acid (GABA) interneurons later in development.[6]

Just as the benefits of organoid research are related to their cell type diversity, so are the drawbacks. For drug screening, having a heterogeneous population of cells confounds efforts to identify drugs affecting specific pathways, since these drugs may produce different effects when targeting different cells. Furthermore, bulk transcriptional and proteomic analysis is challenging when multiple cell types are present, necessitating single cell analysis. Finally, organoids suffer from reproducibility issues.[7] A systematic analysis of organoids following an established protocol revealed a bias in differentiation between experiments (and between culture vessels in an individual experiment), and continuous developmental progression of the organoid population over time.[8] Thus, comparisons between experiments must be done to carefully match the relative developmental timepoint and the relative contribution of different cell types within the organoid(s). Efforts are underway to understand the source of variability, and promising studies have shown improved consistency in organoid derivation.[5,9]

Genetic background differences lead to high variability when comparing lines from different individuals

In hPSC-based neuronal research, as in all biological science, the confidence in experimental design is largely dependent on having proper controls. Genetic diversity in the human population and age-based changes in epigenetic state are confounding factors, and hPSCs from different individuals differ in the ability to generate neuronal cells in vitro.[10] To counteract these variables, early publications primarily focused on using parental or sibling lines as controls for a patient's disease phenotype. However, the genetic diversity even between siblings and age-related differences make these comparisons problematic. Genetic modification of hPSCs allows the repair of monogenic disorders to facilitate the generation of age-matched, isogenic cell lines that only differ at the site of the disease mutation. Scientists can now have full confidence that phenotypic differences are not due to unique genetic components unknown between non-isogenic lines.

Visualizing and analyzing human cells in *in vivo* transplant models is technology challenging

To ensure hPSC neural models are recapitulating in vivo biology, chimeras or transplant models where differentiated human cells are transplanted into non-human models such as primates or rodents are particularly valuable. Numerous studies have shown that transplanted neurons and glia can migrate and integrate with the functional circuitry of these animals to provide a mechanism of studying how human cells behave in an intact nervous system.[11–15] However, it is difficult to assess how these cells are integrating and functioning. Transgenic reporter lines that label the human cells either constitutively or via cell type-specific markers allows tracking of the cells in the brain and information regarding their outcomes. Modulation or monitoring of transplanted cells can be accomplished using transgenes expressing designer receptors exclusively activated by designer drugs (DREADDs), calcium sensing proteins (gCaMP6), or light-activated channels such as channel rhodopsin, among other tools.[15–18] The ability to efficiently generate transgenic cell lines for these purposes has rapidly expanded the toolbox for studying hPSC-derived neural cells following transplant.

Gene editing systems — past, present, and future

Transgenic tools for hPSC-derived neural models require modification of the genome, a process which has evolved rapidly in the past decade. Early genome editing used a variety of approaches including, but not limited to, homologous recombination, recombinases, transposases, transposons, and integrases.[19] These techniques relied on events that occur or are catalyzed in the absence of an induced double-stranded DNA break (DSB). In order to boost efficiency, the field transitioned to using nuclease-based systems to generate a DSB, which is repaired via non-homologous end joining (NHEJ) or homology-directed repair (HDR). Since these mechanisms are well understood, they can be manipulated during the repair process to provide a reproducible gene editing outcome. NHEJ is an error-prone mechanism where the broken DSB is re-attached, typically through the insertion or deletion of nucleotides. This process can be used to disrupt the coding sequence of genes and is often used to generate a knockout of a targeted gene. HDR, on the other hand, utilizes DNA homology surrounding the break site to repair the DNA lesion in a high-fidelity process. By supplying exogenous DNA sequence with homology to the DSB site, scientists can trick the cell into incorporating sequences not usually found at the locus. This method is commonly used for correcting variants identified in patient-derived PSCs.

"Version 1.0" — meganucleases

The technology underlying induction of double-strand breaks (DSBs) has been the focus of intense interest over the past two decades and led to a search for an ideal nuclease containing a trinity of features: ease of implementation, specificity for targeted locus (and associated low off-target effects), and efficiency of DSB induction. As outlined in Fig. 13.1, there are five primary systems that have been developed, each meeting different criteria.[20–22] Meganucleases

Technology	Protein structure	Recognition motif	Nucleotides required for function	Ease of use*	Specificity*
Meganuclease	Monomer	Modified through directed evolution of the DNA binding locus	12-40 nucleotides	-	+++
Zinc Finger Nuclease (ZFN)	Dimer, bind each strand	3-4 zinc fingers/ subunit 3 nt per zinc finger	12 nt per subunit, 24 nt per dimer	+	++
TAL Effector Nuclease (TALEN)	Dimer, bind each strand	15-20 repeats/subunit** 1 nt per repeat	15-20 nt per subunit, 30-40 nt per dimer	++	++
WT CRISPR-Cas9 (SpCas9)	Monomer	gRNA-scaffold facilitates Cas9 adherence	20 nt gRNA, plus 3 nt PAM sequence	+++	+
Nickase CRISPR-Cas9 (Cas9-D10A, Cas9-H840A)	Dimer, bind each strand	gRNA-scaffold facilitates Cas9 adherence	46 nucleotides (dual gRNA+PAM motifs)	+++	+++

*TALENs not limited in length, but in practice utilize this format
**Rankings based on applications in hPSCs, not a quantitative assessment

FIGURE 13.1 Summary of directed endonuclease systems.

were the first site-specific endonuclease to be identified for genome engineering and contain two hallmark functions essential for genome editing: endonuclease activity permitting cleavage of bound DNA, and a specific DNA recognition sequence that can be modulated to alter where the protein can target.[19] Meganucleases are limited by the effort required to alter the specific binding sequence, which is done through directed evolution, a process that involves labor-intensive rounds of mutagenesis and selection, so the field quickly transitioned to new technology with easier methods of modulating DNA binding specificity.

"Version 2.0" – zinc finger nucleases and TAL effector nucleases

The next technology to gain widespread implementation involved the use of chimeric proteins constructed by fusing the nuclease domain from one protein to a well-characterized DNA recognition domain from another. The Zinc Finger Nucleases (ZFNs) contain zinc finger domains that individually recognize specific three to four nucleotide DNA sequences fused to a FokI restriction enzyme nuclease domain. Combining multiple (often 4) zinc finger domains will result in a recognition motif of 12 nucleotides. Since the FokI nuclease domain requires dimerization to function, a second ZFN recognizing 12 nucleotides on the opposite strand of DNA is required to facilitate a DSB at the desired locus.[23,24] Thus, a functional ZFN targeting apparatus requires a 24-nucleotide recognition domain, split between the 12-nucleotide sequences for each monomer.

A similar system, combining transcription activator-like effector (TAL Effector, TALE) DNA binding domains and the same FokI nuclease domain, creates the TALEN class of tool.[25–28] In contrast to ZFNs, TALE repeats are a stretch of 33–35 amino acids separated by two amino acids that confer recognition of a single nucleotide (termed the repeat-variable diresidue, or RVD).[29,30] Combining multiple TALE repeats and RVDs into a single polypeptide chain gives the resulting TALEN DNA sequence-specific binding properties. Although common usage employs 15–20 TALE repeats/RVDs per TALEN protein, there is effectively no limit to the length of DNA sequence that a TALEN can recognize, although specificity may decrease with a longer recognition sequence.[31] As with ZFNs, the FokI domain in a TALEN requires dimerization to function, so two monomeric TALENs are needed to generate a DSB. The total TALEN targeting apparatus typically requires between a 30-nucleotide and 40-nucleotide recognition domain (split between the TALEN monomers), but this motif can be increased or decreased by altering the number of TALE repeat/RVDs.

Both ZFNs and TALENs have relatively high specificity for the designed loci, owing to the combined recognition sequences of 24–30 nucleotides, respectively. The generation of TALEN proteins is easier to adopt due to well-defined TALE repeat/RVD codes and the availability of the individual plasmid components for assembly. The ZFN zinc finger domains are less amenable to cloning, fewer 3-nucleotide sequences have associated zinc finger recognition domains, and assembly of ZFN components does not always correspond to effective nuclease activity at the targeted locus.[32] Although both systems have reasonable efficacy in mammalian cells, their utilization in hPSCs has some limitations.[33] The large coding sequence of these genes necessitates plasmid delivery of the nucleases and are commonly utilized with genetic incorporation of a resistance gene cassette that can leave a genomic scar following removal (discussed in detail later).

"Version 3.0" — CRISPR/cas nucleases

Perhaps the closest system to achieving the desired genome engineering trinity is the Clustered Regularly Interspaced Short Palindromic Repeats (CRISPR) system from bacteria, its utilization for genome editing first reported in 2012–2013.[34–36] These reports characterized a *Streptococcus pyogenes* type II endonuclease called Cas9 (SpCas9) and how it can target DNA sequences for cleavage in many cell types, including hPSCs. Contemporary reports highlight Cas9 proteins from *Neisseria meningitidis*, *Staphylococcus aureus*, and other bacteria with unique sequence binding properties.[37–39] While these systems have their own utilities and have been implemented for some research projects, SpCas9 remains the primary protein utilized in hPSC research.

SpCas9 is directed to DNA through the interaction with two RNA molecules: a CRISPR RNA (crRNA) which serves as a scaffold between the SpCas9 protein and the second RNA molecule, a transactivating crRNA (tracrRNA), which facilitates DNA interaction through complementary base pair recognition. While developing SpCas9 for gene editing, it was discovered that a single chimeric RNA molecule containing a combination of the tracrRNA essential binding region and the crRNA region responsible for DNA recognition, commonly called a guide RNA (gRNA) or single guide RNA (sgRNA), is able to direct SpCas9 to a genomic locus and facilitate a DSB.[34] The specificity of SpCas9 is derived from the tracrRNA/gRNA construct that performs the base pair recognition. This 20-nucleotide stretch of RNA, called a protospacer, must bind to a DNA region followed on the 3′ end by a protospacer-adjacent motif (PAM) of -NGG (where N is any nucleotide). Upon recognition of a compatible locus, the SpCas9 protein can generate a DSB through the activity of its RuvC and HNH nuclease domains.

Much effort has been spent to engineer Cas9 for different genome editing purposes. Removal of a single nuclease domain, such as through a D10A mutation, which abolishes the RuvC nuclease activity, or a H840A mutation, which abolishes the HNH nuclease domain, makes SpCas9 into a "nickase" nuclease where only a single DNA strand is cleaved.[40–42] With the nickase, two gRNAs are required, one to target each strand of DNA simultaneously to generate a DSB, similar to the TALEN and ZFN systems, thus increasing the recognition motif from 23 nucleotides to 46 nucleotides. The nickase system helps alleviate some of the risk of off-target Cas9 activity that has been associated with Cas9 in many cell types, although hPSCs have been shown to be largely resistant to these off-target effects when stringent design criteria are employed.[43–46]

Finally, dual-nuclease mutants (so-called "dead" Cas9, or dCas9) fused with effector proteins form a derivative system that co-opts the RNA-mediated DNA targeting ability of Cas9 for purposes other than genome editing.[47] dCas9 fusion with a fluorescent protein can detail localization of specific regions within the nucleus, fusion with a transcriptional transactivator (such as dCas9-VPR) can increase gene expression at a locus, fusion with a transcriptional repressor can reduce gene expression at a locus, fusion with DNA base editors facilitate conversion of nucleotides in the absence of a DSB to induce SNP modifications, and additional applications that are too numerous to highlight.[47–52]

"Version 4.0" – genome editing systems for translational medicine

The risk of unwanted effects of CRISPR/Cas9 systems has led to alternative approaches with an eye toward translational medicine. The focus of these new technologies begins with altering the targeting capacity of Cas9 itself, either through mutations that increase the on-target binding specificity or through mutations that modify or eliminate the PAM requirement. Systems with different CRISPR/Cas proteins that target RNA (Cas13a) or mechanisms that induce genome modification without DSBs were similarly identified.

Modification of Cas9 binding and specificity was a nearly-immediate focus of research following initial reports of the technology. The first high-fidelity Cas9 orthologs, SpCas9-HF1 and eSpCas9(1.1), reduced off-target effects and remain inactive when bound to mis-matched DNA regions.[53–55] Additional publications on evoCas9 and SniperCas9 report improvements to the first high-fidelity Cas9 proteins, so expectations are that this field will continue to evolve over time.[56,57] In addition, strategies to alter the PAM requirements for SpCas9 were developed. Through directed mutagenesis at the region of Cas9 that interacts with the PAM, mutants were screened that resulted in combinations altering the PAM sequence recognized by Cas9[39,54,58] and in some cases nearly eliminate the PAM requirement altogether (SpCas9-NG)[59] in conjunction with increased on-target specificity (xCas9).[60] While these efforts have reduced the concern of off-target Cas9 activity for therapeutic settings, they are not the only tool researchers are developing to facilitate high-fidelity genome editing in the therapeutic realm.

Another mechanism to prevent potentially-oncogenic DSBs in cells, particularly those destined for transplant, is an enzymatic process that does not introduce DSBs. Base editors facilitate the conversion of nucleotides through enzymatic modification of the bases themselves, followed by altered polymerase recognition during cell division to effectively change the nucleotide at the specific locus.[51,52,61] A more widely applicable system for diverse editing approaches is Prime Editing, which fuses a Cas9 nickase protein with an engineered reverse transcriptase.[62] The protein component is supplied with a pegRNA that contains the gRNA targeting Cas9 to the site of interest, plus additional nucleotides to prime the reverse transcriptase to incorporate the desired edit at the locus, which is also encoded on the pegRNA.

One approach to remove unwanted DNA changes in cells that may be used therapeutically is to forego targeting the DNA altogether. Cas13a is a CRISPR/Cas nuclease that uses a gRNA to target and cleave RNA molecules.[63] Developed from *Leptotrichia wadei* bacteria, the LwaCas13a system has a 28–30 nucleotide recognition motif that was shown to function in mammalian cells to generate 40%–84% knockdown of endogenous mRNA and reasonable specificity for the targeted sequence.[63] The primary benefit of an RNA editor is the reduction of DNA damage that may lead to oncogenesis in a cell. If the DNA is not exposed to DSBs, the likelihood of oncogenesis in a gene edited, transplanted cell line is reduced.

Use of genetic modification to generate isogenic cell lines

The modeling of monogenic disorders with hPSCs is dependent on the ability to identify phenotypes associated with the specific mutation, independent of confounding differences in genetic background. The two primary methods of accomplishing this are to repair a genetic mutation in patient-derived iPSC lines, or to introduce the genetic mutation in a control hPSC line that does not contain any other disease-specific mutations. In a different way, knock out of a gene can provide insight into biology and disease. Isogenic hPSC pairs harboring gene knockouts have shed light on the function of specific genes on neural differentiation and disease states.

Gene editing systems for generating isogenic lines

The TALEN system was used extensively to generate isogenic cell lines, but its usage has gone down dramatically since the introduction of CRISPR/Cas9. TALENs have lost favor due to their lower efficiency necessitating the use of a resistance gene cassette to select targeted clones. Resistance cassettes utilize a loxP-flanked (or similar system) constitutive promoter regulating expression of an antibiotic resistance gene. Following selection with the antibiotic, the resistance cassette is removed to restore the locus. However, removal of most cassettes leaves a scar at the site of the removal, resulting in exogenous DNA remaining at the locus. While this DNA is typically designed to reside within an intron, it could affect splicing or transcription in unexpected ways. The piggybac transposase system can be used to excise the resistance cassette in a scar-free approach, but this system requires a "TTAA" DNA sequence near the site of the SNP modification.[64]

In certain cases, intentional DNA mutations called silent SNPs can be incorporated in the HDR repair construct to prevent unwanted modification of the target locus. This strategy is used when the desired DNA modification does not remove the gRNA binding sequence. In these scenarios, even if the DNA modification is introduced via HDR, Cas9 is able to re-target the locus until the complementary sequence to the gRNA sequence is removed (most commonly by indel formation, which alters the coding sequence and affects protein function).

Silent SNPs are DNA modifications that change the amino acid codon sequence but not the amino acid itself. As such, these mutations change the DNA sequence to prevent subsequent Cas9 activity at the site but do not affect protein function.

The CRISPR/Cas9 system is more efficient and easily introduced to cells, thus precluding the need for resistance cassettes. Cas9 can be delivered to cells through plasmid, mRNA, or protein as part of a Cas9-RiboNucleoProtein (Cas9-RNP) complex.[65–67] Plasmids allow co-delivery of another protein for sorting positively-transduced cells, such as a fluorescent reporter for cell sorting or an antibiotic resistance gene for drug selection.[68,69] sgRNA can be supplied using plasmids containing the RNA polymerase III U6 or similar promoter or through in vitro synthesis. Finally, Cas9-RNP complexes represent the most high-throughput approach. Commercially available Cas9 protein can be preincubated with either in vitro generated or purchased sgRNA molecules to form a Cas9-RNP complex which is subsequently transduced to cells. This system has a limited duration of Cas9 activity, as the presence of Cas9 is limited by the normal degradation rate of this protein in the cell. As such, it is predicted to have lower off-target risks due to limited time of Cas9-sgRNA functionality.

Isogenic hPSC lines for disease modeling

Most of the focus of hPSC research on neurodevelopmental and neurodegenerative disorders centers around monogenic disorders because these are amenable to genetic correction and the affected gene often gives an indication of potential mechanisms to study. Isogenic hPSC pairs have proven useful in identifying mechanisms that may underlie pathogenesis in more common neurodegenerative disorders, such as amyotrophic lateral sclerosis (ALS),[70] as well as in rare developmental disorders, such as Alexander's disease.[71,72] Fig. 13.1 discusses the progression of hPSC research in ALS, which is possibly the most highly-studied neurodegenerative disorder. This timeline shows how gene editing has facilitated research into ALS and allowed isogenic hPSC lines to shed light on this disorder. Although in-depth discussion of isogenic hPSC lines is not feasible for all disorders, the role of gene editing in generating isogenic hPSC lines for neurodevelopmental disorders that are not simple SNP modifications is worth highlighting.

Some disorders that are monogenic are not caused by an SNP. For example, Fragile X syndrome is caused by an expansion of CGG repeats upstream of the *Fragile X Mental Retardation 1 (Fmr1)* gene on the X-chromosome. When the repeats exceed 200, the gene locus becomes highly methylated and silenced, blocking protein expression from this locus. Park et al. used CRISPR/Cas9 to target the repeat locus in Fragile X hESC and iPSC lines and successfully eliminated the repeats.[73] Although they reported additional nucleotides were deleted at the locus, even crude excision of the repeats was sufficient to restore transcription at the *Fmr1* locus in hPSC and neuronal precursors.[73] Similar repeat manipulation has shown promise in studying poly-glutamine expansion in Huntington's disease[74,75] and hexanucleotide expansion at the C9ORF72 locus in ALS,[76,77] among others.

Isogenic knock-out cell lines

Knocking out a single gene is a common method for identifying the function of that specific gene/protein in any biological system. The most common approach to generate knockouts in hPSC lines is to generate DSBs using a site-directed endonuclease and allow the cell to repair the DNA lesion using non-homologous end joining (NHEJ). This process typically results in an insertion/deletion (indel) at the site of the DSB, which can be screened to identify an indel that results in a frameshift that correctly abolishes the coding sequence of the gene of interest. With the development and ease of CRISPR/Cas9, each gene can be targeted with multiple gRNA sequences to increase the efficiency of generating a gene knockout. By localizing gRNAs around the start codon of a gene, a knockout can be generated by eliminating this codon. Additionally, if a specific domain is required for gene function, it can be targeted to abolish protein activity.

In certain scenarios, knockouts can be extended to specific isoforms of a protein. Many genes have multiple isoforms with unique N-terminal or C-terminal sequences that can be specifically targeted, thus knocking out one class of isoforms, but not affecting others. For example, the *MAPT* gene has multiple splice isoforms that give rise to six unique Tau proteins in the human brain.[78] These isoforms differ due to alternative splicing of exons 2, 3, and 10. Many of the mutations in *MAPT* associated with neurological disorders arise from mutations involving exon 10, whose inclusion in the protein gives rise to an additional microtubule binding repeat.[79,80] By individually targeting these exons for knockout, models could be developed to study the role of specific isoforms on normal or pathological neuronal function.

Finally, gene knockouts are useful in studying copy number variants (CNV) found in a subset of autism spectrum disorder (ASD) and other neurological diseases.[81] CNVs are caused by chromosomal rearrangements, can be either deletions or duplications, and often affect multiple genes. Since many CNVs have been mapped, researchers can use targeted knockout of genes from these CNVs to study what effect loss of individual genes or groups of genes within a variant may have on cellular function in an attempt to better understand disease pathology.[81] Indeed, a novel SCORE (Single-guide CRISPR/Cas targeting Of Repetitive Elements) method successfully generated 16p11.2 and 15q13.3 chromosomal deletions in an iPSC line to mimic CNVs found in ASD and schizophrenia.[82–84]

Despite their utility in understanding basic biological processes, gene knockouts have multiple caveats. Genes that are essential to cellular survival cannot be knocked out since their loss is lethal to the cell, and genes that are required for cellular differentiation cannot be knocked out if that lineage is required for analysis. To overcome these limitations, conditional knockout systems have been developed including iCRISPR,[85] iKO,[86] and OPTiKO.[87] These systems require at least one additional round of gene editing, and allow the gene of interest to function normally until the system is turned on. Both the iCRISPR and OPTiKO systems utilize inducible expression of components of the CRISPR/Cas9 machinery to regulate knockout timing. In the iCRISPR system, expression of the Cas9 protein is controlled by a doxycycline-activatable transactivator/tet-on promoter system integrated at the AAVS1 safe harbor locus and requires transfection of the targeting gRNA to function.[85] A benefit of the iCRISPR system is that transfection of multiple gRNA sequences can target multiple loci with a single round of induction. The OPTiKO system utilizes a similar AAVS1 safe harbor transgene with constitutive Cas9 expression, constitutive transactivator expression, and a tet-on promoter regulating gRNA expression.[87] The iKO system will be discussed further in the section on complex transgenic systems.

Safe harbor locus transgenic systems

Early efforts to modulate gene expression in hPSCs utilized lentiviral (or similar) integration of a promoter and transcript randomly in the genome. However, this method came with many downsides, including lentivirus incorporation at different sites and quantities in the genome, deleterious integration of the lentiviral construct, transgene silencing during differentiation, and non-physiological expression levels of the transgene, which may negatively affect the ability to focus analysis of the gene mutation. While lentivirus techniques are still relevant in proof-of-principle studies or to introduce transgenes to hPSC-derived cell types with limited expansion potential, the introduction of site-specific endonucleases has supplanted lentivirus as the gold standard for generating transgenic hPSC cell lines.

Learning from the *Rosa26* locus in the mouse, a so-called "safe harbor" locus is critical for the integration of promoter-gene cassettes in hPSC gene editing. A quality safe harbor locus must be genetically accessible in most, if not all, differentiation capacities and transgene integration should not affect the expression of nearby genes. Other desirable features include an endogenous promoter that can be utilized and unique genomic sequences that makes specific targeting of the locus straightforward.

Three safe harbor loci have been utilized in hPSC studies. A region homologous to the *Rosa26* locus in mouse was tentatively mapped to chromosome 3 in the human genome and was found to be a safe harbor locus.[88] Similarly, the *CCR5* locus in humans, which encodes the co-receptor facilitating HIV infection of human cells, was identified as a potential safe harbor locus since deletions of this region cause no observable phenotype in humans.[89] The third safe harbor site utilized extensively in hPSC research is the AAVS1 locus on chromosome 19, which encodes the *PPP1R12C* gene. This region is a target for adeno-associated viruses whose integration has not shown any deleterious effects following transgene integration and has been shown to facilitate transgene expression in many cell types.[90,91]

Ongoing efforts seek to identify optimal safe harbor loci for therapeutic purposes. While the loci described above have proven effective in hPSC transgenic models, their applicability for cell therapy may not be adequate since expression of *CCR5* and *PPP1R12C* are affected upon transgene integration and their accessibility in all possible cell types has not been explored.[92,93] Therefore, new sites for therapeutic purposes are being identified and characterized using data-driven approaches, rather than piggybacking off safe harbor sites from other organisms or viral integration sites, to identify sites that are fully characterized with regard to oncogenic potential and transgene activity.[94]

Promoters utilized at safe harbor loci

For safe harbor transgenes, the most common approach is to reproduce a promoter-gene-polyA expression cassette. With this approach, the promoters can be expressed constitutively, as a result of a specific differentiation lineage, or inducible through exogenous dosing of a drug or compound.

Constitutive promoters, as the name implies, drive the expression of genes regardless of cellular differentiation status. These are typically derived from essential genes such as actin or ubiquitin, and can be categorized based on the relative strength of the promoter. A "strong" promoter drives high levels of gene expression, and a systematic evaluation of many strong promoters (UBC, PGK, EF1A, CMV, CAG, and SV40) found some variation in promoter activity between different cell lines and organisms.[95] The CAG promoter (aka CAGG) is consistently strong, PGK is consistently medium-strength, and UBC is a lower-strength option. However, strong overexpression of a protein is not always physiologically relevant and can give rise to confounding phenotypes in disease modeling, in addition to promoting protein aggregation and altered subunit stoichiometry.[96] Thus, careful selection and testing of different promoters is needed to identify the appropriate level of gene expression.

An alternative approach to testing promoter strength through generation of multiple lines is to utilize an inducible promoter system. A common example of this is the tetracycline-on (tetON) system and its derivatives.[97–99] One transgene cassette delivers constitutive expression of a drug-responsive transcription factor (rtTA, a fusion between a modified TetR transcription factor from bacteria and the VP16 transcription factor activation domain), and a second cassette with the TRE (tetracycline-regulated promoter, containing multiple repeats of the tetO bacterial promoter sequence fused with a CMV minimal promoter) upstream of the gene of interest. Upon dosing with tetracycline, the rtTA protein will bind to the promoter and drive expression of the gene of interest. By titrating the level of doxycycline, this system can regulate higher or lower levels of gene expression to match the physiologically-relevant level. A recent review highlights other inducible promoter systems that could be applied in safe harbor transgenic systems.[100]

A final class of promoter is a lineage-specific or maturation-specific promoter, identified and isolated from genes specific to a lineage or maturation state. For example, the murine *Hb9* gene promoter is active in hPSC-derived motor neurons and can be utilized with a fluorescent reporter for purifying or labeling this cell type.[101–104] Lineage-specific promoters are becoming less common with the advancement of endogenous locus gene editing, where the desired reporter can be inserted at the location of the gene, and where all possible regulatory mechanisms are present. However, maturation-state promoters such as the synapsin-1 promoter are still widely used. Synapsin-1 expression coincides with functional maturation of a neuron and synapse formation, making this promoter suitable for expressing constructs to manipulate action potential propagation, such as channel rhodopsin or DREADDs.[105]

Effector transgenes utilizing safe harbor loci

Safe harbor transgenes have been developed for modulating synaptic activity through the expression of various genes. Additionally, transgenes that report cellular activity are used to give a read-out of neuronal or glial cell activity. Examples of each system are discussed below.

The ability to control synaptic activity and firing rate of neurons is key to understand function and dysfunction in intellectual and developmental disabilities. Optogenetic tools use proteins that respond to light, such as channel rhodopsin (ChR2) or halorhodopsin, to modulate the membrane potential in neuronal cells. ChR2 encodes a cation channel that opens in response to blue light, prompting membrane depolarization and action potential firing.[106,107] Halorhodopsin is a chloride channel which opens upon light stimulation and hyperpolarizes the membrane, reducing synaptic activity.[108] Weick et al. first applied a lentiviral, *Synapsin-1*-controlled ChR2-mCherry fusion construct to hES-derived neurons and confirmed the system's utility in this model.[109] The *Synapsin-1* promoter restricts ChR2-mCherry expression to neurons that have synapsed with other cells, and the mCherry fusion protein facilitates identification of the functionally-mature cells for electrophysiological recordings via patch clamp. Transplantation of this cell line in a mouse model showed the ability to control the transplanted cells and confirmed functional integration of human neurons into the mouse circuitry.[16] ChR2 has been used as a tool for understanding how human neurons function in vivo following transplantation to host animals,[16,17,110] in regulating induced neuron action potentials upon direct conversion from fibroblasts,[111] in disease modeling and therapeutic development for Parkinson's disease,[112] in hPSC-derived grafts to ameliorate stroke in a rat model,[113] and in *in vitro* long-term potentiation modeling.[114] As these examples indicate, optogenetics represent a diverse tool for understanding human neuronal activity and function.

A second class of transgenic protein that can regulate neuronal activity are the Designer Receptors Exclusively Activated by a Designer Drug (DREADDs) and their related counterparts Receptors Activated Solely by Synthetic Ligands (RASSLs).[115] For simplicity, we will refer to the whole class of proteins as DREADDs. These g-protein coupled receptors (GPCRs) have been mutated to only respond to synthetic compounds instead of their endogenous ligands.[116,117] By mutating the Gq-coupled muscarinic receptor M_3, a protein termed hM3Dq promotes increased neuronal activity via enhanced calcium ion transport.[116,118] Mutating the Gi-coupled muscarinic receptor M_4, yields the hM4Di protein that reduces neuronal activity through reduced cAMP levels.[116] DREADDs have been used in hPSC-based neuronal modeling to regulate the activity of neurons both in vitro to understand the role of calcium dysregulation in frontotemporal lobar degeneration tauopathy[119] and in *in vivo* rodent models of epilepsy[120] and Parkinson's disease.[15]

Other safe harbor transgenes can carry enzymes that regulate cellular activity, dCas9 fusion proteins to promote altered gene expression (discussed previously), and various fluorescent or luciferase constructs to report on physiological processes.

Endogenous locus transgenes

Introduction of transgenes into hPSCs has been pivotal in increasing the speed and complexity of research. Transgenes integrated at the locus of a specific gene can replace, tag, or report the expression of that gene, with each of these mechanisms having a specific experimental use.

Endogenous locus gene replacement transgenes

A gene replacement transgene incorporates the coding sequence of a gene, often times a fluorescent protein such as eGFP paired with a polyA (pA) sequence to promote proper transcription and translation, at the endogenous start codon site for the gene being studied (Fig. 13.2). A benefit of an endogenous gene replacement system over a safe harbor transgene utilizing a gene's promoter sequence is that endogenous promoters retain all of the *cis* and *trans* elements that regulate its activity. However, endogenous gene replacements abolish expression of the gene upon incorporation, so this system is limited to genes where monoallelic expression is sufficient or the gene is not essential to the desired cell lineage. An example of a gene replacement technique utilizes the Neural Retina Leucine zipper gene (NRL) which is important for specifying rod cell development in the eye.[121] Insertion of an eGFP-pA transgene at the endogenous ATG site for one allele of the NRL gene faithfully labels rod cells during optic vesicle differentiation, and expression of NRL protein expression from the other

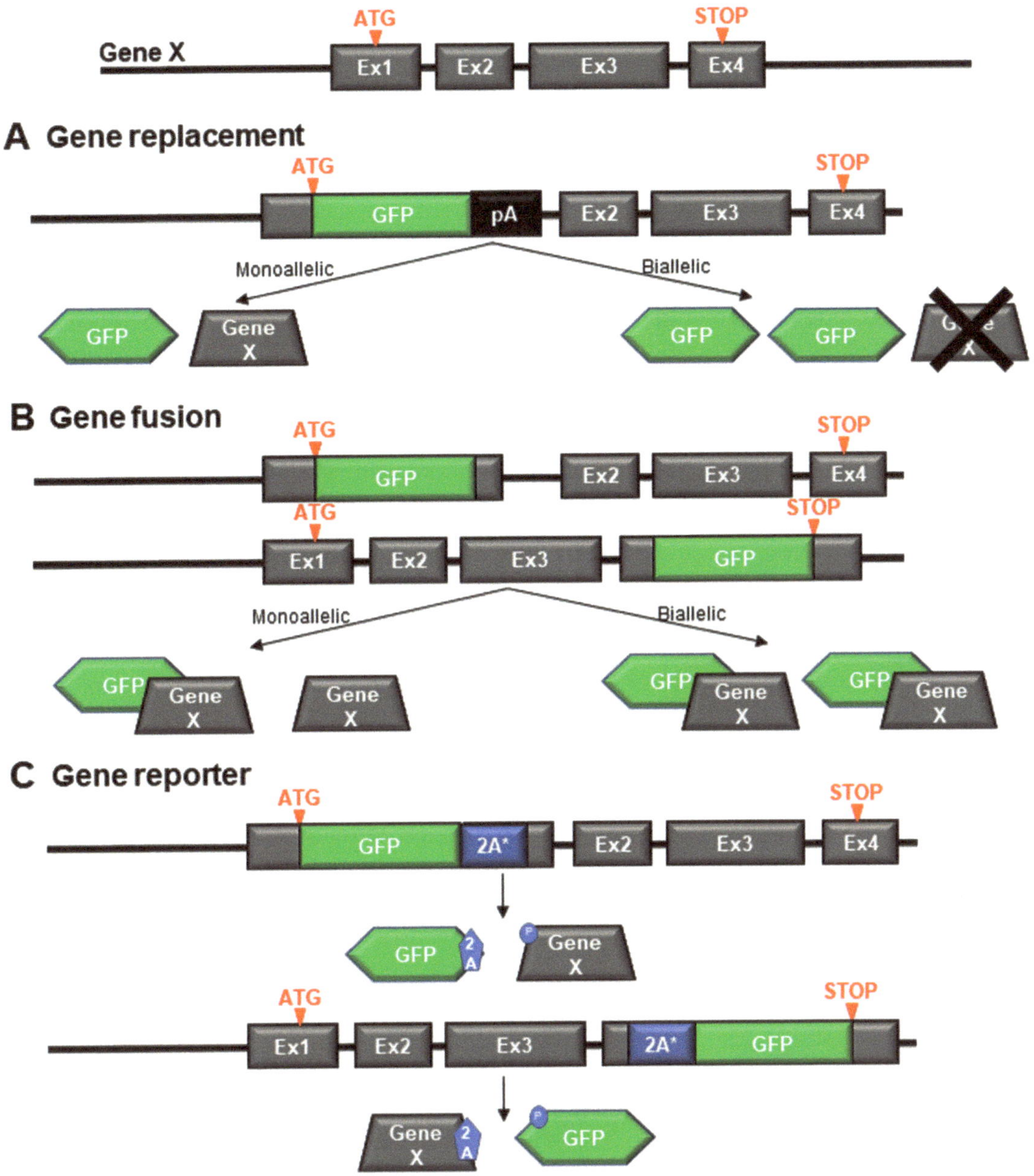

FIGURE 13.2 **Endogenous reporter subtypes.** (A) A gene replacement construct inserts the desired reporter at the endogenous transcription start site. The endogenous promoter elements drive gene expression, and a polyA tail sequence is utilized with the reporter gene to facilitate proper translation. Monoallelic integration of this construct results in one copy of the endogenous gene and one copy of the reporter gene being expressed. Biallelic integration of this construct results in two copies of the reporter gene and effective knockout of the endogenous gene. (B) A gene fusion construct adds the reporter or protein tag to the coding sequence of the endogenous gene. Endogenous promoter and polyA sequences are used to regulate expression, and fusion constructs can be either N-terminal or C-terminal (or interior to the gene in special cases). Monoallelic integration results in both fusion protein and untagged protein, whereas biallelic integration results in only fusion proteins which may affect protein and cellular function. (C) A gene reporter construct utilizes either an IRES or 2A linker to report the expression of a gene. Endogenous promoter and polyA sequences are used to regulate gene expression, and these reporters can be either N-terminal or C-terminal. During translation, the reporter gene forms a different polypeptide than the endogenous gene. With the 2A linker, residual amino acids are retained on the 5′ gene product and a single amino acid is retained on the 3′ gene product.

allele is sufficient to maintain proper rod development. The example discussed earlier, where an OLIG2-GFP gene replacement construct was used to identify altered neuronal progenitor populations in Down syndrome iPSCs, explicitly utilized a heterozygous transgenic cell line in order to allow proper *OLIG2* expression from one allele.[6,122]

Endogenous locus fusion protein or peptide tag transgenes

A fusion protein transgenic system attaches a fluorescent gene or protein tag to the coding sequence of the gene-of-interest to form a single polypeptide product (Fig. 13.2). Since a protein fusion tracks the endogenous gene product throughout the cell, this type of transgene can be used to understand protein dynamics either in normal biology or in a disease state. Recently, the utility of this approach was shown by isogenic iPSCs containing a FUS-eGFP fusion protein to study the effects of a co-knocked-in ALS-causing mutation (FUS^{P525L}) on FUS localization in the cell.[123] The FUS-eGFP fusion protein is retained in the cytoplasm of FUS mutant cells due to FUS retention in large stress granules. Specific small molecules that regulate autophagy were successful in reducing the cumulative load of stress granules in both FUS mutant iPSCs and iPSC-derived neurons, representing a potential therapeutic avenue for ALS.[123]

Protein epitope tags are often utilized when biochemical purification of a protein is needed and there is not a quality antibody for immunostaining or blotting. These tags (FLAG, 6xHis, HA, GST, etc.) have been extensively characterized and, since protein tags are small, have minimal effect on protein function compared to a larger fusion protein such as eGFP. However, most protein tags are limited to either purification or immunocytochemistry using an antibody directed to the tag as they do not harbor inherent fluorescent activity. A multipurpose protein tag seeks to bridge the gap between a fluorescent reporter and a protein tag: HaloTag is a 99 amino acid peptide tag engineered to bind multiple synthetic ligands including fluorescent options and cell-permeable constructs.[124] Thus, a HaloTag fusion protein can be utilized via different ligands to answer diverse questions regarding the endogenous protein's function and localization in the cell.

Bi-cistronic reporter transgenes

The final class of endogenous locus transgene generates a bi-cistronic transcript where the transgene construct is separated from the coding sequence of the gene-of-interest by an IRES (internal ribosome entry site) or a 2A viral linker (Fig. 13.2C). These systems generate two polypeptides from a single transcript (via different mechanisms) which allows the endogenous gene to retain its proper function and creates opportunities to maximize the utility of the transgene. The IRES element was identified as the method RNA viruses use to initiate translation after infecting mammalian cells, and these elements have been adapted in transgenic systems to promote ribosome entry and subsequent translation from within the interior of a transcript.[125] When utilized in a transgenic system, an IRES allows translation of the 5′ coding sequence to be completed independently from the 3′ coding sequence. Thus, instead of fusing the transgenic construct to the gene of interest, two separate polypeptides are generated. Problems with the IRES arise from the relative size of the sequence (typically 400–600 nucleotides in length) and variable levels of expression between the 5′ and 3′ genes. In many cases, such as drug screens where fluorescent readout needs to mirror levels of gene expression, this variability between endogenous gene level and fluorescent gene level can generate confounding effects. IRES sequences were used to generate a NEUROG2-IRES-mCherry-IRES-hygromycin resistance gene reporter line to fluorescently label and allow selection for neural progenitor cells during differentiation.[126] As reported, the IRES sequences allowed independent expression of all three genes from the same transcript, but in-depth characterization of the relative expression levels of the three genes was not performed.

Since they utilize the same translation event to generate the 5′ and 3′ gene products, the 2A class of peptides provide a more stoichiometric ratio of these proteins. 2A peptides are a self-cleaving viral sequence that induces a skipped amino acid during translation, which facilitates the separation of the polypeptide sequence before and after its incorporation.[127] This 2A sequence has a few characterized varieties (e.g., T2A, P2A, E2A, F2A) derived from different viruses and with different cleavage efficiencies in cells, but they share a common 3′ amino acid sequence which appears to underlie the ribosome skip. The full mechanism of action causing this skip is not well understood. Since the most common 2A sequences for mammalian systems (T2A and P2A) minimally require 18 or 19 amino acids, respectively, their molecular cloning is feasible as a component of primer design and incorporating them into a transgene is straightforward. Because translation of a polypeptide chain split by a 2A sequence is continuous other than the skip, the amino acids encoded by the 2A peptide are incorporated into the 5′ and 3′ proteins, with 17 or 18 (respective to T2A or P2A usage) amino acids remaining on the C-terminus of the 5′ gene and the final amino acid (proline) constituting the first amino acid of the 3′ gene. Thus, for endogenous genes whose C-terminus sequence is critical for protein function, utilizing a C-terminal 2A linker may affect function and alternative approaches should be considered. Finally, cleavage of the 2A linker is not always accomplished and can result in a fusion protein consisting of the 5′ and 3′ genes. While this appears rare for the preferred T2A or P2A sequences, the presence of these fusion proteins must be considered. Since the 2A linker has a higher-fidelity expression level between the 5′ and 3′ genes, this approach was utilized in a drug screening platform to identify compounds that could restore FMRP expression from the *Fmr1* locus in Fragile X patient iPSCs.[128] A 2A-nanoluciferase gene was inserted at the 3′ end of the FMRP gene to report on reactivation of the *FMR1* gene locus in response to small molecules.

Complex transgenic systems utilizing multiple gene editing events

Complex transgenic systems, where transgenes are inserted at multiple loci in the genome (commonly an endogenous locus and a safe harbor locus), are necessary to employ a variety of beneficial techniques. The first type of complex transgenic system utilizes an effector protein, such as a transcription factor or DNA recombinase, integrated at the endogenous locus of interest. When that gene is expressed, the effector will mediate its effect on a second transgene at a safe harbor site containing the response element. An example of this is the loxP--stop-loxP or the related dual-floxed inverted open reading frame (DIO) lineage tracing systems. In this system, the effector protein is Cre recombinase, which is integrated at the endogenous locus of a gene expressed in a specific neuronal lineage. A second transgene integrated at a safe harbor locus contains a constitutive promoter, coding sequence of a fluorescent protein, and a polyA tail. To prevent expression of the fluorescent protein until Cre is expressed, the coding sequence can either be preceded by a loxP-stop codon-loxP sequence or inverted and flanked by a specialized dual-loxP orientation (in the DIO system). When the lineage gene is activated, Cre is co-expressed and recombines the loxP sequences, resulting in the final configuration where the constitutive promoter is now expressing the fluorescent protein. This tool permanently labels the resulting daughter cells (and further downstream progeny), facilitating quantitative studies to determine what cell lineages are derived from particular progenitors and whether the developmental process in control versus disease hPSCs affects neurogenesis in a particular lineage.

As discussed in the section on isogenic knockout cell lines, the iKO conditional knockout represents a system where the effector protein is expressed from a safe harbor locus and mediates its effect at an endogenous locus of interest.[86] A safe harbor locus transgene promotes constitutive expression of a Flp recombinase fused with a mutated estrogen receptor (T2) ($FlpeER^{T2}$). At the endogenous locus to be knocked out, FRT sites are introduced which flank the exon for removal. Dosage of the cells with 4-hydroxytamoxifen facilitates $FlpeER^{T2}$ translocation to the nucleus and excision of the FRT-flanked exon.[86] This strategy allows for precise removal of genes of interest at the desired time point when tamoxifen is added.

Future of genetic modification in hPSC-based neuronal research

The use of genetic modification will be a critical component of translating hPSC-derived cells or tissues into therapeutic options. For example, cells can be prepared for transplant with reduced oncogenic risk via expression of suicide genes, such as the herpes simplex virus thymidine kinase (TK), which reacts with ganciclovir to promote apoptosis through DNA replication stress.[129] By regulating TK with a Ki67 promoter, thus expressing TK in cells that are replicating, Tieng et al. demonstrated an approach to reduce oncogenic potential in transplanted hPSC-derived cell types.[130] Restricting TK expression in neural lineages, as demonstrated with a *NANOG* promoter to restrict TK expression to pluripotent cells, can provide a mechanism for purifying cell populations for cell transplant.[131,132] The development of high-fidelity gene editing options will be required to ensure the cells for transplant are safe, but these experiments are well underway and the available tools will continue to improve.

Although it bypasses the hPSC lineage, safe harbor transgenic lines are helpful in promoting the direct conversion of primary cells to neurons. This approach is used to retain epigenetic marks which are believed to be removed during the reprogramming process, thus retaining any age-related attributes that can contribute to disease phenotypes. As discussed in an earlier chapter, direct conversion involves transgene expression of transcription factors that induce neural identity, and extensive work has characterized which factors, alone or in combination, are able to transduce multiple cell types to specific fates.[133] While early techniques utilized viral transduction, new formats are utilizing dCas9-VP64 and related fusion proteins as transcriptional activators that can target specific loci, thus retaining endogenous feedback mechanisms, so sustained transgene activation is not needed.[134,135]

Finally, novel gene editing techniques and applications will continue to play a vital role in disease modeling in vitro. Studying Down syndrome, which is caused by chromosome 21 trisomy, is challenging in any model system. The genes found on chromosome 21 are spread across multiple chromosomes in the mouse, making this model less attractive. Creating isogenic hPSC lines is possible due to mosaicism within the Down syndrome population, but identifying disomy and trisomy hPSC lines from individuals is hampered by proliferation deficiencies for the trisomy cells. By inserting the *XIST* gene, which encodes a long noncoding RNA responsible for the inactivation of X chromosome, onto a single chromosome 21 through genome modification, Jiang et al. were able to silence this extra chromosome and create a functionally isogenic hPSC line.[136] This approach was utilized in a subsequent report to show *XIST*-induced chromosome 21 inactivation at various timepoints during neuronal differentiation, which will help elucidate defects associated with trisomy 21 at different points in development.[137] Creative applications of genome engineering will be essential in developing hPSC models of neural disorders and transitioning hPSC-derived neural cells to clinical applications.

References

1. Kondo T, Imamura K, Funayama M, et al. iPSC-based compound screening and in vitro Trials identify a synergistic anti-amyloid β combination for Alzheimer's disease. *Cell Rep*. 2017;21(8):2304–2312. https://doi.org/10.1016/j.celrep.2017.10.109.
2. Lancaster MA, Renner M, Martin C-A, et al. Cerebral organoids model human brain development and microcephaly. *Nature*. 2013;501(7467):373–379. https://doi.org/10.1038/nature12517.
3. Pasca AM, Sloan SA, Clarke LE, et al. Functional cortical neurons and astrocytes from human pluripotent stem cells in 3D culture. *Nat Methods*. 2015;12(7):671–678. https://doi.org/10.1038/nmeth.3415.
4. Qian X, Jacob F, Song MM, Nguyen HN, Song H, Ming G-L. Generation of human brain region-specific organoids using a miniaturized spinning bioreactor. *Nat Protoc*. 2018;13(3):565–580. https://doi.org/10.1038/nprot.2017.152.
5. Yoon S-J, Elahi LS, Pasca AM, et al. Reliability of human cortical organoid generation. *Nat Methods*. 2019;16(1):75–78. https://doi.org/10.1038/s41592-018-0255-0.
6. Xu R, Brawner AT, Li S, et al. OLIG2 drives abnormal neurodevelopmental phenotypes in human iPSC-based organoid and chimeric mouse models of down syndrome. *Cell Stem Cell*. 2019;24(6):908–926. https://doi.org/10.1016/j.stem.2019.04.014. e8.
7. Di Lullo E, Kriegstein AR. The use of brain organoids to investigate neural development and disease. *Nat Rev Neurosci*. 2017;18(10):573–584. https://doi.org/10.1038/nrn.2017.107.
8. Quadrato G, Nguyen T, Macosko EZ, et al. Cell diversity and network dynamics in photosensitive human brain organoids. *Nature*. 2017;545(7652):48–53. https://doi.org/10.1038/nature22047.
9. Lancaster MA, Corsini NS, Wolfinger S, et al. Guided self-organization and cortical plate formation in human brain organoids. *Nat Biotechnol*. 2017;35(7):659–666. https://doi.org/10.1038/nbt.3906.
10. Hu B-Y, Weick JP, Yu J, et al. Neural differentiation of human induced pluripotent stem cells follows developmental principles but with variable potency. *Proc Natl Acad Sci USA*. 2010;107(9):4335–4340. https://doi.org/10.1073/pnas.0910012107.
11. Guillaume DJ, Johnson MA, Li X-J, Zhang S-C. Human embryonic stem cell-derived neural precursors develop into neurons and integrate into the host brain. *J Neurosci Res*. 2006;84(6):1165–1176.
12. Yang D, Zhang Z-J, Oldenburg M, Ayala M, Zhang S-C. Human embryonic stem cell-derived dopaminergic neurons reverse functional deficit in parkinsonian rats. *Stem Cell*. 2008;26(1):55–63.
13. Han X, Chen M, Wang F, et al. Forebrain engraftment by human glial progenitor cells enhances synaptic plasticity and learning in adult mice. *Cell Stem Cell*. 2013;12(3):342–353. https://doi.org/10.1016/j.stem.2012.12.015.

14. Liu Y, Weick JP, Liu H, et al. Medial ganglionic eminence-like cells derived from human embryonic stem cells correct learning and memory deficits. *Nat Biotechnol*. 2013;31(5):440–447. https://doi.org/10.1038/nbt.2565.
15. Chen Y, Xiong M, Dong Y, et al. Chemical control of grafted human PSC-derived neurons in a mouse model of Parkinson's disease. *Cell Stem Cell*. 2016;18(6):817–826. https://doi.org/10.1016/j.stem.2016.03.014.
16. Weick JP, Liu Y, Zhang S-C. Human embryonic stem cell-derived neurons adopt and regulate the activity of an established neural network. *Proc Natl Acad Sci USA*. 2011;108(50):20189–20194. https://doi.org/10.1073/pnas.1108487108.
17. Piña-Crespo JC, Talantova M, Cho E-G, et al. High-frequency hippocampal oscillations activated by optogenetic stimulation of transplanted human ESC-derived neurons. *J Neurosci*. 2012;32(45):15837–15842. https://doi.org/10.1523/JNEUROSCI.3735-12.2012.
18. Linaro D, Vermaercke B, Iwata R, et al. Xenotransplanted human cortical neurons reveal species-specific development and functional integration into mouse visual Circuits. *Neuron*. 2019;104(5):972–986. https://doi.org/10.1016/j.neuron.2019.10.002. e6.
19. Silva G, Poirot L, Galetto R, et al. Meganucleases and other tools for targeted genome engineering: perspectives and challenges for gene therapy. *Curr Gene Ther*. 2011;11(1):11–27.
20. Musunuru K. Genome editing of human pluripotent stem cells to generate human cellular disease models. *Dis. Models Mech*. 2013;6(4):896–904. https://doi.org/10.1242/dmm.012054.
21. Gaj T, Gersbach CA, Barbas CF. ZFN, TALEN, and CRISPR/Cas-based methods for genome engineering. *Trends Biotechnol*. 2013;31(7):397–405. https://doi.org/10.1016/j.tibtech.2013.04.004.
22. Merkert S, Martin U. Targeted gene editing in human pluripotent stem cells using site-specific nucleases. *Adv Biochem Eng/Biotech*. 2018;163:169–186. https://doi.org/10.1007/10_2017_25.
23. Smith J, Bibikova M, Whitby FG, Reddy AR, Chandrasegaran S, Carroll D. Requirements for double-strand cleavage by chimeric restriction enzymes with zinc finger DNA-recognition domains. *Nucleic Acids Res*. 2000;28(17):3361–3369.
24. Bitinaite J, Wah DA, Aggarwal AK, Schildkraut I. FokI dimerization is required for DNA cleavage. *Proc Natl Acad Sci USA*. 1998;95(18):10570–10575.
25. Li T, Huang S, Jiang WZ, et al. TAL nucleases (TALNs): hybrid proteins composed of TAL effectors and FokI DNA-cleavage domain. *Nucleic Acids Res*. 2011;39(1):359–372. https://doi.org/10.1093/nar/gkq704.
26. Christian M, Cermak T, Doyle EL, et al. Targeting DNA double-strand breaks with TAL effector nucleases. *Genetics*. 2010;186(2):757–761. https://doi.org/10.1534/genetics.110.120717.
27. Miller JC, Tan S, Qiao G, et al. A TALE nuclease architecture for efficient genome editing. *Nat Biotechnol*. 2011;29(2):143–148. https://doi.org/10.1038/nbt.1755.
28. Bogdanove AJ, Voytas DF. TAL effectors: customizable proteins for DNA targeting. *Science*. 2011;333(6051):1843–1846. https://doi.org/10.1126/science.1204094.
29. Moscou MJ, Bogdanove AJ. A simple cipher governs DNA recognition by TAL effectors. *Science*. 2009;326(5959):1501. https://doi.org/10.1126/science.1178817.
30. Boch J, Scholze H, Schornack S, et al. Breaking the code of DNA binding specificity of TAL-type III effectors. *Science*. 2009;326(5959):1509–1512. https://doi.org/10.1126/science.1178811.
31. Guilinger JP, Pattanayak V, Reyon D, et al. Broad specificity profiling of TALENs results in engineered nucleases with improved DNA-cleavage specificity. *Nat Methods*. 2014;11(4):429–435. https://doi.org/10.1038/nmeth.2845.
32. Ramirez CL, Foley JE, Wright DA, et al. Unexpected failure rates for modular assembly of engineered zinc fingers. *Nat Methods*. 2008;5(5):374–375. https://doi.org/10.1038/nmeth0508-374.
33. Hockemeyer D, Wang H, Kiani S, et al. Genetic engineering of human pluripotent cells using TALE nucleases. *Nat Biotechnol*. 2011;29(8):731–734. https://doi.org/10.1038/nbt.1927.
34. Jinek M, Chylinski K, Fonfara I, Hauer M, Doudna JA, Charpentier E. A programmable dual-RNA-guided DNA endonuclease in adaptive bacterial immunity. *Science*. 2012;337(6096):816–821. https://doi.org/10.1126/science.1225829.
35. Cong L, Ran FA, Cox D, et al. Multiplex genome engineering using CRISPR/Cas systems. *Science*. 2013;339(6121):819–823. https://doi.org/10.1126/science.1231143.
36. Mali P, Yang L, Esvelt KM, et al. RNA-guided human genome engineering via Cas9. *Science*. 2013;339(6121):823–826. https://doi.org/10.1126/science.1232033.

37. Hou Z, Zhang Y, Propson NE, et al. Efficient genome engineering in human pluripotent stem cells using Cas9 from Neisseria meningitidis. *Proc Natl Acad Sci USA*. 2013;110(39):15644–15649. https://doi.org/10.1073/pnas.1313587110.
38. Ran FA, Cong L, Yan WX, et al. In vivo genome editing using *Staphylococcus aureus* Cas9. *Nature*. 2015;520(7546):186–191. https://doi.org/10.1038/nature14299.
39. Kleinstiver BP, Prew MS, Tsai SQ, et al. Engineered CRISPR-Cas9 nucleases with altered PAM specificities. *Nature*. 2015;523(7561):481–485. https://doi.org/10.1038/nature14592.
40. Ran FA, Hsu PD, Lin C-Y, et al. Double nicking by RNA-guided CRISPR Cas9 for enhanced genome editing specificity. *Cell*. 2013;154(6):1380–1389. https://doi.org/10.1016/j.cell.2013.08.021.
41. Ran FA, Hsu PD, Wright J, Agarwala V, Scott DA, Zhang F. Genome engineering using the CRISPR-Cas9 system. *Nat Protoc*. 2013;8(11):2281–2308. https://doi.org/10.1038/nprot.2013.143.
42. Shen B, Zhang W, Zhang J, et al. Efficient genome modification by CRISPR-Cas9 nickase with minimal off-target effects. *Nat Methods*. 2014;11(4):399–402. https://doi.org/10.1038/nmeth.2857.
43. Cho SW, Kim S, Kim Y, et al. Analysis of off-target effects of CRISPR/Cas-derived RNA-guided endonucleases and nickases. *Genome Res*. 2014;24(1):132–141. https://doi.org/10.1101/gr.162339.113.
44. Fu Y, Foden JA, Khayter C, et al. High-frequency off-target mutagenesis induced by CRISPR-Cas nucleases in human cells. *Nat Biotechnol*. 2013;31(9):822–826. https://doi.org/10.1038/nbt.2623.
45. Pattanayak V, Lin S, Guilinger JP, Ma E, Doudna JA, Liu DR. High-throughput profiling of off-target DNA cleavage reveals RNA-programmed Cas9 nuclease specificity. *Nat Biotechnol*. 2013;31(9):839–843. https://doi.org/10.1038/nbt.2673.
46. Veres A, Gosis BS, Ding Q, et al. Low incidence of off-target mutations in individual CRISPR-Cas9 and TALEN targeted human stem cell clones detected by whole-genome sequencing. *Cell Stem Cell*. 2014;15(1):27–30. https://doi.org/10.1016/j.stem.2014.04.020.
47. Qi LS, Larson MH, Gilbert LA, et al. Repurposing CRISPR as an RNA-guided platform for sequence-specific control of gene expression. *Cell*. 2013;152(5):1173–1183. https://doi.org/10.1016/j.cell.2013.02.022.
48. Adli M. The CRISPR tool kit for genome editing and beyond. *Nat Commun*. 2018;9(1):1911. https://doi.org/10.1038/s41467-018-04252-2.
49. Chavez A, Scheiman J, Vora S, et al. Highly efficient Cas9-mediated transcriptional programming. *Nat Methods*. 2015;12(4):326–328. https://doi.org/10.1038/nmeth.3312.
50. Chen B, Gilbert LA, Cimini BA, et al. Dynamic imaging of genomic loci in living human cells by an optimized CRISPR/Cas system. *Cell*. 2013;155(7):1479–1491. https://doi.org/10.1016/j.cell.2013.12.001.
51. Gaudelli NM, Komor AC, Rees HA, et al. Programmable base editing of A•T to G•C in genomic DNA without DNA cleavage. *Nature*. 2017;551(7681):464–471. https://doi.org/10.1038/nature24644.
52. Komor AC, Kim YB, Packer MS, Zuris JA, Liu DR. Programmable editing of a target base in genomic DNA without double-stranded DNA cleavage. *Nature*. 2016;533(7603):420–424. https://doi.org/10.1038/nature17946.
53. Chen JS, Dagdas YS, Kleinstiver BP, et al. Enhanced proofreading governs CRISPR-Cas9 targeting accuracy. *Nature*. 2017;550(7676):407–410. https://doi.org/10.1038/nature24268.
54. Kleinstiver BP, Pattanayak V, Prew MS, et al. High-fidelity CRISPR-Cas9 nucleases with no detectable genome-wide off-target effects. *Nature*. 2016;529(7587):490–495. https://doi.org/10.1038/nature16526.
55. Slaymaker IM, Gao L, Zetsche B, Scott DA, Yan WX, Zhang F. Rationally engineered Cas9 nucleases with improved specificity. *Science*. 2016;351(6268):84–88. https://doi.org/10.1126/science.aad5227.
56. Casini A, Olivieri M, Petris G, et al. A highly specific SpCas9 variant is identified by in vivo screening in yeast. *Nat Biotechnol*. 2018;36(3):265–271. https://doi.org/10.1038/nbt.4066.
57. Lee JK, Jeong E, Lee J, et al. Directed evolution of CRISPR-Cas9 to increase its specificity. *Nat Commun*. 2018;9(1):3048. https://doi.org/10.1038/s41467-018-05477-x.
58. Anders C, Bargsten K, Jinek M. Structural plasticity of PAM recognition by engineered variants of the RNA-guided endonuclease Cas9. *Mol Cell*. 2016;61(6):895–902. https://doi.org/10.1016/j.molcel.2016.02.020.
59. Nishimasu H, Shi X, Ishiguro S, et al. Engineered CRISPR-Cas9 nuclease with expanded targeting space. *Science*. 2018;361(6408):1259–1262. https://doi.org/10.1126/science.aas9129.
60. Hu JH, Miller SM, Geurts MH, et al. Evolved Cas9 variants with broad PAM compatibility and high DNA specificity. *Nature*. 2018;556(7699):57–63. https://doi.org/10.1038/nature26155.

61. Anzalone AV, Koblan LW, Liu DR. Genome editing with CRISPR-Cas nucleases, base editors, transposases and prime editors. *Nat Biotechnol*. 2020;38(7):824–844. https://doi.org/10.1038/s41587-020-0561-9.
62. Anzalone AV, Randolph PB, Davis JR, et al. Search-and-replace genome editing without double-strand breaks or donor DNA. *Nature*. 2019;576(7785):149–157. https://doi.org/10.1038/s41586-019-1711-4.
63. Abudayyeh OO, Gootenberg JS, Essletzbichler P, et al. RNA targeting with CRISPR-Cas13. *Nature*. 2017;550(7675):280–284. https://doi.org/10.1038/nature24049.
64. Eggenschwiler R, Moslem M, Fráguas MS, et al. Improved bi-allelic modification of a transcriptionally silent locus in patient-derived iPSC by Cas9 nickase. *Sci Rep*. 2016;6:38198. https://doi.org/10.1038/srep38198.
65. Glass Z, Lee M, Li Y, Xu Q. Engineering the delivery system for CRISPR-based genome editing. *Trends Biotechnol*. 2018;36(2):173–185. https://doi.org/10.1016/j.tibtech.2017.11.006.
66. Lin S, Staahl BT, Alla RK, Doudna JA. Enhanced homology-directed human genome engineering by controlled timing of CRISPR/Cas9 delivery. *Elife*. 2014;3:e04766. https://doi.org/10.7554/eLife.04766.
67. Schumann K, Lin S, Boyer E, et al. Generation of knock-in primary human T cells using Cas9 ribonucleoproteins. *Proc Natl Acad Sci USA*. 2015;112(33):10437–10442. https://doi.org/10.1073/pnas.1512503112.
68. Ding Q, Regan SN, Xia Y, Oostrom LA, Cowan CA, Musunuru K. Enhanced efficiency of human pluripotent stem cell genome editing through replacing TALENs with CRISPRs. *Cell Stem Cell*. 2013;12(4):393–394. https://doi.org/10.1016/j.stem.2013.03.006.
69. Sanjana NE, Shalem O, Zhang F. Improved vectors and genome-wide libraries for CRISPR screening. *Nat Methods*. 2014;11(8):783–784. https://doi.org/10.1038/nmeth.3047.
70. Karagiannis P, Inoue H. ALS, a cellular whodunit on motor neuron degeneration. *Mol Cell Neurosci*. 2020;107:103524. https://doi.org/10.1016/j.mcn.2020.103524.
71. Jones JR, Kong L, Hanna MG, et al. Mutations in GFAP disrupt the distribution and function of Organelles in human astrocytes. *Cell Rep*. 2018;25(4):947–958. https://doi.org/10.1016/j.celrep.2018.09.083. e4.
72. Li L, Tian E, Chen X, et al. GFAP mutations in astrocytes Impair oligodendrocyte progenitor proliferation and myelination in an hiPSC model of alexander disease. *Cell Stem Cell*. 2018;23(2):239–251. https://doi.org/10.1016/j.stem.2018.07.009. e6.
73. Park C-Y, Halevy T, Lee DR, et al. Reversion of FMR1 methylation and silencing by editing the triplet repeats in fragile X iPSC-derived neurons. *Cell Rep*. 2015;13(2):234–241. https://doi.org/10.1016/j.celrep.2015.08.084.
74. Dabrowska M, Ciolak A, Kozlowska E, Fiszer A, Olejniczak M. Generation of new isogenic models of Huntington's disease using CRISPR-cas9 technology. *Int J Mol Sci*. 2020;21(5). https://doi.org/10.3390/ijms21051854.
75. Xu X, Tay Y, Sim B, et al. Reversal of phenotypic Abnormalities by CRISPR/Cas9-Mediated gene correction in Huntington disease patient-derived induced pluripotent stem cells. *Stem Cell Rep*. 2017;8(3):619–633. https://doi.org/10.1016/j.stemcr.2017.01.022.
76. Ababneh NA, Scaber J, Flynn R, et al. Correction of amyotrophic lateral sclerosis related phenotypes in induced pluripotent stem cell-derived motor neurons carrying a hexanucleotide expansion mutation in C9orf72 by CRISPR/Cas9 genome editing using homology-directed repair. *Hum Mol Genet*. 2020;29(13):2200–2217. https://doi.org/10.1093/hmg/ddaa106.
77. Selvaraj BT, Livesey MR, Zhao C, et al. -permeable AMPA receptor-mediated excitotoxicity. *Nat Commun*. 2018;9(1):347. https://doi.org/10.1038/s41467-017-02729-0.
78. Sposito T, Preza E, Mahoney CJ, et al. Developmental regulation of tau splicing is disrupted in stem cell-derived neurons from frontotemporal dementia patients with the 10 + 16 splice-site mutation in MAPT. *Hum Mol Genet*. 2015;24(18):5260–5269. https://doi.org/10.1093/hmg/ddv246.
79. Iovino M, Agathou S, González-Rueda A, et al. Early maturation and distinct tau pathology in induced pluripotent stem cell-derived neurons from patients with MAPT mutations. *Brain: J Neurol*. 2015;138(Pt 11):3345–3359. https://doi.org/10.1093/brain/awv222.
80. Liu F, Gong C-X. Tau exon 10 alternative splicing and tauopathies. *Mol Neurodegener*. 2008;3:8. https://doi.org/10.1186/1750-1326-3-8.
81. Drakulic D, Djurovic S, Syed YA, et al. Copy number variants (CNVs): a powerful tool for iPSC-based modelling of ASD. *Mol Autism*. 2020;11(1):42. https://doi.org/10.1186/s13229-020-00343-4.
82. Maillard AM, Ruef A, Pizzagalli F, et al. The 16p11.2 locus modulates brain structures common to autism, schizophrenia and obesity. *Mol Psychiatr*. 2015;20(1):140–147. https://doi.org/10.1038/mp.2014.145.
83. Sharp AJ, Mefford HC, Li K, et al. A recurrent 15q13.3 microdeletion syndrome associated with mental retardation and seizures. *Nat Genet*. 2008;40(3):322–328. https://doi.org/10.1038/ng.93.

84. Tai DJC, Ragavendran A, Manavalan P, et al. Engineering microdeletions and microduplications by targeting segmental duplications with CRISPR. *Nat Neurosci*. 2016;19(3):517–522. https://doi.org/10.1038/nn.4235.
85. González F, Zhu Z, Shi Z-D, et al. An iCRISPR platform for rapid, multiplexable, and inducible genome editing in human pluripotent stem cells. *Cell Stem Cell*. 2014;15(2):215–226. https://doi.org/10.1016/j.stem.2014.05.018.
86. Chen Y, Cao J, Xiong M, et al. Engineering human stem cell lines with inducible gene knockout using CRISPR/Cas9. *Cell Stem Cell*. 2015;17(2):233–244. https://doi.org/10.1016/j.stem.2015.06.001.
87. Snijders KE, Cooper JD, Vallier L, Bertero A. Conditional gene knockout in human cells with inducible CRISPR/Cas9. *Methods Mol Biol*. 2019;1961:185–209. https://doi.org/10.1007/978-1-4939-9170-9_12.
88. Irion S, Luche H, Gadue P, Fehling HJ, Kennedy M, Keller G. Identification and targeting of the ROSA26 locus in human embryonic stem cells. *Nat Biotechnol*. 2007;25(12):1477–1482.
89. Kim HJ, Lee HJ, Kim H, Cho SW, Kim J-S. Targeted genome editing in human cells with zinc finger nucleases constructed via modular assembly. *Genome Res*. 2009;19(7):1279–1288. https://doi.org/10.1101/gr.089417.108.
90. Hockemeyer D, Soldner F, Beard C, et al. Efficient targeting of expressed and silent genes in human ESCs and iPSCs using zinc-finger nucleases. *Nat Biotechnol*. 2009;27(9):851–857. https://doi.org/10.1038/nbt.1562.
91. Smith JR, Maguire S, Davis LA, et al. Robust, persistent transgene expression in human embryonic stem cells is achieved with AAVS1-targeted integration. *Stem Cell*. 2008;26(2):496–504.
92. Mizutani T, Li R, Haga H, Kawabata K. Transgene integration into the human AAVS1 locus enhances myosin II-dependent contractile force by reducing expression of myosin binding subunit 85. *Biochem Biophys Res Commun*. 2015;465(2):270–274. https://doi.org/10.1016/j.bbrc.2015.08.018.
93. Sadelain M, Papapetrou EP, Bushman FD. Safe harbours for the integration of new DNA in the human genome. *Nat Rev Cancer*. 2011;12(1):51–58. https://doi.org/10.1038/nrc3179.
94. Pellenz S, Phelps M, Tang W, et al. New human chromosomal sites with "safe harbor" potential for targeted transgene insertion. *Hum Gene Ther*. 2019;30(7):814–828. https://doi.org/10.1089/hum.2018.169.
95. Qin JY, Zhang L, Clift KL, et al. Systematic comparison of constitutive promoters and the doxycycline-inducible promoter. *PLoS One*. 2010;5(5):e10611. https://doi.org/10.1371/journal.pone.0010611.
96. Gibson TJ, Seiler M, Veitia RA. The transience of transient overexpression. *Nat Methods*. 2013;10(8):715–721. https://doi.org/10.1038/nmeth.2534.
97. Agha-Mohammadi S, O'Malley M, Etemad A, Wang Z, Xiao X, Lotze MT. Second-generation tetracycline-regulatable promoter: repositioned tet operator elements optimize transactivator synergy while shorter minimal promoter offers tight basal leakiness. *J Gene Med*. 2004;6(7):817–828.
98. Gossen M, Freundlieb S, Bender G, Müller G, Hillen W, Bujard H. Transcriptional activation by tetracyclines in mammalian cells. *Science*. 1995;268(5218):1766–1769.
99. Loew R, Heinz N, Hampf M, Bujard H, Gossen M. Improved Tet-responsive promoters with minimized background expression. *BMC Biotechnol*. 2010;10:81. https://doi.org/10.1186/1472-6750-10-81.
100. Zhang J, Chen L, Zhang J, Wang Y. Drug inducible CRISPR/cas systems. *Comput Struct Biotechnol J*. 2019;17:1171–1177. https://doi.org/10.1016/j.csbj.2019.07.015.
101. Amoroso MW, Croft GF, Williams DJ, et al. Accelerated high-yield generation of limb-innervating motor neurons from human stem cells. *J Neurosci*. 2013;33(2):574–586. https://doi.org/10.1523/JNEUROSCI.0906-12.2013.
102. Di Giorgio FP, Boulting GL, Bobrowicz S, Eggan KC. Human embryonic stem cell-derived motor neurons are sensitive to the toxic effect of glial cells carrying an ALS-causing mutation. *Cell Stem Cell*. 2008;3(6):637–648. https://doi.org/10.1016/j.stem.2008.09.017.
103. Singh Roy N, Nakano T, Xuing L, Kang J, Nedergaard M, Goldman SA. Enhancer-specified GFP-based FACS purification of human spinal motor neurons from embryonic stem cells. *Exp Neurol*. 2005;196(2):224–234.
104. Wichterle H, Lieberam I, Porter JA, Jessell TM. Directed differentiation of embryonic stem cells into motor neurons. *Cell*. 2002;110(3):385–397.
105. Kügler S, Kilic E, Bähr M. Human synapsin 1 gene promoter confers highly neuron-specific long-term transgene expression from an adenoviral vector in the adult rat brain depending on the transduced area. *Gene Ther*. 2003;10(4):337–347.
106. Nagel G, Szellas T, Huhn W, et al. Channelrhodopsin-2, a directly light-gated cation-selective membrane channel. *Proc Natl Acad Sci USA*. 2003;100(24):13940–13945.

107. Sineshchekov OA, Jung K-H, Spudich JL. Two rhodopsins mediate phototaxis to low- and high-intensity light in Chlamydomonas reinhardtii. *Proc Natl Acad Sci USA*. 2002;99(13):8689–8694.
108. Tønnesen J, Sørensen AT, Deisseroth K, Lundberg C, Kokaia M. Optogenetic control of epileptiform activity. *Proc Natl Acad Sci USA*. 2009;106(29):12162–12167. https://doi.org/10.1073/pnas.0901915106.
109. Weick JP, Johnson MA, Skroch SP, Williams JC, Deisseroth K, Zhang S-C. Functional control of transplantable human ESC-derived neurons via optogenetic targeting. *Stem Cell*. 2010;28(11):2008–2016. https://doi.org/10.1002/stem.514.
110. Avaliani N, Sørensen AT, Ledri M, et al. Optogenetics reveal delayed afferent synaptogenesis on grafted human-induced pluripotent stem cell-derived neural progenitors. *Stem Cell*. 2014;32(12):3088–3098. https://doi.org/10.1002/stem.1823.
111. Zhang Y, Pak C, Han Y, et al. Rapid single-step induction of functional neurons from human pluripotent stem cells. *Neuron*. 2013;78(5):785–798. https://doi.org/10.1016/j.neuron.2013.05.029.
112. Steinbeck JA, Choi SJ, Mrejeru A, et al. Optogenetics enables functional analysis of human embryonic stem cell-derived grafts in a Parkinson's disease model. *Nat Biotechnol*. 2015;33(2):204–209. https://doi.org/10.1038/nbt.3124.
113. Daadi MM, Klausner JQ, Bajar B, et al. Optogenetic stimulation of neural grafts enhances neurotransmission and Downregulates the inflammatory response in experimental stroke model. *Cell Trans*. 2016;25(7):1371–1380. https://doi.org/10.3727/096368915X688533.
114. Dong Y, Xiong M, Chen Y, et al. Plasticity of synaptic transmission in human stem cell-derived neural networks. *iScience*. 2020;23(2):100829. https://doi.org/10.1016/j.isci.2020.100829.
115. Roth BL. DREADDs for Neuroscientists. *Neuron*. 2016;89(4):683–694. https://doi.org/10.1016/j.neuron.2016.01.040.
116. Armbruster BN, Li X, Pausch MH, Herlitze S, Roth BL. Evolving the lock to fit the key to create a family of G protein-coupled receptors potently activated by an inert ligand. *Proc Natl Acad Sci USA*. 2007;104(12):5163–5168.
117. Nichols CD, Roth BL. Engineered G-protein coupled receptors are powerful tools to investigate biological processes and Behaviors. *Front Mol Neurosci*. 2009;2:16. https://doi.org/10.3389/neuro.02.016.2009.
118. Alexander GM, Rogan SC, Abbas AI, et al. Remote control of neuronal activity in transgenic mice expressing evolved G protein-coupled receptors. *Neuron*. 2009;63(1):27–39. https://doi.org/10.1016/j.neuron.2009.06.014.
119. Imamura K, Sahara N, Kanaan NM, et al. Calcium dysregulation contributes to neurodegeneration in FTLD patient iPSC-derived neurons. *Sci Rep*. 2016;6:34904. https://doi.org/10.1038/srep34904.
120. Upadhya D, Hattiangady B, Castro OW, et al. Human induced pluripotent stem cell-derived MGE cell grafting after status epilepticus attenuates chronic epilepsy and comorbidities via synaptic integration. *Proc Natl Acad Sci USA*. 2019;116(1):287–296. https://doi.org/10.1073/pnas.1814185115.
121. Phillips MJ, Capowski EE, Petersen A, et al. Generation of a rod-specific NRL reporter line in human pluripotent stem cells. *Sci Rep*. 2018;8(1):2370. https://doi.org/10.1038/s41598-018-20813-3.
122. Liu Y, Jiang P, Deng W. OLIG gene targeting in human pluripotent stem cells for motor neuron and oligodendrocyte differentiation. *Nat Protoc*. 2011;6(5):640–655. https://doi.org/10.1038/nprot.2011.310.
123. Marrone L, Poser I, Casci I, et al. Isogenic FUS-eGFP iPSC reporter lines enable quantification of FUS stress granule pathology that is rescued by drugs inducing autophagy. *Stem Cell Rep*. 2018;10(2):375–389. https://doi.org/10.1016/j.stemcr.2017.12.018.
124. Los GV, Encell LP, McDougall MG, et al. HaloTag: a novel protein labeling technology for cell imaging and protein analysis. *ACS Chem Biol*. 2008;3(6):373–382. https://doi.org/10.1021/cb800025k.
125. Pelletier J, Sonenberg N. Internal initiation of translation of eukaryotic mRNA directed by a sequence derived from poliovirus RNA. *Nature*. 1988;334(6180):320–325.
126. Li S, Xue H, Wu J, et al. Human induced pluripotent stem cell NEUROG2 dual Knockin reporter lines generated by the CRISPR/Cas9 system. *Stem Cell Dev*. 2015;24(24):2925–2942. https://doi.org/10.1089/scd.2015.0131.
127. Kim JH, Lee S-R, Li L-H, et al. High cleavage efficiency of a 2A peptide derived from porcine teschovirus-1 in human cell lines, zebrafish and mice. *PLoS One*. 2011;6(4):e18556. https://doi.org/10.1371/journal.pone.0018556.
128. Li M, Zhao H, Ananiev GE, et al. Establishment of reporter lines for detecting fragile X mental retardation (FMR1) gene reactivation in human neural cells. *Stem Cell*. 2017;35(1):158–169. https://doi.org/10.1002/stem.2463.

129. Fillat C, Carrió M, Cascante A, Sangro B. Suicide gene therapy mediated by the Herpes Simplex virus thymidine kinase gene/Ganciclovir system: fifteen years of application. *Curr Gene Ther*. 2003;3(1):13–26.
130. Tieng V, Cherpin O, Gutzwiller E, et al. Elimination of proliferating cells from CNS grafts using a Ki67 promoter-driven thymidine kinase. *Mol Ther Methods Clin Dev*. 2016;6:16069.
131. Cheng F, Ke Q, Chen F, et al. Protecting against wayward human induced pluripotent stem cells with a suicide gene. *Biomaterials*. 2012;33(11):3195–3204. https://doi.org/10.1016/j.biomaterials.2012.01.023.
132. Rong Z, Fu X, Wang M, Xu Y. A scalable approach to prevent teratoma formation of human embryonic stem cells. *J Biol Chem*. 2012;287(39):32338–32345.
133. Colasante G, Rubio A, Massimino L, Broccoli V. Direct neuronal reprogramming reveals unknown functions for Known transcription factors. *Front Neurosci*. 2019;13:283. https://doi.org/10.3389/fnins.2019.00283.
134. Black JB, Adler AF, Wang H-G, et al. Targeted epigenetic remodeling of endogenous loci by CRISPR/Cas9-Based transcriptional activators directly Converts fibroblasts to neuronal cells. *Cell Stem Cell*. 2016;19(3):406–414. https://doi.org/10.1016/j.stem.2016.07.001.
135. Liu Y, Yu C, Daley TP, et al. CRISPR activation screens systematically identify factors that drive neuronal fate and reprogramming. *Cell Stem Cell*. 2018;23(5):758–771. https://doi.org/10.1016/j.stem.2018.09.003. e8.
136. Jiang J, Jing Y, Cost GJ, et al. Translating dosage compensation to trisomy 21. *Nature*. 2013;500(7462):296–300. https://doi.org/10.1038/nature12394.
137. Czermiński JT, Lawrence JB. Silencing trisomy 21 with XIST in neural stem cells promotes neuronal differentiation. *Dev Cell*. 2020;52(3):294–308. https://doi.org/10.1016/j.devcel.2019.12.015. e3.

CHAPTER

14

Cell therapy and biomanufacturing using hiPSC-derived neurons

Juan Cruz Casabona[1,2]

[1]Instituto de Investigaciones Bioquímicas de Buenos Aires (IIBBA), Fundación Instituto Leloir, Buenos Aires, Argentina; [2]Consejo Nacional de Investigaciones Científicas y Técnicas (CONICET), Buenos Aires, Argentina

Introduction

The utility of stem cells for preclinical research offers one of the great promises of our time, given the huge potential of these cells to be differentiated into different cell types, allowing researchers to use stem cell derivatives to discover new mechanisms of disease as well as designing toxicity assays, drug testing platforms, and even allowing for the development of cell-based therapies for a broad range of diseases that are currently incurable.

The field of stem cell biology has been thriving with new discoveries over the last few decades, especially since the groundbreaking discovery of human reprogrammed stem cells from somatic cells by Shinya Yamanaka in 2006, which led to an explosion of research in this area.[1] This seminal work provided the cornerstone for a new exciting era in stem cell research, in which many past challenges can now be overcome. With the discovery of human induced pluripotent stem cells (hiPSCs) derived from patient-specific somatic cells, issues including stem cell accessibility, ethical considerations, and concerns regarding transplant rejection, could now be left behind.

As the hiPSC field has grown exponentially during this last decade, new challenges have been encountered, and many lie ahead of us. As global science grows steady to build a better future, there are still many diseases that require deeper understanding, and new drugs and therapies to treat them.

This is particularly true for neurological diseases, which have an extreme level of complexity to study, and to treat. For this purpose, hiPSC-derived neurons are an invaluable tool that is pushing forward the understanding of these illnesses every day. New models of diseases related to the central nervous system (CNS) are helping us to understand the complex mechanisms that underlie a plethora of illnesses, and are leading us to the identification of potential treatments in the near future. For many of these disorders, there is also a chance

Phenotyping of Human iPSC-derived Neurons
https://doi.org/10.1016/B978-0-12-822277-5.00009-2

to develop cell replacement therapies, with specific differentiated cells derived from hiPSCs. For the development of cell replacement therapies, the laboratory will have to turn into a manufacturing facility. Such facilities require that protocols should be converted into standard operative procedures (SOPs) and the equipment that the scientists use for their everyday job will have to be validated according to the industry standards and regulations. Every single reagent will have to comply with good manufacturing practices (GMP), and the personnel for these areas will have to be trained according to this new paradigm. Strict quality controls (QC) will have to be standardized and regulatory compliance will be required.

This new world where Clinical-Grade hiPSCs and their derivatives will be flourishing, will rely on setting a network of strategic partnerships, multidisciplinary and multi-institutional collaborations involving many players to achieve this task. Academia and industry will need to join forces, along with funding and regulatory agencies, in a synergistic effort to overcome the multiple challenges that this goal faces.

This chapter will cover some examples of the use of hiPSC-derived neurons to model specific neurological disorders, and the potential clinical relevance of these findings. We will also briefly discuss the historical aspects of biomanufacturing, in order to understand the need for standardization and regulations and we will dig deeper into the manufacturing process needed to make neurons and other cells related to the CNS from hiPSCs. We will cover the technical challenges associated with the process, taking into account the aforementioned topics, and focusing on up to date examples of diseases that are studied with CNS cells differentiated from hiPSC by reviewing current Clinical Trials. At last, we will discuss the potential application of these tools, the multiple challenges that have to be addressed, and the need for collaboration in order to transfer this technology from bench to bedside.

hiPSC-derived neurons to model specific neurological disorders

Due to the complexity of the human brain and the CNS, and the impossibility of obtaining live primary human samples for experimental research, many animal models have been developed for the study of CNS disorders, in order to understand the mechanisms of disease.[2] Despite significant progress using animal models to understand the cellular and molecular mechanisms leading to brain cell dysfunction and degeneration, these tools are limited in revealing some of the most fundamental aspects of development, genetics, pathology, and disease mechanisms that are unique to humans. These differences are particularly augmented with conditions that affect the brain, and lead to poor understanding of the mechanisms that are involved in neurological disorders.

Many models are able to recapitulate the basic pathological features of a particular disease, and occasionally also recapitulate specific molecular mechanisms and degenerative cascades, but they do not fully capture the pathological events or timing that occur in humans and therefore, they have poor predictive power for drug efficacy and fail in the development of clinical interventions.[3,4] Furthermore, even if the final output of a preclinical model recapitulates the main features of a disease, different species can have variable responses to the exact same insult and finding a cure for an animal model could lead to a cure for the animal but not necessarily for humans.

hiPSCs have appeared as a new tool to discover the mechanisms of disease in human cells, complementing animal models with cell-based models, and organoid models derived from human donors.[5] This has been particularly true for diseases that are associated with a strong genetic component, for example, a mutation in a single gene. hiPSCs have the advantage of providing an unlimited supply of cells for experiments as well as circumventing many ethical issues associated with the use of stem cells obtained from human fetal tissue. These cells can be differentiated into various cell types of the CNS, including neurons, astrocytes, and oligodendrocytes,[6] and are currently used for disease modeling, drug discovery or toxicity tests, as well as pre-clinical trials, and could be used for cell replacement therapies in the future.

Most of the early studies that have adopted hiPSCs to study neurological disorders have chosen to model a neurological disease with a known genetic association, for example, a point mutation in a causative single gene. With this starting point, researchers are able to reprogram a somatic cell sample from each of several patients with the same genetic cause of a particular disease. The selection of control lines is critical to the success of any studying using hiPSCs and can include familial controls, a close relative that doesn't carry the particular mutation but has shared genetic background, or isogenic controls, where the genetic mutation is corrected by gene editing techniques in the patient cell line itself.[7] With established hiPSC lines from patients and controls, differentiation can be completed to generate CNS cells of varying types, according to which disease we are studying. For example, if the research study is focused on Parkinson's disease (PD), the researcher will differentiate the hiPSCs to a type of neuron that is known to be failing or absent in that particular disease such as dopaminergic neurons, which are usually derived through a neural stem cell (NSC) intermediate stage. These dopaminergic neurons maintain the genetic background of the patient, including the variant of interest, and therefore should recapitulate the pathological events or mechanisms of the disease. Once the dopaminergic cells are differentiated, experiments can be set up to elucidate what is different between them and healthy control cells. Compound testing can also be implemented to identify new drugs that reverse the pathological phenotypes seen in vitro. Additionally, dopaminergic precursors from healthy individuals can be used to replace the dopaminergic neurons that are lost in PD patient brains. These cell replacement therapies are the ultimate goal of regenerative medicine, because they hold the promise of actually curing the disease, and not only ameliorating its symptoms or delaying its manifestation.

Nowadays, cell banks or repositories have been established to improve access to the originating cell materials. These biobanks often provide access to fibroblasts, peripheral blood mononuclear cells (PBMCs), or even hiPSCs from patients and healthy controls. There are also many companies and academic cores offering services to reprogram hiPSCs from patient-derived samples. For disease modeling, services also exist that provide differentiation services into distinct neuronal or glial subtypes from patient and control hiPSCs. However, these services are primarily for disease modeling and are not held to the same stringency needed for cell therapy indications. Since one of the most common intermediate stages in the differentiation process of a hiPSC into a neuron is the NSC, many repositories also offer NSCs for purchase. However, if a lab wants to reprogram the donor cells themselves, there are ready-to-use media, feeder layers, and reagents to make it possible. There are also a few options of ready-to-use media and additives to differentiate hiPSCs to NSCs, or specific subtypes of neurons, including dopaminergic, glutamatergic or GABAergic neurons, but only for research purposes. Usually the companies selling these reagents use proprietary technology that provides consistency from batch to batch, scalability and quite simple protocols to speed the path to discovery. However, frequently these technologies are quite expensive and are not as flexible and easy to adapt to individual labs or across a variety of hiPSC lines. Some companies even offer ready-to-use neural precursors, which will become fully developed neurons with some extra culture steps. Needless to say, they are limited to produce a very specific final phenotype and may not be made from a hiPSC line from the disease that you want to study, reflecting only the genetic background of the control group or the small cohort of patients that a particular company has access to. In addition, these commercially sourced cells may not be able to be expanded in culture, so a finite amount of cells is provided, instead of a potential unlimited supply of hiPSCs or NSCs.

It is also worth mentioning that despite the great potential and several advantages of hiPSC-derived neurons and glia to elucidate many aspects of diverse diseases, they have limitations, especially for modeling age-related diseases because the cellular aging profile is widely erased during reprogramming of somatic cells to the pluripotent state.[8] Neural cells derived from hiPSCs display a fetal, immature phenotype and may lack the aging signatures of their source cells. This issue can be circumvented by direct differentiation of, for instance, fibroblast into neurons. This was achieved for diseases such as Huntington's or Parkinson's (PD).[9,10] However, direct differentiation protocols are inefficient and fibroblasts do not serve as an infinite source of starting material, a distinctive characteristic that stem cells do have. Maturation of stem cell-derived neuronal cultures has also been achieved using long term culture of 3D organoids, which represents a major step forward for disease modeling of neurodegenerative disorders using these cells. Nevertheless, hiPSC-derived neurons and precursors are a very powerful tool for studying neurodegenerative diseases, despite their limitations.

If the intention is to use hiPSC-derived cells for a clinical trial, the path will be very different, and so far there is only one interventional clinical trial ongoing using CNS cells derived from hiPSCs for PD, which we will discuss deeply in the following sections.

This section will review some of the most advanced and significant discoveries for hiPSC-derived neurons to model specific neurological disorders. Understanding of these models will help us comprehend the current contribution of this field so far, and let us appreciate the need for manufacturing these cells for clinical trials and regenerative medicine.

Differentiating hiPSCs to NSCs

There is no single, unique protocol for differentiating neurons from hiPSCs, but the vast majority of them go through an intermediate differentiation state of NSCs. Not so long ago, NSC differentiation from ESCs or hiPSCs was a complex, expensive and time-consuming process, requiring long time in culture and intensive hands-on labor that made this procedure difficult to reproduce and scale up. This technique would start with stem cell differentiated to embryoid bodies, which are random ensembles of the three germ layers with significant levels of variability, and which grow spontaneously when stem cell factors are removed from growth media. This complex tridimensional culture of these different germ layers would have to be led to ectoderm by the addition of several growth factors and other molecules in order to develop neuroepithelial progenitors that would eventually form neural tube-like rosettes in two-dimensional assemblies. These progenitors would then need to be picked up manually under the microscope based on morphology, in order to obtain a pure culture of NSCs. The entire procedure was very prone to contamination due to the manual picking, and it would take months. Fortunately, now it is possible to differentiate hiPSCs into NSCs without the need for the formation of the three germ layer embryoid bodies, neural rosettes or manual picking,[11] through a protocol that induces dual SMAD signal inhibition by supplementing Noggin and SB431542, inhibiting bone morphogenetic protein (BMP) and transforming growth factor beta (TGFbeta), respectively. Dual SMAD inhibition is an efficient method for the rapid and high efficacy generation of neuroepithelial cells.[12] Many companies offer ready-to-use media and supplements to obtain high quality NSCs in a single week and that can be further expanded, cryopreserved and differentiated into other neural cell types. NSCs are easy to characterize, usually by fluorescence-activated cell sorting (FACS) or immunofluorescence techniques, with commercial antibodies for neuroepithelial stem cell protein (NESTIN), paired box protein 6 (PAX6), and SRY-Box Transcription Factor 1 (SOX1).

Differentiating NSCs to specific neurons for disease modeling

Multiple groups all over the world have been developing protocols for neuron differentiation from hiPSC, for many neurological diseases.[2,13–17] Table 14.1 summarizes most of the diseases which have been modeled using hiPSC derived neurons. Typically, these models involve derivation of a population of mixed neurons, but many subtype specific phenotypes have been developed for a wide range of diseases, including disease-specific mechanisms in dopaminergic, GABAergic, glutamatergic, serotonergic, cholinergic, motor neurons, as well as their precursors. Astrocytes, microglia, and oligodendrocytes have also been differentiated to model the pathology of some of these diseases.

Earlier chapters focused specifically on the development of such protocols and gave examples for how these protocols have been used for the study of neurodevelopmental disorders.

TABLE 14.1 Diseases modeled using hiPSC derived neurons.

Adrenoleukodystrophy	Cockayne's syndrome	Fragile X-associated syndrome	Major depressive disorder	Schizophrenia
Alzheimer's disease	Down syndrome	Friedreich's ataxia	Mucopolysaccharidosis type IIIB	Spinal cord injury
Amyotrophic lateral sclerosis	Dravet syndrome	Gaucher's disease type III	Parkinson's disease (familial)	Spinal muscular atrophy
Angelman syndrome	Duchenne muscular dystrophy	Hirschsprung's disease	Parkinson's disease (sporadic)	Spinocerebellar atrophy
Autism spectrum disorder	Epilepsias	Huntington's disease	Phelan-McDermid syndrome	Timothy syndrome
Bipolar disorder	Familial dysautonomia	Machado-Joseph disease	Rett's syndrome	Zika induced microcephaly

Here, we summarize four neurodegenerative disease models which have made significant advances in the field, focusing on major incidence illnesses for which the science has not yet found a cure.

Alzheimer's disease (AD)

AD is a chronic neurodegenerative disease associated with up to 70% of cases of dementia. Its onset starts with mild symptoms and worsens over time. Losing short term memory, disorientation, speech problems, mood changes, and many other behavioral issues related to normal development of daily activities and communication are reported. The brain pathology in AD is characterized by extensive neuronal loss, particularly of cholinergic neurons, intracellular neurofibrillary tangles composed of the TAU protein (NFTs) and extracellular deposition of plaques composed of β-amyloid (Aβ), a cleavage product of the amyloid precursor protein (APP); however, other causes have been proposed.[18]

hiPSC-derived neural cells, including cholinergic neurons, have been widely used to recapitulate key aspects of AD[19] both for familial AD (fAD), driven by specific genetic mutations that promotes disease onset, or sporadic AD (sAD), with none or less familial evidence. Aβ and TAU pathologies were replicated in mixed neurons derived from hiPSCs, showing increased level of expression when they were cultured for a very long period of time. Increased levels of phosphorylated TAU were also reported for fAD and sAD. Elevated Aβ production due to aberrant APP processing, increased amyloid levels, increased apoptosis of hiPSC-derived GABAergic neurons, and increased susceptibility to Aβ toxicity, have also been reported in cells derived from patients carrying specific mutations, among many other dysfunctions resembling the disease.

Parkinson's disease (PD)

PD is the second most common neurodegenerative disease, after AD, with more than 4 million patients affected worldwide. Similar to AD, PD also has a slow onset, with predominant motor symptoms and as the disease develops, non-motor symptoms become more frequent. Tremor, rigidity, slowness of movement, and difficulty with walking are the most common early manifestations. Dementia, depression, anxiety, sleeping, and emotional problems also occur in many patients in advanced stages of the disease. The main pathological characteristics of the disease is the loss of dopaminergic neurons in the substantia nigra that led to substantial decrease in dopamine level, affecting the circuits linking the cerebrum and the basal ganglia to which dopaminergic neurons project, accounting for the progression of the stereotypical motor symptoms, and the presence of Lewy bodies (accumulations of the protein alpha-synuclein) in many of the remaining neurons. There is no cure for PD, but its symptoms can be delayed by the administration of levodopa or dopamine agonists, which progressively becomes less effective, and produce involuntary movements as a side effect.

There are several models using hiPSC-derived neurons from PD patients with specific mutations,[20] and so far, it is the only disease for which an interventional Clinical Trial is developed with neural precursors derived from hiPSCs, as we will deeply discuss in the following sections. Current work involves several mutations in PD-related genes such as Synuclein Alpha (SNCA), Leucine-Rich Repeat Kinase 2 (LRRK2), PTEN-Induced Kinase 1 (PINK1), and Parkin RBR E3 Ubiquitin Protein Ligase (PARK2), VPS35 Retromer Complex Component

(VPS35), and beta-glucocerebrosidase (GBA), for which hiPSCs were developed as early as 2008.[21] Dopaminergic neurons were differentiated for the first time a year later.[22] Since then, many related dysfunctions have been modeled from these patient-derived cells. From lines with mutations in SNCA, phenotypes and dysfunction have been reported including oxidative stress, nuclear toxicity, mitochondrial toxicity, lysosomal dysfunction and axonal dysfunction. Phenotypes identified related to mutations in the LRRK2 include alpha synuclein pathology, compromised neuronal differentiation, impaired neuronal growth and development, mitochondrial dysfunction, slowed synaptic vesicle transport, reduced autophagy, and several issues involving a pathological role of immunity related to LRRK2 gene. Lines containing mutations in PINK1/PARK2 revealed changes in oxidative stress, mitochondrial dysfunction and autophagy, impaired dopamine regulation, and destabilization of microtubule system. Lines with VP35 mutations revealed alteration of dopamine transportation. And finally, disrupted protein homeostasis, calcium homeostasis, autophagy, and other mechanisms were related to mutations in GBA.

All together, the hiPSC technology provided numerous tools for understanding the underlying mechanisms that lead to disease pathology and progression of PD, offering an unprecedented ability to mimic the disease in vitro with patient-specific disease models. Because of this, many groups have made a number of attempts to use these cell panels for drug screening[23] and an interventional Clinical Trial is currently on its recruiting stage.[24] Even so, numerous scientific and clinical challenges need to be addressed before any clinical application can be safely introduced.

Huntington disease (HD)

HD is a devastating autosomal dominant, progressive neurodegenerative disease characterized by involuntary movements, mood disorders, cognitive impairment and neuropsychiatric symptoms.[25] Lack of coordination and unsteady gait are also present, followed by inability to talk and finally dementia. When it is inherited, it is caused by an unstable CAG triplet in the huntingtin gene (HTT) that affects a wide array of neuronal functions, such as cell signaling, gene transcription, vesicle trafficking, axonal transport, autophagy, cell and mitochondrial metabolism, as well as neurotransmission. There is no treatment to reverse HD neurodegeneration. Although HD affects the whole brain, GABAergic Striatal Medium Spiny Neurons (MSN) are the most vulnerable cell type. These neurons receive excitatory inputs from glutamatergic neurons, which are also compromised. Many groups and a consortium[26] have developed hiPSC models for HD since 2012,[27] differentiating these cells to neural progenitors or to a neuronal subtype similar to MSN that express a certain level of Dopamine- and cyclic-AMP-regulated phosphoprotein of molecular weight 32,000 (DARPP32), a known MSN marker. Current publications show associated changes in electrophysiology, mitochondrial metabolism, cell adhesion, and ultimately cell death and increased susceptibility to BDNF withdrawal for patient lines. These studies also show elevated proteosomal activity, dysregulation of p53 and ERK signaling, superoxide dismutase 1 accumulation, defective calcium signaling, increased caspase activity upon growth factor deprivation, decreased concentrations of glutamate transporters and BDNF of hiPSC differentiated neurons or astrocytes derived from patients.[28] Mutant huntingtin protein aggregation, increased number of lysosomes and autophagosomes and enhanced neuronal death during cell aging was also found in hiPSC patient-derived neurons.

These hiPSC-based models greatly facilitates the study of this disease, but due to the complexity of its mechanism, which involves several types of neurons, along many years of degeneration, it is difficult to recapitulate this entire process aiming to individual phenotypes. This issues can be overcome by two different approaches: the first is the development of patient-specific mixed populations of different neuronal subtypes or with the assembly of 3D structures such as brain organoids, which have also been obtained by a few groups for other diseases, and provide a new exiting opportunity for research and drug testing,[29] and the second involves direct reprogramming of patient cells to striatal neurons, without the need of reprogramming that it is known to erase the age-associated disease phenotypes.[10]

These differentiated cells and organoids provide a great opportunity for high throughput drug screening, but so far there is no successful screening involving these cells.[15,30]

Amyotrophic lateral sclerosis (ALS)

ALS, also known as Lou Gehrig's disease, is a condition that affects motor neurons (MNs) and voluntary movement. Its manifestation includes muscle stiffness, weakness and decreasing muscle size. Gradually, the patient loses the ability to move, speak and swallow, and finally the capacity of breathing, leading to death. Usually, intelligence is mildly affected, or not affected at all, as it happened with the renowned scientist Stephen Hawking. There is no cure for ALS. The average survival from onset is from 2 to 4 years, and medication can only extend this period by a few months.

Although animal models have yielded insight into the mechanisms of disease, more than 50 clinical trials have proven unsuccessful, questioning the validity of preclinical models,[31] and urging for the development of human cell-based models for this disease.

Several groups have managed to obtain hiPSC-derived MNs from healthy individuals, as well as ALS patients carrying specific mutations. It has been reported that ALS patient-derived neurons with a vamp-associated protein B/C (VAPB) mutation, show a significant reduction in the levels of VAPB, particularly in the motor neuron-enriched population, suggesting this could be involved in the initial steps of ALS degeneration.[32] Neurofilament inclusions, mitochondria distribution impairment, and activation of caspase3,[33] along with reduced axon length, lower expression of intermediate neurofilaments, and increased susceptibility to cellular stress were also found in models for the Tar DNA binding protein-43 (TDP-43) mutation. Interesting, ALS-specific phenotypes could be rescued by the addition of anacardic acid.[34] Motor neurons derived from patients with mutations within the FUS gene produced mislocalization and susceptibility to DNA damage.[35] In addition, hiPSC-derived MNs from ALS patients with C9orf72 repeats showed higher glutamate excitotoxicity than MN derived from healthy individuals.[36] Elevated calcium levels, membrane hyperexcitability and cell death were also found in other models. All of these ALS-associated mutation models provide a powerful platform to study this disease with further application for drug screening and a possible cell-based therapy.[37]

Many other models of diverse diseases related to the CNS are under development, and although there is no need for urgent standardization when it comes to understand the mechanism of a disease, once these hiPSC derived neurons or precursors are getting closer to a translational application such as toxicity assays or high throughput drug screening, the production of this cells on a larger scale and with a higher degree of reproducibility is needed. The same applies if we are planning to use them for pre-Clinical Trials in animal models. But, if our goal is to design an interventional Clinical Trial for humans, in order to be able to find a cure for a disease for which there is none, we need to produce these cells under the most strict standards, in a controlled environment, under Good Manufacturing Practices. The following section will briefly go through the history of manufacturing, in order to understand the basic concepts underlying this methodology and the reason why we need to comply with GMP, to be able to produce Clinical-Grade hiPSCs or their derivatives.

Brief history of bio-manufacturing

In order to understand the need and importance of standardization required to manufacture cellular products according to state of the art technology and regulations, we will briefly discuss in the following section, the historical landmarks and the concepts and methods applied by the authors that led to the implementation of good laboratory practices (GLPs) and good manufacturing practices (GMPs).

Historical perspective of the term "quality"

Over time, the term "quality" has had countless definitions regarding its historical context. In certain ways, we can establish some parallelism between how the manufacturing industry has evolved in any particular area and the production of clinical-grade stem cells. From the historical point of view, the definition of quality before the Industrial Revolution was limited to the craftsman's wish to satisfy a client, and make a single product. In that context, quality meant to make the best unique product possible, regardless of the cost or the effort. During the Industrial Revolution, large scale production was identified with quality, in order to satisfy a large demand of goods, and maximize the benefits. During the Second World War, the quality concept was applied to the fast production of military armament, for obvious reasons. After the war, the world needed to produce a large quantity of goods, to satisfy the demand and the loss of goods because of the war. In order to minimize the cost, and be competitive, Japan for instance, applied the concept of quality as to make things right from the beginning.

Nowadays, the concept of quality is defined by the ISO 9000 normative as, *"The totality of features and characteristics of a product or service that bear on its ability to satisfy stated or implied needs"* in other words, a product has *good quality* when it complies with the required specifications. Therefore, current GMPs provide guidelines for systems that assure proper design, monitor and control of manufacturing processes and facilities.

Concept and methods applied to develop quality management

In order to be able to establish a relation between this concept and its practical significance, there are many models and methods applied by several teachers that led the way.

There is a saying at our Laboratory that states: "We never have time to do it right on the first attempt, but we always find time to do it twice." In order to solve this very common problem, many authors established the basis to make a product according to specifications, and doing that perfectly, on the first attempt. Amongst these individuals we can find W.A. Shewhart, W.E. Deming, J.M. Juran, S. Shingo, K. Ishikawa, A.V. Feigenbaum, G. Taguchi, and P. Crosby. Several concepts were introduced by these authors to develop the area, including the classification of variability into random and preventable causes, the pursuit for unreachable perfection and statistical control of quality, introduction of the concept of quality not only for post-production inspection but for the entire process (including planning, sales, and administration), the operation with scientific organization and standards, improvement of statistical methods, implementation of the concept of total quality management, and the pursuit to comply with the requirements from the beginning of the process with the objective of continuous inspections, led to continuous improvement of the process, and assured that the product complies with the specifications.

Quality management applied to GLP

From the beginning of last century, several conceptual and practical implementations of methods for quality assurance were developed in many manufacturing processes. However, some alarming historical landmarks triggered the implementation of normative standards for drug industry, focusing on safety of preclinical studies. Some of these events were substantial birth defects in several thousand babies in Europe because of Talidomide during the 1960s and 1970s, and the tumorigenic effects of Aspartame in the United States. In these cases, fraud and severe deficiencies in animal preclinical studies were the causes of the known consequences that led to the introduction of GLPs regulation. For example, the inspections that took place in 1975 in pharmaceutical laboratories could stratify flaws in several areas, such as personnel training, data management and documentation, deviation from protocols, exclusion of data, flawed statistical procedures, amongst others. By the late 1970s and 1980s, GLP were finally published in order to guaranty the minimal requirements for safety of preclinical studies. These regulations were further adapted to satisfy the standard of different facilities for manufacturing, calibration, and clinical assays, mostly aligned with the ISO9000 norm.

The origins of good manufacturing practices (GMP)

The history of GMP applied to health-related products was also branded by tragic events that led to different regulations in order to assure the safety and purity of drugs. The United States led the way on this path to regulate the drug and food industry, with the development of several administrative acts. The 1902 Biologic Act that precedes the creation of the United States Food and Drug Administration (FDA) in 1906, was the consequence of contaminated batches of diphtheria and smallpox vaccine with tetanus,[38,39] that led to the death of over 20 children in St. Louis and New Jersey. The 1938 Federal Food, Drug, and Cosmetic Act, was the response to another tragedy resulting in over a 100 deaths by the ingestion of sulfanilamide (an antibacterial agent) dissolved in diethylene glycol.[40] Several amendments to this Act have been made over the years until the incident involving over 300 deaths by sulfatiazol contaminated with phenobarbital in 1941,[41] which gave way for what we know now as GMP, and the tragedy of thalidomide in Europe in the 1960s[42] gave place to several regulations that force the manufacturers to demonstrate that their products are safe and effective, before commercialization.

The world tendency over several regulatory acts was taking place until the end of the last millennia and focused on the quality and amount of assays as a way to guaranty quality, until the FDA published a seminal report for the 21st century, with a risk-based approach, to control and reduce variability based on the scientific method. This view is now also applied to the manufacture of biological products, including stem cells.[43]

Different laws and regulations were developed over the world at different times, making a large amount of diverse technical requirements for similar products according to a different region. The ICH International Conference on Harmonization for registration of Pharmaceuticals for Human Use was an effort to rationalize and harmonize the requirements for one country to another to be able to commercialize pharmaceutical products on an international level, with common standards and specifications.

GMP applied to clinical-grade cell manufacturing

The history of hiPSC development started with a robust, but complicated, method of gene transfer in which a set of integrative viruses were transduced into cells in order to convert them into hiPSCs. This method was very useful for research purposes, but viruses were not easy to produce, and each laboratory had to make its own tools for reprogramming, which led to a lot of variability in the process and in the final product. As more efficient tools were developed, many companies started to create ready-to-use solutions for reprogramming. Many different models of disease were established and cell-based therapies in animal models were developed. As the evidence grew stronger, some groups started to use these cells for toxicity assays, drug testing and finally, interventional Clinical Trials. For this purpose, the industry had to develop new reagents to comply with regulations and laboratory grade cells started to be produced to match GMP standards. Since GMP guidelines were written long before the need for standardized cell production, the specific regulatory framework for these cells was directly influenced by the laboratories that were actually making this product. This led to different interpretations and degrees of evolution of these regulations along different jurisdictions, even within members of the International Conference on Harmonization.[44] Nowadays, regular efforts to unify protocols and regulations are made all over the world, and the concepts and principles for these developments are common to all of them.

Implementation of GMP requires a number of different elements depicted in Fig. 14.1 such as quality management, a quality control department and an independent quality assurance department, SOPs regarding all aspects of production, including sanitation, hygiene, sterilization, quality controls, packaging, waste disposal, distribution, and even dressing. All personnel have to be trained rigorously according to the proceedings and training has to be standardized and documented. Production areas have to be physically separated from administrative areas, QC areas, storage and quarantine areas and also bathrooms or dressing areas. All materials have to traceable, and the equipment and premises have to be qualified and validated periodically. There is also need for quarantine and storage separated rooms, as well as refrigerators or freezers if the raw materials demand that. Self-inspection and regular internal and external audits have to be taken into consideration, and it is mandatory to have batch documentation to be able to record every step of the process. The GMP organization

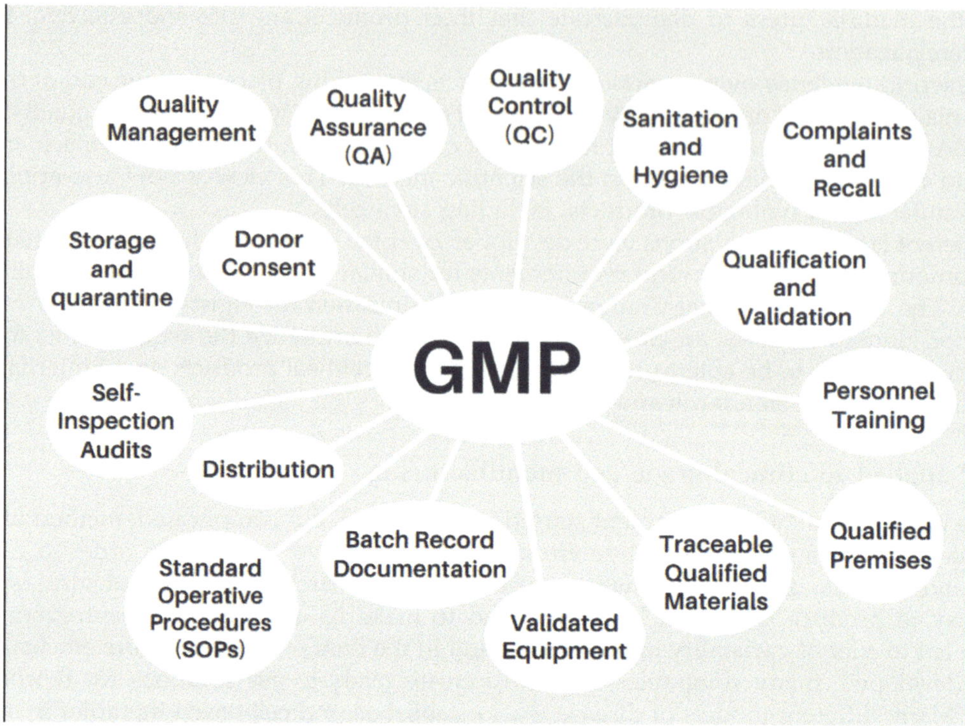

FIGURE 14.1 **GMP elements for stem cell manufacturing.** Basic components of GMP.

also demands a department for complaints and there has to be a way to recall any products that do not comply with customer demands.

For the specific case of cell production, several issues regarding production and QC have to be addressed. A regular facility has an HVAC system (Heat-Ventilation-Air conditioning) that allows us to control room temperature, room pressure and even humidity. Since GMP requires a strict control of personnel, an automatic system will allow us to customize permits and control entry and exit from the premises. Critical electrical devices are required to record several aspects of the facility during the entire process, such as refrigerator temperature, culture incubator temperature and CO_2 level, open or closed doors, and other parameters, and for this purpose many different sensors are needed and a continuous electrical supply has to be provided by the aid of an online UPS that will also be useful to be able to work under sterile conditions under the culture hood uninterruptedly. All these features have to be controlled online by trained personnel, and periodically validated.

Manufacturing hiPSCs and neuronal derivatives

We previously discussed the need for regulation and harmonization, mostly for pharmaceutical products, but stem cells pose an additional challenge, since they could be thought of as a pharmaceutical product, a biological product, and also a human product for transplantation. Regardless of regulations that may differ from country to country, it is mainly agreed upon that biological products for human use are regulated as pharmaceutical products, but it is not taken for granted that a human cell or tissue for transplant would be falling under the same regulations and each country may apply its own rules. In either case, and despite which regulation applies, there are many practical issues that we can take into consideration in order to make Clinical-Grade hiPSC or its derivatives, either for drug testing or transplantation. The first thing that we have to evaluate when we design a project involving hiPSCs, is if we really need to comply with GMP. If the project is about understanding the mechanisms underlying a disease, and neither the hiPSC cells or their derivatives would ever find their way to a clinical trial, neither a patient, the answer will definitely be "no." In terms of cost, it was estimated that production of one line of clinical-grade hiPSCs in compliance with cGMP would cost approximately US$800,000 in comparison with US$10,000 to US$20,000 for a research-grade hiPSC line.[45] The need for GMP compliance is mandatory when we are certain that our biological product will eventually find a way out of the regular basic research laboratory. Because of the complexity and cost of GMP compliance, and also the dedicated infrastructure, human resources and specific reagents, a reasonable advice would be not to go that way, unless you really need it.

For the translational scientists that are convinced to take that big leap forward, there are several issues to take under consideration, and the first of all is how to obtain Clinical-Grade hiPSCs for further differentiation. For this purpose, there are two acceptable methods according to FDA standards: (1) Derivation of cells and cell products under GMP requirements and (2) Conversion of cells or cell products derived under research-grade conditions to match GMP quality standards. For both ways, a GMP-compliant facility is needed and although the latter could seem easier, switching the cells to GMP reagents and rigorous testing for pathogen or animal contamination for non-GMP lines that wish to be converted could be as difficult as starting with a cGMP compatible proceeding from the beginning.[46]

The next section will focus on method (1) the production of clinical-grade hiPSCs and their derivatives, starting from the very first step: the donor.

hiPSC reprogramming and differentiation process overview

The production of clinical-grade hiPSCs and their derivatives is very complex, and requires unique expertise and equipment. If a laboratory decides to go this way, it is because it has years of experience working with Research-Grade hiPSCs, and has the need to standardize its production for a higher goal. Usually, this goal is a clinical trial.

Nowadays, there are several facilities that can provide us with the starting material (i.e., fibroblast lines from patients), hiPSCs for modeling diseases, or even differentiated neural lines, but most of these cell lines are only intended for research use, and would never be allowed for clinical use. There are also a few GMP-qualified hiPSC lines available that may suit your needs,[47] but we will assume that your lab has the need to start from the donor cells.

The first step is to obtain an informed consent from the donors, which needs to be obtained through an approved Institution Review Board (IRB) protocol in the United States. IRB committees have been put in place by the United States government to ensure proper execution of biomedical research involving human patients. After obtaining consent from the donor, one needs to obtain an initial cell to reprogram (usually from a blood draw or skin biopsy), choose the reprogramming method, and a set of quality controls for the obtained hiPSCs to build a Master and a Working bank. Then, we have to find a way to differentiate the hiPSCs to the final desired cellular subtype (i.e., neurons), maybe through an intermediate subtype, such as NSCs that we can also bank, along with other set of quality controls for these cells. The final step includes the differentiation of the NSCs to the final desired neuronal subtype, their characterization, and if this product is intended for human transplantation, rigorous testing for any proliferative cancer cells is also needed.

All of these tasks need to be achieved in a GMP-compliant facility, with adequate validated equipment and reagents, and highly trained personnel working with SOPs, as we briefly described previously.

We will go through each of these steps depicted in Fig. 14.2, in the following section.

Informed consent

hiPSCs have the evident advantage of leaving behind many ethical considerations regarding the use of human embryos as a source of stem cells for clinical trials or treatments, but the hiPSC field and its promise of an unlimited source of cells also face numerous ethical challenges associated with immortalized cell lines, transplantation, human reproduction, and bio-banking for unlimited time. In this context an informed consent (IC) that protects and advises the donors, as well as safeguarding the future of the research and translation is needed.[48] A regular IC will state the Institution, Project and principal investigator that will make use of the donated cells and its derivatives, as well as a brief and clear explanation of how the cells are going to be collected, treated, and used afterward. It will also state the potential benefits and risks of the donation, an explicit declaration of a voluntary nonprofit donation and a disclosure for potential financial benefits. The ISSCR has on its website a " Sample Research Consent Form for Somatic Cell Donation for Induced Pluripotent Stem Cell Research"[49] that could be very useful to take as a model for your own IC that meets the standards of your local ethical committee.

Screening donor samples for contagious disease

Regardless of the starting material, the donor needs to be tested for several diseases, in order to guaranty the safety of the facility personnel and the final hiPSC line. Most common testing includes HIV, CMV, Hep-B and Hep-C, WN (West Nile), HTLV1-2, *Chlamydia, N. gonorrhoeae*, *T-Cruzi* (for Chagas), and *T-Pallidum* (for syphilis). The FDA provides on its webpage a comprehensive list of assays for disease screening.[50]

FIGURE 14.2 **hiPSC derived cells manufacturing pipeline.** Basic Workflow for hiPSC derivation, differentiation, and characterization.

Tissue acquisition, cell isolation, and expansion

There are numerous cell types that could be reprogrammed to hiPSCs, but most of them will likely remain as laboratory-grade material, because of the absence of GMP reagents and operative procedures for sample procurement and reprogramming. Most of the GMP reprogramming has been done with cell types derived from this two tissues: skin and blood. The selected cell types from each tissue are dermal fibroblasts (FBs) and PBMCs because of the simplicity of extraction, cultivation, and reprogramming. FBs are usually taken from a small skin biopsy punch of about 2 or 3 mm, from which the dermal fibroblasts will grow once we plate the dissected, collagenase-treated explant with fibroblast culture media. The morphology of the cells is continuously monitored until the fibroblasts can be passaged, characterized and banked. This technique is quite straightforward and allows us to have a very large supply of frozen cells for future reprogramming, and the cost is relatively low. PBMCs are obtained from whole blood, which make this sample easier to obtain than fibroblasts, but cell isolation is slightly more complicated, expansion is very limited, and the cost is significantly higher than using fibroblasts. Mononuclear cells are separated from whole blood by gradient density centrifugation with Ficoll and enriched for erythroblasts with a media that is supplemented with several factors that support differentiation of the mixed PBMC population toward this particular cell type. After this process, there will be just enough cells for a few reprogramming events, and although you can freeze the original PBMCs (which are also very limited), is not recommended to freeze the final enriched erythroblasts.

Reprogramming method

Reprogramming technology has evolved rapidly since the initial experiments from Yamanaka, and is reviewed in more detail in an earlier chapter. As the original experiments were carried out with retroviruses or lentiviruses, these highly efficient tools had a severe drawback because part of the viral genome, along with the reprogramming genes, were integrated into the host cell genome at random, with unpredictable side effects. Current state of the art technology involves non-integrative methods which are considered safer for clinical use. Among these tools, we can find commercial episomal plasmids, non-integrative viruses and even mRNA transfection kits, which are available for clinical-grade manufacturing. Just to show a few examples of three very different methods, Lonza offers episomal plasmids along with a 4D-nucleofector, which is a key piece of machinery for the reprogramming process due to its higher transfection efficiency than electroporation,[51] Thermo Fischer has the option of using Sendai virus based reprogramming using the CTS CytoTune-iPS 2.1 Reprogramming Kit,[52] and Stemgent has its StemRNA 3rd Gen Reprogramming Kit, that allows for fibroblast, urine and blood reprograming.[53] All of these kits are GMP compliant.

Episomal vectors have the advantage of high efficiency with a single nucleofection, but the nucleofector itself can be an expensive piece of equipment. Reprogramming with Sendai virus doesn't need any specific equipment, but the absence of viral contamination must be proven after reprogramming. mRNA transfection has the advantage of complete clearance of reprogramming material, but the protocols often require multiple transfection steps. In-lab techniques can also be used, but GMP validation would be required, and the reprogramming tool itself has to be made under GMP.

hiPSC banking and quality controls

During the reprogramming process, the initial cells have to go through several drastic changes to be able to finally express the set of genes that makes them hiPSCs. These hiPSCs are the starting material for differentiation to a final cellular subtype that may be used for a Clinical Trial or a therapy. Therefore, it is very important to be able to produce these cells in large quantities and have a set of controls to confirm the purity and identity of this biological product. In order to have a consistent supply of cells with the same characteristics over time, it is recommended to establish a Seed Bank (SB) at a low passage number (10 vials or more), and then from one of these vials, build a Master Cell Bank (MCB) of over 100 vials. From one vial of this MCB, we can further multiply these cells to build a Working Cell Bank (WCB) of over 100 vials, to be able to have an unlimited supply of ready-to-use cells. Both MCB and WCB have to be built under GMP, and it is highly recommended that the SB is also built with the same conditions, but in the case of FDA regulations, GMP compliance for the SB is not required.

Since clinical-grade hiPSCs are starting to be made all over the world, there is a need to establish an agreement of a minimal set of controls to be able to consider a hiPSC line of Clinical-Grade, besides particular regulations and normative for each country. For that purpose, the International Stem Cell Banking Initiative has established a set of guidelines for Quality Control for Clinical-Grade human induced pluripotent stem cell lines.[54] These guidelines include minimal testing requirements concerning identity, microbiological sterility, genetic fidelity and stability, viability, characterization, and potency provided by a consensus from 16 facilities involved in hiPSC manufacturing. Table 14.2 summarizes mandatory testing for clinical-grade hiPSCs.

All testing on seed, master and working cell banks should be performed by accredited and licensed laboratories where available. For most of the assays, standardized methods are available according to the United States, European, and Japanese pharmacopeias (USP, Ph.Eur., and JP). A published detailed description of characterization methods and regulations is also available.[55]

These guidelines also recommend the use of xeno-free reagents in order to exclude the possibility of nonhuman adventitious animal agents or contamination with feeder cells. Reprogramming by episomal vectors it is highly encouraged, to avoid the possibility of viral insertion in the host cell genome.

Although the minimal requirements for characterization and potency (tested by Embryoid Bodies formation) may be as low as a couple of markers for each, there are several genes that could be tested for hiPSCs and EBs, and a routine practice involves rigorous visualization of colony morphology and growth rate, to make sure to have a healthy culture. It is not unusual for the average laboratory to test for several genes in order to characterize the hiPSC lines, by immunofluorescence, flow cytometry, as well as qPCR. Most common genes for hiPSC testing are OCT4, Sox2, Nanog, hKlf4, hMyc, hGdf3, hRex1, hpH34, hECAT1, hECAT2, hFGF4, Nanog, Tra 1-81, Tra 1-60, Sox2, and SSEA. Teratoma formation is not recommended due to it difficult standardization,[56] but EB formation is mandatory. Some of the markers to characterize all three germ layers (Endoderm, Mesoderm, and Ectoderm) of EBs are AFP, CXCR4, Brachyury, Pecam1, MAP2, PAX6, AFP, GATA4, SOX17, TropoTC, Desmin, SMA, DCX, and betaIIItub.

TABLE 14.2 Minimal obligatory testing for clinical-grade hiPSC.

Attribute	Test	Recommended method	Acceptance criteria
Identity	Short tandem repeat	STR profiles performed by accredited laboratory	Identical
Microbiological sterility	Mycoplasma	Qualified qPCR or culture method	Negative
	Bacteriology	Use of pharmacopeial methods	Negative
	Viral testing	Based on risk assessment of starting and raw materials	Negative
Endotoxin	Endotoxin	Use pharmacopeial methods	Negative
Genetic fidelity, stability	Residual vector testing	Appropriate specific assay to be used	Negative
	Karyotype	G banding	Normal, ≥20 metaphases
Viability	Viability	Dye exclusion test or flow cytometry	>60%
Characterization	Flow cytometry	A minimum of two markers from an accepted panel	Positive on 70% of cells in MCB
Potency	Phenotypic	EB formation and/or directed differentiation	Positive cells from all germ layers

Another issue that has to be taken into consideration are genomic aberrations. hiPSC generation involves long-term culturing, from several weeks up to a few months, favoring all sort of genomic abnormalities, such as copy number variations, trisomy, amplifications of chromosomal regions, deletions of chromosomal regions, loss of heterozygosity, and epigenetic changes. Although minimal requirements demand only G-banding karyotyping, whole-genome sequencing, whole-epigenome analyses, as well as single-cell genome sequencing of large cell populations may be useful to identify these changes.[57]

Specific differentiation and characterization of cellular subtypes

As we have discussed previously, GMP-compliant reprogramming, as well as the expansion and banking of hiPSCs has been established in many facilities and there is a reasonable degree of consensus for their characterization. However, as the hiPSC as a biological product itself is relatively easy to define and characterize, when we want to differentiate them to a specific cellular subtype, there are hundreds of different subtypes with numerous stages of development according to stage of maturation over time. Even for the same specific cell type, we can find several different protocols for differentiation and characterization in numerous laboratories around the world. This issue makes standardization and harmonization extremely challenging. Nevertheless, differentiated cells have been made from hiPSCs for interventional Clinical Trials under GMP compliance for cell types including retinal photoreceptors for transplantation,[58–60] cardiomyocytes,[61] natural killer (NK) cells,[62] mesenchymal stem cells (MSC) (at Waisman biomanufacturing center) and hematopoietic stem cells.

If we only focus on the production of neurons, there are also numerous laboratories that make different neurons from hiPSCs under non-GMP conditions, usually through a neural stem cell intermediate stage. There is also evidence that once a GMP-compliant hiPSC cell bank is established and the lines can be differentiated into the three germ layers consistently, they can also be differentiated into neural stem cells, dopaminergic neurons, and astrocytes with non-GMP standardized protocols.[63,64] Neural progenitors[65] and even dopaminergic neurons[66] have been made under GMP conditions from ESCs, and GMP-compatible protocols have also been established for neural progenitor cells[67] and neural stem cells.[68] Although the scientific community has paved the way for the production of several types of fully GMP-compliant neurons from hiPSCs, there are still very few groups that have achieved that milestone. Some of the groups working on GMP-compliant neuronal differentiation protocols have the ultimate goal of curing Parkinson's disease. Clinical-grade midbrain dopaminergic progenitor cells, differentiated in vitro from autologous induced pluripotent stem cells were first tested in an animal model[69] and further implanted into one patient stabilizing or improving PD symptoms after 18–24 months.[70] Also, Kyoto University developed a proceeding for the manufacturing of dopaminergic progenitor cells (DAPs) for treating Parkinson's disease[71] in a Clinical Trial for seven patients. Although these cells are not fully differentiated, these grafted cells survive and function as dopaminergic neurons (DA) for a long time (possibly until the patient dies). Since hiPSC-based therapies are very novel, many regulatory issues are not fully addressed, neither are they universally standardized and each country has its own set of criteria for clinical application. Also, the long-term integration for these grafted cells has to undergo more intensive risk management compared to other trials.[72] For these reasons, Doi and colleagues, who were pioneers in this area, had to run several different quality controls, with parallel methods (i.e., Flow cytometry and qPCR), not only to confirm that these cells were mainly composed of DAPs with the ability to undergo differentiation to DA neurons, but also to prove that no undifferentiated cells were included in the final product, posing the risk of tumorigenicity. They also had to use specific GMP compliant reagents, including culture media and antibodies (one of which they had to develop specifically for this project), and a clinical-grade cell sorter. For further analysis of potential tumorigenicity capacity, they proceeded to repeat DAP induction six times and compared the results of whole-genome sequencing (WGS) and whole-exome sequencing (WES) obtained from original peripheral blood cells, undifferentiated hiPSCs, and differentiated cells on days 12 and 26, to screen for potential cancer-related genes, and also to check for epigenetic modifications in each stage. In order to evaluate the efficacy of these cells, they transplanted the precursors into the striatum of 6-OHDA-lesioned rats, showing behavioral improvement.

Their tests showed that DAPs had the safety and efficacy of clinical-grade cell products and based on these results, they received approval to start a clinical trial for PD from the Pharmaceuticals and Medical Devices Agency and the institutional regulatory board of Kyoto University, in 2018. The details of this clinical trial will be discussed in the following section.

Clinical trials with hiPSC-derived neurons

We have discussed previously the history of GMP, the manufacturing process itself, and about some examples of how hiPSCs could be differentiated to neurons to model disease. We have also mentioned briefly which hiPSC-derived cells were made under GMP conditions. All of them have the final purpose of being used for a clinical trials. But, are there any other Clinical Trials involving hiPSCs and their derivatives?

A very recent systematic multi-data based analysis[73] shows that there are 131 studies involving PSCs (ESCs or hiPSCs), or cells derived from them, but only 30 of those trials are interventional studies, meaning that they involve actual transplantation of derived cells into patients. Out of those 30, only 8 clinical trials involve hiPSC-derived cells, and just one of them involves transplantation of neurons or precursors related to the central nervous system: dopaminergic progenitors (DAPs) derived from hiPSCs for PD.

We have previously discussed about hiPSC-derived DAPs characterization, and safety and efficacy studies for their use into human subjects. This clinical trial (UMIN000033564) consists in transplantation of DAPs into the substantia nigra pars compacta to a group of seven patients with moderate symptoms. The clinical study is being completed by a group led by Ryosuke Takahashi's at Kyoto University and source cells were obtained from third-party HLA homozygous donors to reduce interference from patients' genetic variability, who also received immunosuppressive drugs, for additional safety. As far as 2020, just one patient received a dose of 2.4 million DAPs, with no side effects, and the patient is expected to receive an additional similar dose, if the results hold.[74] The remaining patients were yet to be treated, but the study has been suspended during the 2019–2020 pandemic.

Cell therapy for PD has been practiced since 1987, using several cell sources. Fetal tissue from the ventral midbrain is able to produce viable DA neurons, which also preserve their function, even 20 years after implantation, but this therapy shows little behavioral improvement. hESC-derived neural precursor cells are also used for a Clinical Trial for PD (NCT03119636) involving 50 patients starting in 2017, but the study is still recruiting volunteers, and there are no results available yet. Furthermore, there is also evidence of a case report of autologous DAPs transplantation for PD which showed some degree of improvement,[70] but additional experiments will be needed to be able to evaluate the robustness of this finding. To date, the efficacy of current treatments for neurological disorders using hiPSC derivatives is yet to be determined, but there is a cautious optimism that promising results in this field will be seen in the near future.

Table 14.3 summarizes all clinical interventional studies involving hiPSC registered up to September 2020.

Perspective and challenges for clinical translation

The development of hiPSC reprogramming technology has already revolutionized stem cell research, and is offering us a great opportunity for translation to the Clinic. As we discussed previously, these cells and their derivatives have an enormous potential for toxicity

TABLE 14.3 Interventional clinical trials involving hiPSC derivatives, registered to September 2020.

Id number	Short title	Cell type derived from hiPSC	Disease	Country	Status	Started
JPRNUMIN000011929	A study of transplantation of autologous hiPSC derived RPE cell	Retinal pigment epithelium	Exudative macular degeneration	Japan	Partial results (mandai M)	2013
NCT03222453	Thalassemia treatment based on the hiPSC technology	Hematopoietic stem cells	Beta-thalassemia	China	Results submitted	2015
JPRNUMIN000026003	A study of transplantation of allogenic hiPSC derived RPE	Retinal pigment epithelium	Neovascularmacular degeneration	Japan	Recruitment complete	2017
NCT02923375	A study of CYP-001 for the treatment of steroid-resistant acute graft versus host disease	MSCs	Graft versus host disease	Australia + UK	Recruitment complete	2017
JPRNUMIN000032989	Clinical trial of human hiPSC-derived cardiomyocyte sheet for severe cardiomyopathy	Cardiomyocytes	Ischemic cardiomyopathy	Japan	Recruitment complete	2018
JPRNUMIN000033564	Kyoto trial to evaluate the safety and efficacy of hiPSC-derived dopaminergic progenitors for PD	Dopaminergic progenitors	Parkinson's disease	Japan	Recruiting suspended	2018
NCT03763136	Treating heart failure with hPSCCMs	Cardiomyocytes	Heart failure	China	Recruiting	2019
NCT03841110	FT500 as monotherapy and in combination with inhibitors in subjects with tumors	NK cells	Advanced solid tumors	U.S.A.	Recruiting	2019

testing, drug screening and cell replacement therapies. Nevertheless, there are still numerous challenges that need to be addressed regarding manufacturing, testing, commercialization, regulations, harmonization, ethical issues, and many others.[75]

There are still a great number of diseases for which science has not found a cure, and this is particularly true for neurodegenerative diseases, as the world population increases life expectancy, and for many other diseases which can be treated with stem cells or their derivatives. Therefore, clinical-grade hiPSCs are starting to be manufactured, and the forecast is that demand will grow exponentially. Over time, we can see that this area was first driven by academia, along with medical facilities, and finally GMP-compliant stem cell platforms, mostly with public funding. As the clinical translation becomes a reality, the interest from pharmaceutical companies and private investors is growing. Since this technique is nowadays extremely expensive but has an enormous potential, there will be a question of who is going to pay the cost of these treatments, and if the public funding will lead the development of this area, or the private companies will invest as well, for there is huge profit potential. Any projection makes the area very attractive for public and private investors, who will direct their efforts according to several factors, such as technical capability, location, costs, and most importantly, regulations. Perhaps the most relevant example would be the five-year relaxed regulatory policy introduced by Japan since 2014 that has given the industry a boost, allowing many foreign companies to fast track treatments to the market in that country with lower requirements than in many other countries. This initiative allowed for the registration of almost 4000 therapies over the extension of this policy,[76] and has raised concern about issues regarding efficacy of these therapies and commercialization.

Nevertheless, and despite regulations and whether these initiatives will find public or private funding, one of the most important issues that have to be taken into consideration is whether to give priority to autologous or allogeneic transplant. In autologous transplant, the reprogrammed cells are taken from the same person who will get the transplant. These cells could be corrected for the genetic deficiency in vitro or not, depending on the disease, and transplanted without major rejection issues, because they come from the same patient. In Allogeneic transplantations, the cells are donated from a person other than the patient, either a matched related or unrelated donor. This raises the problem of tissue rejection and the patient will have to be taking immunosuppression drugs for life, if the human leukocyte antigen (HLA) complex is not fully matched. The apparent advantage of autologous transplant, would come very short, considering that the process to develop a hiPSC line and differentiate it to a specific tissue can take over a year, making this tool useless for an urgent therapy. Besides, Clinical-Grade differentiated cells from hiPSC are extremely complex to manufacture and the cost would make this approach almost unfeasible, considering that the total price of a therapy, would be probably over a million USD for one patient. There are also regulatory considerations to be addressed since the agencies worldwide have to ensure that these transplanted cells are safe, pure, potent, and effective, and historically this would apply to large batches for a great number of patients, from which you can take numerous samples to be able to test all of these features. With the case of a single small batch of cells that divide and could change over time, these regulations are very difficult to comply.

Because of the reasons mentioned above, the scientific community is focusing on an alternative approach by building hiPSC haplo-banks, aiming to collect the most frequent HLA homozygous donors to provide HLA-matched material to be able to use it in a significant numbers of patients. One of the great advantages of this approach is that many Blood Cord Banks already have HLA information available all over the world and it is calculated that for a homogenous genetic population like Japan, 80% of the patients could be treated with HLA-matched material from a bank of only 55 donors.[77] In the case of a more diverse populations, such us in the UK, 150 lines would be able to treat 93% of the population,[78] and for a very diverse population, such as the state of California, 80 lines could cover over 50%.[79]

This initiative is doable, since the technical challenges to reprogram such small number of lines is feasible, and is worthy, because with just a few hundreds lines we would be able to cure a very large population with an affordable cost.[80] There are already at least 16 platforms manufacturing Clinical-Grade hiPSC over the world, and most of them are working together to establish a Global Alliance for hiPSC Therapies (GAIT),[81] aiming at harmonization of standards for donor selection and screening, hiPSC manufacture proceedings, QC and regulatory compliance in order to be able to establish a global, Haplobank.[82]

hiPSCs and their derivatives have been developed for over a decade, and have already produced significant advances in understanding the mechanisms of disease. Their unlimited self-renewal potential and ability to differentiate into any cell type of our body is giving us the opportunity of developing new models of disease, cell platforms for toxicity testing and drug screening, and the possibility of cell-based treatments. Their application in neuroscience has been extremely relevant, due to the very limited access to any related tissue and the difficulty in developing animal models that fully address the complexity of these diseases. This field, initially led by academia, is finding its way into the clinic by establishing strategic partnership with companies and platforms that will provide the resources to be able to produce clinical-grade cells under the current standards and regulations. The evolution of this sector will require global efforts from academia, health care professionals, companies and regulators addressing multiple challenges that lie ahead, in order to establish the foundation of this new exciting era of hiPSC-based regenerative medicine.

References

1. Takahashi K, Yamanaka S. Induction of pluripotent stem cells from mouse embryonic and adult fibroblast cultures by defined factors. *Cell*. 2006;126(4):663–676. https://doi.org/10.1016/j.cell.2006.07.024.
2. Zeng X, Hunsberger JG, Simeonov A, Malik N, Pei Y, Rao M. Concise review: modeling central nervous system diseases using induced pluripotent stem cells. *Stem Cell Transl Med*. 2014;3(12):1418–1428. https://doi.org/10.5966/sctm.2014-0102.
3. Zhao X, Bhattacharyya A. Human models are needed for studying human neurodevelopmental disorders. *Am J Hum Genet*. 2018;103(6):829–857. https://doi.org/10.1016/j.ajhg.2018.10.009.
4. Dawson TM, Golde TE, Lagier-Tourenne C. Animal models of neurodegenerative diseases. *Nat Neurosci*. 2018;21(10):1370–1379. https://doi.org/10.1038/s41593-018-0236-8.
5. Hansen W. Modeling neurological diseases with human brain organoids. *Front Synaptic Neurosci*. 2018. https://doi.org/10.3389/fnsyn.2018.00015.
6. Chamberlain SJ, Li XJ, Lalande M. Induced pluripotent stem (iPS) cells as in vitro models of human neurogenetic disorders. *Neurogenetics*. 2008;9(4):227–235. https://doi.org/10.1007/s10048-008-0147-z.
7. Merkert S, Martin U. Targeted gene editing in human pluripotent stem cells using site-specific nucleases. In: *Advances in Biochemical Engineering/Biotechnology*Vol 163.
8. Studer L, Vera E, Cornacchia D. Programming and reprogramming cellular age in the era of induced pluripotency. *Cell Stem Cell*. 2015;16(6):591–600. https://doi.org/10.1016/j.stem.2015.05.004.

9. Miller JD, Ganat YM, Kishinevsky S, et al. Human iPSC-based modeling of late-onset disease via progerin-induced aging. *Cell Stem Cell*. 2013;13(6):691–705. https://doi.org/10.1016/j.stem.2013.11.006.
10. Victor MB, Richner M, Olsen HE, et al. Striatal neurons directly converted from Huntington's disease patient fibroblasts recapitulate age-associated disease phenotypes. *Nat Neurosci*. 2018;21(3):341–352. https://doi.org/10.1038/s41593-018-0075-7.
11. Chambers SM, Fasano CA, Papapetrou EP, Tomishima M, Sadelain M, Studer L. Highly efficient neural conversion of human ES and iPS cells by dual inhibition of SMAD signaling. *Nat Biotechnol*. 2009;188(3).
12. Chambers SM, Fasano CA, Papapetrou EP, Tomishima M, Sadelain M, Studer L. Highly efficient neural conversion of human ES and iPS cells by dual inhibition of SMAD signaling. *Nat Biotechnol*. 2009;27(3):275–280. https://doi.org/10.1038/nbt.1529.
13. Jang J, Quan Z, Yum YJ, Song HS, Paek S, Kang HC. Induced pluripotent stem cells for modeling of pediatric neurological disorders. *Biotechnol J*. 2014;9(7):871–891. https://doi.org/10.1002/biot.201400010.
14. Amin N, Tan X, Ren Q, et al. Recent advances of induced pluripotent stem cells application in neurodegenerative diseases. *Prog Neuro Psychopharmacol Biol Psychiatr*. 2019:95. https://doi.org/10.1016/j.pnpbp.2019.109674.
15. Garcia-Leon J, Vitorica GA. Use of human pluripotent stem cell-derived cells for neurodegenerative disease modeling and drug screening platform. *Future Med Chem*. 2019;11(11). https://doi.org/10.4155/fmc-2018-0520 (Epub).
16. Soliman MA, Aboharb F, Zeltner N, Studer L. Pluripotent stem cells in neuropsychiatric disorders. *Mol Psychiatr*. 2017;22(9):1241–1249. https://doi.org/10.1038/mp.2017.40.
17. Wilson Y. Patient-specific induced pluripotent stem cells as a platform for disease modeling, drug discovery and precision personalized medicine. *J Stem Cell Res Ther*. 2012. https://doi.org/10.4172/2157-7633.S10-010.
18. Ooi L, Sidhu K, Poljak A, et al. Induced pluripotent stem cells as tools for disease modelling and drug discovery in Alzheimer's disease. *J Neural Transm*. 2013;120(1):103–111. https://doi.org/10.1007/s00702-012-0839-2.
19. Leeuw TC. Alzheimer's in a dish - induced pluripotent stem cell-based disease modeling. *Transl Neurodegener*. 2019;8. https://doi.org/10.1186/s40035-019-0161-0. PMID: 31338163; PMCID: PMC6624934.
20. Hu X, Mao C, Fan L, et al. Modeling Parkinson's disease using induced pluripotent stem cells. *Stem Cell Int*. 2020:2020. https://doi.org/10.1155/2020/1061470.
21. Park IH, Arora N, Huo H, et al. Disease-specific induced pluripotent stem cells. *Cell*. 2008;134(5):877–886. https://doi.org/10.1016/j.cell.2008.07.041.
22. Soldner F, Hockemeyer D, Beard C, et al. Parkinson's disease patient-derived induced pluripotent stem cells free of viral reprogramming factors. *Cell*. 2009;136(5):964–977. https://doi.org/10.1016/j.cell.2009.02.013.
23. Xiao B, Hui Ng H, Takahashi R, Tan EK. Induced pluripotent stem cells in Parkinson's disease: scientific and clinical challenges. *J Neurol Neurosurg Psychiatr*. 2016;87(7):697–702. https://doi.org/10.1136/jnnp-2015-312036.
24. *UMIN-CTR Clinical Trial for PD*; 2020. https://upload.umin.ac.jp/cgi-open- bin/ctr_e/ctr_view.cgi?recptno=R000038278 [Accessed September 2020].
25. Gatto PV, Persi G, Parisi V, et al. Clinical and genetic characteristics in patients with Huntington's Disease from Argentina. *Park Relat Disord*. 2011;18(2). https://doi.org/10.1016/j.parkreldis.2011.09.011 (Epub).
26. Mattis VB, Svendsen SP, Ebert A, et al. Induced pluripotent stem cells from patients with huntington's disease show CAG repeat expansion associated phenotypes. *Cell Stem Cell*. 2012;11(2):264–278. https://doi.org/10.1016/j.stem.2012.04.027.
27. Camnasio S, Carri AD, Lombardo A, et al. The first reported generation of several induced pluripotent stem cell lines from homozygous and heterozygous Huntington's disease patients demonstrates mutation related enhanced lysosomal activity. *Neurobiol Dis*. 2012;46(1):41–51. https://doi.org/10.1016/j.nbd.2011.12.042.
28. Csobonyeiova M, Polak S, Danisovic L. Recent overview of the use of iPSCs huntington's disease modeling and therapy. *Int J Mol Sci*. 2020;21(6). https://doi.org/10.3390/ijms21062239.
29. Wu YY, Chiu FL, Yeh CS, Kuo HC. Opportunities and challenges for the use of induced pluripotent stem cells in modelling neurodegenerative disease. *Open Biol*. 2019;9(1). https://doi.org/10.1098/rsob.180177.
30. Zhang N, Bailus BJ, Ring KL, Ellerby LM. iPSC-based drug screening for Huntingtons disease. *Brain Res*. 2016;1638:42–56. https://doi.org/10.1016/j.brainres.2015.09.020.
31. Mitsumoto H, Brooks BR, Silani V. Clinical trials in amyotrophic lateral sclerosis: why so many negative trials and how can trials be improved? *Lancet Neurol*. 2014;13(11):1127–1138. https://doi.org/10.1016/S1474-4422(14)70129-2.

32. Mitne-Neto M, Machado-Costa M, Marchetto MCN, et al. Downregulation of VAPB expression in motor neurons derived from induced pluripotent stem cells of ALS8 patients. *Hum Mol Genet*. 2011;20(18):3642–3652. https://doi.org/10.1093/hmg/ddr284.
33. Xuejiao S, Jianyuan S, Hailong H, Hong C, Kun Q. Modeling hallmark pathology using motor neurons derived from the family and sporadic amyotrophic lateral sclerosis patient-specific iPS cells. *Stem Cell Res Ther*. 2018. https://doi.org/10.1186/s13287-018-1048-1.
34. Egawa N, Kitaoka S, Tsukita K, et al. Drug screening for ALS using patient-specific induced pluripotent stem cells. *Sci Transl Med*. 2012;4(145). https://doi.org/10.1126/scitranslmed.3004052.
35. Higelin J, Demestre M, Putz S, et al. FUS mislocalization and vulnerability to DNA damage in ALS patients derived hiPSCs and aging motoneurons. *Front Cell Neurosci*. 2016;10(2016). https://doi.org/10.3389/fncel.2016.00290.
36. Donnelly CJ, Zhang PW, Pham JT, et al. RNA toxicity from the ALS/FTD C9ORF72 expansion is mitigated by antisense intervention. *Neuron*. 2013;80(2):415–428. https://doi.org/10.1016/j.neuron.2013.10.015.
37. Lee JH, Liu JW, Lin SZ, Harn HJ, Chiou TW. Advances in patient-specific induced pluripotent stem cells shed light on drug discovery for amyotrophic lateral sclerosis. *Cell Transplant*. 2018;27(9):1301–1312. https://doi.org/10.1177/0963689718785154.
38. Borchers AT, Hagie F, Keen CL, Gershwin ME. The history and contemporary challenges of the US food and drug administration. *Clin Therapeut*. 2007;29(1):1–16. https://doi.org/10.1016/j.clinthera.2007.01.006.
39. Lilienfeld DE. The first pharmacoepidemiology investigations: notional drug safety policy in the United States, 1901-1902. *Perspect Biol Med*. 2008;51(2):188–198.
40. Ballentine C. Sulfanilamide disaster. *FDA Consum Mag*. 1981.
41. Swann JP. The 1941 sulfathiazole disaster and the birth of good manufacturing practices. *PDA J Pharm Sci Technol*. 1999;53(3):148–153.
42. Rehman W, Arfons LM, Lazarus HM. The rise, fall and subsequent triumph of thalidomide: lessons learned in drug development. *Ther Adv Hematol*. 2011;2(5):291–308. https://doi.org/10.1177/2040620711413165.
43. Cordoba-Rodriguez R. Raw materials in the manufacture of biotechnology products: regulatory considerations. *PDA J Pharm Sci Technol*. 2010;64(5):445–450.
44. Bedford P, Jy J, Collins L, Keizer S. Considering cell therapy product good manufacturing practice status. *Front Med*. 2018:5. https://doi.org/10.3389/fmed.2018.00118.
45. Bravery CA. Do human leukocyte antigen-typed cellular therapeutics based on induced pluripotent stem cells make commercial sense? *Stem Cell Dev*. 2015;24(1):1–10. https://doi.org/10.1089/scd.2014.0136.
46. Awe JP, Lee PC, Ramathal C, et al. Generation and characterization of transgene-free human induced pluripotent stem cells and conversion to putative clinical-grade status. *Stem Cell Res Ther*. 2013;4(4). https://doi.org/10.1186/scrt246.
47. Baghbaderani BA, Tian X, Neo BH, et al. CGMP-manufactured human induced pluripotent stem cells are available for pre-clinical and clinical applications. *Stem Cell Rep*. 2015;5(4):647–659. https://doi.org/10.1016/j.stemcr.2015.08.015.
48. Lowenthal J, Lipnick S, Rao M, Hull SC. Specimen collection for induced pluripotent stem cell research: harmonizing the approach to informed consent. *Stem Cell Transl Med*. 2012;1(5):409–421. https://doi.org/10.5966/sctm.2012-0029.
49. *ISSCR Informed Consent*; 2020. https://www.isscr.org/policy/guidelines-for-stem-cell-research-and-clinical-translation/guidelines—hesc-research/sample-informed-consent-documents [Accessed September 26, 2020].
50. *FDA sample disease screening*; 2020. https://www.fda.gov/vaccines-blood-biologics/safety-availability-biologics/testing-human-cells-tissues-and-cellular-and-tissue-based-product-hctp-donors-relevant-communicable [Accessed September 26, 2020].
51. Shafa M, Walsh T, Panchalingam KM, et al. Long-term stability and differentiation potential of cryopreserved cGMP-compliant human induced pluripotent stem cells. *Int J Mol Sci*. 2020;21(1). https://doi.org/10.3390/ijms21010108.
52. Haase A, Glienke W, Engels L, et al. GMP-compatible manufacturing of three iPS cell lines from human peripheral blood. *Stem Cell Res*. 2019:35. https://doi.org/10.1016/j.scr.2019.101394.
53. Gagliano O, Luni C, Qin W, et al. Microfluidic reprogramming to pluripotency of human somatic cells. *Nat Protoc*. 2019;14(3):722–737. https://doi.org/10.1038/s41596-018-0108-4.

54. Stephen S, Stacey GN, Akazawa C, et al. Quality control guidelines for clinical-grade human induced pluripotent stem cell lines. *Regen Med*. 2018:859–866. https://doi.org/10.2217/rme-2018-0095.
55. Rehakova D, Souralova T, Koutna I. Clinical-grade human pluripotent stem cells for cell therapy: characterization strategy. *Int J Mol Sci*. 2020;21(7). https://doi.org/10.3390/ijms21072435.
56. Müller FJ, Goldmann J, Löser P, Loring JF. A call to standardize teratoma assays used to define human pluripotent cell lines. *Cell Stem Cell*. 2010;6(5):412–414. https://doi.org/10.1016/j.stem.2010.04.009.
57. Rohani L, Johnson AA, Naghsh P, Rancourt DE, Ulrich H, Holland H. Concise review: molecular cytogenetics and quality control: clinical guardians for pluripotent stem cells. *Stem Cell Transl Med*. 2018;7(12):867–875. https://doi.org/10.1002/sctm.18-0087.
58. Mandai M, Watanabe A, Kurimoto Y, et al. Autologous induced stem-cell-derived retinal cells for macular degeneration. *N Engl J Med*. 2017;376(11):1038–1046. https://doi.org/10.1056/NEJMoa1608368.
59. Wiley LA, Burnight ER, Deluca AP, et al. CGMP production of patient-specific iPSCs and photoreceptor precursor cells to treat retinal degenerative blindness. *Sci Rep*. 2016:6. https://doi.org/10.1038/srep30742.
60. Zhu J, Reynolds J, Garcia T, et al. Generation of transplantable retinal photoreceptors from a current good manufacturing practice-manufactured human induced pluripotent stem cell line. *Stem Cell Transl Med*. 2018;7(2):210–219. https://doi.org/10.1002/sctm.17-0205.
61. Chen VC, Ye J, Shukla P, et al. Development of a scalable suspension culture for cardiac differentiation from human pluripotent stem cells. *Stem Cell Res*. 2015;15(2):365–375. https://doi.org/10.1016/j.scr.2015.08.002.
62. Zhu H, Kaufman DS. An improved method to produce clinical-scale natural killer cells from human pluripotent stem cells. In: *Methods in Molecular Biology*. 2048. United States: Humana Press Inc.; 2019:107–119. https://doi.org/10.1007/978-1-4939-9728-2_12.
63. Rao MS, Pei Y, Garcia TY, et al. Illustrating the potency of current good manufacturing practice–compliant induced pluripotent stem cell lines as a source of multiple cell lineages using standardized protocols. *Cytotherapy*. 2018;20(6):861–872. https://doi.org/10.1016/j.jcyt.2018.03.037.
64. Shafa M, Yang F, Fellner T, Rao MS, Baghbaderani BA. Human-induced pluripotent stem cells manufactured using a current good manufacturing practice-compliant process differentiate into clinically relevant cells from three germ layers. *Front Med*. 2018:5. https://doi.org/10.3389/fmed.2018.00069.
65. Vitillo L, Durance C, Hewitt Z, Moore H, Smith A, Vallier L. GMP-grade neural progenitor derivation and differentiation from clinical-grade human embryonic stem cells. *bioRxiv*. 2020. https://doi.org/10.1101/2020.04.22.043216.
66. Peng J, Liu Q, Rao MS, Zeng X. Survival and engraftment of dopaminergic neurons manufactured by a good manufacturing practice-compatible process. *Cytotherapy*. 2014;16(9):1305–1312. https://doi.org/10.1016/j.jcyt.2014.06.002.
67. Karumbayaram S, Lee P, Azghadi SF, et al. From skin biopsy to neurons through a pluripotent intermediate under good manufacturing practice protocols. *Stem Cell Transl Med*. 2012;1(1):36–43. https://doi.org/10.5966/sctm.2011-0001.
68. Rosati J, Ferrari D, Altieri F, et al. Establishment of stable iPS-derived human neural stem cell lines suitable for cell therapies. *Cell Death Dis*. 2018;9(10). https://doi.org/10.1038/s41419-018-0990-2.
69. Song B, Cha Y, Ko S, et al. Human autologous iPSC-derived dopaminergic progenitors restore motor function in Parkinson's disease models. *J Clin Invest*. 2020;130(2):904–920. https://doi.org/10.1172/JCI130767.
70. Schweitzer JS, Song B, Herrington TM, et al. Personalized iPSC-derived dopamine progenitor cells for Parkinson's disease. *N Engl J Med*. 2020;382(20):1926–1932. https://doi.org/10.1056/NEJMoa1915872.
71. Doi D, Magotani H, Kikuchi T, et al. Pre-clinical study of induced pluripotent stem cell-derived dopaminergic progenitor cells for Parkinson's disease. *Nat Commun*. 2020;11(1). https://doi.org/10.1038/s41467-020-17165-w.
72. Trounson A, McDonald C. Stem cell therapies in clinical trials: progress and challenges. *Cell Stem Cell*. 2015;17(1):11–22. https://doi.org/10.1016/j.stem.2015.06.007.
73. Deinsberger J, Reisinger D, Weber B. Global trends in clinical trials involving pluripotent stem cells: a systematic multi-database analysis. *NPJ Regen Med*. 2020;5(1):15. https://doi.org/10.1038/s41536-020-00100-4.
74. Stoddard-Bennett T, Pera R. Stem cell therapy for Parkinson's disease: safety and modeling. *Neural Regen Res*. 2020;15(1):36–40. https://doi.org/10.4103/1673-5374.264446.
75. Allsopp TE, Ebneth A, Cabrera-Socorro A. Deploying human pluripotent stem cells to treat central nervous system disorders: facts, challenges and realising the potential. *Stem Cell Res*. 2019:41. https://doi.org/10.1016/j.scr.2019.101581.

76. Cyranoski D. The potent effects of Japan's stem-cell policies. *Nature*. 2019;573(7775):482–485. https://doi.org/10.1038/d41586-019-02847-3.
77. Nakajima F, Tokunaga K, Nakatsuji N. Human leukocyte antigen matching estimations in a hypothetical bank of human embryonic stem cell lines in the Japanese population for use in cell transplantation therapy. *Stem Cell*. 2006;25. https://doi.org/10.1634/stemcells.2006-0566 (Epub).
78. Okita K, Matsumura SY, Yoshiko S, et al. A more efficient method to generate integration-free human iPS cells. *Nat Methods*. 2011;8(5). https://doi.org/10.1038/nmeth.1591.
79. Pappas DJ, Gourraud PA, Le Gall C, et al. Proceedings: human leukocyte antigen haplo-homozygous induced pluripotent stem cell haplobank modeled after the California population: evaluating matching in a multiethnic and admixed population. *Stem Cell Transl Med*. 2015;4(5):413–418. https://doi.org/10.5966/sctm.2015-0052.
80. Solomon PF, Rao MS. Banking on iPSC—is it doable and is it worthwhile. *Stem Cell Rev Rep*. 2015;11(1). https://doi.org/10.1007/s12015-014-9574-4. PMID: 25516409.
81. Turner M, Leslie S, Martin NG, et al. Toward the development of a global induced pluripotent stem cell library. *Cell Stem Cell*. 2013;13(4):382–384. https://doi.org/10.1016/j.stem.2013.08.003.
82. Barry J, Hyllner J, Stacey G, Taylor CJ, Turner M. Setting up a haplobank: issues and solutions. *Curr Stem Cell Rep*. 2015;1(2):110–117. https://doi.org/10.1007/s40778-015-0011-7.

CHAPTER

15

Ethical considerations for the use of stem cell-derived therapies

Kevin T. FitzGerald[1,2], *Samantha Wu*[2] *and James Giordano*[2,3]

[1]Department of Medical Humanities, Creighton University School of Medicine, Omaha, NE, United States; [2]Pellegrino Center for Clinical Bioethics, Georgetown University Medical Center, Washington, DC, United States; [3]Departments of Neurology and Biochemistry, Georgetown University Medical Center, Washington, DC, United States

Overview of induced pluripotent stem cell (iPSC) therapies for neurological application

The field of neurological therapeutics has shifted from a focus on pharmacological and structural approaches intended to prevent and reduce inflammation and provide stability[1–3] to cell-based approaches that seek to restore functionality via cell-replacement and regeneration in cases of neurological injury and disease.[1,2,4–6] The last two decades have seen stem cell transplant studies in animal models for application to traumatic brain injury (TBI),[7,8] stroke,[9] Parkinson's disease (PD),[10–13] Huntington disease (HD),[14,15] and other conditions of the central nervous system (CNS).[1,16–19] In recent years, several stem cell therapies have advanced to clinical trials.[2]

Embryonic stem cells (ESCs), fetal stem cells (FSCs), adult somatic stem cells (ASCs), and induced pluripotent stem cells (iPSCs) are among the stem cells that have been explored to date. Research and applications involving ESCs and FSCs have been limited by ethical, medical, technical, and financial challenges and concerns.[2,20,21] As potential alternatives to ESCs and FSCs, ASCs are derived from adult tissues, such as skeletal muscle, bone marrow, adipose tissue, and central nervous system tissue.[2,21] Certain ASCs – most notably mesenchymal stem cells (MSCs)[22,23] —have garnered attention as potential therapeutics due to advantages such as autologous transplantation and mitigation of the ethical issues associated with ESCs. This is not to say that ASCs are without their challenges; preclinical and early clinical trials highlight concerns over limited, inefficient, and inconsistent differentiation and proliferation of different

Phenotyping of Human iPSC-derived Neurons
https://doi.org/10.1016/B978-0-12-822277-5.00012-2

ASCs, as well as quality and safety concerns associated with transplantation and clinical application.[21,24,25] Neural stem cells (NSCs) derived from the adult central nervous system have multipotent capabilities to differentiate into oligodendrocytes, astrocytes, or neurons, and thus, have also shown promise for tissue regeneration and repair.[2,21]

Innovation in cellular reprogramming techniques and a renewed interest in regenerative medicine have led to the exploration of iPSC technology[26–28] for neurological therapeutic applications. Any iPSC treatment development requires optimization and establishment of safety, effectiveness, efficiency, and quality at three primary stages: (1) selection of a donor cell, (2) manipulation and manufacture of stem cells, and (3) delivery of stem cells into the patient.[2] Specific scientific, technical, financial, and ethical considerations, however, can vary at each stage depending upon the disease or condition in question, and the patient in need. Currently, decisions regarding the selection of donors is likely to be determined primarily by biological considerations. However, it is not too early to begin drawing up selection criteria that will include diversity and inclusion characteristics so an equitable selection process can be put in place once biological characteristics no longer dictate selection of donor cells.

Ethical and social issues

The lack of established and reliable therapies for regeneration and repair of the injured CNS to date, coupled with the promise of iPSC-derived neurons in basic and preclinical studies has led to rapid expansion of research and development into this area.[29,30] Alongside the promise of such iPSC-derived treatments are uncertainties that remain with regard to safety, effectiveness, efficiency, and quality. The field of regenerative neurological therapeutics is entering into early phase clinical trials to assess and establish safety and efficacy of therapies that are not only palliative, but also potentially curative (see[2]). Scientific and regulatory communities must now face questions about the ethical, legal, and social implications, as well as the potential social value and utility of these approaches.

Quality control (QC) and quality assurance (QA) issues

Translation of iPSC therapies to the clinic raises unique concerns related to quality control and quality assurance. These include considerations, such as the timeline for the manufacture and testing of the iPSC product, as well as three notable areas of concern: therapeutic cell product characterization, potency assays, and tissue specificity.

Characterization of the therapeutic cell product

iPSCs pose challenges to cell characterization techniques, which are intended to define and characterize phenotypic, biological and functional properties of the cell population in question.[31] Such characterization is key to the development of relevant and effective potency assays. Quality control of this process necessitates consensus regarding the defining of required stem cell characteristics (e.g., specific surface markers, differentiation into specific cell types) of the therapeutic cell product and an appropriate bioassay. Variations in cell source, reprogramming, isolation, culture, and maintenance techniques, transplant procedures, as well as characterization bioassays make it difficult to define clear criteria for the consistent characterization of therapeutic iPSC products.[31,32]

Potency assays

Potency, as defined in 21 CFR 600.3(s),[33] is "the specific ability or capacity of the product, as indicated by appropriate laboratory tests or by adequately controlled clinical data obtained through the administration of the product in the manner intended, to effect a given result" (21 CFR 600.3(s)).[33] Potency tests and measurements are used to provide assurance of product identity, strength, stability, and purity – during its manufacture, as well as, prior to and during administration. During clinical trial investigations and with market approval, potency testing is also required to ensure the quality of the product lot and its compliance with applicable quality standards and regulations prior to release.[34]

For cell therapy products, the FDA notes several key challenges to the development of potency assays: inherent variability of starting materials, limited lot size and material for testing, limited stability, lack of appropriate reference standards, the potential for interference or synergy between multiple active ingredients, complex mechanisms of action, and the in vivo fate of the cell product.[34]

Currently, potency assays for dopaminergic precursor cell therapies for Parkinson's measure the following: that the cells are capable of dendritic sprouting; the cells express dopamine (DA); and, the cells are engrafting.[35–37] Measuring dendritic sprouting is relatively easy but may be insufficient to demonstrate potency alone. Development of assays intended to measure potency via DA expression can be technically challenging, and potentially inappropriate for measuring the strength of the therapeutic cell product in achieving the desired effect. Dopaminergic precursor cells do not express DA like their differentiated derivatives; measuring DA expression in a given lot prior to release may not represent the DA expression levels of precursor cells (or their derivatives) following in vivo transplantation. Lastly, measuring cell engraftment requires in vivo transplantation in either animal or human models – neither of which is cost or time-effective in the manufacture of iPSCs. These issues are not unique to dopaminergic precursor cell therapies; this provides one example of the complex issues raised in the development of iPSCs, and the need for further development of appropriate and effective potency assays as iPSCs move toward clinical application.

Tissue specificity

Tissue specificity has become an area of growing concern in the development and regulation of iPSCs. In part, this is due to the concerns raised by some scientists regarding the potential for c-kit^{+} cells (hematopoietic stem cells derived from the bone-marrow) to differentiate into cardiomyocytes and contribute to regeneration of heart tissue.[38,39] Research that originally suggested that bone-marrow derived stem cells could repair damaged heart tissue has now been called into question and retracted, revealing false and fabricated data.[40] As a result, the scientific community and the NIH are becoming increasingly vigilant with regard to tissue specificity issues in the development of any stem cell treatments. Considering the risks involved in possible brain and spinal cord iPSC interventions, this increased scrutiny is warranted even if treatment development is slowed.

Risks, benefits, and safety for participants and patients

The literature indicates a renewed interest and shift in the field of regenerative medicine toward stem cell therapies that alleviate at least some of the ethical and regulatory concerns that have hampered research involving ESCs and FSCs. However, the ethical benefits of any single approach do not provide sufficient grounds to preclude any approach from the need to demonstrate its safety and efficacy (i.e., through sufficient basic and preclinical evidence);[41] nor does it circumvent the need to consider the associated risks and benefits for patients and research participants as the field moves into clinical trials.

A patient-centered assessment of the risks of stem cell-based approaches must include both known and potentially unknown effects − especially for interventions that have not yet been tested in humans. For iPSCs that involve in vitro manipulation, for example, a risk assessment should include defined risks with regard to cell differentiation patterns (consistency and stability), any uncontrolled or aberrant activity of expression of transgenes, and any potential for mutagenesis during cell manipulation and manufacture. If such iPSCs are also intended for transplant into humans, the known risks associated with the chosen delivery method (injury to surrounding nervous tissue, infection) and the risks associated with the stem cell-based therapy (i.e., potential for tumorigenicity, migration, immunogenicity, and interaction with the host substrate) post-transplant within the human body should be evaluated and disclosed to prospective research participants.[42] With regard to the proposed benefits of an intervention, the iPSC's efficacy for its intended purpose within the human body should be made clear to participants, such that patients have a reasonable expectation about any potential benefit from the trial.[17] These claimed benefits and known risks of a therapy should be reinforced by basic and preclinical evidence, all of which should contribute to an assessment of the safety of the proposed therapy. Indeed, some of the published literature suggests the safety and efficacy of particular iPSC interventions may be adequately achieved according to these standards, though these reports are predominantly using animal models.[2,17]

Given that many of these iPSCs are novel developments that have not yet been tested in human subjects, it is difficult to account for all potential risks and benefits to patients. Requirements and guidelines for the conduct of clinical trials involving new technologies and techniques should encourage investigators to consider such uncertainties in the design of their trial, as well as in communications and informed consent processes with potential research participants. The unanticipated consequences that may be considered for novel iPSC therapies, particularly those intended for transplantation, stem from risks defined in basic and preclinical studies and from technological and medical precedents. As Giordano (2012) describes in detail elsewhere (with regard to neurogenetic and neuroimplantation technologies), there may be, for instance, (1) unanticipated changes in the local genotypic and phenotypic environment at the site of transplantation and the possibility that such changes would have larger physiopsychosocial effects in the host, (2) runaway effects, in which the transplanted stem cells or progenitor cells effect detrimental outcomes that are unmanageable given current medical and technological capabilities, (3) domino effects, in which the transplanted cells interact with neural networks through previously unknown mechanisms, and (4) synergistic effects that may occur when the stem cells are combined with other technologies or techniques (see Giordano, 2012 for a more complete discussion of how to estimate benefits, burdens, and risks for neurogenetic and neural tissue-implantation technology).[42,43]

These uncertainties must be considered in the design and implementation of clinical trials, as well as in the recruitment and consent of research participants, with regard to the specific stem cell therapy in question, the delivery method of choice, the intended effect, and the patient's context.

Informed consent and patient vulnerability

Informed consent forms and processes for clinical trials involving investigational iPSCs must adequately communicate the risks – again, both known and unknown – and benefits of participation in the study. This begs the question: what constitutes "adequate" communication of risks and benefits? And how should study leaders decide that this information has been properly understood and synthesized by consenting parties?

Valid informed consent requires participant/patient autonomy in the consent process. This necessitates participants understanding the potential risks and benefits – often demonstrated by participants stating their understanding of the risks and benefits in their own words – and an ability to weigh said risks and benefits in their decision to participate. It also requires that the participant provide consent free of undue coercion, and with a reasonable expectation of how/what they stand to benefit from the study (i.e., avoidance of the "therapeutic misconception").[17] These mandates require primary investigators and their research teams to consider the relevant risks and benefits, the amount, type, and quality of information being provided in the consent process, and the physiopsychosocial context of their potential participants.

Target populations for regenerative iPSCs in neurological injury and disease often involve particularly vulnerable patients; many of them may suffer from chronic and debilitating neurological disorders or conditions – brought on by injury or stroke – that can impair mental, behavioral, and/or physical well-being.[17] Questions of oversight, risk-benefit ratios, and safety and efficacy becoming increasingly important in preserving patient autonomy and minimizing risks to these vulnerable populations.

For investigational stem cell therapies, it is worth consulting recently released guidelines and recommendations (e.g., from the ISSCR, MRCT, FDA) that address regulatory gaps and post-trial responsibilities, in order to plan for further potentially unanticipated risks to research participants both during and following clinical trials.

Access/provision

As Giordano (2012) raises in his discussion of neurogenetic and neural tissue-implantation technology, even at this stage in the development of novel iPSCs, we must begin addressing the question of how such therapies will be accessed and distributed once validated and approved.[42] Allocation of novel medical technologies is often such that distribution and access will be limited to those who can afford it. Furthermore, health insurance plans frequently do not fully cover the costs of these technologies or treatments. This would imply that iPSCs that are currently in development may not be available to a majority of the population. What, then is the goal of advancing such technologies and who stands to benefit? As stem cell therapies continue to advance to clinical trial stages, these questions will remain pertinent and will ultimately come to bear as the therapies reach marketability and clinical application.

Ethical translation of promising stem cell-based neurological therapeutics

Scientific and regulatory communities have recognized the shortcomings of the regulatory frameworks currently in place – namely, the gaps that have allowed for the development of "unproven" therapies and the possibility that the current frameworks are ill-equipped to sufficiently evaluate proposed stem cell and stem cell-based therapies for INDs and market approval – as well as the detriment that such shortcomings can have in terms of patient safety and scientific progress. In response, researchers and agencies have developed guidelines and frameworks for IRBs, researchers, manufacturers, and/or sponsors to help them assess (1) whether their product is ready (safe and effective) for clinical trial testing in humans, and (2) whether their proposed trial design shows consideration for relevant regulatory and ethical concerns (see[17,41,44] for examples of guidelines). This is certainly a step in the right direction; however, this addresses only a fraction of the relevant issues and questions. While the issues presented in this chapter are categorized as ethical, regulatory, or social (for the purposes of emphasizing the importance of considering each of them in the regulation and translation of iPSCs), this is not to imply that these issues are unrelated to one another or that they should be addressed in isolation.

Giordano (2016) and Shook and Giordano (2019) suggest that uses of scientific, technological, and medical approaches in the public sphere (e.g., in the clinic and on the market) are determined by the interactive relationships amongst science, ethics, guidelines, and policy (see Fig. 15.1[42,45,46]). An approach that recognizes these interactive relationships may ensure (1) that the ethics and regulations keep pace with the science and (2) the ethical translation of these novel stem cell and stem cell-based therapies.

As Shook and Giordano advocate, the relative value (i.e., as a "social good") of a scientific/technological approach is determined by an assessment of these interacting factors and a thorough analysis of the relevant practical and ethical concerns. The relevant practical and ethical concerns that accompany scientific, technological and/or medical approaches include those that are inherent to the scientific/technical aspects and those that are socially derived or brought about by "social factors, circumstances and concerns"[45,46] (see Fig. 15.2).

Specific scientific and technological issues are closely tied to and inform questions of safety (as evaluated by defined parameters, long-term replicable studies, and peer-reviewed data[47]), efficacy and clinical effectiveness, as well as cost.[46] Safety, efficacy, and clinical effectiveness, in turn, are key in determining the relative risks, benefits and burdens to patients. Together, these considerations inform the viability (as a "social good") of the approach in question. The safety, effectiveness, cost, and viability of the approach all contribute to an assessment of the relative value of said approach.[46] Questions of access and provision are also implicated in the consideration of costs and the assessment of value, adding to the merit and completeness of this form of value assessment.[46]

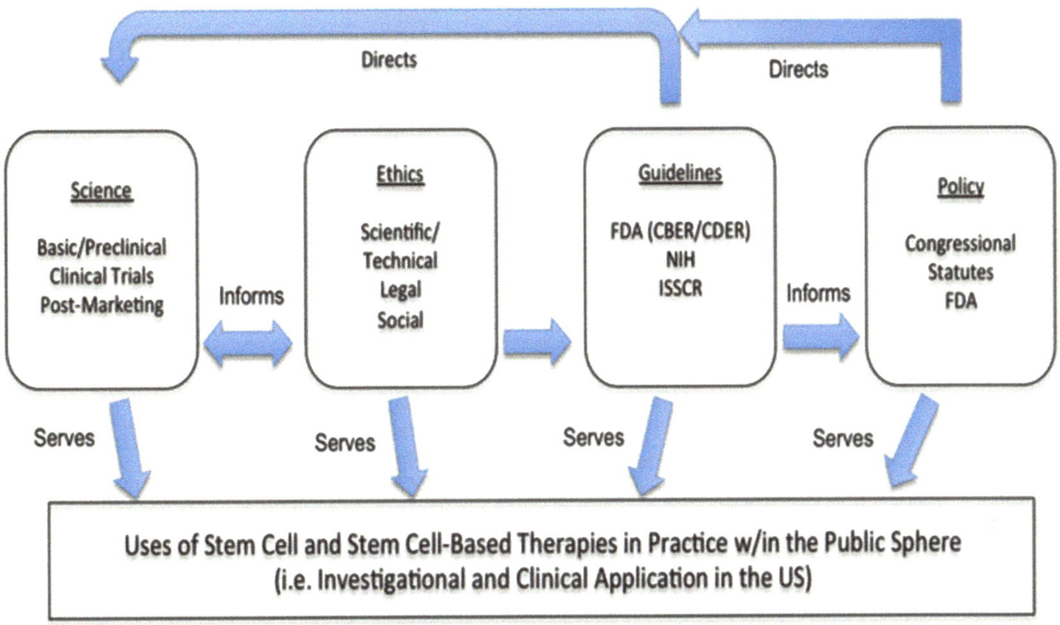

FIGURE 15.1 Interactive relationships amongst science, ethics, guidelines, and policies within the regulatory framework that indicates the use of stem cell and stem cell-based therapies in the US. *From Giordano J. Neurogenetic and neural tissue-implantation technology: Neuroethical, legal, and social issues. In:* Neurotechnology: Premises, Potential, and Problems *CRC Press. 2012:59–67. https://doi.org/10.1201/b11861.*

Ethical translation in practice

To demonstrate the utility of frameworks that recognize the interactive relationships amongst science, ethics, and policy, we apply the frameworks in Figs. 15.1 and 15.2 to assess the value of iPSC therapies.

As Fig. 15.1 suggests, translation of any of these approaches also implicates careful consideration of current policies, regulatory gaps (i.e., vague definitions of "homologous use" and "minimal manipulation"; evaluation of early phase clinical trials involving novel iPSCs; post trial access and responsibilities), and how these might influence the existing literature, ongoing preclinical and clinical research, and early applications of the iPSC.

Ethical issues implicated in the use of stem cell and stem cell-derived therapies are closely related to and, in some cases, overlap with scientific and regulatory considerations.[42,46] In turn, the ability to discern the relative risks, benefits, and burdens is compromised, as are the veridicality of the approach and the informed consent process. Broadly speaking, a valid informed consent process should satisfy the previously discussed criteria (see "Informed consent and patient vulnerability" section), but for emerging stem cell and stem cell-based therapies, should also communicate the (a) cell type, cell source, the intended use; (b) associated potential risks, benefits, and burdens of the therapy as informed by consideration of its safety, efficacy, and cost; and (c) communication of plans for continuity of care (see post-trial access and responsibilities).

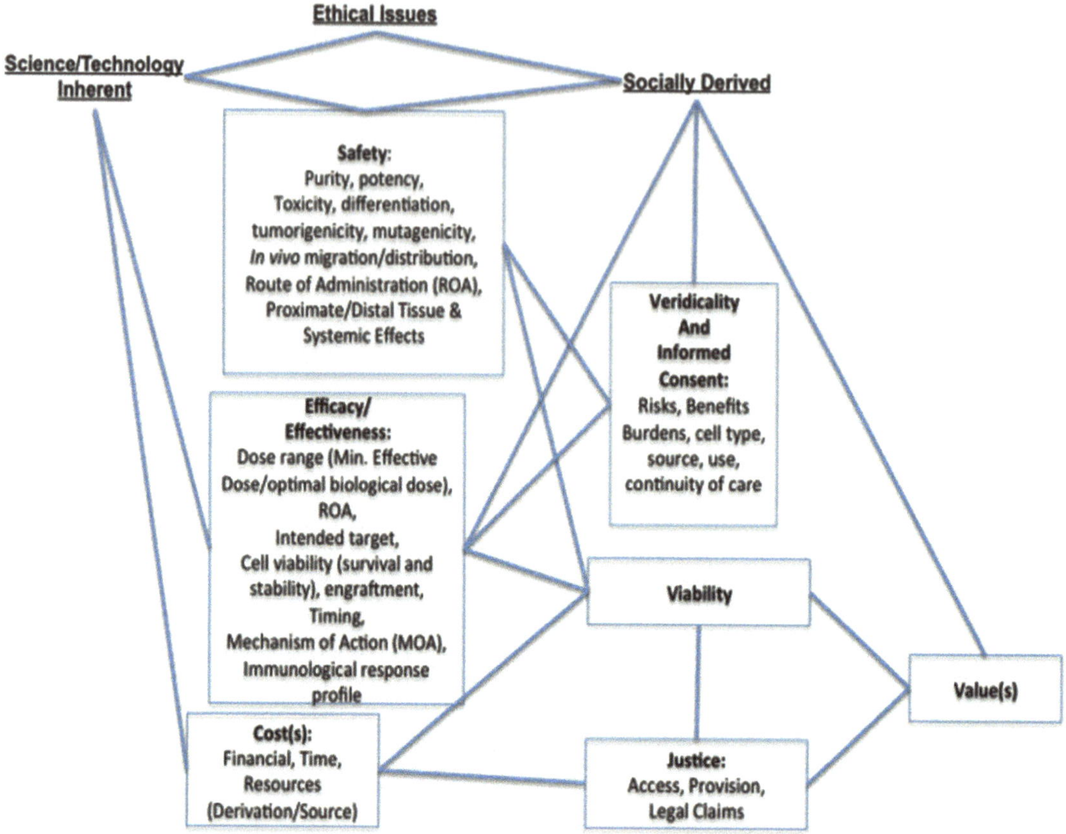

FIGURE 15.2 Ethical and practical issues for consideration in the translation of stem cell therapies. *From Shook JR, Giordano J. Ethical contexts for the future of neuroethics.* AJOB Neuroscience *2019;10(3):134–136. https://doi.org/10.1080/21507740.2019.1632969, Fig. 1: FDA, Nov 2013.*

ASCs can be reprogrammed into iPSCs that are capable of differentiating into different types of neural precursor cells, glial cells, or other neuronal cell types.[20,27,48] Ethical concerns related to safety and efficacy are important to address and consider in translation of therapies involving iPSCs. All the safety and efficacy concerns related to the use of iPSCs necessitate greater consideration for cell stability, differentiation, mutagenicity and tumorigenicity both in vitro during manufacturing and in vivo, once delivered into the patient.[2,27,49,50] Again, ethical translation of any approach necessitates a proper informed consent process, as previously described.

In sum, each of these approaches necessitates consideration of ethical issues related to safety and efficacy; cost (cell source, supply, derivation); and, the relative risks, benefits and burdens to be communicated to patients and research participants in a valid informed consent process.

We have herein assessed the state-of-the-science and primary ethical concerns pertinent to iPSCs. Overall, such an analysis (Figs. 15.1 and 15.2) provides the user with a comprehensive overview of the issues within each of these fields that must be considered in the ethical translation of stem cell or stem cell-based therapies for neurological application. It also allows for a comparative analysis to assess the relative viability and value of each of these "promising" stem cell approaches as the field moves into early phase clinical trials.

Conclusions

Research involving iPSC therapies is set to continue its advance toward clinical application, driven by the influence of medical, humanitarian, technological, and financial incentives. Appropriate and sufficient consideration of these issues necessitates an assessment of the current landscape, inclusive of gaps in the science, policies, and ethical frameworks and the interactive relationships among these fields. Such an assessment assists those who manufacture, apply, or regulate these emerging iPSCs in ensuring that their therapies are valid and viable. It also brings attention to gaps in the science, policy, or ethical frameworks that need to be addressed moving forward, to ensure the ethical translation of emerging iPSCs to clinics and patients.

References

1. Tsintou M, Dalamagkas K, Seifalian AM. Advances in regenerative therapies for spinal cord injury: a biomaterials approach. *Neural Regen Res.* 2015;10(5):726–742. https://doi.org/10.4103/1673-5374.156966.
2. Wu S, FitzGerald KT, Giordano J. On the viability and potential value of stem cells for repair and treatment of central neurotrauma: overview and speculations. *Front Neurol.* 2018;9. https://doi.org/10.3389/fneur.2018.00602.
3. Yiu G, He Z. Glial inhibition of CNS axon regeneration. *Nat Rev Neurosci.* 2006;7(8):617–627. https://doi.org/10.1038/nrn1956.
4. Bliss TM, Andres RH, Steinberg GK. Optimizing the success of cell transplantation therapy for stroke. *Neurobiol Dis.* 2010;37(2):275–283. https://doi.org/10.1016/j.nbd.2009.10.003.
5. Miller FD, Gauthier-Fisher A. Home at last: neural stem cell niches defined. *Cell Stem Cell.* 2009;4(6):507–510. https://doi.org/10.1016/j.stem.2009.05.008.
6. Mothe AJ, Tator CH. Advances in stem cell therapy for spinal cord injury. *J Clin Invest.* 2012;122(11):3824–3834. https://doi.org/10.1172/JCI64124.
7. Berkowitz AL, Miller MB, Mir SA, et al. Glioproliferative lesion of the spinal cord as a complication of stem-cell tourism. *N Engl J Med.* 2016;375(2):196–198. https://doi.org/10.1056/NEJMc1600188.
8. Teng YD, Lavik EB, Qu X, et al. Functional recovery following traumatic spinal cord injury mediated by a unique polymer scaffold seeded with neural stem cells. *Proc Natl Acad Sci USA.* 2002;99(5):3024–3029. https://doi.org/10.1073/pnas.052678899.
9. Veizovic T, Beech JS, Stroemer RP, Watson WP, Hodges H. Resolution of stroke deficits following contralateral grafts of conditionally immortal neuroepithelial stem cells. *Stroke.* 2001;32(4):1012–1019. https://doi.org/10.1161/01.STR.32.4.1012.
10. Kordower JH, Freeman TB, Snow BJ, et al. Neuropathological evidence of graft survival and striatal reinnervation after the transplantation of fetal mesencephalic tissue in a patient with Parkinson's disease. *N Engl J Med.* 1995;332(17):1118–1124. https://doi.org/10.1056/NEJM199504273321702.
11. Piccini P, Brooks DJ, Björklund A, et al. Dopamine release from nigral transplants visualized in vivo in a Parkinson's patient. *Nat Neurosci.* 1999;2(12):1137–1140. https://doi.org/10.1038/16060.
12. Studer L, Tabar V, McKay RDG. Transplantation of expanded mesencephalic precursors leads to recovery in parkinsonian rats. *Nat Neurosci.* 1998;1(4):290–295. https://doi.org/10.1038/1105.

13. de Melo-Martín I, Hellmers N, Henchcliffe C. First-in-human cell transplant trials in Parkinson's disease: the need for an improved informed consent process. *Park Relat Disord*. 2015;21(8):829–832. https://doi.org/10.1016/j.parkreldis.2015.05.011.
14. Bachoud-Lévi AC, Rémy P, Nǵuyen JP, et al. Motor and cognitive improvements in patients with Huntington's disease after neural transplantation. *Lancet*. 2000;356(9246):1975–1979. https://doi.org/10.1016/S0140-6736(00)03310-9.
15. Freeman TB, Cicchetti F, Hauser RA, et al. Transplanted fetal striatum in Huntington's disease: phenotypic development and lack of pathology. *Proc Natl Acad Sci USA*. 2000;97(25):13877–13882. https://doi.org/10.1073/pnas.97.25.13877.
16. Batista CEM, Mariano ED, Marie SKN, et al. Stem cells in neurology - current perspectives. *Arq Neuro Psiquiatr*. 2014;72(6):457–465. https://doi.org/10.1590/0004-282X20140045.
17. Cote DJ, Bredenoord AL, Smith TR, et al. Ethical clinical translation of stem cell interventions for neurologic disease. *Neurology*. 2017;88(3):322–328. https://doi.org/10.1212/WNL.0000000000003506.
18. Grisolía JS. CNS stem cell transplantation: clinical and ethical perspectives. *Brain Res Bull*. 2002;57(6):823–826. https://doi.org/10.1016/S0361-9230(01)00766-3.
19. Rice CM, Halfpenny CA, Scolding NJ. Stem cells for the treatment of neurological disease. *Transfus Med*. 2003;13(6):351–361. https://doi.org/10.1111/j.1365-3148.2003.00463.x.
20. Hockemeyer D, Jaenisch R. Induced pluripotent stem cells meet genome editing. *Cell Stem Cell*. 2016;18(5):573–586. https://doi.org/10.1016/j.stem.2016.04.013.
21. Nam H. Adult human neural stem cell therapeutics: current developmental status and prospect. *World J Stem Cell*. 2015;7(1):126. https://doi.org/10.4252/wjsc.v7.i1.126.
22. Lazarus HM, Haynesworth SE, Gerson SL, Rosenthal NS, Caplan AI. Ex vivo expansion and subsequent infusion of human bone marrow-derived stromal progenitor cells (mesenchymal progenitor cells): implications for therapeutic use. *Bone Marrow Transplant*. 1995;16(4):557–564.
23. Otto WR, Wright NA. Mesenchymal stem cells: from experiment to clinic. *Fibrogenesis Tissue Repair*. 2011;4(1). https://doi.org/10.1186/1755-1536-4-20.
24. Alagesan S, Griffin MD. Autologous and allogeneic mesenchymal stem cells in organ transplantation: what do we know about their safety and efficacy? *Curr Opin Organ Transplant*. 2014;19(1):65–72. https://doi.org/10.1097/MOT.0000000000000043.
25. Sharma RR, Pollock K, Hubel A, McKenna D. Mesenchymal stem or stromal cells: a review of clinical applications and manufacturing practices. *Transfusion*. 2014;54(5):1418–1437. https://doi.org/10.1111/trf.12421.
26. Chang YJ, Su HL, Hsu LF, et al. Isolation of human neural stem cells from the amniotic fluid with diagnosed neural tube defects. *Stem Cell Dev*. 2015;24(15):1740–1750. https://doi.org/10.1089/scd.2014.0516.
27. Khazaei M, Ahuja CS, Fehlings MG. Induced pluripotent stem cells for traumatic spinal cord injury. *Front Cell Dev Biol*. 2017;4. https://doi.org/10.3389/fcell.2016.00152.
28. Takahashi K, Yamanaka S. Induction of pluripotent stem cells from mouse embryonic and adult fibroblast cultures by defined factors. *Cell*. 2006;126(4):663–676. https://doi.org/10.1016/j.cell.2006.07.024.
29. Boer GJ. Transplantation and xenotransplantation. In: *Scientific and Philosophical Perspectives in Neuroethics*. Giordano and B Gordijn; 2010:190–216.
30. Meloni R, Mallet J, Faucon Biguet N. Brain gene transfer and brain implants. *Stud Ethics Law Technol*. 2011;4(3). https://doi.org/10.2202/1941-6008.1141.
31. Burk J, Brehm W, Smith RK, Dudhia J. Editorial to the special issue "stem cell characterization across species.". *Cytometry*. 2018;93(1):16–18. https://doi.org/10.1002/cyto.a.23306.
32. Dominici M, Le Blanc K, Mueller I, et al. Minimal criteria for defining multipotent mesenchymal stromal cells. The International Society for Cellular Therapy position statement. *Cytotherapy*. 2006;8(4):315–317. https://doi.org/10.1080/14653240600855905.
33. Regulatory Information. Code of Federal Regulations (annual edition). https://www.govinfo.gov/app/details/CFR-2011-title21-vol7/CFR-2011-title21-vol7-sec600-3.
34. FDA. *Guidance for Industry: Potency Tests for Cellular and Gene Therapy Products*. Center for Biologics Evaluation and Research; 2011. https://www.fda.gov/regulatory-information/search-fda-guidance-documents/potency-tests-cellular-and-gene-therapy-products.
35. Barker RA, Barrett J, Mason SL, Björklund A. Fetal dopaminergic transplantation trials and the future of neural grafting in Parkinson's disease. *Lancet Neurol*. 2013;12(1):84–91. https://doi.org/10.1016/S1474-4422(12)70295-8.

36. Kefalopoulou Z, Politis M, Piccini P, et al. Long-term clinical outcome of fetal cell transplantation for Parkinson disease: two case reports. *JAMA Neurol.* 2014;71(1):83–87. https://doi.org/10.1001/jamaneurol.2013.4749.
37. Moon J, Schwarz SC, Lee HS, et al. Preclinical analysis of fetal human mesencephalic neural progenitor cell lines: characterization and safety in vitro and in vivo. *Stem Cells Transl Med.* 2017;6(2):576–588. https://doi.org/10.5966/sctm.2015-0228.
38. Balsam LB, Wagers AJ, Christensen JL, Kofidis T, Weissmann IL, Robbins RC. Haematopoietic stem cells adopt mature haematopoietic fates in ischaemic myocardium. *Nature.* 2004;428(6983):668–673. https://doi.org/10.1038/nature02460.
39. Van Berlo JH, Kanisicak O, Maillet M, et al. C-kit+ cells minimally contribute cardiomyocytes to the heart. *Nature.* 2014;509(7500):337–341. https://doi.org/10.1038/nature13309.
40. Johnson CY. *Heart Failure Stem Cell Trial to be Paused After Calls for Retraction.* Wash Post; 2018. https://www.washingtonpost.com/science/2018/10/29/heart-failure-stem-cell-trial-be-paused-after-calls-retractions/?utm_term=.7ee4c4f20edd. Accessed 29 Oct 2018.
41. Barker RA, Parmar M, Kirkeby A, Björklund A, Thompson L, Brundin P. Are stem cell-based therapies for Parkinson's disease ready for the clinic in 2016? *J Parkinsons Dis.* 2016;6(1):57–63. https://doi.org/10.3233/JPD-160798.
42. Giordano J. Neurogenetic and neural tissue-implantation technology: Neuroethical, legal, and social issues. In: *Neurotechnology: Premises, Potential, and Problems.* United States: CRC Press; 2012:59–67. https://doi.org/10.1201/b11861.
43. Giordano J, Benedikter R, Boswell MV. Pain medicine, biotechnology and market effets: tools, tekne and moral responsibility. *J Ethics Biol Engineer Med.* 2010;1(2):135–142.
44. Stem cell-based clinical trials: practical advice for physicians and ethics/institutional review boards. *Int Soc Stem Cell Res;* 2018. http://www.isscr.org/docs/default-source/clinical-resources/isscr-stem-cell-based-clnical-trials-practical-advice_final_23jan2018.pdf?sfvrsn=2.
45. Giordano J. Commentary: the value of patient benefit: consideration of framing contingencies to guide the ethical use of DBS-a case analysis. *Camb Q Healthc Ethics.* 2016;25(4):755–758. https://doi.org/10.1017/S0963180116000530.
46. Shook JR, Giordano J. Ethical contexts for the future of Neuroethics. *AJOB Neurosci.* 2019;10(3):134–136. https://doi.org/10.1080/21507740.2019.1632969.
47. Weiss DJ, Rasko JEJ, Cuende N, et al. Part 2: making the unproven proven. *Cytotherapy.* 2016;18(1):120–123. https://doi.org/10.1016/j.jcyt.2015.11.005.
48. Brouwer M, Zhou H, Nadif Kasri N. Choices for induction of pluripotency: recent developments in human induced pluripotent stem cell reprogramming strategies. *Stem Cell Rev Rep.* 2016;12(1):54–72. https://doi.org/10.1007/s12015-015-9622-8.
49. Costa M, Dottori M, Sourris K, et al. A method for genetic modification of human embryonic stem cells using electroporation. *Nat Protoc.* 2007;2(4):792–796. https://doi.org/10.1038/nprot.2007.105.
50. Ji J, Ng S, Sharma V, et al. Elevated coding mutation rate during the reprogramming of human somatic cells into induced pluripotent stem cells. *Stem Cell.* 2012;30(3):435–440. https://doi.org/10.1002/stem.1011.

Index

Note: 'Page numbers followed by "f" indicate figures and "t" indicate tables.'

Kolibri 360 GmbH
Ferdinand-Jühlke-Straße 7
99095 Erfurt, Deutschland
produktsicherheit@kolibri360.de